Legionella: Infections and Pathology

Legionella: Infections and Pathology

Edited by Charlotte Ortiz

STATES
ACADEMIC PRESS
www.statesacademicpress.com

States Academic Press,
109 South 5th Street,
Brooklyn, NY 11249, USA

Visit us on the World Wide Web at:
www.statesacademicpress.com

ISBN: 978-1-63989-324-9

Cataloging-in-Publication Data

Legionella : infections and pathology / edited by Charlotte Ortiz.
p. cm.
Includes bibliographical references and index.
ISBN 978-1-63989-324-9
1. Legionella. 2. Legionnaires' disease. 3. Legionella pneumophila.
4. Legionnaires' disease--Diagnosis. I. Ortiz, Charlotte.
QR201.L44 L44 2022
616.241--dc23

Table of Contents

Preface

Legionella refers to a genus of pathogenic bacteria which can cause diseases such as legionellosis, Pontiac fever and Legionnaires' disease. Their symptoms are similar to pneumonia. Complicated cases might also lead to issues related to the gastrointestinal tract and nervous system. The infections generally spread by the inhalation of water droplets from a source in which the bacteria has spread. It has not been observed to spread from person-to-person. The incubation period of these bacteria inside a host organism is up to two weeks. It usually causes severe complications in individuals who are aged or immunocompromised. This book contains some path-breaking studies in the research on legionella. Also included herein is a detailed explanation of the various infections and pathology of these bacteria. The extensive content of this book provides the readers with a thorough understanding of the subject.

After months of intensive research and writing, this book is the end result of all who devoted their time and efforts in the initiation and progress of this book. It will surely be a source of reference in enhancing the required knowledge of the new developments in the area. During the course of developing this book, certain measures such as accuracy, authenticity and research focused analytical studies were given preference in order to produce a comprehensive book in the area of study.

This book would not have been possible without the efforts of the authors and the publisher. I extend my sincere thanks to them. Secondly, I express my gratitude to my family and well-wishers. And most importantly, I thank my students for constantly expressing their willingness and curiosity in enhancing their knowledge in the field, which encourages me to take up further research projects for the advancement of the area.

Editor

1

Beyond Paralogs: The Multiple Layers of Redundancy in Bacterial Pathogenesis

*Soma Ghosh and Tamara J. O'Connor**

Department of Biological Chemistry, Johns Hopkins University School of Medicine, Baltimore, MD, United States

Edited by:
Matthias P. Machner, National Institutes of Health (NIH), United States

Reviewed by:
Anne-Marie Krachler, University of Texas Health Science Center at Houston, United States
Michael L. Vasil, University of Colorado Denver School of Medicine, United States

***Correspondence:**
Tamara J. O'Connor
toconno7@jhmi.edu

Redundancy has been referred to as a state of no longer being needed or useful. Microbiologists often theorize that the only case of true redundancy in a haploid organism would be a recent gene duplication event, prior to divergence through selective pressure. However, a growing number of examples exist where an organism encodes two genes that appear to perform the same function. For example, many pathogens translocate multiple effector proteins into hosts. While disruption of individual effector genes does not result in a discernable phenotype, deleting genes in combination impairs pathogenesis: this has been described as redundancy. In many cases, this apparent redundancy could be due to limitations of laboratory models of pathogenesis that do not fully recapitulate the disease process. Alternatively, it is possible that the selective advantage achieved by this perceived redundancy is too subtle to be measured in the laboratory. Moreover, there are numerous possibilities for different types of redundancy. The most common and recognized form of redundancy is functional redundancy whereby two proteins have similar biochemical activities and substrate specificities allowing each one to compensate in the absence of the other. However, redundancy can also exist between seemingly unrelated proteins that manipulate the same or complementary host cell pathways. In this article, we outline 5 types of redundancy in pathogenesis: molecular, target, pathway, cellular process, and system redundancy that incorporate the biochemical activities, the host target specificities and the impact of effector function on the pathways and cellular process they modulate. For each type of redundancy, we provide examples from *Legionella* pathogenesis as this organism employs over 300 secreted virulence proteins and loss of individual proteins rarely impacts intracellular growth. We also discuss selective pressures that drive the maintenance of redundant mechanisms, the current methods used to resolve redundancy and features that distinguish between redundant and non-redundant virulence mechanisms.

Keywords: redundancy, pathogenesis, effector, functional redundancy, genetic redundancy, *Legionella*

REDUNDANCY—BIOLOGY'S CONTINGENCY PLAN

Bacteria are one of nature's ultimate survivalists, able to adapt to extreme and dynamic environmental conditions. One of the reasons for their robustness is redundancy, contingency plans for a given process that enhances their fitness. Genetic redundancy describes two copies of the same gene whereby the protein encoded by one can function in place of the other.

A classic example of genetic redundancy occurs in metabolism, where two genes encode proteins that catalyze the same reaction (Toda et al., 1987). However, redundancy extends beyond gene duplication. Two proteins or sets of proteins with different catalytic activities can generate the same product (Wagner, 2000). The ability to synthesize a molecule *de novo* and the ability to acquire that molecule from the environment is also a form redundancy. In this case, the proteins and their functions are completely unrelated but they serve a common goal. Thus, redundancy can occur at multiple levels within a system and is largely defined by what a bacterium is trying to accomplish.

REDUNDANCY IN MICROBIAL PATHOGENESIS

Koch's postulates outline a set of criteria to define causal relationships between pathogens and disease (Koch, 1891; Evans, 1976). With advances in molecular biology techniques and bacterial genetics, Stanley Falkow proposed a molecular version of Koch's postulates to define virulence factors responsible for the pathogenesis of an individual microorganism (Falkow, 1988). The postulate sets an exclusive condition where disruption of a gene should result in a virulence defect and that phenotype should be reversed upon allelic replacement of the gene. For decades, the postulate has been used to identify many virulence factors in numerous pathogens (Isberg et al., 1987; Hersh et al., 1999). At the same time however, a growing number of genes that failed Falkow's criteria but played important roles in disease began to emerge (Falkow, 2004; Choy et al., 2012; Gaspar and Machner, 2014). The lack of phenotypes associated with genetic mutations was attributed to redundancy amongst virulence factors. While redundancy is not the only explanation for this phenomenon (discussed below), it is becoming a common feature in microbial pathogenesis with examples from *Legionella* (Luo and Isberg, 2004; Belyi et al., 2006), *Pseudomonas* (Kvitko et al., 2009; Cunnac et al., 2011), *Yersinia* (Ratner et al., 2016), *Chlamydia* (Cocchiaro and Valdivia, 2009), *Salmonella* (Zhou et al., 2001), and *Mycobacterium* (Downing et al., 2005; Ganapathy et al., 2015). While an exciting challenge for microbiologists, redundancy is a major obstacle in identifying virulence factors, deciphering their roles in disease and developing new therapeutic agents to combat infection.

REDUNDANCY IN *LEGIONELLA* PATHOGENESIS

Legionella pneumophila is an intracellular bacterial pathogen with a broad host range spanning over 15 species of amoebae and ciliated protozoa (Rowbotham, 1980) to mammalian macrophages (Horwitz and Silverstein, 1980). Intracellular growth of *L. pneumophila* requires a number of key events be accomplished. *L. pneumophila* must disrupt endocytic and autophagic targeting of its membrane-bound compartment, termed the *Legionella*-containing vacuole (LCV) to avoid digestion in the lysosome (Horwitz, 1983; Berger et al., 1994; Swanson and Isberg, 1995; Wiater et al., 1998; Choy et al., 2012); transform the phagosome into a replication-permissive compartment (Kagan and Roy, 2002; Derre and Isberg, 2004; Kagan et al., 2004); acquire nutrients to grow (Sauer et al., 2005; Allard et al., 2009; Isaac et al., 2015); expand and maintain the integrity of the replication vacuole to accommodate increasing bacterial numbers (Laguna et al., 2006; Creasey and Isberg, 2012); avoid detection by host innate immune recognition (Laguna et al., 2006; Zamboni et al., 2006; Coers et al., 2007; Fontana et al., 2011; Pereira et al., 2011; Creasey and Isberg, 2012; Barry et al., 2013); inhibit host cell death to maintain an intracellular environment that supports replication (Losick and Isberg, 2006; Abu-Zant et al., 2007); and eventually, exit from the host cell (Horwitz and Silverstein, 1980). As it turns out, *L. pneumophila* employs multiple strategies to accomplish each of these tasks.

With Falkow's molecular Koch's postulates in mind, several genetic screens to correlate gene disruptions with virulence defects have been employed to identify *L. pneumophila* virulence genes (Berger and Isberg, 1993; Sadosky et al., 1993; VanRheenen et al., 2004; Laguna et al., 2006). Parallel genetic screens independently identified a collection of 26 genes encoding components of a Type IVb secretion system, subsequently named Icm/Dot (Marra et al., 1992; Berger and Isberg, 1993; Brand et al., 1994). Mutations in *icm*/*dot* genes abolish *L. pneumophila* intracellular growth in macrophages (Berger and Isberg, 1993; Brand et al., 1994) and amoebal hosts (Segal and Shuman, 1999) demonstrating a critical role for the Icm/Dot complex in *L. pneumophila* pathogenesis. The identification of Icm/Dot was not surprising as numerous pathogens employ secretion systems to deploy proteins, termed effectors to the host cell to establish growth. Yet the search for Icm/Dot translocated substrates (IDTS) using similar genetic screening strategies was relatively unsuccessful, identifying only a small handful of IDTS-encoding genes that were important for *L. pneumophila* pathogenesis (VanRheenen et al., 2004; Laguna et al., 2006; Isaac et al., 2015). As a consequence, more creative genetic screening strategies were implemented (Luo and Isberg, 2004; Campodonico et al., 2005): not only did this lead to the identification of the first set of IDTS but also the presence of multiple paralogs of many IDTS in the *L. pneumophila* genome (Luo and Isberg, 2004). As a result, the lack of phenotypes associated with genetic mutations in a single IDTS was attributed to redundancy.

The presence of multiple IDTS paralogs was the first evidence of redundancy in *L. pneumophila* pathogenesis. However, the simultaneous deletion of all paralogs from a single family of IDTS did not impair *L. pneumophila* intracellular growth (Bardill et al., 2005). The simplest explanation was that these genes were dispensable under the experimental conditions tested. However, the subsequent use of biochemical and bioinformatics-based approaches had begun to define a collection of 270 translocated proteins (de Felipe et al., 2005; Huang et al., 2011; Zhu et al., 2011). In the process, pairs of IDTS that modulate the same host protein via different mechanisms or different components of the same pathway were identified (Nagai et al., 2002; Machner and Isberg, 2006; Murata et al., 2006; Belyi et al., 2008). In parallel, genetic screens in host cells to identify host factors important for *L. pneumophila* pathogenesis demonstrated that while depletion of a single host factor rarely impaired *L. pneumophila* replication, the combined deletion of pairs of host proteins that function in common processes significantly disrupted intracellular growth of the bacterium (Dorer et al., 2006). Collectively, these results suggested that redundancy extends beyond paralogs to more complex mechanisms that function at pathway and system levels, providing an explanation for the lack of phenotypes for mutants lacking all members of a paralogous family of IDTS. In support of this, it was subsequently shown that deleting specific combinations of unrelated IDTS impairs *L. pneumophila* intracellular growth while deletion of each gene individually failed to elicit a phenotype (O'Connor et al., 2011, 2012). Thus, redundancy appeared to be a multi-tiered phenomenon integrating many different forms.

TYPES OF REDUNDANCY

Redundancy in microbial pathogenesis manifests in many forms that encompass a broad spectrum of functional relationships and multiple levels of biological systems: this complexity necessitates a structured nomenclature to define the different types of redundancy. Genetic and functional redundancy are often used interchangeably, defining compensatory roles for two proteins with the same biochemical activities that allow one to substitute in place of the other. However, the use of function can be somewhat subjective, as it can refer to a precise biochemical activity or more generally, to the impact of that activity on a particular pathway or cellular process. As a consequence, the term functional redundancy has been omitted here as it could be used to describe more than one type of redundancy outlined below. Instead, we propose 5 types of redundancy (**Figure 1A**): molecular, target, pathway, cellular process, and system redundancy that incorporate the biochemical activities of effectors, their host target specificities, their impact on host cell biology and their contributions to pathogenesis. In many cases, virulence strategies are multi-tiered encompassing several types of redundancy.

Molecular Redundancy

Molecular redundancy defines two or more effectors that modify the same host target using the same molecular mechanism (**Figure 1A**). In this case, one effector can function in place of the other because it has the same activity and target specificity as its counterpart(s). In some cases, molecular redundancy is likely to be a byproduct of gene duplication however, this is not the sole source with examples of horizontal gene transfer and convergent evolution leading to the presence of molecularly redundant proteins.

Molecular redundancy is exemplified by the *L. pneumophila* SidE family of IDTS, SidE, SdeA, SdeB, and SdeC (Luo and Isberg, 2004). Individual paralogs consist of a mono-ADP-ribosyltransferase domain, a deubiquitylation domain, and a phosphohydrolase domain that collectively catalyze the ubiquitination of the host proteins Reticulon 4 (Rtn4) (Kotewicz et al., 2016) and Rab33b (Qiu et al., 2016) (**Figure 1B**). While each member of this family is individually dispensable for intracellular replication, the simultaneous deletion of all four members impairs growth of *L. pneumophila* in the amoebal hosts *Acanthamoebae castellanii* (Bardill et al., 2005) and *Dictyostelium discoideum* (Qiu et al., 2016). The virulence defects can be rescued by SdeA alone (Bardill et al., 2005; Qiu et al., 2016) demonstrating that, at least in these two hosts, a single paralog is sufficient for *L. pneumophila* intracellular replication.

A second example of molecular redundancy in *L. pneumophila* is the Lgt family of proteins consisting of Lgt1, Lgt2/LegC8, and Lgt3/LegC5 (de Felipe et al., 2005; Belyi et al., 2006) (**Figure 1B**). Each paralog is a functional glucosyltransferase that covalently modifies the host protein elongation factor 1A (eEF1A) at serine 53 via mono-O-glycosylation (Belyi et al., 2006, 2008). Modification of eEF1A by any of the three paralogs impairs host protein synthesis (Belyi et al., 2006, 2008). The simultaneous deletion of all three 3 paralogs does not impair *L. pneumophila* intracellular growth nor does it completely abolish host protein translation in infected cells (Belyi et al., 2006, 2008) suggesting that *L. pneumophila* encodes additional IDTS that modulate this process.

Effectors with Similar Activities Can Be Misinterpreted as Redundant

Effector paralogs are most readily identified by sequence and/or structural similarities. However, homology does not necessarily indicate that two proteins perform the same function and therefore have molecular redundancy. For example, the IDTS VipD is targeted to early endosomes through its interaction with the host protein Rab5 (Gaspar and Machner, 2014). Binding to Rab5 activates VipD phospholipase activity resulting in dephosphorylation of phosphoinositol 3-phosphate on endosomes (Gaspar and Machner, 2014). Sequence homology comparisons identified three paralogs of VipD encoded in the *L. pneumophila* genome: VdpA, VpdB, and VpdC, each consisting of a functional phospholipase domain based on the conservation of all active site and catalytic residues (VanRheenen et al., 2006; Gaspar and Machner, 2014). However, *in vitro* binding assays demonstrated that unlike VipD, neither VpdA nor VpdB bind Rab5 (Gaspar and Machner, 2014). Thus, while it appears that the catalytic activities of these four proteins are conserved and they are all likely to alter host phosphoinositide pools, their respective binding partners and the host pathways they modulate may vary significantly.

The unlikelihood of redundancy amongst larger families of effectors with conserved activities or domains is more apparent, as demonstrated by the five F-box domain-containing proteins of the E3 ubiquitin ligase family in *L. pneumophila* (Ensminger and Isberg, 2010). The F-box protein provides substrate specificity to the E3 ubiquitin ligase complex, typically Skp-Cullin-F-box (SCF) (Zheng et al., 2002). Despite their common F-box domain, only LegU1, LegAU13/AnkB, and LicA interact with Skp1 while only LegU1 and LegAU13/AnkB interact with CUL1 (Ensminger and Isberg, 2010). Moreover, the pattern of host protein ubiquitination varies significantly between the five IDTS (Ensminger and Isberg, 2010). Thus, despite similar biochemical activities, the host proteins they target for ubiquitination and the corresponding host processes they impact are likely to differ. Indeed, many families of IDTS with common functional

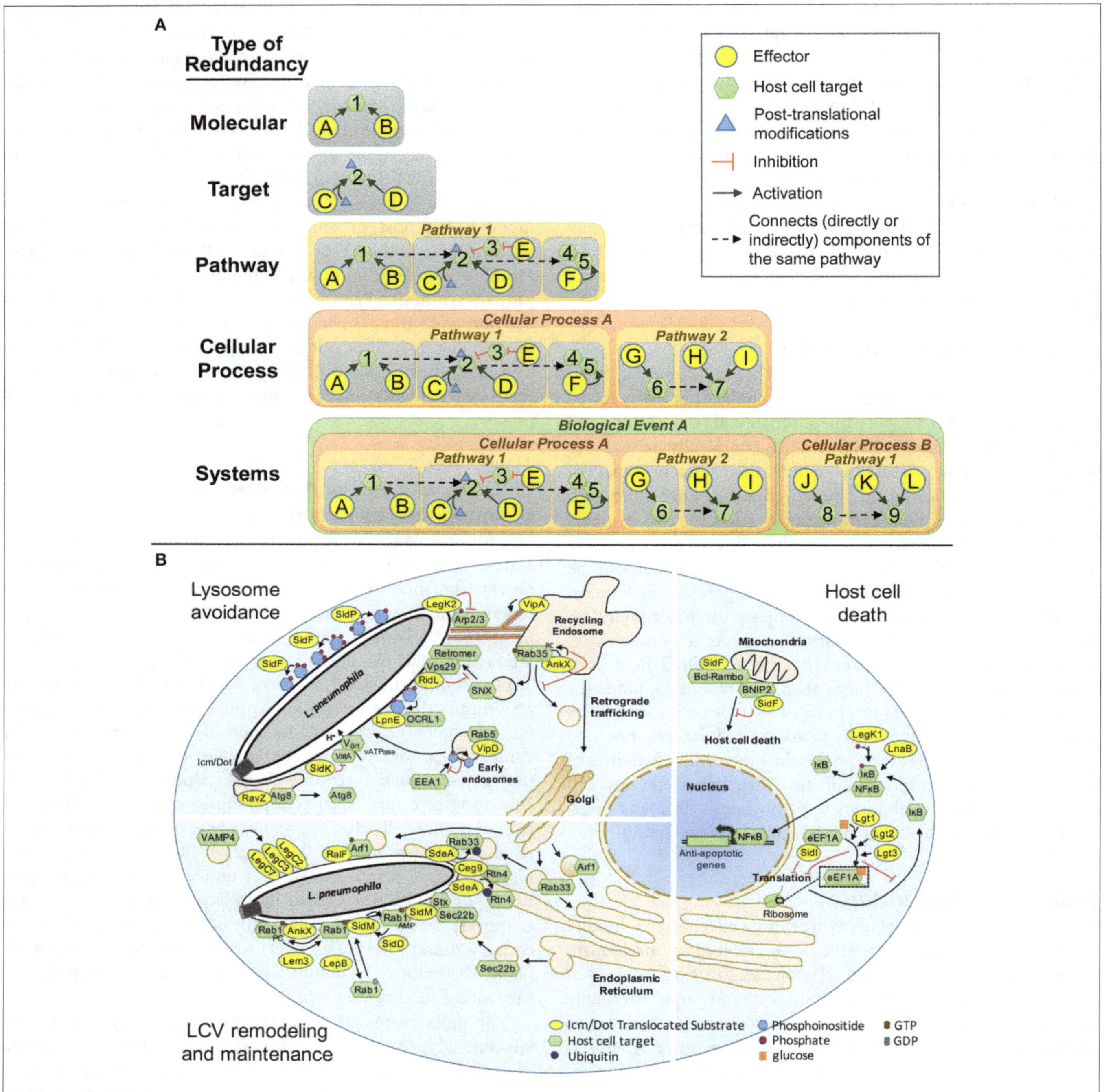

FIGURE 1 | Types of Redundancy. **(A)** Schematic representations of the 5 classes of redundancy: *Molecular*, two or more effectors that modify the same host target using the same molecular mechanism; *Target*, effectors that modulate the same host protein using different molecular mechanisms; *Pathway*, effectors that modulate a single host pathway but target different components of that pathway; *Cellular Process*, effectors that target redundant or complementary host pathways that collectively govern a single cellular process; *System*, effectors that modulate more than one host cellular process to accomplish a common goal. **(B)** Redundant Icm/Dot translocated substrates that modulate lysosomal trafficking, vacuole remodeling and maintenance and host cell death in *Legionella* pathogenesis.

domains such as kinases, phosphatases, ankyrin-repeat, or coil-coil domains (de Felipe et al., 2005) typically have additional, unrelated functional domains that set them apart.

Effectors that exhibit similar activities or target specificities *in vitro* or *in vivo* outside the context of an infection can also be misinterpreted as redundant, as these similarities may not translate to redundancy in the context of a host under native conditions. Additionally, enzymatic functions and/or target specificities defined *in vitro* could be biased based on the substrates and/or assays used to investigate them and thus, misleading as to their true functions within a host. As a consequence, *in vitro* studies may suggest redundancy between two proteins that are in fact quite distinct in the context of an infection.

Target Redundancy

Target redundancy defines effectors that modulate the same host protein using different molecular mechanisms (**Figure 1A**). In this case, the activity of one effector cannot replace the other but can have a similar impact on the function of the targeted host protein, and its component pathway. Thus, contrary to molecular redundancy, target redundancy defines redundant strategies rather than redundant activities. Effectors that are redundant at the target level are more difficult to identify because they typically lack sequence, structural and functional similarity.

Target redundancy is exemplified by the IDTS SidM/DrrA (Machner and Isberg, 2006; Murata et al., 2006) and AnkX/LegA8/AnkN (Pan et al., 2008; Mukherjee et al., 2011; Allgood et al., 2017), herein after referred to as AnkX (**Figure 1B**). Both effectors modulate the activity of the host small GTPase Rab1 to remodel the LCV but do so by different molecular mechanisms. SidM/DrrA AMPylates the GTP-bound form of Rab1 preventing GTP to GDP exchange by its cognate GAP protein (Muller et al., 2010), whereas AnkX phosphocholinates GTP-bound Rab1, locking it in the active state (Mukherjee et al., 2011). Two additional effectors SidD and Lem3 reverse Rab1 constitutive activation by de-AMPylation and de-phosphocholination, respectively (Neunuebel et al., 2011; Tan and Luo, 2011). Thus, *L. pneumophila* encodes two sets of IDTS, SidM-SidD and AnkX-Lem3 that are both able to regulate Rab1 activity but do so through different mechanisms.

A second example of target redundancy is observed between the glucosyltransferases Lgt1, Lgt2, Lgt3 (Belyi et al., 2008), and SidI (Shen et al., 2009) (**Figure 1B**). While the Lgt proteins and SidI both impair host protein synthesis by targeting eEFA1, SidI appears to do so by an alternative mechanism. Similar to Lgt1, Lgt2, and Lgt3, SidI directly interacts with eEFA1 to impair its function however, direct binding is not solely responsible for this effect (Shen et al., 2009). If SidI inactivates eEF1A through modification, the lack of a glycosyltransferase domain suggests it is likely to differ from glycosylation. In addition to Lgt1, Lgt2, Lgt3, and SidI, a fifth effector, SidL has been implicated in impairing host protein synthesis (Fontana et al., 2011), although the mechanism has yet to be elucidated including whether this occurs through eEF1A or another component of the translation machinery. Moreover, the deletion of all five of these IDTS only partially restores host protein synthesis (Fontana et al., 2011), suggesting that additional IDTS regulate this process. Thus, this example encompasses multiple types of redundancy from molecular and target redundancy to pathway and possibly, cellular process redundancy (see below). We predict that many virulence strategies will similarly consist of more than one type of redundancy.

Effectors with Similar Targets but Different Activities Can Be Misinterpreted as Redundant

It is important to distinguish between effectors that modulate the activity of common host proteins but do not achieve the same effect on the component host pathway: these types of effectors are not redundant. An example of non-redundant IDTS with a common target is SidM/DrrA and LepB. SidM/DrrA functions as a GDI displacement factor to recruit Rab1 to the *Legionella* vacuole, then constitutively activates Rab1 by locking it in the GTP bound form via covalent modification (Machner and Isberg, 2007; Muller et al., 2010). Upon de-AMPylation of Rab1 by SidD (Neunuebel et al., 2011), LepB acts as a Rab1 GTPase activating protein (GAP) promoting GTP hydrolysis and release of Rab1 from the LCV (Ingmundson et al., 2007). Although, SidM and LepB both target Rab1, they have opposite effects on its activity and distribution which differentially impacts Rab1-mediated vesicle trafficking events. Thus, SidM/DrrA and LepB are not target redundant.

Pathway Redundancy

Pathway redundancy defines effectors that modulate a single host pathway but target different components of that pathway (**Figure 1A**). Sets of effectors that belong to this category can manipulate different proteins in a single complex, different components at various steps along the pathway or regulators of the pathway. However, while the mechanisms and the host proteins used to modulate the pathway differ, the outcome of that modulation is the same and these effectors collectively serve to achieve a common goal.

Pathway redundancy is illustrated by the IDTS VipD and SidK that both modulate the endocytic pathway but do so by targeting different components at different stages of LCV maturation on the way to the lysosome (**Figure 1B**). On early endosomes, VipD dephosphorylates PI3P, which functions as an anchor for the tethering protein EEA1 (Gaspar and Machner, 2014). The lack of EEA1 at endosomal surfaces prevents endosome fusion with the LCV (Gaspar and Machner, 2014). Vacuole acidification occurs downstream of early endosome fusion events and is mediated by vATPases, multi-component proton pumps (Forgac, 2007). SidK directly binds VatA, a component of the vATPase to inhibit its function (Xu et al., 2010). While VipD can impair early endosome fusion with the LCV, it is not sufficient to avoid endosomal fusion completely as 40% of LCVs containing wild type bacteria stain positive for the early endosomal marker Rab5 (Gaspar and Machner, 2014). SidK acts as part of a contingency plan when endocytic maturation of the LCV is not completely thwarted. Moreover, while vacuoles containing the Δ*vipD* mutant are more likely to accumulate Rab5 than those containing wild type bacteria, the frequency is significantly lower than that observed for a *dot-* mutant (Gaspar and Machner, 2014): this suggests that other effectors function to modulate the endocytic pathway. Several effectors including VipA, VipF, SetA, and Ceg19 are likely candidates based on their ability to disrupt trafficking along the vacuole sorting pathway in yeast (Shohdy et al., 2005; Franco et al., 2012).

A second example of pathway redundancy is observed between the effectors RidL, LpnE, and AnkX which target separate mediators of retrograde trafficking between endosomes and the *trans*-Golgi network to alter the fate of the LCV (**Figure 1B**). RidL directly interacts with the retromer complex subunit Vps29 to compete with endosome sorting nexins for retromer and PI3P binding (Finsel et al., 2013). LpnE directly interacts with OCRL1 (Weber et al., 2009), a phosphoinositol 5-phosphatase that regulates retrograde trafficking by altering phosphoinositide phosphate pools. AnkX phosphocholinates Rab35 (Mukherjee et al., 2011), a regulator of cargo sorting and recycling from recycling endosomes. Modification of Rab35 prevents microtubule-dependent endosomal vesicle transport to the LCV (Pan et al., 2008). Loss of RidL, LpnE, or AnkX moderately increases the frequency of LAMP1 staining of LCVs demonstrating that the all three effectors independently contribute to disrupting maturation of the LCV along the endocytic pathway (Newton et al., 2007; Pan et al., 2008; Finsel et al., 2013).

Cellular Process Redundancy

Cellular process redundancy occurs when sets of effectors compensate for one another by targeting redundant or complementary host pathways that collectively govern a single cellular process (**Figure 1A**). An example of a cellular process that is mediated by multiple pathways is the unfolded protein response (UPR). The UPR is activated through three separate sensory pathways: inositol requiring enzyme-1 (IRE1), protein kinase RNA-like ER kinase (PERK), and activating transcription factor 6 (ATF6) (Walter and Ron, 2011). Distinct pathways allow the cell to respond to multiple signs of ER stress enhancing the sensitivity and breadth of the sensory system but all pathways lead to a common response that includes global translation inhibition, upregulation of ER stress proteins, ER membrane expansion and under extreme conditions, activation of pro-apoptotic pathways (Walter and Ron, 2011). While multiple pathways provide robustness to the host, it affords pathogens multiple ways to hijack a cellular process and when necessary, the ability to do so without completely abolishing the cellular process, which can have negative, even detrimental effects on the pathogen itself.

A critical event in *Legionella* pathogenesis is remodeling and maintenance of the LCV to support bacterial replication: this is accomplished through the recruitment of ER-derived membrane material. Three parallel mechanisms by which *Legionella* achieves this have been described (**Figure 1B**). The SdeA, SdeB, and SdeC family of IDTS drives rearrangement of tubular ER and its association with the LCV through ubiquitination of Rtn4 (Kotewicz et al., 2016), a regulator of tubular ER dynamics (Zurek et al., 2011) (**Figure 1B**). SidM/DrrA and RalF target components of the early secretory pathway to redirect vesicles trafficking between the ER and the Golgi to the LCV. SidM/DrrA does so by recruiting and activating Rab1 at the LCV and promoting non-canonical functional pairing between the plasma membrane tSNARE syntaxins at the LCV and the ER-derived vesicle vSNARE Sec22b (Arasaki et al., 2012). RalF does so by recruiting and activating the host protein ARF1 at the LCV (Nagai et al., 2002) (**Figure 1B**). The loss of either SidM/DrrA or RalF alters the timing and efficiency of ER protein accumulation at the LCV (Nagai et al., 2002; Ingmundson et al., 2007) demonstrating redundant roles for these proteins in LCV remodeling.

A second example of redundancy at the level of cellular processes is SidF and SidP (**Figure 1B**). Each effector contributes to modulation of host lipid metabolism to modulate the relative abundance of phosphoinositides (PIs) at the LCV, specifically conversion from a PI(3)P rich environment to a PI(4)P rich environment. SidF is a phosphoinositide 3-dephosphatase with specificity for $PI(3,4)P_2$ and $PI(3,4,5)P_3$ preventing PI(3)P accumulation at the LCV (Banga et al., 2007) while SidP is a phosphoinositide 3-phosphatase that hydrolyzes PI(3)P and $PI(3,5)P_2$ removing PI(3)P from the LCV (Toulabi et al., 2013). The host protein ORCL1, a $PI(4,5)P_2$ 5-phosphatase also localizes to the LCV and thus may also promote PI(4)P accumulation at the surface (Weber et al., 2009) (**Figure 1B**). OCRL1 targeting to the LCV is Icm/Dot-dependent but the specific IDTS required for this has yet to be determined (Weber et al., 2009). Phosphoinositides distinguish individual organelle membranes in the host cell and serve as anchors for organelle-specific host proteins. Several IDTS exploit PIs decorating the LCV to anchor themselves to the surface. Many of these IDTS have been implicated in LCV remodeling including SidM/DrrA, SidC, LidA, and RidL (Machner and Isberg, 2006; Murata et al., 2006; Finsel et al., 2013; Hsu et al., 2014). Loss of SidF impairs Rab1 recruitment to the LCV (Toulabi et al., 2013), likely as a consequence of the inability of SidM to attach itself to the LCV surface. Thus, PI dynamics play a central role determining the repertoires of IDTS at the vacuole surface and thus the fate of the *Legionella* vacuole.

System Redundancy

System redundancy defines effectors that modulate more than one host cellular process to accomplish a single task (**Figure 1A**). An example in biology of a single event that is governed by multiple host cellular processes is cell death. Cell death can be achieved through apoptosis, necrosis, pyroptosis, or autophagy. Each process may be triggered by different cues and the mechanisms by which the cell is terminated may vary but the result is the same—death. In some cases, components mediating these pathways are completely distinct; in other cases they may overlap. For a pathogen, the more options at its disposal for manipulating the host cell to accomplish a specific goal, the greater the likelihood of its success. System redundancy provides yet another layer of insurance by allowing a pathogen to tap into multiple cellular processes to ensure completion of a critical event.

Intracellular growth of *L. pneumophila* requires the viability of the host cell but cell death is induced by host cells when bacteria cannot be eradicated through lysosomal targeting. *L. pneumophila* regulates host cell death by targeting host signal transduction, translation, and apoptosis (**Figure 1B**). The IDTS LnaB and LegK1 activate the host transcription factor, nuclear factor κB (NFκB) causing upregulation of anti-apoptotic pathway-associated genes (Losick and Isberg, 2006; Ge et al., 2009). While the mechanism of action of LnaB is unknown, *in vitro* studies suggest that LegK1 promotes the degradation of the NFκB inhibitor IκB through direct phosphorylation (Ge et al., 2009). Lgt1, Lgt2, Lgt3, SidI, and SidL promote prolonged NFκB signaling by blocking host protein synthesis and thus cellular levels of IκB from being replenished (Fontana et al., 2011). SidF promotes host cell survival by inhibiting the activity of the pro-apoptotic proteins BNIP3 and Bcl-Rambo through direct binding (Banga et al., 2007). Thus, *L. pneumophila* orchestrates the induction of host cell survival mechanisms while simultaneously obstructing host cell death pathways by targeting distinct cellular processes.

Lysosomal avoidance by *L. pneumophila* is orchestrated through four separate cellular processes: the endocytic pathway using VipD (Gaspar and Machner, 2014) and SidK (Xu et al., 2010); retrograde transport via RidL (Finsel et al., 2013), LpnE (Weber et al., 2009), and AnkX (Mukherjee et al., 2011); actin cytoskeleton dynamics through LegK2 (Michard et al., 2015) and VipA (Franco et al., 2012); and autophagy by RavZ (Choy et al., 2012) (**Figure 1B**). The mechanisms of action of VipD, SidK, RidL, LpnE, and AnkX have been discussed previously (see section Pathway Redundancy). Altering endosome transport to the LCV is also achieved by manipulating the actin cytoskeleton. LegK2 phosphorylates the Arp2/3 complex subunits ARPC1B and ARP3 (Michard et al., 2015): this prevents actin nucleation at the site of the LCV thus perturbing endosome trafficking to the LCV (Michard et al., 2015). In contrast, VipA localizes to endosomes and promotes actin polymerization by directly binding to actin. In yeast, VipA impairs vacuole sorting and thus is predicted to similarly alter organelle trafficking during infection (Shohdy et al., 2005; Franco et al., 2012). Host cells are not without their own forms of redundancy. When bacteria fail to be delivered to the lysosome, host cells can also target pathogens to the lysosome via autophagy (Xie and Klionsky, 2007). The IDTS RavZ localizes to the LCV where it irreversibly deconjugates the autophagic protein Atg8 thereby preventing autophagasome membrane nucleation at the site of the LCV (Choy et al., 2012).

System redundancy is also exemplified by SidM/DrrA, SdeA, SdeB, SdeC, RalF, and a functional complex formed by LegC2/YlfB, LegC3, and LegC7 (**Figure 1B**). SidM/DrrA, SdeABC, and RalF modulate LCV remodeling by hijacking tubular ER dynamics and vesicle trafficking along the early secretory pathway (see section Cellular Process Redundancy). LegC2/YlfB, LegC3, and LegC7 collectively mimic Q-SNARE proteins and directly bind the R-SNARE protein VAMP4 (Shi et al., 2016). LegC2/YlfB-LegC3-LegC7/YlfA complex pairing with VAMP4 diverts VAMP4-containing vesicle trafficking along the retrograde transport pathway between endosomes and the *trans*-Golgi network to the LCV (Shi et al., 2016). *L. pneumophila* mutants lacking LegC2/YlfB and LegC7/YlfA show reduced accumulation of the ER marker calnexin at the LCV but do not exhibit an increase in LAMP1 staining (Campodonico et al., 2016). Thus, recruitment of VAMP4-containing vesicles serves to remodel and maintain the LCV but does not impact LCV trafficking to the lysosome (Campodonico et al., 2016; Shi et al., 2016). Differential targeting of endosomes to the LCV suggests the existence of distinct populations of endosomal vesicles, some of which are actively recruited to the LCV to enable *L. pneumophila* replication while others are actively excluded because they promote *L. pneumophila* trafficking to the lysosome.

Effectors and Host Target Specificity

Several studies have identified a number of effectors capable of interacting with more than one host target that often function in more than one host pathway or host cellular process. The most common example in *L. pneumophila* pathogenesis is IDTS that target host Rab proteins, the gatekeepers of membrane transport and trafficking. In addition to Rab1, SidM/DrrA also binds Rab8B, Rab10, and Rab27A (Machner and Isberg, 2006; Yu et al., 2015). Similarly, the IDTS LidA binds activated Rab1 and Rab6A (Machner and Isberg, 2006; Murata et al., 2006; Chen and Machner, 2013) but has also been shown to interact with Rab8B, Rab10, and Rab27A (Yu et al., 2015). Lpg0393 is a guanine nucleotide exchange factor for Rab5, Rab21, and Rab22, all of which are associated with endosomal trafficking (Sohn et al., 2015) while PieE can interact with Rab1, Rab2, Rab5c, Rab6a, and Rab7 (Mousnier et al., 2014) which encompass various stages of secretory, endocytic, and endosome recycling pathways as well as late endosome- and autophagosome-lysosome fusion events (Stenmark, 2009). While overlapping functions between IDTS may provide a source of redundancy and thus, insurance against failure to complete critical events in the infection cycle, it can also be a potential source of decreased specificity. In cases where *L. pneumophila* has to exploit subtle differences in host cellular pathways, for instance to discriminating between subpopulations of endosomal vesicles, redundancy may be less beneficial. Importantly, many of the Rab protein targets were identified using *in vitro* systems or *in vivo* systems outside the context of infection. In the case of SidM/DrrA, initial screening experiments identified seven putative Rab protein targets but subsequent validation experiments narrowed the list down to only two (Yu et al., 2015). Thus, extreme caution has to be exercised in assigning redundant functions before the biological relevance of effector-host target interactions is determined.

SELECTIVE PRESSURES DRIVING THE MAINTENANCE OF REDUNDANT VIRULENCE PROTEINS

Genetic redundancy is unstable over time. Genes performing similar functions tend to experience genetic drift, unless each gene undergoes independent selective pressure (Clark, 1994; Force et al., 1999; Bergthorsson et al., 2007). So how are redundant proteins maintained? The simplest explanation is that so-called redundant effectors have both overlapping and distinct functions and that selection for their independent activities drives the maintenance of their redundant functions. For example, within the Lgt1/Lgt2/Lgt3/SidI/SidL family of IDTS that inhibit protein synthesis by targeting eEF1A, SidI also interacts with eEF1Bγ (Shen et al., 2009) another component of the translation machinery (Browne and Proud, 2002). Similarly, Lgt1 has a second putative binding partner, Hsb1 that plays a role in mRNA surveillance during translation (Belyi et al., 2009). In addition, members of this family vary in their ability to block the unfolded protein response during *L. pneumophila* infection (Hempstead and Isberg, 2015; Treacy-Abarca and Mukherjee, 2015). While the significance of these differences has not been elucidated, the independent activities of individual members of this group may be responsible for their maintenance in the genome despite their apparent redundant functions.

Redundancy amongst effectors may compensate for temporal or regulatory differences in gene expression. For instance, Lgt1 is expressed early in the infection cycle while Lgt3 is expressed at later stages prior to bacterial egress (Belyi et al., 2008). The overlapping functions of effectors may allow a specific host processes to be modulated throughout the infection cycle despite differences in their individual expression patterns. Differences in gene expression between redundant effectors may correlate with requirements for their non-overlapping functions at different stages of the infection cycle or differences in the regulatory mechanisms controlling their expression. While redundancy resulting from gene duplication is likely to establish common regulatory networks for individual paralogs, this is unlikely for independently acquired redundant effector genes that are dispersed throughout the genome. Indeed, the mechanisms by which newly acquired effector genes are integrated into existing regulatory networks are not well established. Conservation of redundant effectors may ensure their functions are fulfilled despite variations in their respective gene expression patterns.

Redundancy between effectors may drive the maintenance of redundant virulence strategies when a host protein, pathway or process is impaired by a single effector but not completely abolished. For instance, while VipD can impair fusion of early endosomes with the LCV, it is not sufficient to avoid it completely (Gaspar and Machner, 2014). Variations in the numbers of endosomes in a host cell, the timing and amount of VipD translocated into the host cell, the efficiency of VipD targeting to endosomes, variations in substrate abundance and/or rates of catalysis or the efficiency of endosome fusion with the LCV may render VipD insufficient to avoid downstream events of the endocytic pathway. SidK (Xu et al., 2010) is part of a back-up plan when inhibition of endosome fusion with the LCV is incomplete. As many IDTS, including VipD are toxic when expressed at high levels, the need to limit effector abundance may restrict the ability of any one effector to completely control a particular event. Additional IDTS like LegK2 (Michard et al., 2015), AnkX (Mukherjee et al., 2011), RidL (Finsel et al., 2013), and VipA (Franco et al., 2012) allow *L. pneumophila* to impair endocytic maturation of the LCV at different points without obliterating major cellular processes.

Redundancy can provide an advantage when enhanced fidelity is required for critical functions (Thomas, 1993). Variations in the host cell type or fluctuations in their external environment may necessitate redundant virulence strategies. In its natural habitat, *L. pneumophila* is destined to encounter many amoebal species thus, the greater the number of amoebae *L. pneumophila* can survive and replicate within, the greater its fitness. The importance of individual IDTS could be impacted by multiple factors: differences in amoebal cell biology, nutrient availability, variations in host targets that impact their recognition or manipulation by IDTS or differences in the components or pathways governing cellular processes targeted by *L. pneumophila*. Maintaining a large cohort of IDTS arms the bacterium with the specific combinations of IDTS necessary for optimal growth in multiple hosts but as a consequence may indirectly result in the accumulation of IDTS that perform overlapping or redundant functions under certain circumstances. Redundancy may also provide a means for pathogens to evolve virulence strategies without compromising fitness. For pathogens that are subject to dynamic and unpredictable environments, have broad host ranges, or find themselves in a perpetual co-evolutionary arms race with their host, this is particularly important.

WHEN REDUNDANCY IS NOT REDUNDANCY AT ALL

Redundancy is not the only explanation for the absence of phenotypes associated with genetic mutations. Whether a gene is required for pathogenesis can vary depending on the host examined, the conditions under which gene requirements are assessed or the type and sensitivity of the assay used. As a consequence, what may be perceived as redundancy is instead an inability to detect phenotypes using a particular experimental system. For example, the IDTS SdhA is essential for *L. pneumophila* replication in bone marrow-derived primary macrophages but not in cultured U937 cells, a monocyte-derived macrophage cell line (Laguna et al., 2006). Loss of SdhA causes the induction of host cell death in response to *L. pneumophila* challenge (Laguna et al., 2006; Creasey and Isberg, 2012), which is likely to differ between primary and immortalized cells. Similarly, the SidE family of IDTS is important for *L. pneumophila* growth in amoebal hosts but is dispensable in primary macrophages (Bardill et al., 2005; Qiu et al., 2016). In these two cases, the host cell type greatly impacts whether a gene is designated as important for *L. pneumophila* pathogenesis. While redundancy is becoming the default justification for a lack of phenotypes, it is not always the culprit. As more effectors and the host processes they modulate are characterized, key differences between effectors that appear to be redundant will most certainly be revealed.

METHODS TO RESOLVE REDUNDANCY

A significant body of work has focused on defining the role of individual virulence factors in isolation, yet understanding how these components coordinately contribute to pathogenesis is necessary to define key determinants of disease. This is particularly important for pathogens that employ compensatory virulence strategies, as redundancy can greatly impact the ability to define what a pathogen requires to survive and grow within a host. A number of strategies have emerged to address redundancy in bacterial pathogenesis that encompass genetic, biochemical and bioinformatics-based techniques. While many of the approaches do not specifically determine redundant mechanisms at a molecular level, they do define functional relationships between individual proteins, and in some cases the host pathways they target, enabling targeted analyses to decipher the basis of redundancy at multiple levels.

Brute-Force Characterization One Effector at a Time

Combined biochemical, molecular, and cell biological characterization of effectors is the most comprehensive way to identify redundant proteins. It provides detailed information about their mechanism of action, their host cell targets and their direct impact on host cellular processes. Moreover, deciphering the intricate details of an effector's function can define subtle distinctions between effectors with overlapping functions and thus circumvent their improper classification as redundant. However, this method is not without its drawbacks. The amount of time required to exhaustively characterize protein function can be lengthy, especially if the techniques to do so are not available or the function of the host target protein or its component pathway have yet to be characterized. For a pathogen like *L. pneumophila* that employs at least 270 IDTS, such an endeavor would be an arduous one. In addition, for many effectors, sequence homology and structure prediction tools are not always informative. For *L. pneumophila*, as many as one third of all IDTS lack domain homology to any other protein characterized to date and often very little is learned from structural predictions. Finally, while there are several methods to define host targets, including more recently adapted high throughput methods (Mousnier et al., 2014; Yu et al., 2015), this can be challenging as host targets can range from proteins to lipids to small molecules (Machner and Isberg, 2006; Toulabi et al., 2013; Isaac et al., 2015). Thus, while characterizing individual effector functions can be highly informative, the road to defining redundant virulence mechanisms can be bumpy and painstakingly slow.

Insertional Mutagenesis and Depletion (iMAD)

iMAD is a genetic screening strategy developed to resolve redundancy amongst effectors by defining sets of bacterial proteins that target common host pathways and parallel pathways exploited by a pathogen to accomplish a single task (O'Connor et al., 2012; O'Connor and Isberg, 2014). To do so, iMAD integrates bacterial mutagenesis and host RNA interference to systematically identify genetic interactions between a pathogen gene and a host gene based on impaired replication of the pathogen (O'Connor et al., 2012; O'Connor and Isberg, 2014). In the case of *L. pneumophila*, a library of transposon mutants were assessed for their ability to replicate within host cells depleted of one of five early secretory proteins that promote *L. pneumophila* intracellular growth (O'Connor et al., 2012). Hierarchical clustering of bacterial gene mutations in IDTS with similar behavioral patterns across all host conditions examined revealed several important functional relationships: (1) Common phenotypic signatures identified sets of bacterial proteins that target common host pathways: these sets of proteins defined distinct functional groups; (2) Deleting pairs of bacterial genes from separate functional groups impaired intracellular growth of *L. pneumophila*: these functional groups defined separate but redundant host pathways targeted by *L. pneumophila* to generate a replication vacuole; (3) Specific defects in host cell biology resulting from loss of bacterial proteins could be predicted for genes based on the characterization of other members of its group: this identified three sets of proteins that independently contribute to the maintenance of replication vacuole membrane integrity; (4) Different combinations of bacterial genes were required for optimal growth in different hosts defining sources of adaptation to host variation. By grouping individual effectors that commonly manipulate a single host pathway and redundant pathways that contribute to a single process, iMAD defines functional relationships between effectors at the target, pathway, cellular process, and system levels. With more efficient methods for generating arrayed bacterial mutant libraries, commercially available RNAi libraries and the development of CRISPR technology to facilitate host cell protein depletion, and the replacement of DNA microarrays with massively parallel sequencing techniques to monitor bacterial mutant populations, more comprehensive, high-throughput iMAD screens are now possible.

Genome Reduction and Minimal Effector Repertoires

Genome reduction followed by effector repertoire reconstitution is another strategy used to identify redundant effectors and the host pathways they target (Cunnac et al., 2011). Progressive removal of all 28 effector genes from the plant pathogen *Pseudomonas syringae Pto* DC3000 determined 15 of the effector genes are collectively dispensable for growth in the plant host *Nicotiana benthamiana* (Kvitko et al., 2009). The subsequent reintroduction of different combinations of effectors defined several redundant-effector groups (REGs) that promote *P. syringae* growth in *N. benthamiana*, two of which were determined to mediate resistance to independent arms of plant innate immunity (Block and Alfano, 2011; Cunnac et al., 2011). By analyzing correlates between effector combinations and rescued *P. syringae* growth during infection, a minimal set of 8 effectors was defined that was sufficient to promote growth of *P. syringae* to near wild type levels. The effector repertoire reconstitution linked individual REGs with the host pathways they target, identified specific combinations of effectors that are sufficient to cause disease and demonstrated the ability to swap different members of individual REGs and still achieve robust *P. syringae* growth. For genetically amenable pathogens with manageable sizes of effector repertoires, preferably clustered in a minimal set of genetic loci to facilitate combinatorial effector reintroduction into the genome, effector repertoire reduction, and reconstitution strategies provides a comprehensive method to define redundant effectors, the cellular processes they modulate and the minimum set of effector functions required for pathogenesis.

Effector Interactome Mapping

Proteome-based analyses that use high throughput mass spectrometry allow the interactomes of entire effector subfamilies to be mapped. Affinity purified-mass spectrometry has been used to define the host interacting partners of 58 secreted virulence factors in *Chlamydia trachomatis* called inclusion membrane proteins (Incs) (Mirrashidi et al., 2015). The results not only allowed groups of effectors to be assigned to specific cellular processes but identified sets of Inc proteins that target the same host proteins or different members of the same multiprotein complex. Mapping the interactome network of all 58 Inc proteins revealed sets of Inc proteins that converge on common targets, pathways and cellular processes: this defined focal points of host modification by *C. trachomatis* and thus, potential sources of redundancy. Moreover, the *C. trachomatis* Inc-human interactome had significant overlap with that of other pathogens. Comparisons with three viral-human interactomes (Jager et al., 2011; Davis et al., 2015; Ramage et al., 2015) identified 98 shared host targets between *C. trachomatis* and at least one of the three viruses. Similarly, a number of the Inc host protein targets are also common targets of *L. pneumophila* IDTS including Rtn4 (Kotewicz et al., 2016), vATPases (Xu et al., 2010), and the retromer complex (Finsel et al., 2013). The lack of similarly between the respective *C. trachomatis* Inc proteins and *L. pneumophila* IDTS and differences in the host protein complex subunits targeted demonstrates that each bacterial pathogen has acquired or evolved different mechanisms to modulate the same host proteins, pathways and/or cellular processes. Comparing effector functions between pathogens not only allows additional redundant virulence mechanisms to be defined but establishes a critical set of events central to microbial pathogenesis.

Comparative and Functional Genomics Can Predict Redundant Virulence Mechanisms

Comparative genomics combined with phenotypic analyses provide a means to define correlates between effector conservation and redundant virulence mechanisms (Baltrus et al., 2011). The *P. syringae* pan-genome effector repertoire consists of 57 effectors but different subsets of effectors are sufficient for growth in the same plant host. Computational analyses that correlate specific combinations of effectors and host tropism (Baltrus et al., 2011) allow redundant virulence mechanisms to be elucidated on a global scale. Alternatively, comparative genomes can be used as a more targeted approach. For example, in *L. pneumophila* SdhA is a critical virulence determinant in macrophages (Laguna et al., 2006). While the precise function of SdhA is still unclear, the severe growth defect of the *ΔsdhA* mutant is due to loss of vacuole integrity that leads to a robust host innate immune response and consequently either bacterial or host cell death (Laguna et al., 2006; Creasey and Isberg, 2012). *L. pneumophila* encodes two paralogs of SdhA, SidH, and SdhB, but their deletion only moderately enhances the already severe intracellular growth defect of the Δ*sdhA* mutant (Laguna et al., 2006). *Legionella feeleii* lacks a *sdhA* paralog but grows almost as well as the wild type strain of *L. pneumophila* in macrophages (**Figure 2**). While the presence of *sidH* in *L. feeleii* may compensate for the absence of *sdhA*, deletion of *sidH* does not impair *L. feeleii* growth in macrophages (**Figure 2**). The dispensability of SdhA (and SidH) in *L. feeleii* is not due to the lack of *plaA* and/or *traI*, which suppresses the Δ*sdhA* mutant phenotype in *L. pneumophila* (Creasey and Isberg, 2012). Thus, while SdhA plays a critical role in *L. pneumophila* pathogenesis, the entire family of SidH paralogs is dispensable in *L. feeleii*. While there are a number of explanations for this discrepancy, *L. feeleii* encodes 27 additional putative IDTS that are not conserved in *L. pneumophila* (Burstein et al., 2016), one or more of which may compensate for the absence of SdhA despite their lack of homology. As more genomes of pathogen isolates are sequenced, correlates between effector conservation and phenotypes will allow alternate virulence mechanisms employed by pathogens to be defined.

FUTURE DIRECTIONS

Much of the research in microbial pathogenesis employs a reductionist's approach, where the individual components are investigated in isolation. While this strategy has proven extremely

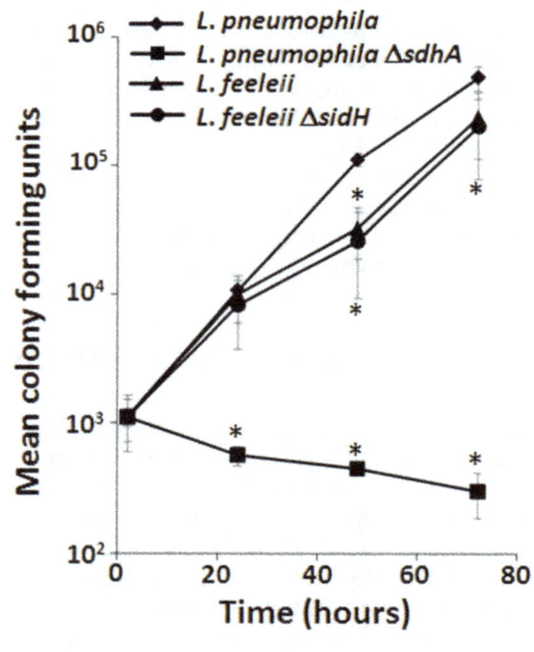

FIGURE 2 | Lack of the SidH family of Dot/Icm translocated substrates does not impair growth of *L. feeleii* in macrophages despite being indispensable in *L. pneumophila*. Growth of wild type *L. pneumophila*, *L. pneumophila* Δ*sdhA*, wild type *L. feeleii* and *L. feeleii* Δ*sidH* in A/J mouse bone marrow-derived macrophages, based on recovered colony forming units (CFU) on solid media from lysed host cells, was monitored over 72 h encompassing 3 consecutive rounds of infection (Supplemental Material). Plotted is the total bacterial yield at the indicated time points normalized to the *L. pneumophila* wild-type strain by the number of intracellular bacteria 2 h post infection. Data are representative of at least 2 independent experiments ± standard deviation of 3 replicates. An asterisk indicates a $P < 0.05$ based on a Student's *t*-test relative to the *L. pneumophila* wild type strain.

useful in identifying key players and their functions, it does not offer tremendous insight into the complex interactions that exist at the systems level. Pathogens invest an incredible amount of resources to build a robust virulence strategy. Redundancy allows pathogens to rapidly adapt to frequently changing environments and the elaborate, multi-tiered antimicrobial strategies employed by their hosts. As more and more effectors are characterized, a striking pattern of redundancy is beginning to emerge. In this review, we establish a structured nomenclature for the different forms of redundancy observed across multiple levels of biological organization. The types of redundancy defined here are not mutually exclusive nor are they expected to be exhaustive as more virulence factors are characterized. Instead, we offer a framework to generate a broader, more dynamic view of the mechanisms governing microbial pathogenesis.

AUTHOR CONTRIBUTIONS

SG generated data presented in the manuscript. SG and TO wrote the manuscript.

ACKNOWLEDGMENTS

We thank Joseph Vogel, Kim Davis, Mohammad Hossain, Sara Rego, and Jason Park for thoughtful review of the manuscript. This work was supported by the National Institutes of Health grant 1R01AI125402 (TO).

REFERENCES

Abu-Zant, A., Jones, S., Asare, R., Suttles, J., Price, C., Graham, J., et al. (2007). Anti-apoptotic signalling by the Dot/Icm secretion system of *L. pneumophila*. *Cell Microbiol.* 9, 246–264. doi: 10.1111/j.1462-5822.2006.00785.x

Allard, K. A., Dao, J., Sanjeevaiah, P., McCoy-Simandle, K., Chatfield, C. H., Crumrine, D. S., et al. (2009). Purification of Legiobactin and importance of this siderophore in lung infection by *Legionella pneumophila*. *Infect. Immun.* 77, 2887–2895. doi: 10.1128/IAI.00087-09

Allgood, S. C., Romero Dueñas, B. P., Noll, R. R., Pike, C., Lein, S., and Neunuebel, M. R. (2017). Legionella effector AnkX disrupts host cell endocytic recycling in a phosphocholination-dependent manner. *Front. Cell. Infect. Microbiol.* 7:397. doi: 10.3389/fcimb.2017.00397

Arasaki, K., Toomre, D. K., and Roy, C. R. (2012). The *Legionella pneumophila* effector DrrA is sufficient to stimulate SNARE-dependent membrane fusion. *Cell Host Microbe* 11, 46–57. doi: 10.1016/j.chom.2011.11.009

Baltrus, D. A., Nishimura, M. T., Romanchuk, A., Chang, J. H., Mukhtar, M. S., Cherkis, K., et al. (2011). Dynamic evolution of pathogenicity revealed by sequencing and comparative genomics of 19 *Pseudomonas syringae* isolates. *PLoS Pathog.* 7:e1002132. doi: 10.1371/journal.ppat.1002132

Banga, S., Gao, P., Shen, X., Fiscus, V., Zong, W. X., Chen, L., et al. (2007). *Legionella pneumophila* inhibits macrophage apoptosis by targeting pro-death members of the Bcl2 protein family. *Proc. Natl. Acad. Sci. U.SA.* 104, 5121–5126. doi: 10.1073/pnas.0611030104

Bardill, J. P., Miller, J. L., and Vogel, J. P. (2005). IcmS-dependent translocation of SdeA into macrophages by the *Legionella pneumophila* type IV secretion system. *Mol. Microbiol.* 56, 90–103. doi: 10.1111/j.1365-2958.2005.04539.x

Barry, K. C., Fontana, M. F., Portman, J. L., Dugan, A. S., and Vance, R. E. (2013). IL-1alpha signaling initiates the inflammatory response to virulent *Legionella pneumophila in vivo*. *J. Immunol.* 190, 6329–6339. doi: 10.4049/jimmunol.1300100

Belyi, Y., Niggeweg, R., Opitz, B., Vogelsgesang, M., Hippenstiel, S., Wilm, M., et al. (2006). *Legionella pneumophila* glucosyltransferase inhibits host elongation factor 1A. *Proc. Natl. Acad. Sci. U.S.A.* 103, 16953–16958. doi: 10.1073/pnas.0601562103

Belyi, Y., Stahl, M., Sovkova, I., Kaden, P., Luy, B., and Aktories, K. (2009). Region of elongation factor 1A1 involved in substrate recognition by *Legionella pneumophila* glucosyltransferase Lgt1: identification of Lgt1 as a retaining glucosyltransferase. *J. Biol. Chem.* 284, 20167–20174. doi: 10.1074/jbc.M109.008441

Belyi, Y., Tabakova, I., Stahl, M., and Aktories, K. (2008). Lgt: a family of cytotoxic glucosyltransferases produced by *Legionella pneumophila*. *J. Bacteriol.* 190, 3026–3035. doi: 10.1128/JB.01798-07

Berger, K. H., and Isberg, R. R. (1993). Two distinct defects in intracellular growth complemented by a single genetic locus in *Legionella pneumophila*. *Mol. Microbiol.* 7, 7–19. doi: 10.1111/j.1365-2958.1993.tb01092.x

Berger, K. H., Merriam, J. J., and Isberg, R. R. (1994). Altered intracellular targeting properties associated with mutations in the *Legionella pneumophila* dotA gene. *Mol. Microbiol.* 14, 809–822. doi: 10.1111/j.1365-2958.1994.tb01317.x

Bergthorsson, U., Andersson, D. I., and Roth, J. R. (2007). Ohno's dilemma: evolution of new genes under continuous selection. *Proc. Natl. Acad. Sci. U.S.A.* 104, 17004–17009. doi: 10.1073/pnas.0707158104

Block, A., and Alfano, J. R. (2011). Plant targets for *Pseudomonas syringae* type III effectors: virulence targets or guarded decoys? *Curr. Opin. Microbiol.* 14, 39–46. doi: 10.1016/j.mib.2010.12.011

Brand, B. C., Sadosky, A. B., and Shuman, H. A. (1994). The *Legionella pneumophila* icm locus: a set of genes required for intracellular multiplication in human macrophages. *Mol. Microbiol.* 14, 797–808. doi: 10.1111/j.1365-2958.1994.tb01316.x

Browne, G. J., and Proud, C. G. (2002). Regulation of peptide-chain elongation in mammalian cells. *Eur. J. Biochem.* 269, 5360–5368. doi: 10.1046/j.1432-1033.2002.03290.x

Burstein, D., Amaro, F., Zusman, T., Lifshitz, Z., Cohen, O., Gilbert, J. A., et al. (2016). Genomic analysis of 38 Legionella species identifies large and diverse effector repertoires. *Nat. Genet.* 48, 167–175. doi: 10.1038/ng.3481

Campodonico, E. M., Chesnel, L., and Roy, C. R. (2005). A yeast genetic system for the identification and characterization of substrate proteins transferred into host cells by the *Legionella pneumophila* Dot/Icm system. *Mol. Microbiol.* 56, 918–933. doi: 10.1111/j.1365-2958.2005.04595.x

Campodonico, E. M., Roy, C. R., and Ninio, S. (2016). *Legionella pneumophila* type IV effectors YlfA and YlfB are SNARE-like proteins that form homo- and heteromeric complexes and enhance the efficiency of vacuole remodeling. *PLoS ONE* 11:e0159698. doi: 10.1371/journal.pone.0159698

Chen, Y., and Machner, M. P. (2013). Targeting of the small GTPase Rab6A' by the *Legionella pneumophila* effector LidA. *Infect. Immun.* 81, 2226–2235. doi: 10.1128/IAI.00157-13

Choy, A., Dancourt, J., Mugo, B., O'Connor, T. J., Isberg, R. R., Melia, T. J., et al. (2012). The Legionella effector RavZ inhibits host autophagy through irreversible Atg8 deconjugation. *Science* 338, 1072–1076. doi: 10.1126/science.1227026

Clark, A. G. (1994). Invasion and maintenance of a gene duplication. *Proc. Natl. Acad. Sci. U.S.A.* 91, 2950–2954. doi: 10.1073/pnas.91.8.2950

Cocchiaro, J. L., and Valdivia, R. H. (2009). New insights into Chlamydia intracellular survival mechanisms. *Cell. Microbiol.* 11, 1571–1578. doi: 10.1111/j.1462-5822.2009.01364.x

Coers, J., Vance, R. E., Fontana, M. F., and Dietrich, W. F. (2007). Restriction of *Legionella pneumophila* growth in macrophages requires the concerted action of cytokine and Naip5/Ipaf signalling pathways. *Cell. Microbiol.* 9, 2344–2357. doi: 10.1111/j.1462-5822.2007.00963.x

Creasey, E. A., and Isberg, R. R. (2012). The protein SdhA maintains the integrity of the Legionella-containing vacuole. *Proc. Natl. Acad. Sci. U.S.A.* 109, 3481–3486. doi: 10.1073/pnas.1121286109

Cunnac, S., Chakravarthy, S., Kvitko, B. H., Russell, A. B., Martin, G. B., and Collmer, A. (2011). Genetic disassembly and combinatorial reassembly identify a minimal functional repertoire of type III effectors in *Pseudomonas syringae*. *Proc. Natl. Acad. Sci. U.S.A.* 108, 2975–2980. doi: 10.1073/pnas.1013031108

Davis, Z. H., Verschueren, E., Jang, G. M., Kleffman, K., Johnson, J. R., Park, J., et al. (2015). Global mapping of herpesvirus-host protein complexes reveals a transcription strategy for late genes. *Mol. Cell* 57, 349–360. doi: 10.1016/j.molcel.2014.11.026

de Felipe, K. S., Pampou, S., Jovanovic, O. S., Pericone, C. D., Ye, S. F., Kalachikov, S., et al. (2005). Evidence for acquisition of Legionella type IV secretion substrates via interdomain horizontal gene transfer. *J. Bacteriol.* 187, 7716–7726. doi: 10.1128/JB.187.22.7716-7726.2005

Derré, I., and Isberg, R. R. (2004). *Legionella pneumophila* replication vacuole formation involves rapid recruitment of proteins of the early secretory system. *Infect. Immun.* 72, 3048–3053. doi: 10.1128/IAI.72.5.3048-3053.2004

Dorer, M. S., Kirton, D., Bader, J. S., and Isberg, R. R. (2006). RNA interference analysis of legionella in drosophila cells: exploitation of early secretory apparatus dynamics. *PLoS Pathogens* 2:e34. doi: 10.1371/journal.ppat.0020034

Downing, K. J., Mischenko, V. V., Shleeva, M. O., Young, D. I., Young, M., Kaprelyants, A. S., et al. (2005). Mutants of Mycobacterium tuberculosis lacking three of the five rpf-like genes are defective for growth *in vivo* and for resuscitation *in vitro*. *Infect. Immun.* 73, 3038–3043. doi: 10.1128/IAI.73.5.3038-3043.2005

Ensminger, A. W., and Isberg, R. R. (2010). E3 ubiquitin ligase activity and targeting of BAT3 by multiple *Legionella pneumophila* translocated substrates. *Infect. Immun.* 78, 3905–3919. doi: 10.1128/IAI.00344-10

Evans, A. S. (1976). Causation and disease: the Henle-Koch postulates revisited. *Yale J. Biol. Med.* 49, 175–195.

Falkow, S. (1988). Molecular Koch's postulates applied to microbial pathogenicity. *Rev. Infect. Dis.* 10 (Suppl. 2), S274–S276. doi: 10.1093/cid/10.Supplement_2.S274

Falkow, S. (2004). Molecular Koch's postulates applied to bacterial pathogenicity–a personal recollection 15 years later. *Nat. Rev. Microbiol.* 2, 67–72. doi: 10.1038/nrmicro799

Finsel, I., Ragaz, C., Hoffmann, C., Harrison, C. F., Weber, S., van Rahden, V. A., et al. (2013). The Legionella effector RidL inhibits retrograde trafficking to promote intracellular replication. *Cell Host Microbe* 14, 38–50. doi: 10.1016/j.chom.2013.06.001

Fontana, M. F., Banga, S., Barry, K. C., Shen, X., Tan, Y., Luo, Z. Q., et al. (2011). Secreted bacterial effectors that inhibit host protein synthesis are critical for induction of the innate immune response to virulent *Legionella pneumophila*. *PLoS Pathog.* 7:e1001289. doi: 10.1371/journal.ppat.1001289

Force, A., Lynch, M., Pickett, F. B., Amores, A., Yan, Y. L., and Postlethwait, J. (1999). Preservation of duplicate genes by complementary, degenerative mutations. *Genetics* 151, 1531–1545.

Forgac, M. (2007). Vacuolar ATPases: rotary proton pumps in physiology and pathophysiology. *Nat. Rev. Mol. Cell Biol.* 8, 917–929. doi: 10.1038/nrm2272

Franco, I. S., Shohdy, N., and Shuman, H. A. (2012). The *Legionella pneumophila* effector VipA is an actin nucleator that alters host cell organelle trafficking. *PLoS Pathog.* 8:e1002546. doi: 10.1371/journal.ppat.1002546

Ganapathy, U., Marrero, J., Calhoun, S., Eoh, H., de Carvalho, L. P., Rhee, K., et al. (2015). Two enzymes with redundant fructose bisphosphatase activity sustain gluconeogenesis and virulence in Mycobacterium tuberculosis. *Nat. Commun.* 6, 7912. doi: 10.1038/ncomms8912

Gaspar, A. H., and Machner, M. P. (2014). VipD is a Rab5-activated phospholipase A1 that protects *Legionella pneumophila* from endosomal fusion. *Proc. Natl. Acad. Sci. U.S.A.* 111, 4560–4565. doi: 10.1073/pnas.1316376111

Ge, J., Xu, H., Li, T., Zhou, Y., Zhang, Z., Li, S., et al. (2009). A Legionella type IV effector activates the NF-kappaB pathway by phosphorylating the IkappaB family of inhibitors. *Proc. Natl. Acad. Sci. U.S.A.* 106, 13725–13730. doi: 10.1073/pnas.0907200106

Hempstead, A. D., and Isberg, R. R. (2015). Inhibition of host cell translation elongation by *Legionella pneumophila* blocks the host cell unfolded protein response. *Proc. Natl. Acad. Sci. U.S.A.* 112, E6790–E6797. doi: 10.1073/pnas.1508716112

Hersh, D., Monack, D. M., Smith, M. R., Ghori, N., Falkow, S., and Zychlinsky, A. (1999). The Salmonella invasin SipB induces macrophage apoptosis by binding to caspase-1. *Proc. Natl. Acad. Sci. U.S.A.* 96, 2396–2401. doi: 10.1073/pnas.96.5.2396

Horwitz, M. A. (1983). The Legionnaires' disease bacterium (*Legionella pneumophila*) inhibits phagosome-lysosome fusion in human monocytes. *J. Exp. Med.* 158, 2108–2126. doi: 10.1084/jem.158.6.2108

Horwitz, M. A., and Silverstein, S. C. (1980). Legionnaires' disease bacterium (*Legionella pneumophila*) multiplies intracellularly in human monocytes. *J. Clin. Invest.* 66, 441. doi: 10.1172/JCI109874

Hsu, F., Luo, X., Qiu, J., Teng, Y. B., Jin, J., Smolka, M. B., et al. (2014). The Legionella effector SidC defines a unique family of ubiquitin ligases important for bacterial phagosomal remodeling. *Proc. Natl. Acad. Sci. U.S.A.* 111, 10538–10543. doi: 10.1073/pnas.1402605111

Huang, L., Boyd, D., Amyot, W. M., Hempstead, A. D., Luo, Z. Q., O'Connor, T. J., et al. (2011). The E Block motif is associated with *Legionella pneumophila* translocated substrates. *Cell. Microbiol.* 13, 227–245. doi: 10.1111/j.1462-5822.2010.01531.x

Ingmundson, A., Delprato, A., Lambright, D. G., and Roy, C. R. (2007). *Legionella pneumophila* proteins that regulate Rab1 membrane cycling. *Nature* 450, 365–369. doi: 10.1038/nature06336

Isaac, D. T., Laguna, R. K., Valtz, N., and Isberg, R. R. (2015). MavN is a *Legionella pneumophila* vacuole-associated protein required for efficient iron acquisition during intracellular growth. *Proc. Natl. Acad. Sci. U.S.A.* 112, E5208–E5217. doi: 10.1073/pnas.1511389112

Isberg, R. R., Voorhis, D. L., and Falkow, S. (1987). Identification of invasin: a protein that allows enteric bacteria to penetrate cultured mammalian cells. *Cell* 50, 769–778. doi: 10.1016/0092-8674(87)90335-7

Jäger, S., Cimermancic, P., Gulbahce, N., Johnson, J. R., McGovern, K. E., Clarke, S. C., et al. (2011). Global landscape of HIV-human protein complexes. *Nature* 481, 365–370. doi: 10.1038/nature10719

Kagan, J. C., and Roy, C. R. (2002). Legionella phagosomes intercept vesicular traffic from endoplasmic reticulum exit sites. *Nat. Cell Biol.* 4, 945–954. doi: 10.1038/ncb883

Kagan, J. C., Stein, M. P., Pypaert, M., and Roy, C. R. (2004). Legionella subvert the functions of Rab1 and Sec22b to create a replicative organelle. *J. Exp. Med.* 199, 1201–1211. doi: 10.1084/jem.20031706

Koch, R. (1891). *Über bakteriologische Forschung Verhandlung des X Internationalen Medichinischen Congresses*. Berlin.

Kotewicz, K. M., Ramabhadran, V., Sjoblom, N., Vogel, J. P., Haenssler, E., Zhang, M., et al. (2016). A single legionella effector catalyzes a multistep ubiquitination pathway to rearrange tubular endoplasmic reticulum for replication. *Cell Host Microbe.* 21, 169–181. doi: 10.1016/j.chom.2016.12.007

Kvitko, B. H., Park, D. H., Velásquez, A. C., Wei, C. F., Russell, A. B., Martin, G. B., et al. (2009). Deletions in the repertoire of *Pseudomonas syringae* pv. tomato DC3000 type III secretion effector genes reveal functional overlap among effectors. *PLoS Pathog.* 5:e1000388. doi: 10.1371/journal.ppat.1000388

Laguna, R. K., Creasey, E. A., Li, Z., Valtz, N., and Isberg, R. R. (2006). A *Legionella pneumophila*-translocated substrate that is required for growth within macrophages and protection from host cell death. *Proc. Natl. Acad. Sci. U.S.A.* 103, 18745–18750. doi: 10.1073/pnas.0609012103

Losick, V. P., and Isberg, R. R. (2006). NF-kappaB translocation prevents host cell death after low-dose challenge by *Legionella pneumophila*. *J. Exp. Med.* 203, 2177–2189. doi: 10.1084/jem.20060766

Luo, Z.-Q., and Isberg, R. R. (2004). Multiple substrates of the *Legionella pneumophila* Dot/Icm system identified by interbacterial protein transfer. *Proc. Natl. Acad. Sci. U.S.A.* 101, 841–846. doi: 10.1073/pnas.0304916101

Machner, M. P., and Isberg, R. R. (2006). Targeting of host Rab GTPase function by the intravacuolar pathogen *Legionella pneumophila*. *Dev. Cell* 11, 47–56. doi: 10.1016/j.devcel.2006.05.013

Machner, M. P., and Isberg, R. R. (2007). A bifunctional bacterial protein links GDI displacement to Rab1 activation. *Science* 318, 974–977. doi: 10.1126/science.1149121

Marra, A., Blander, S. J., Horwitz, M. A., and Shuman, H. A. (1992). Identification of a *Legionella pneumophila* locus required for intracellular multiplication in human macrophages. *Proc. Natl. Acad. Sci. U.S.A.* 89, 9607–9611. doi: 10.1073/pnas.89.20.9607

Michard, C., Sperandio, D., Baïlo, N., Pizarro-Cerdá, J., LeClaire, L., Chadeau-Argaud, E., et al. (2015). The legionella kinase LegK2 targets the ARP2/3 complex to inhibit actin nucleation on phagosomes and allow bacterial evasion of the late endocytic pathway. *MBio* 6, e00354–e00315. doi: 10.1128/mBio.00354-15

Mirrashidi, K. M., Elwell, C. A., Verschueren, E., Johnson, J. R., Frando, A., Von Dollen, J., et al. (2015). Global mapping of the inc-human interactome reveals that retromer restricts chlamydia infection. *Cell Host Microbe* 18, 109–121. doi: 10.1016/j.chom.2015.06.004

Mousnier, A., Schroeder, G. N., Stoneham, C. A., So, E. C., Garnett, J. A., Yu, L., et al. (2014). A new method to determine *in vivo* interactomes reveals binding of the *Legionella pneumophila* effector PieE to multiple Rab GTPases. *MBio* 5:e01148-14. doi: 10.1128/mBio.01148-14

Mukherjee, S., Liu, X., Arasaki, K., McDonough, J., Galán, J. E., and Roy, C. R. (2011). Modulation of Rab GTPase function by a protein phosphocholine transferase. *Nature* 477, 103–106. doi: 10.1038/nature10335

Müller, M. P., Peters, H., Blümer, J., Blankenfeldt, W., Goody, R. S., and Itzen, A. (2010). The Legionella effector protein DrrA AMPylates the membrane traffic regulator Rab1b. *Science* 329, 946–949. doi: 10.1126/science.1192276

Murata, T., Delprato, A., Ingmundson, A., Toomre, D. K., Lambright, D. G., and Roy, C. R. (2006). The *Legionella pneumophila* effector protein DrrA is a Rab1 guanine nucleotide-exchange factor. *Nat. Cell Biol.* 8, 971–977. doi: 10.1038/ncb1463

Nagai, H., Kagan, J. C., Zhu, X., Kahn, R. A., and Roy, C. R. (2002). A bacterial guanine nucleotide exchange factor activates ARF on Legionella phagosomes. *Science* 295, 679–682. doi: 10.1126/science.1067025

Neunuebel, M. R., Chen, Y., Gaspar, A. H., Backlund, P. S. Jr., Yergey, A., and Machner, M. P. (2011). De-AMPylation of the small GTPase Rab1 by the pathogen *Legionella pneumophila*. *Science* 333, 453–456. doi: 10.1126/science.1207193

Newton, H. J., Sansom, F. M., Dao, J., McAlister, A. D., Sloan, J., Cianciotto, N. P., et al. (2007). Sel1 repeat protein LpnE is a *Legionella pneumophila* virulence determinant that influences vacuolar trafficking. *Infect. Immun.* 75, 5575–5585. doi: 10.1128/IAI.00443-07

O'Connor, T. J., Adepoju, Y., Boyd, D., and Isberg, R. R. (2011). Minimization of the *Legionella pneumophila* genome reveals chromosomal regions involved in host range expansion. *Proc. Natl. Acad. Sci. U.S.A.* 108, 14733–14740. doi: 10.1073/pnas.1111678108

O'Connor, T. J., Boyd, D., Dorer, M. S., and Isberg, R. R. (2012). Aggravating genetic interactions allow a solution to redundancy in a bacterial pathogen. *Science* 338, 1440–1444. doi: 10.1126/science.1229556

O'Connor, T. J., and Isberg, R. R. (2014). iMAD, a genetic screening strategy for dissecting complex interactions between a pathogen and its host. *Nat. Protoc.* 9, 1916–1930. doi: 10.1038/nprot.2014.133

Pan, X., Lührmann, A., Satoh, A., Laskowski-Arce, M. A., and Roy, C. R. (2008). Ankyrin repeat proteins comprise a diverse family of bacterial type IV effectors. *Science* 320, 1651–1654. doi: 10.1126/science.1158160

Pereira, M. S., Morgantetti, G. F., Massis, L. M., Horta, C. V., Hori, J. I., and Zamboni, D. S. (2011). Activation of NLRC4 by flagellated bacteria triggers caspase-1-dependent and -independent responses to restrict *Legionella pneumophila* replication in macrophages and *in vivo*. *J. Immunol.* 187, 6447–6455. doi: 10.4049/jimmunol.1003784

Qiu, J., Sheedlo, M. J., Yu, K., Tan, Y., Nakayasu, E. S., Das, C., et al. (2016). Ubiquitination independent of E1 and E2 enzymes by bacterial effectors. *Nature* 533, 120–124. doi: 10.1038/nature17657

Ramage, H. R., Kumar, G. R., Verschueren, E., Johnson, J. R., Von Dollen, J., Johnson, T., et al. (2015). A combined proteomics/genomics approach links hepatitis C virus infection with nonsense-mediated mRNA decay. *Mol. Cell* 57, 329–340. doi: 10.1016/j.molcel.2014.12.028

Ratner, D., Orning, M. P., Starheim, K. K., Marty-Roix, R., Proulx, M. K., Goguen, J. D., et al. (2016). Manipulation of interleukin-1beta and interleukin-18 production by Yersinia pestis effectors YopJ and YopM and redundant impact on virulence. *J. Biol. Chem.* 291, 9894–9905. doi: 10.1074/jbc.M115.697698

Rowbotham, T. J. (1980). Preliminary report on the pathogenicity of *Legionella pneumophila* for freshwater and soil amoebae. *J. Clin. Pathol.* 33, 1179–1183. doi: 10.1136/jcp.33.12.1179

Sadosky, A. B., Wiater, L. A., and Shuman, H. A. (1993). Identification of *Legionella pneumophila* genes required for growth within and killing of human macrophages. *Infect. Immun.* 61, 5361–5373.

Sauer, J. D., Bachman, M. A., and Swanson, M. S. (2005). The phagosomal transporter A couples threonine acquisition to differentiation and replication *Legionella pneumophila* in macrophages. *Proc. Natl. Acad. Sci. U.S.A.* 102, 9924–9929. doi: 10.1073/pnas.0502767102

Segal, G., and Shuman, H. A. (1999). *Legionella pneumophila* utilizes the same genes to multiply within Acanthamoeba castellanii and human macrophages. *Infect. Immun.* 67, 2117–2124.

Shen, X., Banga, S., Liu, Y., Xu, L., Gao, P., Shamovsky, I., et al. (2009). Targeting eEF1A by a *Legionella pneumophila* effector leads to inhibition of protein synthesis and induction of host stress response. *Cell. Microbiol.* 11, 911–926. doi: 10.1111/j.1462-5822.2009.01301.x

Shi, X., Halder, P., Yavuz, H., Jahn, R., and Shuman, H. A. (2016). Direct targeting of membrane fusion by SNARE mimicry: convergent evolution of Legionella effectors. *Proc. Natl. Acad. Sci. U.S.A.* 113, 8807–8812. doi: 10.1073/pnas.1608755113

Shohdy, N., Efe, J. A., Emr, S. D., and Shuman, H. A. (2005). Pathogen effector protein screening in yeast identifies Legionella factors that interfere with membrane trafficking. *Proc. Natl. Acad. Sci. U.S.A.* 102, 4866–4871. doi: 10.1073/pnas.0501315102

Sohn, Y. S., Shin, H. C., Park, W. S., Ge, J., Kim, C. H., Lee, B. L., et al. (2015). Lpg0393 of *Legionella pneumophila* is a guanine-nucleotide exchange factor for Rab5, Rab21 and Rab22. *PLoS ONE* 10:e0118683. doi: 10.1371/journal.pone.0118683

Stenmark, H. (2009). Rab GTPases as coordinators of vesicle traffic. *Nat. Rev. Mol. Cell Biol.* 10, 513–525. doi: 10.1038/nrm2728

Swanson, M. S., and Isberg, R. R. (1995). Association of *Legionella pneumophila* with the macrophage endoplasmic reticulum. *Infect. Immun.* 63, 3609–3620.

Tan, Y., and Luo, Z. Q. (2011). *Legionella pneumophila* SidD is a deAMPylase that modifies Rab1. *Nature* 475, 506–509. doi: 10.1038/nature10307

Thomas, J. H. (1993). Thinking about genetic redundancy. *Trends Genet.* 9, 395–399. doi: 10.1016/0168-9525(93)90140-D

Toda, T., Cameron, S., Sass, P., Zoller, M., and Wigler, M. (1987). Three different genes in *S. cerevisiae* encode the catalytic subunits of the cAMP-dependent protein kinase. *Cell* 50, 277–287. doi: 10.1016/0092-8674(87)90223-6

Toulabi, L., Wu, X., Cheng, Y., and Mao, Y. (2013). Identification and structural characterization of a Legionella phosphoinositide phosphatase. *J. Biol. Chem.* 288, 24518–24527. doi: 10.1074/jbc.M113.474239

Treacy-Abarca, S., and Mukherjee, S. (2015). Legionella suppresses the host unfolded protein response via multiple mechanisms. *Nat. Commun.* 6, 7887. doi: 10.1038/ncomms8887

VanRheenen, S. M., Dumenil, G., and Isberg, R. R. (2004). IcmF and DotU are required for optimal effector translocation and trafficking of the *Legionella pneumophila* vacuole. *Infect. Immun.* 72, 5972–5982. doi: 10.1128/IAI.72.10.5972-5982.2004

VanRheenen, S. M., Luo, Z. Q., O'Connor, T., and Isberg, R. R. (2006). Members of a *Legionella pneumophila* family of proteins with ExoU (phospholipase A) active sites are translocated to target cells. *Infect. Immun.* 74, 3597–3606. doi: 10.1128/IAI.02060-05

Wagner, A. (2000). Robustness against mutations in genetic networks of yeast. *Nat. Genet.* 24, 355–361. doi: 10.1038/74174

Walter, P., and Ron, D. (2011). The unfolded protein response: from stress pathway to homeostatic regulation. *Science* 334, 1081–1086. doi: 10.1126/science.1209038

Weber, S. S., Ragaz, C., and Hilbi, H. (2009). The inositol polyphosphate 5-phosphatase OCRL1 restricts intracellular growth of Legionella, localizes to the replicative vacuole and binds to the bacterial effector LpnE. *Cell. Microbiol.* 11, 442–460. doi: 10.1111/j.1462-5822.2008.01266.x

Wiater, L. A., Dunn, K., Maxfield, F. R., and Shuman, H. A. (1998). Early events in phagosome establishment are required for intracellular survival of *Legionella pneumophila*. *Infect. Immun.* 66, 4450–4460.

Xie, Z., and Klionsky, D. J. (2007). Autophagosome formation: core machinery and adaptations. *Nat. Cell Biol.* 9, 1102–1109. doi: 10.1038/ncb1007-1102

Xu, L., Shen, X., Bryan, A., Banga, S., Swanson, M. S., and Luo, Z. Q. (2010). Inhibition of host vacuolar H+-ATPase activity by a *Legionella pneumophila* effector. *PLoS Pathog.* 6:e1000822. doi: 10.1371/journal.ppat.1000822

Yu, X., Decker, K. B., Barker, K., Neunuebel, M. R., Saul, J., Graves, M., et al. (2015).

Host-pathogen interaction profiling using self-assembling human protein arrays. *J. Proteome Res.* 14, 1920–1936. doi: 10.1021/pr5013015

Zamboni, D. S., Kobayashi, K. S., Kohlsdorf, T., Ogura, Y., Long, E. M., Vance, R. E., et al. (2006). The Birc1e cytosolic pattern-recognition receptor contributes to the detection and control of *Legionella pneumophila* infection. *Nat. Immunol.* 7, 318–325. doi: 10.1038/ni1305

Zheng, N., Schulman, B. A., Song, L., Miller, J. J., Jeffrey, P. D., Wang, P., et al. (2002). Structure of the Cul1-Rbx1-Skp1-F boxSkp2 SCF ubiquitin ligase complex. *Nature* 416, 703–709. doi: 10.1038/416703a

Zhou, D., Chen, L. M., Hernandez, L., Shears, S. B., and Galán, J. E. (2001). A Salmonella inositol polyphosphatase acts in conjunction with other bacterial effectors to promote host cell actin cytoskeleton rearrangements and bacterial internalization. *Mol. Microbiol.* 39, 248–259. doi: 10.1046/j.1365-2958.2001.02230.x

Zhu, W., Banga, S., Tan, Y., Zheng, C., Stephenson, R., Gately, J., et al. (2011). Comprehensive identification of protein substrates of the Dot/Icm type IV transporter of *Legionella pneumophila*. *PLoS ONE* 6:e17638. doi: 10.1371/journal.pone.0017638

Zurek, N., Sparks, L., and Voeltz, G. (2011). Reticulon short hairpin transmembrane domains are used to shape ER tubules. *Traffic* 12, 28–41. doi: 10.1111/j.1600-0854.2010.01134.x

2

Legionella pneumophila Strain 130b Evades Macrophage Cell Death Independent of the Effector SidF in the Absence of Flagellin

Mary Speir[1], Adam Vogrin[1], Azadeh Seidi[1], Gilu Abraham[1], Stéphane Hunot[2,3], Qingqing Han[2], Gerald W. Dorn II[4], Seth L. Masters[5,6], Richard A. Flavell[2], James E. Vince[5,6]* and Thomas Naderer[1]*

[1] Department of Biochemistry and Molecular Biology, Biomedicine Discovery Institute, Monash University, Clayton, VIC, Australia, [2] Department of Immunobiology, Howard Hughes Medical Institute, Yale University School of Medicine, New Haven, CT, USA, [3] Centre National de la Recherche Scientifique, Institut National de la Santé et de la Recherche Médicale, Institut du Cerveau et la Moelle - Hôpital Pitié-Salpêtrière, Boulevard de l'hôpital, Sorbonne Universités, UPMC Univ Paris 06, Paris, France, [4] Department of Medicine, Center for Pharmacogenomics, Washington University School of Medicine, St. Louis, MO, USA, [5] Walter and Eliza Hall Institute of Medical Research, Parkville, VIC, Australia, [6] Department of Medical Biology, University of Melbourne, Parkville, VIC, Australia

Edited by:
Matthias P. Machner,
National Institutes of Health, USA

Reviewed by:
Sunny Shin,
University of Pennsylvania, USA
John-Demian Sauer,
University of Wisconsin-Madison, USA

***Correspondence:**
James E. Vince
vince@wehi.edu.au
Thomas Naderer,
thomas.naderer@monash.edu

The human pathogen *Legionella pneumophila* must evade host cell death signaling to enable replication in lung macrophages and to cause disease. After bacterial growth, however, *L. pneumophila* is thought to induce apoptosis during egress from macrophages. The bacterial effector protein, SidF, has been shown to control host cell survival and death by inhibiting pro-apoptotic BNIP3 and BCL-RAMBO signaling. Using live-cell imaging to follow the *L. pneumophila*-macrophage interaction, we now demonstrate that *L. pneumophila* evades host cell apoptosis independent of SidF. In the absence of SidF, *L. pneumophila* was able to replicate, cause loss of mitochondria membrane potential, kill macrophages, and establish infections in lungs of mice. Consistent with this, deletion of BNIP3 and BCL-RAMBO did not affect intracellular *L. pneumophila* replication, macrophage death rates, and *in vivo* bacterial virulence. Abrogating mitochondrial cell death by genetic deletion of the effectors of intrinsic apoptosis, BAX, and BAK, or the regulator of mitochondrial permeability transition pore formation, cyclophilin-D, did not affect bacterial growth or the initial killing of macrophages. Loss of BAX and BAK only marginally limited the ability of *L. pneumophila* to efficiently kill all macrophages over extended periods. *L. pneumophila* induced killing of macrophages was delayed in the absence of capsase-11 mediated pyroptosis. Together, our data demonstrate that *L. pneumophila* evades host cell death responses independently of SidF during replication and can induce pyroptosis to kill macrophages in a timely manner.

Keywords: infection, bacteria, pyroptosis, caspases, pneumonia, mitochondria, live-cell imaging

INTRODUCTION

Legionella pneumophila is the aetiological agent of Legionnaires' Disease, a potentially life-threatening form of pneumonia in the elderly and immuno-compromised individuals (Cunha et al., 2016). Infection is initiated by inhaling aerosols derived from *L. pneumophila* contaminated water sources, such as cooling towers. Within the lungs, *L. pneumophila* establishes a specialized niche, termed the *Legionella*-containing vacuole, in resident alveolar macrophages, which promotes immune protection and bacterial growth. Virulence is absolutely dependent on the Type IV secretion system (T4SS). Many of the over 300 effector proteins translocated by the T4SS hijack host cell processes, including apoptotic cell death pathways, important for intracellular survival (Isberg et al., 2009; Speir et al., 2014).

Apoptosis is a tightly regulated program of cellular suicide depending on the activation of cytosolic cysteine-dependent aspartic acid-specific proteases, such as caspase-3. In the case of intrinsic apoptosis, the pro- and anti-apoptotic members of the BCL-2 protein family control the activity of the sentinel cell death regulators, BAX, and BAK (Czabotar et al., 2014). Activation of BAX and/or BAK leads to the loss of mitochondrial membrane integrity and release of cytochrome-*c*, which nucleates apoptosome formation to activate caspase-9 (Youle and Strasser, 2008). Active caspase-9 then cleaves caspases-3 and -7, which initiate dismantling of the cell via proteolysis of essential proteins (Elmore, 2007). Cellular stresses, including bacterial infections, can promote activation of the pro-apoptotic BH3-only proteins that either directly, or indirectly, induce BAX/BAK-mediated apoptosis (Chipuk et al., 2010). This effectively controls intracellular pathogens by compromising their replicative niche and triggering bacterial clearance in a cell autonomous manner (Chow et al., 2016).

L. pneumophila primarily replicates in macrophages and, thus, depends critically upon the health of its host cell for survival. For example, detection of flagellin results in the rapid induction of caspase-1 dependent pyroptotic cell death, which prevents *L. pneumophila* replication and infection in mice (Molofsky et al., 2006; Ren et al., 2006; Zamboni et al., 2006; Miao et al., 2010; Zhao et al., 2011). Similarly, *L. pneumophila* must also prevent mitochondrial apoptosis to promote replication, as loss of pro-survival BCL-2 family members, BCL-XL and MCL-1, induces cell death of infected macrophages and *Legionella* clearance in lungs (Speir et al., 2016). How *Legionella* evades apoptosis remains unresolved as *L. pneumophila* can trigger caspase-3 activation during macrophage invasion without inducing immediate cell death (Molmeret et al., 2004; Abu-Zant et al., 2005). *L. pneumophila* induces the transcriptional up-regulation of several pro-survival BCL-2 family members in a T4SS dependent manner, but, paradoxically, also pro-apoptotic factors, such as BNIP3 (Losick and Isberg, 2006b; Abu-Zant et al., 2007). While BNIP3 activity can be blocked by BCL-2, it can target mitochondria directly and induce cell death independently of BCL-2 (Zhang and Ney, 2009). SidF is the only *L. pneumophila* effector identified to bind and inhibit the activity of BNIP3, as well as BCL-RAMBO, which may similarly induce death in a BCL-2-dependent and -independent manner (Kataoka et al., 2001; Banga et al., 2007). Consistent with this, loss off SidF was reported to result in increased apoptotic cell death of *L. pneumophila* infected macrophages, raising the possibility that BNIP3 and BCL-RAMBO are important host factors that control *L. pneumophila* (Banga et al., 2007). In late stages of *L. pneumophila* infections, macrophages are characterized by nucleic acid fragmentation and activated apoptotic caspases, suggesting that *L. pneumophila* induces BAX/BAK-dependent apoptosis during escape (Abu-Zant et al., 2005; Santic et al., 2005; Fischer et al., 2006). Also, *L. pneumophila* infected lungs of susceptible mice show apoptosis associated phenotypes (Santic et al., 2007). While at least five effectors have been identified that can activate mitochondria-mediated apoptosis in macrophages, combined deletion of these effectors did not abrogate growth in macrophage or resulted in reduced caspase-3 activity (Nogueira et al., 2009; Zhu et al., 2013).

Besides apoptosis and flagellin/caspase-1 mediated pyroptosis, cytosolic contamination with *L. pneumophila* lipopolysaccharide triggers the activation of caspase-11, independent of flagellin, which subsequently cleaves Gasdermin D to form pores in the plasma membrane (Case et al., 2013; Casson et al., 2013; Shi et al., 2015). Caspase-11 can also activate the NLRP3/caspase-1 inflammasome and caspase-1 the apoptotic caspase-7 to control cell death-independent mechanisms as observed in *L. pneumophila* infection (Akhter et al., 2009; Case et al., 2013; Casson et al., 2013; Cerqueira et al., 2015), suggesting that during *L. pneumophila* infections caspase activation can affect multiple cellular events.

To gain a better understanding of the role of apoptosis in *L. pneumophila* infection, we have established a novel imaging method that allows for single-cell analysis of *L. pneumophila*-infected macrophages in real-time. This enables the identification of subtle and transient host-pathogen interactions, which may be overlooked in traditional methods that extrapolate from only a small number of isolated data points, or only analyze cells at the population level. With this technique, we now show that *L. pneumophila* is able to replicate and induce normal macrophage killing rates in the absence of SidF, as well as BNIP3 and BCL-RAMBO. Consistent with this, loss of SidF, BNIP3, or BCL-RAMBO did not affect *L. pneumophila* lung infections in mice. Moreover, we show that mitochondrial apoptosis itself is not essential for the induction of host macrophage cell death, nor for bacterial replication.

MATERIALS AND METHODS

Ethics Statement

Animal experiments were performed in accordance with the National Health and Medical Research Council Australian Code of Practice for the Care and Use of Animals and were approved by the Monash University Animal Ethics Committee (approval number 2011/086), and by the Walter and Eliza Hall Institute Animal Ethics Committee. All mice were maintained under specific pathogen-free conditions. Age- and sex-matched mice were chosen to be included in different treatment groups without randomization.

Legionella pneumophila Strains

Legionella pneumophila 130b serogroup 1 (ATCC BAA-74) is a spectinomycin-resistant clinical isolate from the Wadsworth Veterans Administration Hospital, Los Angeles, CA (Edelstein, 1986). The avirulent Δ*dotA* and the flagellin-deficient Δ*flaA* strains are deletion mutants of *L. pneumophila* 130 b. To generate the Δ*flaA*/Δ*sidF* deletion strain, ~500 bp fragments from upstream and downstream of sidF (LPW28321) were amplified and fused together using overlap extension PCR. The construct was cloned into the SalI site of the plasmid pSR47S, and Δ*flaA* *L. pneumophila* were transformed to select for kanamycin resistant clones. The second integration to delete the entire *sidF* coding region was selected for on 10% sucrose plates and individual colonies were verified by PCR for loss of SidF. The complemented Δ*flaA*/Δ*sidF* strains were generated by cloning the full-length sidF gene into the plasmid pMMB207C. All *L. pneumophila* strains were grown from −80°C frozen stocks on buffered charcoal-yeast extract (BCYE) agar at 37°C for 48 h before each infection. To determine bacterial numbers, *L. pneumophila* were re-suspended in PBS to determine optical density at 600 nm (OD_{600}), whereby an OD_{600} of 1 equaled 10^9 bacteria/mL. Based on this, macrophages were infected with multiplicity of infections (MOI) of 10, unless otherwise indicated.

Cell Culture

Murine bone marrow-derived macrophages (BMDMs) were obtained from femora and tibiae of female 6–8 week-old C57BL/6 mice, or from mice of the indicated genotypes. Macrophages were cultured in RPMI 1640 medium supplemented with 15% fetal bovine serum (Serana), 20% L-cell-conditioned medium (containing macrophage colony-stimulating factor), and 100 U/mL of penicillin-streptomycin (Sigma) in bacteriological dishes for 7 days, at 37°C + 5% CO_2. For infections, BMDMs were gently scraped from plates using a cell scraper (BD Falcon) and washed three times in PBS, before seeding into tissue culture-treated plates.

Live-Cell Imaging To Determine Macrophage Viability

To follow *Legionella* infection in real-time using live-cell imaging, macrophages (2.5 × 10^5 cells/mL) were seeded into 96-well tissue culture-treated plates. Before infection, BMDMs were stained with 1 μM Cell Tracker Green (CTG) (Invitrogen) for 20 min in serum-free RPMI 1640. Medium was then replaced with RPMI 1640 supplemented with 15% FBS and 10% L-cell-conditioned medium containing 50 nM tetramethylrhodamine (TMRM) and 600 nM Draq7 (Abcam). Cells were infected with *L. pneumophila* strains at a MOI of 10. In some experiments, the CellEvent Casaspase-3/7 detection reagent (Invitrogen) was added to measure caspase activity by time-lapse imaging. Before imaging, 50 μL of mineral oil (Sigma) was added to each well to prevent evaporation.

Experiments were performed on a Leica AF6000 LX epi-fluorescence microscope equipped with an incubator chamber set at 37°C + 5% CO_2 and an inverted, fully-motorized stage driven by Leica Advanced Suite Application software. Time-lapse images were acquired with bright-field, GFP, TxRed, and Y5 filters every hour for up to 72 h using a 10 × /0.8-A objective. To determine the percentage of dead cells, images were analyzed in ImageJ and in MetaMorph (Molecular Devices) using a custom-made journal suite incorporating the count nuclei function to segment and count the number of CTG, TMRM, caspase active and Draq7-positive cells (adapted from Croker et al., 2011). The data was analyzed in Excel and GraphPad Prism.

Measurement of Colony-Forming Units (CFUs)

To determine bacterial burdens, macrophages were seeded at a density of 2.5 × 10^5 cells/mL into 12-well tissue culture plates and infected with *L. pneumophila* strains at an MOI of 5. After 2 h, cells were washed 3× in PBS and the medium replaced. For analysis, cells were lysed in 0.05% digitonin for 5 min at room temperature and serial dilutions of the cell lysates and the corresponding culture media were plated on BCYE agar plates. Bacterial colonies were counted after 72 h at 37°C.

Mice Infections

C57BL/6 mice were obtained from Monash Animal Research Platform (MARP). BNIP3$^{-/-}$ (Diwan et al., 2007), BAK$^{-/-}$, BAX/BAK$^{-/-}$ (Willis et al., 2007), Ppif$^{-/-}$ (Baines et al., 2005), Casp1/11$^{-/-}$ and Casp-11$^{-/-}$ (Kayagaki et al., 2013) mice have been characterized previously. A targeting vector for BCL-RAMBO was generated from a 129/Sv genomic library, linearized and transfected into TC-1 embryonic stem (ES) cells by electroporation. Southern blotting was used to identify ES cell clones with homologous recombination, which were used for injection into blastocysts to generate chimeric mice. Chimeric mice were bred at least 10 generations onto the C57BL/6 background and were deficient in BCL-RAMBO expression (**SI Figure 1**).

Six to eight week-old male or female mice, in groups of five or more, were anesthetized by 4% isofluorane inhalation and infected intra-nasally with 2.5 × 10^6 *L. pneumophila* in 50 μL of sterile PBS. For CFUs, at 6 or 48 h following infection, both lung lobes were removed and homogenized for 30 s in PBS at 30,000 rpm using the Omni Tissue Master homogenizer. Serial dilutions of the lung homogenates were plated onto BCYE agar plates and bacterial colonies were counted after 72 h at 37°C to determine CFUs.

Immunoblot Analysis

2.5 × 10^5 cells were lysed in 120 μL SDS-loading dye, boiled for 5 min, and samples analyzed by 12% SDS-PAGE. After transfer to nitrocellulose membranes (Millipore), membranes were blocked with 5% skim milk in T-BST (Tween-20, Tris-buffer) for 1 h at room temperature. Membranes were probed with anti-cleaved caspase-3 antibody (CST #9964) or anti-β-actin antibody (Millipore #04-1116) (loading control) and re-suspended in T-BST + 5% skim milk, overnight at 4 °C. After washing, membranes were probed with secondary goat anti-rabbit IgG (Life Technologies) and goat anti-mouse IgG (Life Technologies) antibodies conjugated to HRP (1:20,000 dilution in T-BST + 5% skim milk). Membranes were developed with the luminol-based enhanced chemiluminescence (ECL) and exposed to film (Kodak). Scanned images were processed in Photoshop Adobe.

Statistical Analyses

For all *in vitro* data, two-way analysis of variance was performed before using Tukey's *post hoc* test for pairwise comparisons. For mice infections, data were analyzed by the Mann-Whitney U-test. In all experiments, $p \leq 0.05$ were taken to be significant.

RESULTS

Live-Cell Imaging of *L. pneumophila* Infected Macrophages

To examine *L. pneumophila* infection of bone marrow-derived macrophages (BMDMs) in real-time and to monitor their viability, infected cells were stained with the cell-permeable fluorescent dye, tetramethylrhodamine methyl ester (TMRM), which is sequestered by active mitochondria, depending on the inner membrane potential. In addition, the macrophage culture media contained the membrane impermeable DNA fluorophore Draq7 to specifically stain dead cells (**Figure 1A**). Importantly, more than 90% of BMDMs left uninfected, or infected with the avirulent Δ*dotA* strain, which lacks a functional T4SS, did not show uptake of Draq7 and remained viable for up to 72 h, demonstrating that it is possible to follow host-pathogen interactions over extended time periods (**Figure 1B**). As expected, the BMDMs infected with WT *L. pneumophila* died more rapidly than those infected with the flagellin-deficient strain, Δ*flaA*, consistent with a flagellin/caspase-1-mediated pyroptotic cell death. For example, at 30 h post infection more than 60% of WT-infected BMDMs were Draq7-positive, whereas <30% of the Δ*flaA* infected BMDMs (**Figure 1C**). Over time, 80% of BMDMs infected with WT or Δ*flaA L. pneumophila* were killed by 72 h post infection, consistent with repeated rounds of bacterial infection, egress, and re-infection. However, only Δ*flaA L. pneumophila* is able to replicate in BMDMs, demonstrating that *L. pneumophila* critically depends on evading macrophage death during early stages of infections for growth. In addition to Draq7 staining, we monitored mitochondrial membrane potential (ΔΨm) over time. The Δ*dot*-infected BMDMs showed little change in TMRM fluorescence, similar to that of uninfected BMDMs (**Figure 1D**). In contrast, in WT- and Δ*flaA*-infected BMDMs the ΔΨm decreased by more that 50% relative to uninfected BMDMs (**Figure 1D**). Comparable to the Draq7 uptake, loss of ΔΨm occurred more quickly in the BMDMs infected with WT *L. pneumophila* than in those infected with Δ*flaA L. pneumophila* (**Figure 1D**). These results demonstrate that live-cell fluorescent imaging, to follow Draq7- and TMRM-staining to quantify cell death and mitochondrial integrity, respectively, is able to distinguish between the different cell death kinetics involved in *L. pneumophila* infection.

Loss of the Bacterial Effector SidF Does Not Lead to Increased Apoptosis of Infected Macrophages

The *L. pneumophila* effector protein SidF is the only effector reported to target and inhibit host cell pro-apoptotic factors (Banga et al., 2007). To investigate its role in *Legionella* induced killing of BMDMs and to avoid rapid pyroptotic cell death, we

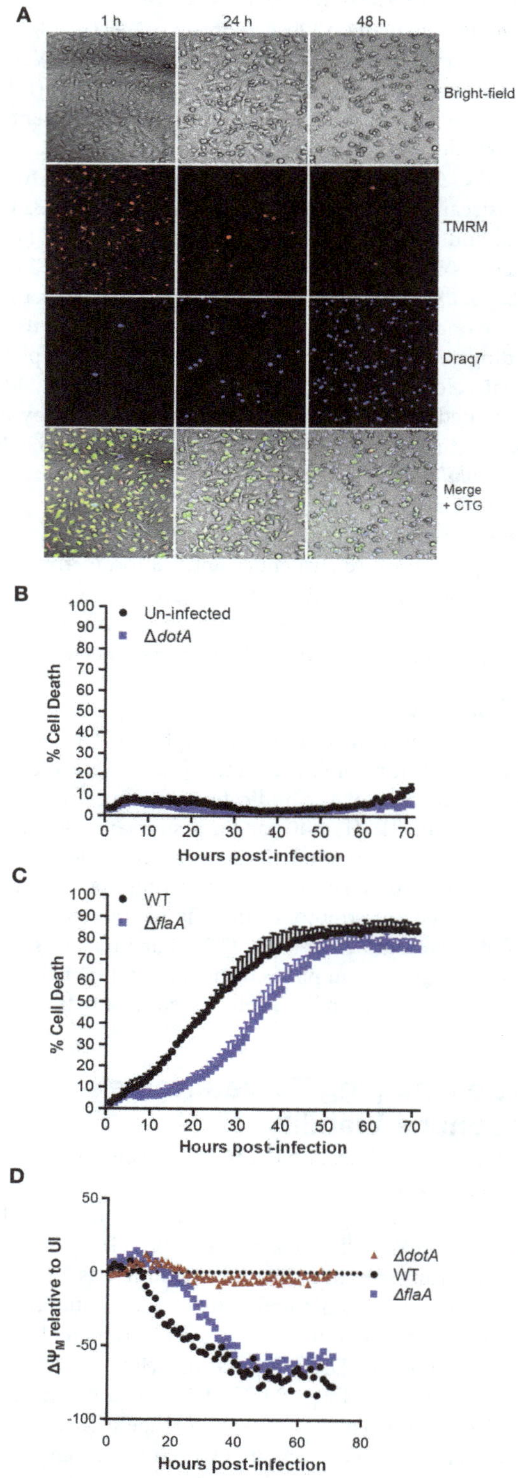

FIGURE 1 | Live cell imaging of macrophage and mitochondrial health in *L. pneumophila* infections. (A) Wild type C57BL/6 bone marrow-derived macrophages (BMDMs) labeled with the fluorescent dyes tetramethylrhodamine methyl ester (TMRM) to stain active mitochondria and Draq7 to detect dead cells were infected at an MOI of 10 with

(Continued)

FIGURE 1 | Continued
Δ*flaA L. pneumophila*. Live-cell images from 1, 24, and 48 h post-infection are shown. **(B)** Draq7 positive (dead) uninfected and Δ*dotA L. pneumophila* infected BMDMs determined by live-cell imaging over 72 h. **(C)** Draq7 positive (dead) WT and Δ*flaA L. pneumophila* infected BMDMs. **(D)** TMRM fluorescence ($\Delta\Psi_M$, mitochondrial membrane potential) of WT, Δ*dotA*, and Δ*flaA L. pneumophila* treated BMDMs over 72 h. TMRM fluorescence intensity is relative to that in uninfected BMDMs (dotted line). Mean and SD. of three independent biological replicates are shown.

generated a Δ*flaA*/Δ*sidF Legionella* mutant. Surprisingly, we did not observe any significant increase in the rate or extent of cell death in BMDMs infected with the Δ*flaA*/Δ*sidF* strain compared to Δ*flaA* or the complemented Δ*flaA*/Δ*sidF* strain (**Figure 2A**). Although loss of SidF did result in increased BMDM death at 20 h post infection, this was not significantly different to the Δ*flaA*- and complemented Δ*flaA*/Δ*sidF*-induced killing (**Figure 2A**). Furthermore, <2% of the Δ*flaA*/Δ*sidF*-infected BMDMs underwent apoptotic cell death during the first 24 h of infection, as judged by cell shrinkage and membrane blebbing, which was similar to the number of Δ*flaA*-infected BMDMs (**Figure 2B**). Consistent with this finding, there was no detectable caspase-3 cleavage, indicative of apoptotic caspase activation, in either the Δ*flaA*- or Δ*flaA*/Δ*sidF*-infected BMDMs after 8, 12, and 24 h of infection, as determined by immunoblotting of the caspase-3 p17/p19 fragment (**Figure 2C**). Finally, infection with the Δ*flaA*/Δ*sidF L. pneumophila* strain did not result in increased mitochondrial damage compared to Δ*flaA* over 72 h (**Figure 3D**). Taken together, these data demonstrate that loss of SidF does not result in a dramatic induction of apoptosis or increased BMDM death in *L. pneumophila* infections.

SidF is Dispensable for *L. pneumophila* Replication in Macrophages

Given our observations that SidF does not play a major role in preventing death of BMDMs, we next tested whether it is required for replication of *L. pneumophila*, as previously reported (Banga et al., 2007). We determined bacterial burdens (CFU/mL) from BMDMs infected with Δ*flaA*, Δ*flaA*/Δ*sidF*, or Δ*flaA*/Δ*sidF* + SidF *L. pneumophila* at 48 h post-infection (**Figure 3A**). After 48 h of infection, there was a small but significant difference (≃3-fold; $p < 0.01$) between the CFU/mL recovered from the Δ*flaA*/Δ*sidF* strain compared to the Δ*flaA* or Δ*flaA*/Δ*sidF* + SidF *L. pneumophila* strains in BMDMs, which was less obvious in immortalized BMDMs that support rapid bacterial growth (**SI Figure 2**). Given this observed, albeit marginal, growth defect in BMDMs, the role of SidF was further examined in *L. pneumophila* infections *in vivo*. Bacterial burdens from the lungs of WT C57BL/6 mice 48 h after infection with either Δ*flaA* or Δ*flaA*/Δ*sidF L. pneumophila* were not significantly different (**Figure 3B**). This demonstrates that, while SidF promotes bacterial replication to a small degree *in vitro*, it is dispensable for *L. pneumophila* survival and burdens during lung infections in mice.

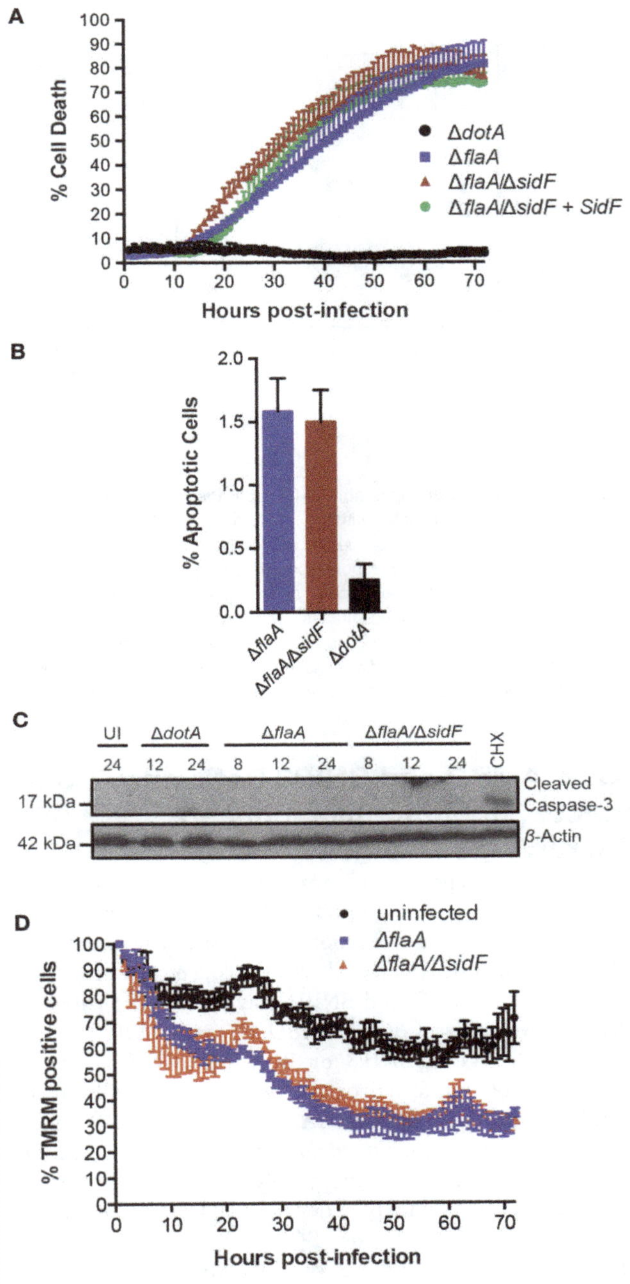

FIGURE 2 | Loss of SidF does not induce apoptotic cell death of infected BMDMs. (A) Draq7 positive (dead) BMDMs infected at a MOI of 10 with Δ*dotA*, Δ*flaA*, Δ*flaA*/Δ*sidF*, and Δ*flaA*/Δ*sidF* + SidF *L. pneumophila*. Data are representative of three independent experiments. Mean and S.D. of three independent biological replicates shown. **(B)** Percentage apoptotic cells as determined by membrane blebbing in BMDMs infected with Δ*dotA*, Δ*flaA*, and Δ*flaA*/Δ*sidF L. pneumophila*. >800 cells were scored from live-cell images taken every 30 min for 48 h. Mean and S.D. shown. **(C)** Time course immuno-blot analysis for cleaved (indicative of active) caspase-3 in BMDMs infected with Δ*flaA* or Δ*flaA*/Δ*sidF L. pneumophila*. BMDMs treated with 10 μM cycloheximide (CHX) were used as a positive control. Actin blot is a loading control. **(D)** TMRM fluorescence of uninfected BMDMs or infected with Δ*flaA* and Δ*flaA*/Δ*sidF L. pneumophila* over 72 h. Mean and SEM from two independent experiments containing three biological repeats shown.

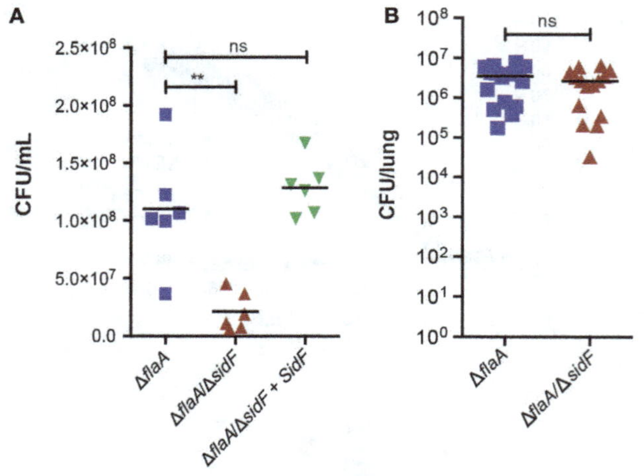

FIGURE 3 | *L. pneumophila* is able to infect lungs in the absence of SidF. (A) BMDMs were infected with the *ΔflaA*, *ΔflaA/ΔsidF*, and *ΔflaA/ΔsidF* + SidF *L. pneumophila* and bacterial burdens (CFU/mL) calculated 48 post infection. Symbols represent biological repeats from two independent experiments. Line indicates mean. $^{**}p < 0.01$, ns, not significant **(B)** C57BL/6 mice were infected intranasal with 2.5×10^6 *ΔflaA* and *ΔflaA/ΔsidF L. pneumophila* and lung bacterial burdens (CFU/lung) determined 48 h post infection. Symbols represent data from individual mice from two independent experiments. Line indicates mean. ns = not significant.

BCL-RAMBO and BNIP3 Deficiency Do Not Affect *L. pneumophila* Induced Macrophage Death

SidF has been reported to target and inhibit BCL-RAMBO and BNIP3 (Banga et al., 2007), which may act independently to modulate cell death in both a caspase-dependent and -independent manner (Kim et al., 2011; Rikka et al., 2011). To test whether BCL-RAMBO and BNIP3 play role in *L. pneumophila* infection, we utilized *BCL-RAMBO*$^{-/-}$ and *BNIP3*$^{-/-}$ mice. As expected, loss of BNIP3 or BCL-RAMBO did not affect the viability of uninfected or *ΔdotA-L. pneumophila* infected BMDMs over 72 h (**Figures 4A,B**). Loss of BNIP3 did not influence *ΔflaA L. pneumophila*-induced killing of BMDMs (**Figure 4C**), which was marginally, but not significantly, increased in BCL-RAMBO deficient BMDMs (**Figure 4C**). Finally, loss of BNIP3 or BCL-RAMBO did not affect *ΔflaA*/*ΔsidF*-induced killing of BMDMs, but reduced the increased death rates observed in *ΔflaA*-infected BCL-RAMBO deficient BMDMs (**Figure 4D**). This demonstrates that even in the absence of SidF, BNIP3, and BCL-RAMBO are dispensable for *L. pneumophila*-mediated killing of macrophages. To confirm that BCL-RAMBO or BNIP3 do not contribute to *L. pneumophila* infection *in vivo*, bacterial burdens were calculated from the lungs of WT C57BL/6, *BCL-RAMBO*$^{-/-}$, and *BNIP3*$^{-/-}$ mice 48 h after infection with *ΔflaA L. pneumophila.* As shown in **Figure 4E**, there was no significant difference in bacterial numbers recovered from the lungs of the different mouse genotypes ($p > 0.05$). In agreement with this, *ΔflaA* and *ΔflaA*/*ΔsidF* replication in *BCL-RAMBO*$^{-/-}$ macrophages was similar compared to WT macrophages (**SI Figure 2**).

L. pneumophila Induces Macrophage Cell Death Independently of Cyclophilin-D and BAX/BAK

Although BNIP3 and BCL-RAMBO do not influence *Legionella* infection, alternate regulators of mitochondria-mediated cell death signaling may be targeted by *Legionella* to promote bacterial replication or egress. These include host cell death mediated by formation of the mitochondrial permeability transition pore (mPT) (Khemiri et al., 2008), or intrinsic (mitochondrial) apoptosis, mediated by BAX and BAK oligomerisation on mitochondrial membranes. The *Ppif* gene product, Cyclophilin-D (CycD), is a critical component of the mPT, can modulate apoptosis independent of BCL-2 and may be targeted by BNIP3 (Carneiro et al., 2009; Gutiérrez-Aguilar and Baines, 2015). We therefore tested the role of CycD in *L. pneumophila* infection. As expected, *ΔdotA* infected CycD-deficient BMDMs (*Ppif*$^{-/-}$) remained viable for 72 h (**Figure 5A**). In the absence of CycD, *ΔflaA*-infected BMDMs remained viable for the first 10 h post infection and then showed increased cell death that were indistinguishable from infected WT BMDMs (**Figure 5B**). This result demonstrates that *L. pneumophila* can still induce cell death normally in the absence of a functional mitochondrial permeability transition pore complex.

To test whether *Legionella* infection is influenced by a loss of intrinsic (mitochondrial) apoptosis, we utilized BMDMs deficient in BAK alone, or both BAX and BAK, which has been demonstrated to completely prevent intrinsic apoptotic cell death (van Delft et al., 2006). As expected, in the absence of BAK, or BAX and BAK together, *ΔdotA L. pneumophila*-infected BMDMs remained viable (**Figure 5C**), indicating that neither BAX nor BAK were essential to cell survival under these conditions. Infection with *ΔflaA L. pneumophila* resulted in the death of similar numbers of *BAK*$^{-/-}$ and *BAX*$^{-/-}$*BAK*$^{-/-}$ BMDMs by 72 h post infection (**Figure 5D**). Deletion of both BAX and BAK resulted in a 15–20% decrease in the rate of cell death compared to WT, or deletion of BAK alone, between 30 and 40 h post infection, but not during the initial killing phase (10–25 h). In addition, the rate loss of TMRM signal was similar between WT, *BAK*$^{-/-}$, and *BAX*$^{-/-}$*BAK*$^{-/-}$ BMDMs, suggesting that BAX and BAK do not significantly contribute to loss of mitochondrial membrane potential in *L. pneumophila* infections (**Figure 5E**). Consistent with this, we and other have recently shown that *ΔflaA L. pneumophila* replicates normally in *BAX*$^{-/-}$*BAK*$^{-/-}$ BMDMs (Nogueira et al., 2009; Speir et al., 2016). Together, these data suggest that while *ΔflaA L. pneumophila* can induce BAX/BAK-mediated apoptosis in late stage infections, it is not critical for bacterial replication.

Inhibition of Host Protein Synthesis by *L. pneumophila* Does Not Lead to Apoptosis, but Pyroptosis

We have recently shown that *L. pneumophila* limits host cell protein synthesis to reduce levels of the short-lived pro-survival BCL-2 family member MCL-1, akin to chemically inhibiting protein synthesis by cycloheximide (CHX) treatment

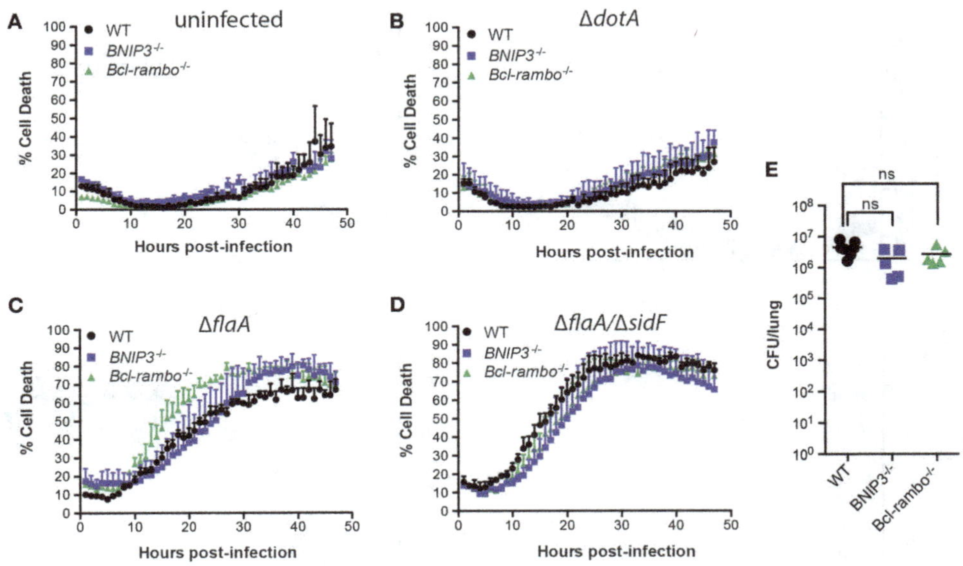

FIGURE 4 | *L. pneumophila* infects macrophages and mice independent of BNIP3 and BCL-RAMBO. Draq7 positive (dead) uninfected **(A)**, Δ*dotA* **(B)**, Δ*flaA* **(C)**, Δ*flaA*/Δ*sidF* **(D)** infected C57BL/6 wild-type (WT), BCL-RAMBO$^{-/-}$, and BNIP3$^{-/-}$ BMDMs. Data are representative of three independent experiments. Mean and S.D. of three independent biological replicates shown. **(E)** Bacterial lung burdens (CFU/lung) of C57BL/6 wild-type (WT), BNIP3$^{-/-}$, and BCL-RAMBO$^{-/-}$ mice infected intranasal with 2.5 × 10^6 Δ*flaA L. pneumophila* 48 h post infection. Data from individual mice and mean (line) are shown. Ns, not significant.

(Speir et al., 2016). Δ*flaA L. pneumophila* infection or CHX treatment also reduced the mitochondrial membrane potential and induced cell death with similar kinetics at the MOI and concentrations used (**Figure 6A**). To follow the activation of apoptotic caspases on a single cell level over extended periods, BMDMs were incubated with a fluorescent probe to detect caspase-3/7 activity. As expected, CHX treatment caused activation of caspase-3/7 at around 20 h post treatment, coinciding with cell death (**Figure 6B**, **SI Video 1**). In contrast, Δ*flaA L. pneumophila* induced cell death with minimal caspase-3/7 activity (**Figure 6B**, **SI Video 2**, consistent with western blot analysis (**Figure 2C**). Only at late stage infections did Δ*flaA L. pneumophila* cause caspase3/7 activation (**Figure 6B**), at which point most BMDMs were stained by Draq7 and thus contained compromised membranes (**Figure 6A**). Similar results were obtained with Δ*flaA/sidF L. pneumophila*, whereas uninfected BMDMs remained viable with little evidence of caspase-3/7 activity (**Figure 6B**). WT *L. pneumophila* caused sustained low levels of caspase-3/7 activity immediately after infections (**Figure 6B**, **SI Video 3**).

CHX induces mitochondrial apoptosis, which is delayed by the pan-caspase inhibitor, QVD-ph (QVD) for at least for 30 h (**Figure 6C**). In contrast, QVD treatment had only a marginal effect on Δ*flaA L. pneumophila* induced killing of BMDMs (**Figure 6D**). This demonstrates that *L. pneumophila* kills macrophages independent of the activity of apoptotic caspases, despite inhibition of host protein synthesis and the loss of mitochondrial membrane potential. Given that *L. pneumophila* can trigger pyroptosis, which can consequently activate apoptotic caspases, we finally tested whether pyroptosis is induced in late stage *L. pneumophila* infections. Δ*flaA* *L. pneumophila* caused delayed (~10 h) death in caspase-1/11 double deficient BMDMs compared to WT BMDMs (**Figure 6E**). We observed the same delayed death response in caspase-11 deficient BMDMs, suggesting that in the absence of flagellin, *L. pneumophila* triggers caspase-11 mediated pyroptosis in late stage infections (**Figure 6E**). Caspase-1 and -11 were not required for efficient replication (**Figure 6F**) or macrophage killing at higher infection rates (**SI Figure 3**), suggesting that besides pyroptosis *L. pneumophila* can utilize other mechanisms to induce macrophage death during egress.

DISCUSSION

The role of programmed host cell death signaling in *Legionella* infections has been studied extensively over the past decade. This has mainly been in the context of the caspase-1-dependent inflammatory cell death, termed pyroptosis, during invasion of macrophages. We have now utilized both host cell and bacterial genetic approaches combined with live cell imaging to more accurately define the role of apoptotic cell death in *Legionella* infections. We show that loss of the critical intrinsic apoptotic proteins, BAK and BAX, or the mPT pore component, cyclophilin D, does not significantly alter *L. pneumophila* replication or the killing of macrophages. Moreover, we also demonstrate that the genetic deletion of BCL-RAMBO or BNIP3, reported host cell pro-apoptotic molecules inhibited by the bacterial effector SidF, have no impact on *in vivo Legionella* replication. Therefore, although several *Legionella* effectors may target mitochondria and activate apoptotic caspase activity (Zhu et al., 2013), our data suggest that key mitochondrial cell death signaling pathways do not facilitate bacterial replication or retard

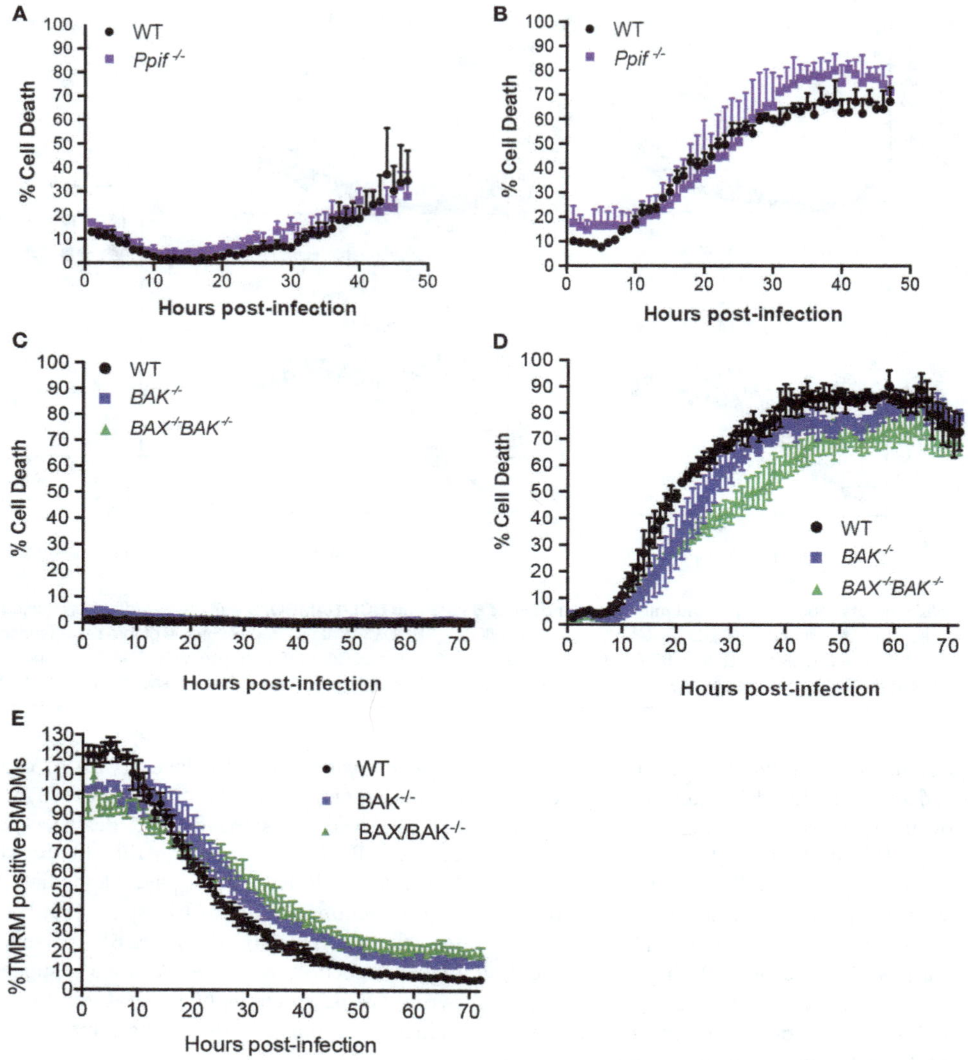

FIGURE 5 | *L. pneumophila* induces macrophage death independent of Cyclophilin-D, BAX, and BAK. Draq7 positive (dead) Δ*dotA* **(A)** and Δ*flaA* *L. pneumophila* **(B)** infected wild-type (WT) and *Ppif*−/− BMDMs. Draq7 positive (dead) Δ*dotA* **(C)** and Δ*flaA* *L. pneumophila* **(D)** infected (MOI of 10) wild-type (WT), *BAK*−/− and *BAX*−/− *BAK*−/− BMDMs. Mean and S.D. of three biological replicates shown. **(E)** TMRM positive BMDMs infected with Δ*flaA* *L. pneumophila* (relative to Δ*dotA* infected BMDMs) over 72 h. Mean and SD from three biological repeats, representative of at least two independent experiments, shown.

bacterial egress, and therefore do not significantly alter *Legionella* infectivity in mice.

Several T4SS effectors have been identified that trigger sustained NF-κB signaling and consequent transcriptional up-regulation of BCL-2 family members, that includes pro-survival BCL-2 and A1 (Losick and Isberg, 2006a; Abu-Zant et al., 2007). In agreement with this, infections with virulent *L. pneumophila* renders macrophages resistant to apoptosis inducing agents (Abu-Zant et al., 2005). However, protein levels of the major pro-survival factors in infected macrophages remain stable, or are reduced, consistent with the notion that *L. pneumophila* prevents translation of most host proteins (Speir et al., 2016). Thus, *L. pneumophila* may rely on other mechanisms to sustain macrophage viability during infection. In part, this may depend on effectors that directly inhibit pro-death factor which can be activated by post-translational processes. So far, however, only one effector, SidF, has been reported to directly block the pro-apoptotic activities of BCL-RAMBO and BNIP3 (Banga et al., 2007). Genetic deletion of SidF was reported to more than double the number of apoptotic BMDMs in late stage *L. pneumophila* infections (Banga et al., 2007). To define the role of SidF and apoptosis in *L. pneumophila*-infected BMDMs, we used live-cell imaging to follow the entire infection cycle and to measure macrophage health in real time by determining plasma membrane rupture and loss of mitochondrial membrane potential. Using this technique, we did not detect significantly increased apoptosis in BMDMs infected with an *L. pneumophila* strain lacking SidF and,

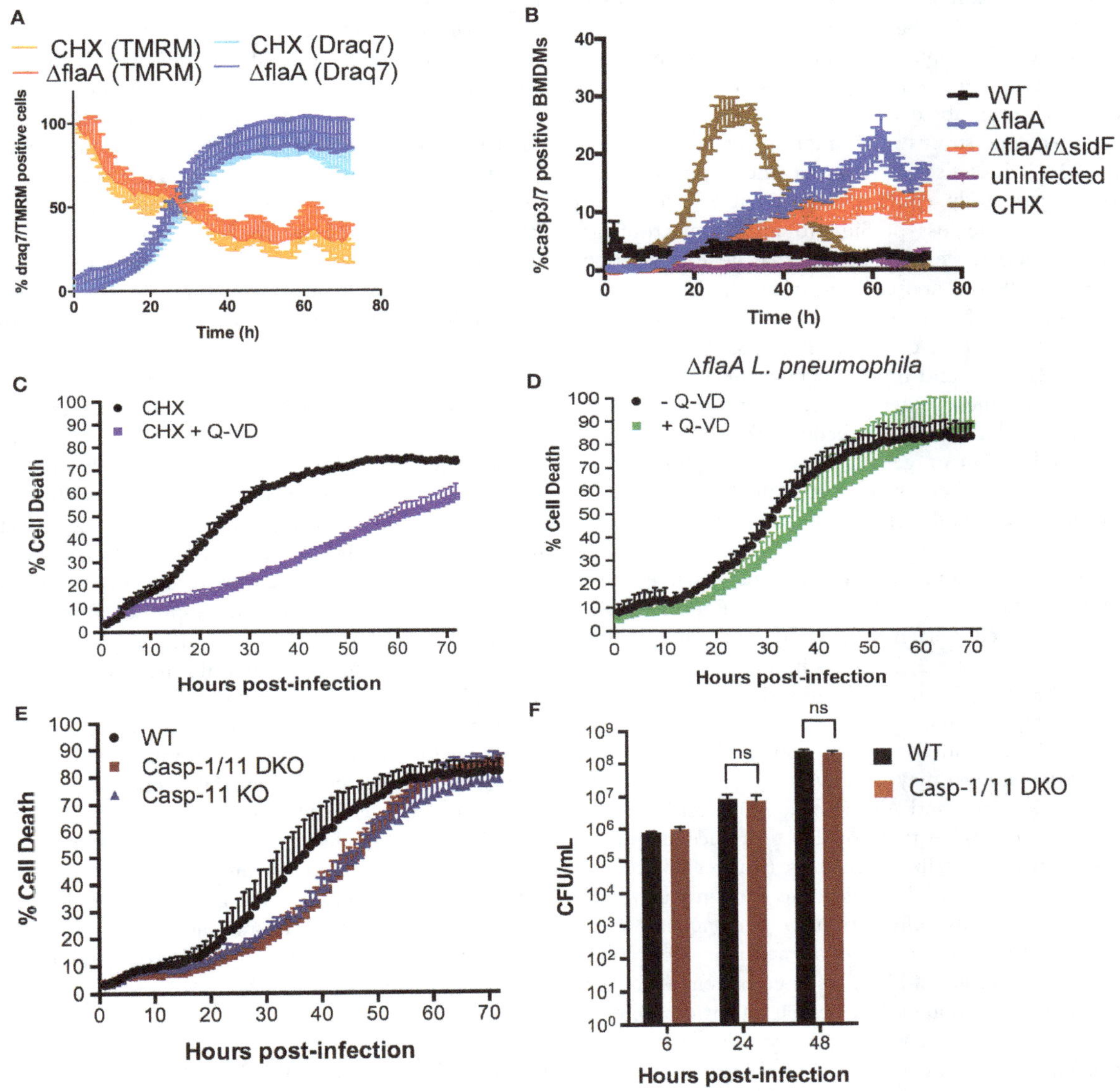

FIGURE 6 | *L. pneumophila* mediated macrophage death is independent of apoptotic caspases, but induced by pyroptotic caspase-11. (A) Draq7 and TMRM positive BMDMs treated with cycloheximide (2 μg/ml) or infected with Δ*flaA L. pneumophila*. **(B)** Caspase-3/7 activity in uninfected, WT, Δ*flaA* and Δ*flaA*/Δ*sdiF L. pneumophila* infected or cycloheximide (CHX, 2 μg/ml) treated BMDMs. **(C)** Draq7 positive (dead) BMDMs treated with cycloheximide (CHX) with or without Q-VD (20 μM). **(D)** Draq7 positive (dead) BMDMs infected with Δ*flaA L. pneumophila* and treatment with or without Q-VD (20 μM). Data are representative of three independent experiments. Mean and S.D. of three independent biological replicates shown. **(E)** Draq7 positive (dead) WT, Caspase-1/11 DKO and Caspase-11 KO BMDMs infected with Δ*flaA L. pneumophila.* Data are representative of two independent experiments. Mean and S.D. of three independent biological replicates shown. **(F)** Bacterial burdens (CFU/mL) from Δ*flaA L. pneumophila*-infected WT and Caspase-1/11 DKO BMDMs at 6, 24, and 48 h post-infection. Mean and S.E.M of three independent experiments shown.

furthermore, show that SidF is dispensable in lung infections in mice.

BMDMs derived from C57BL/6 mice readily detect flagellin present in wild-type *L. pneumophila* and induce caspase-1-mediated pyroptotic cell death. To specifically investigate the role of SidF and apoptosis in *L. pneumophila*-infected mice, we generated a Δ*flaA*/Δ*sidF* mutant, which evades caspase-1 detection. Because the Δ*sidF* strain in the original report was on a flagellated *L. pneumophila* background, it is possible that, in the absence of SidF, flagellin reached the cytosolic sensors of NAIP5 and NLRC4 to activate caspase-1 even in permissive macrophages, albeit at reduced rates (Zamboni et al., 2006; Lamkanfi et al., 2007). The establishment and integrity of the *Legionella* containing vacuole may directly dependent on SidF and its phosphoinositide phosphatase activity which thus likely promotes efficient bacterial growth (Hsu et al., 2012). Besides triggering pyroptosis, caspase-1 may also cleave apoptotic caspases, such as caspase-3 and 7 (Amer, 2010). Our study now shows that in the absence of flagellin-mediated pyroptosis, the loss of SidF does not significantly affect the ability of *L. pneumophila* to evade apoptosis. Of note, this and the previous study used genetically different *Legionella* strains, which can affect the degree of apoptotic death as not all of the effectors are conserved and as some strains infect more efficiently (Gomez-Valero et al., 2011). It is also possible that the *L. pneumophila* strain used in this study contains additional effectors that can compensate for the loss of SidF to inhibit macrophage cell death. Nevertheless, we demonstrate that genetic deletion of the proposed SidF pro-apoptotic host cell target proteins, BCL-RAMBO, or BNIP3, also does not alter *L. pneumophila* replication, infection or host cell death kinetics. While the co-deletion of BCL-RAMBO and BNIP3 together may be required to reveal a role for these potential pro-apoptotic factors in modulating intracellular bacterial infections, the limited effects of pan-caspase inhibition or genetic loss of cyclophilin D or BAX and BAK, argue that abrogating mitochondrial death signaling does not significantly influence bacterial replication and infectivity.

L. pneumophila is able to directly manipulate host cell apoptotic signaling during infection. For example, *L. pneumophila*-infected cells have been reported to contain high levels of active caspase-3, but only induce host cell death with apoptotic features, such as chromosome condensation and nucleic acid fragmentation, in late stages of infection (Gao and Abu Kwaik, 1999; Abu-Zant et al., 2005). This led to the notion that *L. pneumophila* may selectively trigger apoptosis to facilitate egress (Molmeret and Abu Kwaik, 2002). At least five effectors have been identified that are able to induce apoptosis when expressed in immortalized cell lines (Zhu et al., 2013). However, their exact roles during infections remain elusive, as the co-deletion of these five effectors in *L. pneumophila* does not influence bacterial infection and intracellular replication in macrophages (Zhu et al., 2013). This is consistent with our findings showing that deletion of the essential intrinsic apoptotic executioners, BAX and BAK, does not overtly reduce the ability of *L. pneumophila* to kill macrophages or to abrogated bacterial growth (Nogueira et al., 2009). Similarly, loss of BAX/BAK and caspase-3 did not affect *L. pneumophila* growth *in vitro* (Nogueira et al., 2009). In contrast to macrophages, the above effectors are able to induce apoptosis in dendritic cells and, thus, prevent bacterial survival (Zhu et al., 2013). Bacterial growth in dendritic cells can also be restored by overexpressing pro-survival BCL-2, or by loss of BAX/BAK, suggesting that at least some effectors may act upstream of BAX/BAK (Nogueira et al., 2009). This also highlights that host cell death signaling following *L. pneumophila* infection is likely to be cell-type specific, and depend on the host cell expression levels of different cell death components.

In the absence of apoptosis, it is possible that *L. pneumophila* may induce other forms of programmed cell death to facilitate bacterial egress. While the expression of flagellin is up-regulated in late stage infections (Molmeret et al., 2010), flagellin-deficient *Legionella* species (e.g., Δ*flaA L. pneumophila* or WT *L. longbeachae*, which is naturally deficient in flagellin; Cazalet et al., 2010), are still able to efficiently kill macrophages in the late stages of infection, suggesting that this is not mediated via flagellin/caspase-1-dependent pyroptosis. Furthermore, deletion of extrinsic apoptosis (Caspase-8 deficient BMDMs) and necroptosis (RIPK3 and MLKL deficient BMDMs) did not abrogate killing of macrophages by virulent *L. pneumophila* (Speir et al., 2016). *L. pneumophila* also activates a caspase-11-dependent form of pyroptotic cell death, particularly in LPS-primed cells, to induce caspase-11 expression (Case et al., 2013; Casson et al., 2013). Cytosolic caspase-11 recognizes and binds LPS directly to either induce NLRP3/caspase-1 dependent or caspase-1 independent pyroptosis (Hagar et al., 2013; Kayagaki et al., 2013). Although caspase-11 is not required for NLRC4-dependent pyroptosis, nor for the restriction of flagellated *Legionella* infection (Cerqueira et al., 2015), there is some evidence that *Legionella* complete their terminal rounds of proliferation within the cytosol (Molmeret et al., 2007) and, thus, may activate caspase-11 upon escaping its vacuole, in order to facilitate egress. Other vacuolar pathogens also induce caspase-11-mediated death to effectively escape, as is the case in *Salmonella* Typhimurium. Caspase-11 activation is detrimental to the host as it expedites bacterial egress, allowing S. Typhimurium to replicate extracellularly in the absence of a caspase-1-mediated immune response (Broz et al., 2012). However, the delay in cell death after *Legionella* infections in the absence of caspase- 11 is only detectable at low infection levels. Higher numbers of bacteria must be able to trigger escape independently of, and more quickly than, caspase-11 activation alone. Furthermore, even at a low infection rates, there is no corresponding defect in bacterial replication, indicating that this delay in cell death does not limit bacterial replication. It is formally possible that multiple programmed cell death pathways are activated during *L. pneumophila* infection to facilitate bacterial egress. Alternatively, bacterial-induced killing may include other mechanisms such as the expression of lytic enzymes or overwhelming bacterial burden that lead to host cell rupture (Molmeret et al., 2002).

AUTHOR CONTRIBUTIONS

MS and AV designed and performed experiments and interpreted data; SH, QH, GD, SM, RF generated knock out mice and revised the manuscript; JV and TN conceived the work and analyzed the data. MS, JV, and TN wrote the manuscript. All authors approved the final version of the manuscript.

ACKNOWLEDGMENTS

We thank Dr P. Bouillet (WEHI) for providing BAX/BAK deficient macrophages and members of the Monash Micro Imaging facility for their technical support.

SUPPLEMENTARY MATERIAL

SI Figure 1 | Genetic deletion of BCL-RAMBO. (A) Gene targeting strategy. The targeting construct replaces exon 2 (black bar on the WT locus) with a neomycin resistance cassette (neo) resulting also in a frame-shift. Restriction sites are indicated (H, Hind III; X, XbaI; S, SacI; B, BamHI). A 3′ probe was designed to recognize a 5.9 kb and 8.7 kb fragment from Sac1-digested wild-type and mutated genomic DNA alleles, respectively. **(B)** Southern blot analysis of Sac I digested mouse tail DNA using the 3′ probe, showing the wild-type (lower, 5.9 kb) and targeted (upper; 8.7 kb) alleles of *Bcl-rambo* gene. **(C)** Western blot analysis of embryonic fibroblast cell lysates using anti-Bcl-rambo antibodies (Abcam) and anti-Actin as internal control.

SI Figure 2 | *L. pneumophila* replicates in BCL-RAMBO deficient macrophages. WT and BCL-RAMBO deficient immortalized macrophages were infected with Δ*flaA* and Δ*flaA*/Δ*sidF* (MOI 10) for 2 h and the colony forming units (CFUs) determined at 6 and 48 h post infection. Mean and SD (from three independent colonies) are shown.

SI Figure 3 | Cell death of caspase-11 deficient BMDMs at high MOI. Draq7 positive (dead) WT, caspase-1/11 DKO and caspase-11 KO BMDMs infected at a MOI of 20 with ΔflaA L. pneumophila. Data are representative of two independent experiments. Mean and S.D. of three independent biological replicates shown.

SI Video 1 | Detection of caspase-3/7 activity in cycloheximide treated macrophages. The mitochondria of BMDMs were labeled with TMRM (red), treated with cycloheximide and incubated with Draq7 (blue DNA stain) and caspase-3/7 fluorescent substrate (green). Fluorescent and bright field images were acquired every 60 min.

SI Video 2 | Detection of caspase-3/7 activity in Δ*flaA L. pneumophila* infected macrophages. The mitochondria of BMDMs were labeled with TMRM (red), infected with Δ*flaA L. pneumophila* and incubated with Draq7 (blue DNA stain) and caspase-3/7 fluorescent substrate (green). Fluorescent and bright field images were acquired every 60 min.

SI Video 3 | Detection of caspase-3/7 activity in WT *L. pneumophila* infected macrophages. The mitochondria of BMDMs were labeled with TMRM (red), infected with WT *L. pneumophila* and incubated with Draq7 (blue DNA stain) and caspase-3/7 fluorescent substrate (green). Fluorescent and bright field images were acquired every 60 min.

REFERENCES

Abu-Zant, A., Jones, S., Asare, R., Suttles, J., Price, C., Graham, J., et al. (2007). Anti-apoptotic signalling by the Dot/Icm secretion system of *L. pneumophila*. *Cell Microbiol.* 9, 246–264. doi: 10.1111/j.1462-5822.2006.00785.x

Abu-Zant, A., Santic, M., Molmeret, M., Jones, S., Helbig, J., and Abu Kwaik, Y. (2005). Incomplete activation of macrophage apoptosis during intracellular replication of *Legionella pneumophila*. *Infect. Immun.* 73, 5339–5349. doi: 10.1128/IAI.73.9.5339-5349.2005

Akhter, A., Gavrilin, M. A., Frantz, L., Washington, S., Ditty, C., Limoli, D., et al. (2009). Caspase-7 activation by the Nlrc4/Ipaf inflammasome restricts *Legionella pneumophila* infection. *PLoS Pathog.* 5:e1000361. doi: 10.1371/journal.ppat.1000361

Amer, A. O. (2010). Modulation of caspases and their non-apoptotic functions by *Legionella pneumophila*. *Cell. Microbiol.* 12, 140–147. doi: 10.1111/j.1462-5822.2009.01401.x

Baines, C. P., Kaiser, R. A., Purcell, N. H., Blair, N. S., Osinska, H., Hambleton, M. A., et al. (2005). Loss of cyclophilin D reveals a critical role for mitochondrial permeability transition in cell death. *Nature* 434, 658–662. doi: 10.1038/nature03434

Banga, S., Gao, P., Shen, X., Fiscus, V., Zong, W. X., Chen, L., et al. (2007). *Legionella pneumophila* inhibits macrophage apoptosis by targeting pro-death members of the Bcl2 protein family. *Proc. Natl. Acad. Sci. U.S.A.* 104, 5121–5126. doi: 10.1073/pnas.0611030104

Broz, P., Ruby, T., Belhocine, K., Bouley, D. M., Kayagaki, N., Dixit, V. M., et al. (2012). Caspase-11 increases susceptibility to Salmonella infection in the absence of caspase-1. *Nature* 490, 288–291. doi: 10.1038/nature11419

Carneiro, L. A., Travassos, L. H., Soares, F., Tattoli, I., Magalhaes, J. G., Bozza, M. T., et al. (2009). Shigella induces mitochondrial dysfunction and cell death in nonmyleoid cells. *Cell Host Microbe* 5, 123–136. doi: 10.1016/j.chom.2008.12.011

Case, C. L., Kohler, L. J., Lima, J. B., Strowig, T., de Zoete, M. R., Flavell, R. A., et al. (2013). Caspase-11 stimulates rapid flagellin-independent pyroptosis in response to *Legionella pneumophila*. *Proc. Natl. Acad. Sci. U.S.A.* 110, 1851–1856. doi: 10.1073/pnas.1211521110

Casson, C. N., Copenhaver, A. M., Zwack, E. E., Nguyen, H. T., Strowig, T., Javdan, B., et al. (2013). Caspase-11 activation in response to bacterial secretion systems that access the host cytosol. *PLoS Pathog.* 9:e1003400. doi: 10.1371/journal.ppat.1003400

Cazalet, C., Gomez-Valero, L., Rusniok, C., Lomma, M., Dervins-Ravault, D., Newton, H. J., et al. (2010). Analysis of the legionella longbeachae genome and transcriptome uncovers unique strategies to cause Legionnaires' disease. *PLoS Genet.* 6:e1000851. doi: 10.1371/journal.pgen.1000851

Cerqueira, D. M., Pereira, M. S., Silva, A. L., Cunha, L. D., and Zamboni, D. S. (2015). Caspase-1 but not Caspase-11 is required for NLRC4-mediated pyroptosis and restriction of infection by flagellated legionella species in mouse macrophages and *in vivo*. *J. Immunol.* 195, 2303–2311. doi: 10.4049/jimmunol.1501223

Chipuk, J. E., Moldoveanu, T., Llambi, F., Parsons, M. J., and Green, D. R. (2010). The BCL-2 family reunion. *Mol. Cell* 37, 299–310. doi: 10.1016/j.molcel.2010.01.025

Chow, S. H., Deo, P., and Naderer, T. (2016). Macrophage cell death in microbial infections. *Cell Microbiol.* 18, 466–474. doi: 10.1111/cmi.12573

Croker, B. A., O'Donnell, J. A., Nowell, C. J., Metcalf, D., Dewson, G., Campbell, K. J., et al. (2011). Fas-mediated neutrophil apoptosis is accelerated by Bid, Bak, and Bax and inhibited by Bcl-2 and Mcl-1. *Proc. Natl. Acad. Sci. U.S.A.* 108, 13135–13140. doi: 10.1073/pnas.1110358108

Cunha, B. A., Burillo, A., and Bouza, E. (2016). Legionnaires' disease. *Lancet.* 23, 376–385. doi: 10.1016/S0140-6736(15)60078-2

Czabotar, P. E., Lessene, G., Strasser, A., and Adams, J. M. (2014). Control of apoptosis by the BCL-2 protein family: implications for physiology and therapy. *Nat. Rev. Mol. Cell Biol.* 15, 49–63. doi: 10.1038/nrm3722

Diwan, A., Krenz, M., Syed, F. M., Wansapura, J., Ren, X., Koesters, A. G., et al. (2007). Inhibition of ischemic cardiomyocyte apoptosis through targeted ablation of Bnip3 restrains postinfarction remodeling in mice. *J. Clin. Invest.* 117, 2825–2833. doi: 10.1172/JCI32490

Edelstein, P. H. (1986). Control of Legionella in hospitals. *J. Hosp. Infect.* 8, 109–115. doi: 10.1016/0195-6701(86)90037-X

Elmore, S. (2007). Apoptosis: a review of programmed cell death. *Toxicol. Pathol.* 35, 495–516. doi: 10.1080/01926230701320337

Fischer, S. F., Vier, J., Müller-Thomas, C., and Häcker, G. (2006). Induction of apoptosis by *Legionella pneumophila* in mammalian cells requires the mitochondrial pathway for caspase activation. *Microbes Infect.* 8, 662–669. doi: 10.1016/j.micinf.2005.08.016

Gao, L. Y., and Abu Kwaik, Y. (1999). Activation of caspase 3 during *Legionella pneumophila*-induced apoptosis. *Infect. Immun.* 67, 4886–4894.

Gomez-Valero, L., Rusniok, C., Cazalet, C., and Buchrieser, C. (2011). Comparative and functional genomics of legionella identified eukaryotic like proteins as key players in host-pathogen interactions. *Front. Microbiol.* 2:208. doi: 10.3389/fmicb.2011.00208

Gutiérrez-Aguilar, M., and Baines, C. P. (2015). Structural mechanisms of cyclophilin D-dependent control of the mitochondrial permeability transition pore. *Biochim. Biophys. Acta* 1850, 2041–2047. doi: 10.1016/j.bbagen.2014.11.009

Hagar, J. A., Powell, D. A., Aachoui, Y., Ernst, R. K., and Miao, E. A. (2013). Cytoplasmic LPS activates caspase-11: implications in TLR4-independent endotoxic shock. *Science* 341, 1250–1253. doi: 10.1126/science.1240988

Hsu, F., Zhu, W., Brennan, L., Tao, L., Luo, Z. Q., and Mao, Y. (2012). Structural basis for substrate recognition by a unique Legionella phosphoinositide phosphatase. *Proc. Natl. Acad. Sci. U.S.A.* 109, 13567–13572. doi: 10.1073/pnas.1207903109

Isberg, R. R., O'Connor, T. J., and Heidtman, M. (2009). The *Legionella pneumophila* replication vacuole: making a cosy niche inside host cells. *Nat. Rev.* 7, 13–24. doi: 10.1038/nrmicro1967

Kataoka, T., Holler, N., Micheau, O., Martinon, F., Tinel, A., Hofmann, K., et al. (2001). Bcl-rambo, a novel Bcl-2 homologue that induces apoptosis *Rev.* 7, 13–24. doi: 10.1038/nrmicro1967

Kataoka, T., Holler, N., Micheau, O., Martinon, F., Tinel, A., Hofmann, K., et al. (2001). Bcl-rambo, a novel Bcl-2 homologue that induces apoptosis via its unique C-terminal extension. *J. Biol. Chem.* 276, 19548–19554. doi: 10.1074/jbc.M010520200

Kayagaki, N., Wong, M. T., Stowe, I. B., Ramani, S. R., Gonzalez, L. C., Akashi-Takamura, S., et al. (2013). Noncanonical inflammasome activation by intracellular LPS independent of TLR4. *Science* 341, 1246–1249. doi: 10.1126/science.1240248

Khemiri, A., Jouenne, T., and Cosette, P. (2008). Presence in *Legionella pneumophila* of a mammalian-like mitochondrial permeability transition pore? *FEMS Microbiol. Lett.* 278, 171–176. doi: 10.1111/j.1574-6968.2007.00985.x

Kim, J. Y., Kim, Y. J., Lee, S., and Park, J. H. (2011). BNip3 is a mediator of TNF-induced necrotic cell death. *Apoptosis* 16, 114–126. doi: 10.1007/s10495-010-0550-4

Lamkanfi, M., Amer, A., Kanneganti, T. D., Muñoz-Planillo, R., Chen, G., Vandenabeele, P., et al. (2007). The Nod-like receptor family member Naip5/Birc1e restricts *Legionella pneumophila* growth independently of caspase-1 activation. *J. Immunol.* 178, 8022–8027. doi: 10.4049/jimmunol.178.12.8022

Losick, V. P., and Isberg, R. R. (2006a). NF-κB translocation prevents host cell death after low-dose challenge by *Legionella pneumophila*. *J. Exp. Med.* 203, 2177–2189. doi: 10.1084/jem.20060766

Losick, V. P., and Isberg, R. R. (2006b). NF-κB translocation prevents host cell death after low-dose challenge by *Legionella pneumophila*. *J. Exp. Med.* 203, 2177–2189. doi: 10.1084/jem.20060766

Miao, E. A., Leaf, I. A., Treuting, P. M., Mao, D. P., Dors, M., Sarkar, A., et al. (2010). Caspase-1-induced pyroptosis is an innate immune effector mechanism against intracellular bacteria. *Nat. Immunol.* 11, 1136–1142. doi: 10.1038/ni.1960

Molmeret, M., and Abu Kwaik, Y. (2002). How does *Legionella pneumophila* exit the host cell? *Trends Microbiol.* 10, 258–260. doi: 10.1016/S0966-842X(02)02359-4

Molmeret, M., Alli, O. A., Zink, S., Flieger, A., Cianciotto, N. P., and Kwaik, Y. A. (2002). icmT is essential for pore formation-mediated egress of *Legionella pneumophila* from mammalian and protozoan cells. *Infect. Immun.* 70, 69–78. doi: 10.1128/IAI.70.1.69-78.2002

Molmeret, M., Jones, S., Santic, M., Habyarimana, F., Esteban, M. T., and Kwaik, Y. A. (2010). Temporal and spatial trigger of post-exponential virulence-associated regulatory cascades by *Legionella pneumophila* after bacterial escape into the host cell cytosol. *Environ. Microbiol.* 12, 704–715. doi: 10.1111/j.1462-2920.2009.02114.x

Molmeret, M., Santic, M., Asare, R., Carabeo, R. A., and Abu Kwaik, Y. (2007). Rapid escape of the dot/icm mutants of *Legionella pneumophila* into the cytosol of mammalian and protozoan cells. *Infect. Immun.* 75, 3290–3304. doi: 10.1128/IAI.00292-07

Molmeret, M., Zink, S. D., Han, L., Abu-Zant, A., Asari, R., Bitar, D. M., et al. (2004). Activation of caspase-3 by the Dot/Icm virulence system is essential for arrested biogenesis of the Legionella-containing phagosome. *Cell. Microbiol.* 33–48. doi: 10.1046/j.1462-5822.2003.00335.x

Molofsky, A. B., Byrne, B. G., Whitfield, N. N., Madigan, C. A., Fuse, E. T., Tateda, K., et al. (2006). Cytosolic recognition of flagellin by mouse macrophages restricts *Legionella pneumophila* infection. *J. Exp. Med.* 203, 1093–1104. doi: 10.1084/jem.20051659

Nogueira, C. V., Lindsten, T., Jamieson, A. M., Case, C. L., Shin, S., Thompson, C. B., et al. (2009). Rapid pathogen-induced apoptosis: a mechanism used by dendritic cells to limit intracellular replication of *Legionella pneumophila*. *PLoS Pathog.* 5:e1000478. doi: 10.1371/journal.ppat.1000478

Ren, T., Zamboni, D. S., Roy, C. R., Dietrich, W. F., and Vance, R. E. (2006). Flagellin-deficient Legionella mutants evade caspase-1- and Naip5-mediated macrophage immunity. *PLoS Pathog.* 2:e18. doi: 10.1371/journal.ppat.0020018

Rikka, S., Quinsay, M. N., Thomas, R. L., Kubli, D. A., Zhang, X., Murphy, A. N., et al. (2011). Bnip3 impairs mitochondrial bioenergetics and stimulates mitochondrial turnover. *Cell Death Differ.* 18, 721–731. doi: 10.1038/cdd.2010.146

Santic, M., Asare, R., Doric, M., and Abu Kwaik, Y. (2007). Host-dependent trigger of caspases and apoptosis by *Legionella pneumophila*. *Infect. Immun.* 75, 2903–2913. doi: 10.1128/IAI.00147-07

Santic, M., Molmeret, M., and Abu Kwaik, Y. (2005). Maturation of the *Legionella pneumophila*-containing phagosome into a phagolysosome within gamma interferon-activated macrophages. *Infect. Immun.* 73, 3166–3171. doi: 10.1128/IAI.73.5.3166-3171.2005

Shi, J., Zhao, Y., Wang, K., Shi, X., Wang, Y., Huang, H., et al. (2015). Cleavage of GSDMD by inflammatory caspases determines pyroptotic cell death. *Nature* 526, 660–665. doi: 10.1038/nature15514

Speir, M., Lawler, K. E., Glaser, S. P., Abraham, G., Chow, S., Vogrin, A., et al. (2016). Eliminating *Legionella* by inhibiting BCL-XL to induce macrophage apoptosis. *Nat. Microbiol.* 1:15034. doi: 10.1038/nmicrobiol.2015.34

Speir, M., Vince, J., and Naderer, T. (2014). Programmed cell death in *Legionella infection*. *Future Microbiol.* 9, 107–118. doi: 10.221/fmb.13.139

van Delft, M. F., Wei, A. H., Mason, K. D., Vandenberg, C. J., Chen, L., Czabotar, P. E., et al. (2006). The BH3 mimetic ABT-737 targets selective Bcl-2 proteins and efficiently induces apoptosis via Bak/Bax if Mcl-1 is neutralized. *Cancer Cell* 10, 389–399. doi: 10.1016/j.ccr.2006.08.027

Willis, S. N., Fletcher, J. I., Kaufmann, T., van Delft, M. F., Chen, L., Czabotar, P. E., et al. (2007). Apoptosis initiated when BH3 ligands engage multiple Bcl-2 homologs, not Bax or Bak. *Science* 315, 856–859. doi: 10.1126/science.1133289

Youle, R. J., and Strasser, A. (2008). The BCL-2 protein family: opposing activities that mediate cell death. *Nat. Rev. Mol. Cell Biol.* 9, 47–59. doi: 10.1038/nrm2308

Zamboni, D. S., Kobayashi, K. S., Kohlsdorf, T., Ogura, Y., Long, E. M., Vance, R. E., et al. (2006). The Birc1e cytosolic pattern-recognition receptor contributes to the detection and control of *Legionella pneumophila* infection. *Nat. Immunol.* 7, 318–325. doi: 10.1038/ni1305

Zhang, J., and Ney, P. A. (2009). Role of BNIP3 and NIX in cell death, autophagy, and mitophagy. *Cell Death Differ.* 16, 939–946. doi: 10.1038/cdd.2009.16

Zhao, Y., Yang, J., Shi, J., Gong, Y. N., Lu, Q., Xu, H., et al. (2011). The NLRC4 inflammasome receptors for bacterial flagellin and type III secretion apparatus. *Nature* 477, 596–600. doi: 10.1038/nature10510

Zhu, W., Hammad, L. A., Hsu, F., Mao, Y., and Luo, Z. Q. (2013). Induction of caspase 3 activation by multiple *L. pneumophila* Dot/Icm substrates. *Cell Microbiol.* 15, 1783–1795. doi: 10.1111/cmi.12157

3

Yersinia Pestis and *Yersinia Pseudotuberculosis* Infection: A Regulatory RNA Perspective

*Luary C. Martínez-Chavarría[1] and Viveka Vadyvaloo[2]**

[1] Departamento de Patología, Facultad de Medicina Veterinaria y Zootecnia, Universidad Nacional Autónoma de México, México, Mexico, [2] Paul G. Allen School for Global Animal Health, Washington State University, Pullman, WA, USA

Edited by:
Beiyan Nan,
Texas A&M University, USA

Reviewed by:
Wyndham W. Lathem,
Northwestern University Feinberg School of Medicine, USA
Yanping Han,
Beijing Institute of Microbiology and Epidemiology, China

***Correspondence:**
Viveka Vadyvaloo,
Paul G. Allen School for Global Animal Health, Washington State University,
240 SE Ott Road, Pullman,
WA 99163, USA
viveka@vetmed.wsu.edu

Yersinia pestis, responsible for causing fulminant plague, has evolved clonally from the enteric pathogen, *Y. pseudotuberculosis*, which in contrast, causes a relatively benign enteric illness. An ~97% nucleotide identity over 75% of their shared protein coding genes is maintained between these two pathogens, leaving much conjecture regarding the molecular determinants responsible for producing these vastly different disease etiologies, host preferences and transmission routes. One idea is that coordinated production of distinct factors required for host adaptation and virulence in response to specific environmental cues could contribute to the distinct pathogenicity distinguishing these two species. Small non-coding RNAs that direct posttranscriptional regulation have recently been identified as key molecules that may provide such timeous expression of appropriate disease enabling factors. Here the burgeoning field of small non-coding regulatory RNAs in *Yersinia* pathogenesis is reviewed from the viewpoint of adaptive colonization, virulence and divergent evolution of these pathogens.

Keywords: post-transcriptional regulation, RNA-binding proteins, sRNAs, riboswitch, Hfq

Introduction

Yersinia pestis and *Yersinia pseudotuberculosis*: so Similar Yet so Different

The Gram-negative genus *Yersinia* comprises 17 different species, but only three have been shown to be virulent to humans and animals: *Yersinia enterocolitica*, *Y. pseudotuberculosis* and *Y. pestis*. Although both *Y. pseudotuberculosis* and *Y. enterocolitica* are enteric pathogens, they are much less closely related at a DNA level than *Y. pestis* is with *Y. pseudotuberculosis*. In fact, *Y. pseudotuberculosis* diverged from *Y. enterocolitica* between 41 and 186 million years ago, while *Y. pestis* diverged from *Y. pseudotuberculosis* within the last 1500–20 000 years, which implies that these two species are more closely genetically related (Achtman et al., 1999).

Interestingly, however, *Y. pestis* and *Y. pseudotuberculosis* produce very different clinical diseases in their hosts and they use both a radically different mechanism of transmission as well as arsenal of virulence factors. While the latter is an enteropathogen which causes self-limiting food-borne enteric diseases that rarely lead to death, the former is a zoonotic pathogen that causes plague, a vector-borne disease transmitted by fleas and one of the most deadly diseases.

Yersinia pseudotuberculosis can be hosted in various animal reservoirs, e.g.: dogs, cats, cattle, horses, rabbits, deer, turkey, ducks, and many others, as well as in soil, plants insects and amoeba in the environment. *Y. pseudotuberculosis* infection generally occurs after ingestion of contaminated food or water, after which it colonizes the gastrointestinal tract, first at the Peyer's patches of the distal small intestine and then it disseminates to the liver and spleen either directly or via the mesenteric lymph nodes. Infections of *Y. pseudotuberculosis* in humans usually lead to gastroenteritis, characterized by a self-limiting mesenteric lymphadenitis and diarrhea. Generally antibiotics are not even required; however, in people with underlying chronic liver ailments or that are immunocompromised, infection has been associated with septic complications and they can develop severe and potentially fatal systemic infections. Enteric *Yersinia* can spend significant portions of their lives outside the mammalian host in soil and water environments and within free-living amoeba (Lambrecht et al., 2013; Santos-Montanez et al., 2015) with varied sources of nutrients, and therefore they retain metabolic capabilities that have been lost in *Y. pestis*.

On the other hand, plague is acquired mainly by the bite of an infected flea, but also by contact with infected tissues or inhalation of respiratory droplets or aerosols. Fleas acquire *Y. pestis* from the blood of a highly bacteremic host and the bacteria multiply within the flea proventriculus and midgut forming a thick coherent biofilm that occludes, and eventually blocks, the foregut proventriculus and esophagus (Hinnebusch et al., 1996; Vadyvaloo et al., 2010). Transmission of *Y. pestis* from the fleas occurs mainly through the proventricular blockage regurgitation mechanism after a period of extrinsic incubation during which the bacteria have to adapt to the flea gut environment. This blockage impedes fresh bloodmeal ingestion and facilitates reflux from the midgut to the flea feeding mouthparts, thus causing the flea to starve. The "blocked" fleas will often regurgitate a mixture of blood and *Y. pestis*-bearing blockage material back into the flea bite site (Jarrett et al., 2004). After flea bite, *Y. pestis* transiently colonizes the dermis of the host, interacts mainly with macrophages which are permissive to its survival and replication (Pujol and Bliska, 2003; Shannon et al., 2015), and then rapidly migrates to the regional lymph node. The bacteria multiply to high numbers in the lymph node causing necrosis and inflammation that characterize bubo formation (Sebbane et al., 2005). From the bubo, hematogenous spread of the bacteria occurs to deeper organs like the spleen and liver causing septicemia. It is via this route that *Y. pestis* can also reach the lungs, to cause secondary pneumonic plague. Person-to-person spread of primary pneumonic plague can then ensue by direct inhalation of infectious droplets or aerosols through coughing.

Primary pneumonic plague is the deadliest form of plague because of its rapid progression to death in 3–4 days (Lathem et al., 2005; Bubeck et al., 2007; Price et al., 2012). Once the bacteria enter the lung tissues via the respiratory route they multiply at an accelerated rate within the first 24–36 h. This time period of the infection is characterized by absence of histopathological changes in lung tissue architecture, and anti-inflammatory molecules, thus marking an anti-inflammatory phase (Lathem et al., 2005; Bubeck et al., 2007; Price et al., 2012). Beyond the 36 h, the bacteria continue to multiply to up to 10^{10} cfus, within 3 days, at which time the bacteria disseminate from the lungs to peripheral tissues like the spleen and continue to multiply there. This latter stage of infection is characterized by raised levels of cytokines and chemokines and leads to a purulent multifocal severe exudative bronchopneumonia (Lathem et al., 2005) and death. Plague infection can be fatal without antibiotic treatment early during infection.

Surprisingly *Y. pseudotuberculosis* and *Y. pestis* differ radically in their pathogenesis despite sharing >97% identity in 75% of their chromosomal genes (Chain et al., 2004). This fact has arisen much interest in elucidating what factors are responsible for such virulence differences (Achtman et al., 2004; Pouillot et al., 2008). There is, however, one critical plasmid-encoded virulence factor that is conserved between *Y. pestis* and *Y. pseudotuberculosis*, and this is the Yop-Ysc Type three secretion system (T3SS; Cornelis et al., 1998). Most *ysc* genes encode proteins that form an injectisome structure that is required to deliver Yop effector proteins into the cytosol of host cells, e.g., immune cells. Carefully orchestrated synthesis of the T3SS Yop effector proteins is required during infection to mediate virulence. This enables the bacteria to subvert host immune function.

At the DNA level the differences between these two strains is being revealed by whole-genome sequencing based phylogenetic studies, which shows that genomes of *Y. pestis* isolates are constantly changing, acquiring and losing genetic elements and undergoing genomic rearrangements when compared with its ancestor *Y. pseudotuberculosis* (Achtman et al., 1999; Morelli et al., 2010). Some of these gene content changes between *Y. pestis* and *Y. pseudotuberculosis* have been able to explain the differences in host colonization and virulence of these pathogens (Chouikha and Hinnebusch, 2014; Sun et al., 2014; Zimbler et al., 2015). Regulatory changes and constant development of new regulatory networks are, however, also well-established ways of evolution of virulence characteristics among pathogens.

Transcriptional regulation is perhaps the most well studied form of controlling gene expression and several global transcriptional regulators play major roles in regulating virulence/pathogenesis in the yersiniae, including Crp (Zhan et al., 2008; Heroven et al., 2012b; Lathem et al., 2014), RovA (Cathelyn et al., 2006; Heroven and Dersch, 2006) and RovM (Heroven and Dersch, 2006). Few comparative studies to understand evolution of the regulons of these transcriptional factors have been undertaken in *Y. pestis* and *Y. pseudotuberculosis*. Beyond transcriptional regulation, post-transcriptional regulation which provides a powerful way for the bacteria to rapidly fine tune gene expression to the needs of the cell, at a more localized level, has recently been appreciated in *Yersinia* species.

Small Non-Coding RNAs (sRNAs): a Mechanism of Post-Transcriptional Regulation

A major manner in which post-transcriptional regulation can be accomplished is through sRNA regulation. These are small molecules of RNA that are not translated into proteins (Gottesman and Storz, 2011). In general these molecules carry out their regulatory function by base-pairing to a limited complementary sequence (6–8 contiguous base-pairs) in the mRNAs of their cognate target gene. This interaction leads to modification of mRNA translation or stability or both, thereby influencing the target gene expression and protein activity (Narberhaus and Vogel, 2009; Waters and Storz, 2009). Small RNAs are commonly known to repress gene expression as the base-pairing sequester mRNA ribosome binding sites, resulting in translational repression and accelerated transcript degradation. However, it is emerging that they can act as translational activators and mRNA stabilizers, for which the underlying mechanisms differ considerably from repression. Translational activation occurs through interactions of sRNAs with the 5′ untranslated region (UTR), the coding sequence, or the 3′-UTR of the target mRNAs (Papenfort and Vanderpool, 2015).

Post-transcriptional regulation mediated through sRNA molecules can currently be characterized into three categories (Waters and Storz, 2009; Oliva et al., 2015): (1) Trans-encoded sRNAs: small RNAs that are distally located from, and interact through limited complementarity with their target mRNAs. These sRNAs usually bind to the Shine-Dalgarno (SD) sequence thereby occluding the ribosome-binding site (RBS), or bind to the coding region of the mRNA. Both base-pairing interactions result in inhibition of translation, and can be coupled with enhanced RNAse activity that facilitates increased rate of mRNA cleavage and degradation. The loose base-pairing interaction between the *trans*-encoded sRNA and target mRNA is often stabilized by an RNA chaperone protein, Hfq. (2) *Cis*-acting sRNAs are a second type of sRNA that is transcribed from the antisense strand of its target mRNA. Being usually encoded in the 5′-UTR region of the mRNA, it mediates its interaction by forming a duplex that contorts into a secondary structure which interferes with ribosome binding or mRNA stability. Besides the sRNA-mRNA interaction, sRNAs can directly interact with regulatory proteins to interfere with their function. This is best exemplified by the interaction of the CsrB and CsrC sRNAs with the global regulator protein CsrA which is a mechanism described for *Y. pseudotuberculosis* (Oliva et al., 2015). This will be discussed in greater detail below. (3) An alternate mechanism of post-transcriptional regulation that results in activation or repression of translation, can be achieved by conformational alterations of complex RNA structures that occurs via binding of small metabolites/cofactors (called riboswitches; Oliva et al., 2015) or thermo-modulation (called thermosensors; Krajewski and Narberhaus, 2014).

With regards to post-transcriptional regulation in the *Yersinia* species, there are several studies that have identified the arsenal of sRNAs expressed by *Y. pseudotuberculosis* and *Y. pestis* under different growth conditions. This has revealed both similarities and unexpected differences, not only limited to the presence or absence of sRNAs genes, but also related to the spatial and temporal expression patterns, and dependence on RNA binding proteins. In this review our aim is to take stock of our current understanding of the role of sRNA-mediated post-transcriptional regulation in these pathogens, and how this may have influenced the unique disease manifestations that define each species.

sRNA Identification: What We Know in *Yersinia*

sRNA Identification

Interest in finding small RNAs (sRNAs) in bacteria has significantly increased in recent years due to their important regulatory functions. Identification of sRNAs has been undertaken in diverse pathogenic bacterial species, e.g., *Salmonella* (Sittka et al., 2008), *Vibrio cholerae* (Liu et al., 2009), Group A *Streptococcus* (Perez et al., 2009), *Helicobacter pylori* (Sharma et al., 2010), *Clostridium difficile* (Soutourina et al., 2013), *Acinetobacter baumannii* (Sharma et al., 2014), and *Porphyromonas gingivalis* (Phillips et al., 2014) amongst numerous others. The identification and characterization of these molecules in the pathogenesis of *Yersinia*, has also in the past decade been gaining importance.

Initially sRNAs were discovered only by computational analysis using gene homology with closely related bacterial species; but without experimental data, erroneous determination of gene start sites or incorrect annotations can be made. For example, Livny et al. (2006) predicted 1478 sRNAs encoded in *Y. pestis* using sRNAPredict2 which searched for intergenic regions (IGs) with a conserved sequence and adjacent Rho-independent terminator that had to be present in 3–7 related bacterial species. Recent development of high-throughput methods and more sophisticated computational algorithms has allowed rapid identification of sRNA candidates in different species. However, given their varying sizes (50–500 nucleotides [nt]) and their potential genomic locations in the 5′- or 3′-UTRs as well as in IGs, identification and validation of true sRNAs remains challenging.

Several studies have identified multiple sRNAs expressed by *Y. pestis* and *Y. pseudotuberculosis* under various *in vitro* conditions (Koo et al., 2011; Qu et al., 2012; Beauregard et al., 2013; Yan et al., 2013; Schiano et al., 2014; Nuss et al., 2015); one study included identification of sRNAs expressed *in vivo*, in infected mouse lung and spleen organ tissue in the virulence -restricted *Y. pestis* biovar (bv.) microtus strain (Yan et al., 2013). The acronym Ysr (representing ***Y****ersinia* **s**mall **R**NAs) was initially coined by the Lathem group to name identified *Yersinia* specific sRNAs (Koo et al., 2011). Two subsequent studies assigned different names to their sRNAs (Qu et al., 2012; Yan et al., 2013), but more recently the norm has been to assign a continuous number to each newly identified *Yersinia* sRNA and prefix this with 'Ysr.' Moving forward the 'Ysr' name should be maintained to achieve standardization of *Yersinia* sRNA nomenclature.

Y. pseudotuberculosis IP32953 and *Y. pestis* CO92 sRNAs

Two pioneering studies by the Lathem group used a deep sequencing approach that enabled a comprehensive identification, validation, and partial functional characterization of a global set of sRNAs in *Y. pseudotuberculosis and Y. pestis* (Koo et al., 2011; Schiano et al., 2014). The first study, used the *Y. pseudotuberculosis* IP32953 strain, and identified 150 previously unannotated Ysrs. One hundred and eighteen of these were specific to *Y. pseudotuberculosis* and *Y. pestis* and 32 were orthologs to *E. coli* and *Salmonella* Typhimurium sRNAs (Koo and Lathem, 2012). The second study was performed using the *Y. pestis* CO92 strain and this supported and extended their previous study, as they found 144 sRNAs previously identified in *Y. pseudotuberculosis* and 63 new potential sRNAs, 10 of which were further validated by Northern blot analysis (Schiano et al., 2014). Both studies were performed essentially identically which permitted direct comparisons of the data sets. Both *Y. pseudotuberculosis* and *Y. pestis* were grown at 26 and 37°C to mimic the environment/flea and mammalian host infection temperatures, respectively, and analyzed at early log, mid log, late log and stationary phases of growth in the rich medium, brain heart infusion (BHI). The total number of potential sRNAs that they identified between *Y. pestis* and *Y. pseudotuberculosis* was 216 (Koo et al., 2011; Schiano et al., 2014). Although the majority of these sRNAs are conserved between both strains, the timing of sRNA expression and functional dependence on the post-transcriptional chaperone protein Hfq was shown to differ.

In general, in *Y. pseudotuberculosis* most Ysrs start to accumulate at later time points at both temperatures. An exception is Ysr45/GcvB which is expressed highly at the start of growth after which its expression declines over time. This was also the most abundantly expressed Ysr at 26°C, while at 37°C, Ysr7/MicA and Ysr149/OmrA/B were the most highly expressed Ysrs (Koo et al., 2011).

The sRNAs examined when *Y. pestis* grows at 26°C, show mostly steady state levels over time and little dependence on Hfq, which may explain why Hfq is not essential for *Y. pestis* survival in fleas (Rempe et al., 2012). At 37°C, however, these Ysrs are expressed differently than in *Y. pseudotuberculosis*, as most show stable levels or accumulation over time that peak at late-log phase but are almost undetectable when *Y. pestis* reaches stationary phase. Additionally, all the sRNAs tested in *Y. pestis* require Hfq for their stability/expression which may explain the severe growth defect displayed by a *hfq* mutant of *Y. pestis* at 37°C relative to that in *Y. pseudotuberculosis* (Bai et al., 2010; Koo et al., 2011; Schiano et al., 2014). This suggests that subtle evolutionary differences in post-transcriptional gene regulation exist between the two pathogenic *Yersinia* species, which may influence differential temporal regulation of targets that finally contribute to the production of divergent clinical diseases.

A group of known sRNAs (MicA/Ysr7, FnrS/Ysr11, RprA/Ysr40, GcvB/Ysr45, RybB/Ysr48, MicM/Ysr145, RyhB/Ysr146.1 and Ysr146.2, GlmY/Ysr147, GlmZ/Ysr148, and OmrA/B/Ysr149) were amongst those identified in *Yersinia* that were previously annotated and functionally characterized in *E. coli* and *Salmonella*, confirming the conservation of these sRNAs in enteric bacteria. On the other hand, a significant number of the newly discovered sRNAs contained mismatches or were absent, from the genomes of *Y. enterocolitica*, *E. coli*, and *Salmonella* Typhimurium. Many of the Ysrs encoded in both *Y. pseudotuberculosis* and *Y. pestis* contained subtle single or multiple variances in sequence. However expression of six unique sRNAs (Ysr29, Ysr53, Ysr70, Ysr84, Ysr94, and Ysr118) was noted for *Y. pseudotuberculosis* IP32953 (Koo et al., 2011), and five (Ysr142, Ysr143, Ysr144, Ysr163, and Ysr185) for *Y. pestis* CO92 (Schiano et al., 2014); these sRNAs may reflect specific and unique regulatory adaptations to host environments, and disease states caused by each of the mentioned *Yersinia* species (Supplementary Table S1).

Y. pestis KIM6+ sRNAs

Another study using a deep sequencing approach in *Y. pestis* was reported by Beauregard et al. (2013), where they identified 31 sRNAs of which only 17 matched previously identified putative sRNAs. They mapped the 5′ ends of 18 and the 3′ ends of 28 sRNAs, finding that some of them overlap an annotated protein-coding gene. All of these were conserved between *Y. pestis* and *Y. pseudotuberculosis* and all but two are conserved in *Y. enterocolitica*. Only 14 were conserved in *E. coli* but several were only partially conserved, suggesting that even when sRNAs are conserved, their functions could have diverged. Some sRNAs are conserved only in the region required for base-pairing with targets identified in *E. coli* which suggests that these Ysrs share some mRNA targets besides their species-specific targets. Similar to the findings of Koo et al. (2011), a wide variety of sRNA expression patterns differed between *Y. pestis* and *Y. pseudotuberculosis* depending upon temperature and the presence of Hfq. Comparative Northern blot analysis of all 31 sRNAs in both *Y. pestis* and *Y. pseudotuberculosis* showed that most of them were constitutively expressed in both species regardless of temperature or the presence of Hfq (Supplementary Table S1).

Y. pestis biovar Microtus strain 201 sRNAs

Two other sRNA identification analyses were undertaken in the enzootic *Y. pestis* strain 201, which is avirulent in humans, but highly virulent to mice and belongs to the newly established, bv. microtus (Zhou et al., 2004; Qu et al., 2012; Yan et al., 2013). The first study used a supposed improved cDNA cloning approach to find novel sRNAs expressed in *Y. pestis* in chemically defined TMH medium in exponential and stationary growth phases, and under stressful conditions that *Y. pestis* might encounter during infection, e.g., iron starvation, Ca^{2+} deprivation and low Mg^{2+}. For stressful conditions early exponential grown cultures at 26°C were transferred to 37°C. They identified a total of 43 sRNAs. Six of these were previously annotated, 25 were encoded on the antisense strand of annotated genes or non-coding RNAs, 12 were located in IGs, and 8 were not reported previously (Qu et al., 2012). However, sRNAs were not categorized under their condition of expression because even though bacteria were grown in five separate treatment conditions, the RNA isolated from each condition was pooled in equimolar ratios for the subsequent cDNA library construction. However, four sRNAs were individually detected in the five growth conditions by Northern blot analysis: Yp-sR1, Yp-sR2, Yp-sR16 and Yp-sR38 (Qu et al., 2012). Comparisons with the Rfam database (http://rfam.sanger.ac.uk), revealed identification of six homologs of known sRNAs of other enteric bacteria, including 6S RNA, SsrA, 4.5S RNA, CyaR, CopA and STnc490. Of these, 6S RNA, which is highly abundant in *E. coli* (Hindley, 1967), was found under all five tested conditions but reached maximal abundance at stationary phase. Four sRNAs that were discovered, are absent in *Y. pseudotuberculosis* and unique to the *Y. pestis* genome, of which the Yp-sR33 is specific to *Y. pestis* bv. microtus and the CO92 genomes. Unfortunately none of these were analyzed by Northern blot or RT-qPCR to further confirm their presence (Qu et al., 2012). However, Northern blot analysis verified the presence and size of six unique sRNAs not previously identified (Supplementary Table S1). Additionally, RT-qPCR analysis of expression of a collective set of 24 sRNAs shows they are highly abundant upon entry into stationary growth phase.

In the chromosome of the *Y. pestis* 201 strain, 36 identical copies and 63 highly homologous sequences of a sRNA called Yp-sR27 was found. These highly repetitive sequences possess all the features of typical transposons, except that they do not encode a transposase and are thus termed non-inserting sequences. Non-inserting sequences are present in the ancestors of *Y. pestis*, *Y. pseudotuberculosis,* and *E. coli*, indicating that they might have evolved conservatively among bacteria. The process of the replication of these sequences remains unknown.

The second study undertaken investigated sRNA transcriptome profiles of the *Y. pestis* bv. microtus strain, grown in TMH (using the same five conditions tested above), BHI and from infected mouse lungs and spleens (Yan et al., 2013). One hundred and four sRNAs were identified, 26 already annotated and 78 representing a novel set of sRNA candidates in *Y. pestis*, 62 of which were intergenic and 16 located on the antisense of annotated ORFs. Sixty two sRNAs were identified in all four conditions tested. Four known sRNAs were found to be conserved in *E. coli* and *Salmonella* species too. However, 93 out of 104 were conserved in *Y. pseudotuberculosis* and *Y. pestis* and only 7 sRNAs were specific to *Y. pestis*. Five of the *Y. pestis* specific sRNAs were located on the *Y. pestis* pMT1 and pPCP1 plasmids indicating that acquisition of sRNA may have an important function during the evolution of *Y. pestis* from *Y. pseudotuberculosis.*

Recently, several of these sRNAs (e.g., sR028, sR041, sR050, sR066, and sR070) that were adjacently located to open-reading frames were re-annotated as 5′-UTRs (Nuss et al., 2015) in a study employing comparative RNA-seq-based high nucleotide resolution transcriptomic profiling. Because sR066 was detected as a short transcript by Northern blotting (Yan et al., 2013), it is presumed that these putative sRNAs are processed forms of a premature 5′-UTR transcript.

Y. pseudotuberculosis YPIII sRNAs

A most recent study, identified sRNAs in *Y. pseudotuberculosis* YPIII and its derivative *crp* mutant. This study employed a more rigorous and comprehensive methodology than previous studies, and determined at single nucleotide resolution the global gene expression of these bacteria grown in LB to exponential and stationary phases at 25 and 37°C (Nuss et al., 2015). A total of 78 putative *trans*-encoded sRNAs, of which 42 were new annotations, and 80 putative antisense RNAs were identified, making this the largest account of antisense sRNAs in the *Yersinia*. Nineteen antisense sRNAs were located on the virulence plasmid supporting the prevailing idea that finely controlled synchronous expression of the T3SS delivery and effector system is required during infection. A comprehensive set of 36 of these sRNAs that comprised 12 known and conserved sRNAs in the *Enterobacteriaceae*, two non-validated sRNAs (Ysr100 and Ysr103), a previously validated sRNA (Ysr164), 11 new sRNAs and 10 antisense sRNAs were confirmed by Northern blot. Most *trans*-encoded sRNAs were largely temperature and growth phase responsive in keeping with studies described above and reflect the transition of the bacteria between heterothermic environments during transmission. This study has to date advanced identification of post-transcriptional regulation involving sRNAs in *Yersinia* and allowed clear categorization of sRNAs into trans-encoded or antisense sRNAs, or as 5′-UTR regulatory elements.

Factors Influencing sRNA Identification

Taking the five studies together, we determine the number of sRNAs identified in *Y. pestis* and *Y. pseudotuberculosis* thus far to be ~354, of which about 105 have been alternately validated by Northern or RT-qPCR (Supplementary Table S1). Of all these identified Ysrs, 13 have been tested in mice, but only three (Ysr29, Ysr35, and tmRNA/ssrA) have been shown to be attenuated (Supplementary Table S1).

While there exists some overlap in the sRNA sets discovered in each of these studies, there are a large number of potential sRNAs that are unique to each study as exemplified in a comparison of *Y. pestis* sRNAs identified in the four studies above (**Figure 1**, Supplementary Table S2). The variable sRNA expression profiles in the studies are likely a consequence of several experimental factors: (1) specific experimental culture conditions, e.g., medium composition, temperature, phase of growth, *in vivo* vs. *in vitro*, (2) species and strains, e.g., *Y. pestis* vs. *Y. pseudotuberculosis* or *Y. pestis* CO92 vs. *Y. pestis* KIM6+ (3) methodologies used, e.g., deep sequencing vs. cDNA cloning, (4) bioinformatics analysis and pipelines applied to data, e.g., expression cut-off thresholds. However other intrinsic factors could also impact variable expression of sRNAs. Take for instance a scenario in which *Y. pestis* and *Y. pseudotuberculosis* encode the same sRNA but that this sRNAs controls different mRNAs targets in the two species, which in turn are additionally differentially regulated by distinct global regulators or growth conditions. In this case, the expression of the sRNA will vary according to the availability of its targets, as in the absence of the mRNA target the sRNA is subject to destabilization and rapid degradation. Indeed, it has been demonstrated that conserved sRNAs between *Y. pseudotuberculosis* and *Y. pestis* can differ in stability and/or expression (Koo et al., 2011; Beauregard et al., 2013). Furthermore, many of the sRNAs that are encoded and expressed by both species contain single nucleotide variations, mismatches, insertions or deletions which could alter the RNA secondary structure and result in distinct interactions with target mRNAs between the species. Some other sRNAs are duplicated, like RyhB (Ysr146.1 and Ysr146.2) in *Y. pseudotuberculosis*, which is regulated by the iron level. It would be interesting to analyze if those duplicated sRNAs have different roles in response to different environmental conditions during the infection and accordingly, if they control the expression of different sets of genes. An added complication

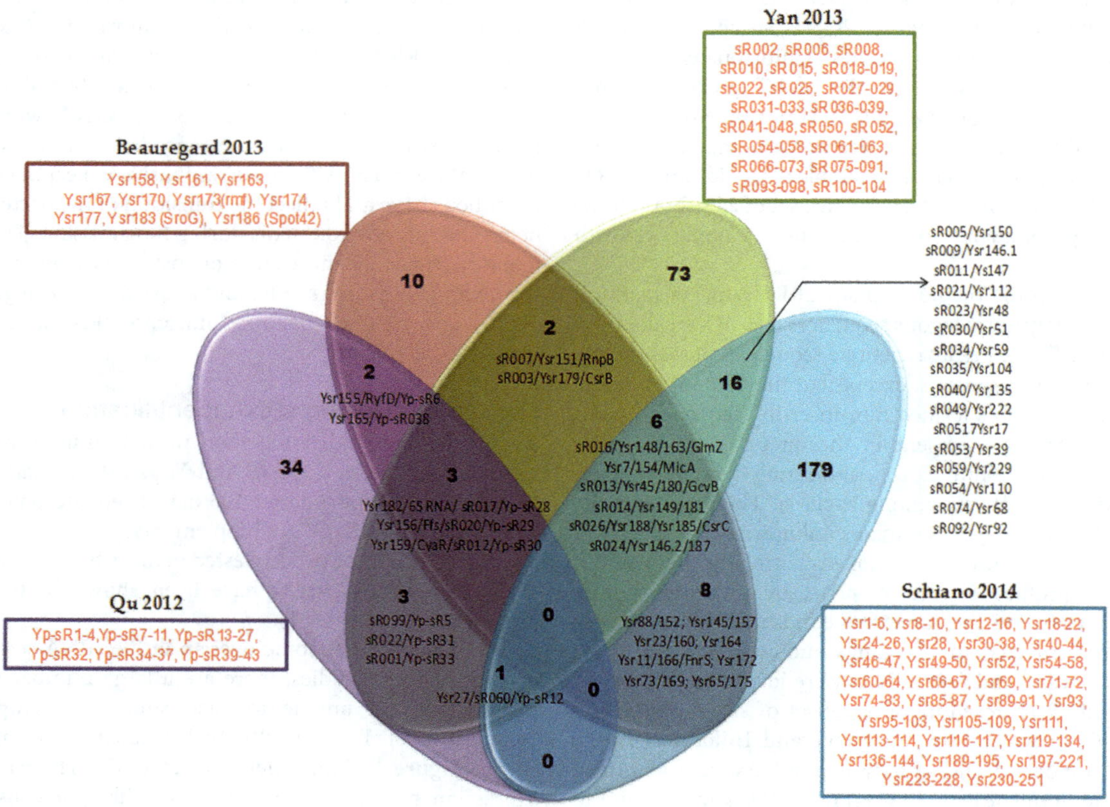

FIGURE 1 | Common (black font) and exclusively expressed (red font) sRNA repertoires are found in *Yersinia pestis* strains of genetically distinct backgrounds in four separate studies using different growth conditions. Common sRNAs are represented by all their annotated names (these are separated by a slash) from different *Yersinia* and *E. coli* studies. Exclusively expressed consecutive sRNAs are not separately named, instead an encompassing dash indicates that they compose the list. Those non-consecutive exclusively expressed sRNAs are named singly and separated by commas from other sRNAs. Each study identifying a sRNA repertoire is represented in the Venn diagram by a different color and denoted according to the first author and year of publication of the study.

can occur during infection if only a fraction of the cell population expresses a particular sRNA. For example, it has been shown that only a small fraction of either *Y. pseudotuberculosis* or *Y. pestis* bacterial populations express the sRNAs Ysr35 or Ysp8, at low copy numbers of 0 to 10 transcripts per cell. Such low quantities can be difficult to detect, limiting identification of such sRNAs (Shepherd et al., 2013). Certainly these differences in sRNA expression between the two pathogens are responsible for some of the divergent clinical disease outcomes.

Although, alternate validation of sRNA expression using techniques such as Northern blotting to avoid false positive identification of sRNAs is important, one is cautioned that such analysis can be tenuous especially for sRNAs expressed at low abundance. Additionally sRNAs may be expressed under conditions not examined thus far, such as upon host-cell contact, inside the cells, during animal infection or even inside the flea in case of *Y. pestis*.

Overall, studies by the Lathem group (Koo et al., 2011; Schiano et al., 2014) proved highly informative because their identification of sRNAs was accompanied by further verification of their expression by Northern analysis, investigation of sRNA Hfq-dependency, identification of cognate mRNAs targets and functional analysis of newly identified sRNAs that revealed their roles in virulence. On the other hand the Nuss et al. (2015) study has used improved techniques to identify and categorize sRNAs, as well, they managed to pinpoint Crp as another major sRNA regulator and verify its direct targeting of several sRNAs.

Functional Characterization of sRNAs in *Yersinia*

Currently >350 Ysrs have been identified but the functional roles that most of these molecules play are yet to be discovered. The functional roles for known *Yersinia* sRNAs conserved in *E. coli* and other related pathogens, that have been subject to thorough investigation of their function, mechanism of action, mRNA targets and conditions of expression are usually inferred. However, the physiological and virulence-inducing processes have become the focus of functional studies on Ysrs, and such roles have been established for a small number of these Ysrs thus far. In this section we will discuss some of these findings.

Virulence

SsrA/sR022: a Vaccine Candidate for Pneumonic Plague?

Small stable RNA A (SsrA RNA), also known as tmRNA or 10S, and small protein B (SmpB) are components of a unique bacterial translational control system (Karzai et al., 2000). This system is important for maintaining cellular homeostasis and bacterial survival in adverse conditions, and as such may be advantageous for efficient response to adverse infection conditions. The SsrA-SmpB system helps maintain the bacterial translational machinery in a fully operational state by dealing with ribosomes that are stalled on defective mRNAs that lack stop codons (Karzai et al., 2000; Dulebohn et al., 2007; Keiler, 2008). SsrA RNA functions both as a tRNA and mRNA. Whereas SmpB is essential for recognition and delivery of SsrA to target stalled ribosomes, as it binds specifically to the tRNA-like domain of SsrA and thus stabilizes SsrA tertiary structure (Karzai et al., 2000; Barends et al., 2001).

The role of SsrA was first evaluated in *Y. pseudotuberculosis* but among deep sequence analyses that searched for new *Yersinia* sRNAs, it was identified in the *Y. pestis* bv. microtus strain and referred to as sR022 or Yp-sR31 (Yan et al., 2013). In *Y. pseudotuberculosis*, expression of VirF, a key TTSS transcriptional activator, and secretion of Yops (YopB, YopD, YopM, LcrH, and LcrV), are reduced in the *smpB-ssrA* mutant. Furthermore this mutant exhibits delayed host cell cytotoxicity and is more sensitive to oxidative and nitrosative stresses, low pH, and sublethal concentrations of translation-specific antibiotics, and is non-motile (Okan et al., 2006, 2010). Consequently, the *smpB-ssrA* mutant strain is avirulent to mice via the orogastric route, as well as defective in survival and replication in macrophages (Okan et al., 2006). It also showed partial protection when challenged with a lethal dose of *Y. pestis*.

In *Y. pestis*, a *smpB-ssrA* mutant is severely attenuated in a mouse model of infection via both the intranasal and intravenous routes. This mutant exhibits a slower growth rate at 37°C (Okan et al., 2010). In agreement, the transcriptional level of SsrA is higher at 37°C than that at 26°C in *Y. pestis* bv. microtus (Qu et al., 2012), which may help *Y. pestis* adapt to natural temperature alterations during its transmission from fleas to mammals. Most significantly, it has been demonstrated that intranasal vaccination of mice with the *ssrA* mutant induced a strong IgG antibody response, and vaccinated animals were well protected against pulmonary *Y. pestis* infection (Okan et al., 2010). Taken together, these characteristics present this strain as a favorable candidate for a live attenuated cell-based vaccine against pneumonic plague.

Ysr141

Ysr141 is present in both *Y. pseudotuberculosis* and *Y. pestis* as identified by sequencing analysis. Its genomic location is mapped to the T3SS-carrying plasmid, pCD1 and it is encoded on the opposite strand within the IG between *yopH* and the gene YPCD1.68c (Koo et al., 2011; Schiano et al., 2014).

Ysr141 is an unstable sRNA which stimulates production of T3SS-associated effector proteins (YopE, YscF, YopK, and LcrF) and regulates *yopJ*, posttranscriptionally, by base-pairing to the 5′-UTR of *yopJ* (Schiano et al., 2014). It has been suggested that Ysr141 may be a mechanism to regulate the T3SS, conserved among the pathogenic *Yersinia* species and it may link environmental cues to the modulation of the T3SS in such species. An investigation into the environmental and temporal cues that trigger expression of Ysr141 and its activation of the T3SS could potentially provide further insight into the differing clinical disease manifestations in *Y. pseudotuberculosis* and *Y. pestis*.

Ysr35

This sRNA was identified from deep sequence analysis (Koo et al., 2011; Schiano et al., 2014). It has been suggested that this sRNA could be required for *Yersinia* adaptation to the host because a Ysr35 mutant showed significantly compromised survival in a mouse model in both *Y. pseudotuberculosis* and *Y. pestis* (Koo et al., 2011). The expression of Ysr35 has been evaluated by measurement of gene expression in *Y. pseudotuberculosis* using single-molecule fluorescence hybridization (smFISH) and its expression was demonstrated to increase upon a temperature upshift from 25 to 37°C, further supporting its importance in pathogenesis (Shepherd et al., 2013).

CsrB and CsrC

These two sRNAs were first described in *Y. pseudotuberculosis* where they were found to be essential for the initial phase of the infection in *Yersinia* species (Heroven et al., 2008, 2012a). CsrC has also been identified by deep sequencing analysis in *Y. pseudotuberculosis* (called Ysr188, Koo et al., 2011), in *Y. pestis* strain 201 (called sR026, Yan et al., 2013), and in *Y. pestis* KIM6+ (called Ysr185, Beauregard et al., 2013), whereas CsrB has been only identified in *Y. pestis* by Beauregard et al. (2013) and Yan et al. (2013) and called Ysr179 and sR003, respectively. These sRNAs has been widely studied in many bacteria as they are part of the Csr system, along with the RNA binding protein CsrA, as such they will be described in the next section below.

Metabolism

RyhB/Ysr48: a Key Regulator of Bacterial Iron Metabolism

The RyhB/Ysr48 sRNA, conserved in *E. coli* and in other *Enterobacteriaceae*, is involved in the post-transcriptional regulation of numerous genes during iron depletion (Masse and Gottesman, 2002; Masse et al., 2003, 2005; Murphy and Payne, 2007; Alice et al., 2008). It is also involved in bacterial growth, biofilm formation, chemotaxis, acid resistance and intracellular growth (Mey et al., 2005; Bollinger and Kallio, 2007; Murphy and Payne, 2007; Boughammoura et al., 2008; Guillemet and Moreau, 2008).

Yersinia pestis encodes two RyhB homologs, RyhB1 and RyhB2, that are located some distance from each other on the chromosome. RyhB1 and RyhB2 share a sequence with *E. coli* and *Salmonella* Typhi that ranges from 61 to 72% (Deng et al., 2012). These sRNAs were identified in deep sequencing studies of *Y. pseudotuberculosis* (Ysr146.1 and Ysr146.2, Koo et al., 2011), virulence-restricted *Y. pestis* bv. microtus (sR009/sR024, Qu et al., 2012) and *Y. pestis* C092 (Ysr146.1/Ysr146.2, Schiano et al., 2014), whereas only RyhB2 was detected in *Y. pestis* KIM (Ysr146.2/187, Beauregard et al., 2013). Both sRNAs are induced by iron deficiency and are negatively regulated by the ferric uptake regulator, Fur (Deng et al., 2012).

The stabilization of RyhB1 and RyhB2 is differentially dependent on Hfq, with RyhB1 stabilization being mediated by Hfq, and that of RyhB2 not (Deng et al., 2012). Both sRNAs are strongly expressed in lungs of mice infected intranasally with *Y. pestis* which indicates that signals such as iron depletion may be present in the infected lungs (Deng et al., 2012). However it was shown that a *ryhb1-ryhB2* double mutant has no discernable defect in survival and dissemination in the host after intranasal inoculation, which suggests that other iron uptake systems have compensatory effects in this environment.

GlmY and GlmZ: Cell Wall Synthesis

These sRNAs regulate synthesis of the enzyme glucosamine-6-phosphate (GlcN6P) synthase, GlmS, in *Enterobacteriaceae*, which catalyzes formation of GlcN6P, the initial building block in the pathway that generates precursors of cell wall synthesis (Urban et al., 2007; Reichenbach et al., 2008; Urban and Vogel, 2008). The GlmY and GlmZ sRNAs are the only known direct targets of the GlrR/GlrK two component system, where GlrK is the sensor kinase that phosphorylates GlrR, which is the response regulator that directly binds to their promoters. GlrR induces *glmY* expression through activation of the σ54-promoter, when cells enter the stationary growth phase (Reichenbach et al., 2009). It has been shown that overlapping σ54- and σ70-promoters direct expression of *glmY* gene, while expression of *glmZ* is achieved from a single constitutively active σ70-promoter (Gopel et al., 2011).

Homologous sRNAs to GlmY and GlmZ have been identified in both *Y. pestis* and *Y. pseudotuberculosis* (Koo et al., 2011; Beauregard et al., 2013; Yan et al., 2013; Schiano et al., 2014). Although their role in virulence has not been evaluated, in *Y. pseudotuberculosis*, these sRNAs are transcribed from σ54-dependent promoters and are induced by the two component system GlrR/GlrK, through direct binding of GlrR to sites located upstream of their promoters (Gopel et al., 2011). In addition, it was shown that putative binding sites for the integration host factor, IHF, are present in the *glmY* and *glmZ* promoter regions, suggesting that IHF could be involved in their regulation, maybe facilitating interaction of GlrR with the σ54-RNA polymerase by binding-induced bending of critical promoter sequences, consistent with the usual role of IHF (Swinger and Rice, 2004).

SraG

First reported in *E. coli*, SraG located between *pnp* (PNPase) and *rpsO* (30S ribosomal protein S15), is expressed preferentially at late-logarithmic phase, and activated by heat and cold shock treatments (Argaman et al., 2001; Sridhar et al., 2009). A comparative sequence analysis with *E. coli* revealed a SraG homolog in *Yersinia* species (Sridhar et al., 2009). In *Y. pseudotuberculosis* 16 proteins were identified as potential regulatory targets of SraG (Lu et al., 2012). However, only *pnp* and YPK_1205 showed significantly different mRNA levels when a RT-PCR validation of these targets was performed. It was shown that SraG negatively regulates the YPK_1206-1205 operon post-transcriptionally by likely targeting the coding region of YPK_1206. The YPK_1206-1205 operon is present only in *Y. pseudotuberculosis* and *Y. enterocolitica* and they share 90% similarity. Unfortunately the role of this operon has not been elucidated but YPK_1206 is predicted to have roles in DNA bending, therefore SraG could be acting as a regulatory element in this process (Lu et al., 2012). As three of the 16 potential targets correspond to proteins associated with maltose metabolism, it has been suggested that SraG might also be involved in regulation of maltose metabolism. This is a classic example of a sRNA that is not conserved in *Y. pestis* but may contribute to adaptation of *Y. pseudotuberculosis* to its host, e.g., maltose is found in the human gut when it is broken down from grains like wheat.

GcvB/Ysr45

Initially predicted using a bioinformatics search of the *Y. pestis* genome. The *gcvB* gene is adjacent to, and divergent from *gcvA* that shares considerable sequence homology (77%) with the *E. coli gcvB* sequence (McArthur et al., 2006). Deep sequence analysis of both *Y. pestis* and *Y. pseudotuberculosis* have also identified GcvB (where it has been called Ysr45, Ysr45/180 or sR013; Koo et al., 2011; Beauregard et al., 2013; Yan et al., 2013; Schiano et al., 2014).

Yersinia pestis gcvB encodes two sRNAs that repress expression of *dppA* that encodes a periplasmic-binding protein component of the dipeptide transport system (McArthur et al., 2006). Deletion of the *gcvB* gene in *Y. pestis*, results in altered growth rate and colony morphology, and due to the pleiotropic nature of these effects it has been suggested that this sRNA is a global regulator of multiple downstream genes in addition to *dppA*, similar to its function in *E. coli* (Urbanowski et al., 2000; McArthur et al., 2006). These target genes remain to be identified.

In *Y. pestis*, transcription of *gcvB* is activated by the GcvA protein and repressed by the GcvR protein. A comparison of the *gcvB* regulatory regions in *Yersinia* species have shown that the putative GcvA binding sites for activation of *gcvB*, are 100% identical in all *Y. pestis* strains, and >92% identical in other *Yersinia* species, which suggests that the regulatory mechanisms of the GcvB RNAs are possibly similar in all *Yersinia* species.

SgrS

SgrS is an Hfq-dependent sRNA that has been widely studied in *E. coli*, where it regulates the *ptsG* mRNA, which encodes the major glucose transporter PtsG, by occluding the *ptsG* RBS leading to degradation of *ptsG* (Kawamoto et al., 2006).

This sRNA also functions as an mRNA, as it encodes the protein SgrT which is ectopically produced under glucose-phosphate metabolic stress conditions when cells are unable to appropriately metabolize phosphorylated sugars (Wadler and Vanderpool, 2007). SgrS prevents new sugar transporters from being produced under conditions where the accumulated sugar-phosphates have become toxic; in so doing it maintains continued cell growth under such conditions (Vanderpool and Gottesman, 2004; Wadler and Vanderpool, 2007).

In *Yersinia* this sRNA was found by bioinformatics analysis aimed at finding homologs of *E. coli* SgrS. The *Y. pestis* SgrS homolog is, however, truncated at the 5′ end such that it is devoid of the *sgrT* coding sequence and retains only typical sRNA base-pairing function (Horler and Vanderpool, 2009). This sRNA has been detected by deep sequencing in both *Y. pestis* and *Y. pseudotuberculosis*, where it has been called Ysr150 or sR005 (Koo et al., 2011; Yan et al., 2013; Schiano et al., 2014).

The *Y. pestis sgrS* ortholog is able to base pair with the *E. coli* K12 *ptsG* mRNA and inhibit its translation (Wadler and Vanderpool, 2007), however, it fails to inhibit growth on minimal glucose medium because the *sgrT* CDS is missing. It is still able to promote recovery of an *E. coli sgrS* mutant strain from glucose-phophate stress, which indicates it conserves the function of regulating target gene expression; the base pairing function of *E. coli* and *Y. pestis* SgrS homologs is critical for rescue from glucose-phosphate stress (Wadler and Vanderpool, 2007).

YsrH

YsrH is a novel *cis*-encoded sRNA identified as Yp-sR16 by Qu et al. (2012) in the avirulent *Y. pestis* bv. microtus strain. This sRNA is conserved in all *Yersinia* species and located on the opposite strand to *fabH2*, which encodes β-ketoacyl-acyl carrier protein synthase III, an enzyme essential for fatty acid biosynthesis in bacteria (Qu et al., 2012; Lu et al., 2014). YsrH is expressed in the early-exponential growth phase and this expression is maintained at the same level in later stages of growth (Lu et al., 2014). It has been reported that YsrH negatively regulates the fatty acid synthesis post-transcriptionally by specifically targeting *fabH2* mRNA transcripts for degradation and this mechanism also involves the PNPase and RNase E-associated processing pathways (Lu et al., 2014).

Stress Response

Ysr29

Ysr29 was identified in reports by Koo and Schiano (Koo et al., 2011; Schiano et al., 2014) but is specific to the *Y. pseudotuberculosis* IP32953 strain. It is expressed much better at 26°C, where its expression depends on the chaperone Hfq. Ysr29 seems to negatively regulate GST and positively regulate RpsA, OmpA and GroEL at a posttranscriptional level because the absence of the sRNA does not affect their transcript levels but instead alters the levels of protein (Koo et al., 2011). GST participates in protecting cells against the damage of oxidative stress (Allocati et al., 2009), and Ysr29 repression of GST levels may prevent an aberrant response to this stress. Interestingly, reactive oxidative species produced by the flea upon infection, is a known stress that *Y. pestis* has to defend itself against during the early stages of infection (Zhou et al., 2012). Therefore, the fact that Ysr29 is not conserved in *Y. pestis* could be as a result of evolutionary selection, as loss of a negative regulator of GST, would lead to expression of GST to cope with the oxidative stress it finds in a new niche, e.g., the flea.

Y. pestis Transmission from Fleas

HmsB: the Biofilm Regulator

Biofilm is a population of bacterial cells embedded in a self-produced exopolysaccharide (EPS) matrix. Biofilm formation results in blockage of the flea proventriculus, which facilitates *Y. pestis* transmission to new mammalian hosts (Jarrett et al., 2004; Hinnebusch and Erickson, 2008).

The second messenger 3′,5′-cyclic diguanosine monophosphate (c-di-GMP), is central to the biofilm formation as it promotes EPS production (Simm et al., 2004). In *Y. pestis* EPS biosynthesis is encoded by the *hmsHFRS* operon, whereas HmsT and HmsD are the two sole diguanylate cyclases responsible of biosynthesis of c-di-GMP, while HmsP is the sole phosphodiesterase responsible for c-di-GMP degradation (Kirillina et al., 2004; Bobrov et al., 2005, 2011; Sun et al., 2011).

HmsB is an Hfq-dependent sRNA (originally called sRNA035) identified by RNA-seq study of *Y. pestis* bv. microtus (Yan et al., 2013). This is the first sRNA to be reported as a biofilm regulator. An *hmsB* deletion reduces biofilm formation both *in vitro* and on nematodes, which is a result of the decreased production of c-di-GMP (Fang et al., 2014). These effects are linked to the fact that HmsB positively regulates *hmsCDE, hmsT, hmsHFRS* and itself, thereby enhancing biofilm production while it negatively regulates *hmsP*, encoding the biofilm-inhibiting phosphodiesterase (Fang et al., 2014). However, the exact molecular interaction of HmsB with *hmsCDE, hmsT, hmsHFRS, hmsP* and other direct targets that could be involved in this process, remain to be elucidated.

Globally Acting RNA Binding Proteins: The Hfq and CsrA Paradigms in *Yersinia*

Post-transcriptional activation of transcripts can occur indirectly, through mechanisms that do not require direct base pairing of sRNAs with their targets. In these cases, globally acting RNA-binding proteins Hfq and CsrA often play an important role. These thoroughly studied RNA binding proteins and their roles in sRNA function in the *Yersinia* genus will be discussed below.

Hfq: the sRNA Chaperone *par Excellence*

Hfq, a close relative of the Sm-like (Lsm) family of eukaryotic proteins, is a well-known RNA chaperone widely recognized for being required for the proper functioning of many *trans*-acting sRNAs. It promotes stable sRNA:mRNA base-paring interactions as in *trans*-acting sRNAs the contact on the target is typically short and imperfect (Jousselin et al., 2009) and it also protects the sRNA from degradation (Brennan and Link, 2007). Most sRNAs characterized to date need binding of Hfq as a RNA chaperone to stabilize the sRNA-target mRNA duplexes (Han et al., 2013). Due to its global post-transcriptional regulatory impact, *hfq* mutation typically results in pleiotropic phenotypes affecting outer membrane biogenesis, quorum sensing, virulence factor synthesis, protein secretion, virulence gene expression and general stress response pathways (Chao and Vogel, 2010; Papenfort and Vogel, 2010). Hfq binding sites have only been mapped for a limited number of transcripts; but all appear to contain a consensus AU-rich single stranded region following an RNA stem loop (Schumacher et al., 2002; Gottesman and Storz, 2011), that are able to interact with the poly-U tail of certain Rho-independent terminator containing RNAs (Otaka et al., 2011).

Hfq was first identified in yersiniae in the *Y. enterocolitica* species, as the protein Yrp which regulates production of the heat stable enterotoxin, Y-ST (Nakao et al., 1995); a subsequent global identification of *Y. enterocolitica* sRNAs has not to date been reported. The Hfq regulatory protein in *Yersinia* was first studied by Geng et al. (2009) using the enzootic *Y. pestis* strain 201. They examined the role of this sRNA chaperone in virulence and performed whole-genome transcriptomic profiling of an *hfq* mutant. Their transcriptional analyses revealed that *hfq* mutation results in dysregulation of ~243 genes in *Y. pestis*. Among the upregulated genes are those involved in macromolecule metabolism, heat shock response and virulence regulation, whereas genes belonging to classes of degradation of small molecules, or energy metabolism and oxidative stress, were downregulated. Numerous hypothetical genes were also dysregulated in the *hfq* mutant.

In general, the Hfq mutant in both *Y. pestis* and *Y. pseudotuberculosis* exhibit growth defects, at ~26 and 37°C, however, the growth defect in *Y. pestis* appears to be more severe compared to that in *Y. pseudotuberculosis* at 37°C (Bai et al., 2010; Schiano et al., 2010). Northern blot analysis of *hfq* shows that *hfq* transcript levels are substantially lower at 37°C than at 28°C in both *Y. pestis* and *Y. pseudotuberculosis* (Beauregard et al., 2013). Nevertheless Hfq is required for the full virulence of both *Y. pestis* and *Y. pseudotuberculosis* in mouse models of infection (Bai et al., 2010; Schiano et al., 2010; Koo et al., 2011). In the lower temperature flea host, an *hfq* mutant of *Y. pestis* is also compromised in its ability to form a biofilm blockage in the flea proventriculus and is outcompeted by the wild type strain during coinfection in fleas (Rempe et al., 2012). In *Y. pseudotuberculosis*, Hfq plays other important roles in sensitivity to heat, oxidative stress resistance, tolerance to long-term nutrient-limitation and antibacterial peptides, phagocytosis resistance, survival and persistence within phagocytes (Schiano et al., 2010). The effects on virulence are likely associated with the role of Hfq in post-transcriptionally regulating multiple components of the T3SS, the major virulence factors in both *Y. pseudotuberculosis* and *Y. pestis* (Schiano et al., 2010, 2014).

From the various sRNA studies in *Yersinia* it can clearly be gleaned that a complex relationship between temperature and Hfq dependence is in place for sRNA regulation between the *Yersinia* strains. So far, about 41 sRNAs have been shown to be regulated by Hfq in both *Y. pestis* and *Y. pseudotuberculosis*. For instance, the sRNA Ysr48/RyhB is Hfq-dependent in *Y. pestis* at both 37 and 26°C but requires Hfq in *Y. pseudotuberculosis* only at 37°C (Supplementary Table S1, Koo et al., 2011). The Ysr48/RyhB mutant is slightly attenuated in *Y. pseudotuberculosis* yet remains virulent in *Y. pestis* in a pneumonic plague infection model. As Hfq is 100% identical between *Y. pestis* and *Y pseudotuberculosis*, regulation by Hfq-dependent sRNAs rather than Hfq itself is what likely contributes to the differences in virulence in both species at 37°C.

Some mechanistic aspects of Hfq regulation of the sRNA HmsB, which controls biofilm formation, and RyhB which is involved in the iron metabolism, have been described in *Y. pestis*. In enzootic *Y. pestis* strain 201, the Hfq-regulated sRNA, HmsB, which controls biofilm formation, is dramatically degenerated in the absence of *hfq* (Fang et al., 2014). Yet, in the presence of *hfq* enhanced expression of HmsB occurs, which leads to increases in expression of the genes enhancing biofilm formation, which are the *hmsCDE* (Bobrov et al., 2015), *hmsT* and *hmsHFRS* genes (Hinnebusch et al., 1996); simultaneously, inhibition of expression of *hmsP* which negatively regulates biofilm formation occurs (Fang et al., 2014), resulting in increases in biofilm formation. However, this effect on biofilm formation is the opposite in the epidemic *Y. pestis* CO92 strain, where Hfq acts as a repressor of biofilm formation through inhibiting expression of *hmsT*, and *hmsHFRS* but stimulating that of *hmsP* (Bellows et al., 2012). This difference in Hfq-dependent biofilm regulation may be explained by the different conditions under which biofilm formation was tested between the two studies. However, it does not rule out that other strain-specific Hfq-regulated sRNAs involved in biofilm control may play a role here. This idea would be consistent with a role for sRNA regulation in the evolution of the distinct pathologies between ancestral and newly evolved strains because the mouse virulent *Y. pestis* strain 201 is less recently evolved than the human virulent *Y. pestis* strains (Morelli et al., 2010).

In *Y. pestis*, Hfq dependence of RyhB1 and RyhB2, the two sRNA encoded by RhyB, is different, as RyhB1 stabilization is mediated by Hfq, whereas RyhB2 does not require Hfq for stability (Deng et al., 2012). In absence of Hfq, rapid degradation of both sRNAs occurs. The ribonuclease PNPase is the main enzyme that degrades Hfq-free RyhB (Deng et al., 2014). This is different from the general mechanism of PnPase activity which usually involves PnPase formation of multi-enzyme ribonucleolytic complexes with RNase E and/or RNA helicase, RhlB, to mediate degradation of the structured RNA (Kaberdin et al., 2011; Silva et al., 2011). Further work is required to explain the mechanism by which RyhB2 maintains stability in absence of Hfq and if RyhB1 and RyhB2 have distinct targets based on their requirement for Hfq.

In spite of the well accepted role of Hfq in *Yersinia* virulence, not all Hfq-dependent sRNAs are important for virulence. For instance, 50 and 10% of mice survived after intragastric infection by the Hfq-dependent sRNAs, Ysr29 and Ysr48/RybB mutants, respectively (Koo et al., 2011), whereas 50% of mice survived after inoculation with the Hfq-independent Ysr35 sRNA mutant. Further, in *Y. pestis* neither the Hfq-independent Ysr23 nor the Hfq-dependent Ysr48 affect the ability of *Y. pestis* to cause disease and death in a pneumonic plague infection model (Koo et al., 2011). These data indicate that not all the sRNAs that depend on Hfq, play a direct role in virulence. Even for those sRNAs whose deletion resulted in a decrease of virulence, the attenuation was not as dramatic as the loss of virulence displayed by mutants lacking Hfq. Multiple Hfq-dependent sRNAs contribute to this Hfq virulence phenotype and an approach in which multiple sRNAs are deleted may be worthwhile in producing a phenotype resembling an *hfq* mutant.

The fact that the overexpression of *hfq* causes substantial decreases of some sRNAs even when the deletion of *hfq* has no substantial effect on sRNA levels suggests that Hfq expression levels itself need to be finely controlled to avoid aberrant effects. Hfq-dependence for the same sRNA can vary between bacteria.

For example, MicA and GcvB whose expression/stability is known to rely on Hfq in *E. coli* and *Salmonella*, are Hfq-independent in *Y. pseudotuberculosis* (Koo et al., 2011).

The distinct expression patterns of conserved sRNAs between *Y. pestis* and *Y. pseudotuberculosis* could be what influences their different pathologies. However, alignments between the sRNAs have shown that almost all of them share >95% identity and Hfq is 100% identical between these strains. Thus the difference in expression patterns may be due to the variance in abundance of target mRNAs, as their absence or abundance would influence base-pairing interactions which would determine stability and degradation of the cognate sRNAs. This in turn, may affect the amounts of available Hfq. At the same time, the abundance of the target mRNAs is subject to first order transcriptional regulation and different environmental conditions that also influence sRNA expression.

Hfq contributes to the regulation of the global transcriptional regulator Crp in a post-transcriptional manner in *Y. pestis* at mammalian physiological temperatures (Lathem et al., 2014); but not in *Y. pseudotuberculosis* (Nuss et al., 2015). Both *crp* and *hfq* mutants of both *Y. pestis* (Bai et al., 2010; Qu et al., 2013; Lathem et al., 2014) and *Y. pseudotuberculosis* (Schiano et al., 2010; Heroven et al., 2012b) are attenuated during mouse infection. In *Y. pestis*, attenuation of the *hfq* mutant is partially attributable to Crp (Lathem et al., 2014). These findings imply that disparate Hfq-dependent sRNA regulation, factors into the different diseases between *Y. pestis* and *Y. pseudotuberculosis*.

CsrA Control is Mediated by sRNAs, CsrC and CsrB

The global translational regulator CsrA has been discovered and studied in many bacteria and some species encode multiple paralogous regulators (Lapouge et al., 2008). This protein typically binds to GGA-rich elements in RNAs. CsrA binding to the SD sequence of bacterial mRNAs results in reduced translation and subsequent mRNA decay. Although it acts primarily as a repressor, in few cases CsrA also activates target mRNA translation (Patterson-Fortin et al., 2013; Yakhnin et al., 2013). CsrA is a major component of the Csr system along with the sRNAs CsrB and CsrC (Babitzke and Romeo, 2007). These sRNAs contain multiple binding sites for CsrA and can therefore bind and titrate CsrA away from its repressed mRNA targets – amounting to indirect activation of these mRNA targets. As a consequence of sequestration, the usual target mRNAs of the CsrA proteins are mostly upregulated (Liu et al., 1997; Weilbacher et al., 2003; Babitzke and Romeo, 2007; Lapouge et al., 2008; Duss et al., 2014). CsrB and CsrC are often found in multiple copies and are present in many bacteria where they can function redundantly.

The Csr system was identified during a search for regulators that influence the expression of the global virulence transcriptional factor RovA in *Y. pseudotuberculosis*, which activates the primary cell entry factor invasion, InvA (Nagel et al., 2001; Heroven et al., 2004; Tran et al., 2005). There are no studies to date reporting the effect of the Csr system in *Y. pestis* therefore our knowledge of CsrA is based on studies in *Y. pseudotuberculosis*. It has been elucidated that CsrA induces the expression of the LysR regulator RovM, which in turn represses *rovA*, thus CsrA represses RovA synthesis indirectly through control of RovM. CsrB and CsrC titrate and sequester CsrA which leads to the activation of *rovA* (Heroven et al., 2008). In agreement, overexpression of CsrB and CsrC induces *rovA* (Heroven et al., 2008); however, neither a *csrC* or *csrB* single mutant affects expression of *rovA*, suggesting that these sRNAs are redundant in the manner in which they regulate *rovA*. Their function in other pathogens is the same, where it has been demonstrated that the loss of CsrB causes a compensatory increase in CsrC levels and vice versa (Weilbacher et al., 2003; Fortune et al., 2006).

The global influence of the Csr system in *Yersinia* is reflected by many different physiological changes for which it is responsible. A *csrA* mutant in *Y. pseudotuberculosis* has a growth defect whereas CsrA overexpression alters cell morphology (Heroven et al., 2008, 2012a); it inhibits glycogen synthesis and about 20% of the CsrA-dependent genes are involved in metabolism (Heroven et al., 2012a). On the other hand, CsrA positively controls motility in *Y. pseudotuberculosis*. It induces flagella biosynthesis by binding directly to the transcript of *flhDC*, which encodes the master regulator of flagellum biosynthesis (Heroven et al., 2008). Absence of CsrA also affects the expression of genes involved in resistance to stress (Heroven et al., 2008). Similar to what has been reported in other microorganisms, intracellular levels of CsrA have to be tightly regulated, as both the loss and the overexpression of CsrA negatively affect invasion to cells (Heroven et al., 2008).

Orthologs of *csrB* and *csrC* genes as well as related Csr (Rsm)-type RNA genes in other bacteria, are activated by the two-component signal transduction system BarA/UvrY which regulates the expression of genes associated with virulence, secondary metabolism, motility, exoenzyme production, quorum sensing or biofilm formation (Goodier and Ahmer, 2001; Lapouge et al., 2008). In *Y. pseudotuberculosis* this scenario seems to be similar only for *csrB* expression, which is positively induced by UvrY (Heroven et al., 2008). However, only when CsrB is present, UvrY seems to repress the expression of CsrC (Heroven et al., 2008).

Recently it has been demonstrated that CsrC, but not CsrB, is regulated by the global regulator PhoP. PhoP induces the expression of CsrC from two different promoters by directly binding to two distinct sites located within the *csrC* regulatory region (Nuss et al., 2014).

Expression of CsrB and CsrC is medium dependent; *csrC* is mostly detected during growth in nutrient rich media, where *csrB* expression is very low, whereas only very low levels of CsrC are observed in minimal medium (Nagel et al., 2001; Heroven et al., 2008). Therefore the CsrA-mediated control of the RovM-RovA-InvA virulence cascade is strongly affected by changes in carbon source availability through alterations of the Csr RNAs, in particular CsrC (Nagel et al., 2001; Heroven and Dersch, 2006; Heroven et al., 2008). This links the virulence CsrABC-RovM-RovA-InvA cascade with the cAMP receptor protein Crp, a crucial global regulator that controls the transcription of multiple genes and operons in bacteria by catabolic repression in response to the glucose supply (Saier, 1998). Crp regulates the synthesis of both Csr RNAs in an opposite manner, as it activates CsrC and represses CsrB transcription (Heroven et al., 2012b). More details of role of Crp in *Yersinia* are given below.

Crp Commands a sRNA Repertoire

More recently it is becoming apparent that the global transcriptional regulator cyclic AMP (cAMP) receptor protein, Crp, has an important role in regulating sRNAs in *Y. pseudotuberculosis* and *Y. pestis* (Lathem et al., 2014; Nuss et al., 2015). Within the *Enterobacteriaceae* family, Crp is a well-studied first order transcriptional regulator that regulates numerous target genes and operons in response to carbon sugar availability (Zhan et al., 2008). Transcriptional regulation by Crp is exacted following activation of the Crp protein by cAMP and once the cAMP-Crp complex binds to a symmetrical recognition site (TGTGA-N_6-TCACA) in the promoter region of target genes and operons.

Importantly, mutants in the *crp* gene of *Y. pseudotuberculosis* (Heroven et al., 2012b) and *Y. pestis* (Zhan et al., 2008; Lathem et al., 2014) are attenuated in virulence. The discerning feature of the attenuated phenotype exhibited by a *crp* mutant is the inability of the bacteria to disseminate from early infection sites to deeper tissues (Zhan et al., 2008; Heroven et al., 2012b). Although the *crp Y. pestis* attenuation mainly results from decreases in expression of genes encoding the major virulence factors like Pla protease and the T3SS and associated Yop effector proteins, alterations in stress adaptation and numerous metabolic genes notably contribute to this phenotype (Zhan et al., 2008, 2009). Similarly, in *Y. pseudotuberculosis* a large proportion of genes involved in primary metabolism, stress and virulence were dysregulated in a *Y. pseudotuberculosis crp* mutant (Nuss et al., 2015). Alongside this, extensive metabolome and fluxome changes to central carbon metabolism related to the pyruvate-tricarboxylic acid cycle were noted in a *crp* mutant of *Y. pseudotuberculosis* (Bucker et al., 2014). Collectively, this data emphasized that Crp plays a role in linking nutritional status and virulence.

Expression of Crp appears to be increased at stationary phase in *Y. pseudotuberculosis* and this was demonstrated to be a function of post-transcriptional control. In *Y. pestis,* Hfq is able to post-transcriptionally control expression of Crp at mammalian host temperature through interaction with a 79 nt long 5′-UTR of Crp (Lathem et al., 2014), however, expression of Crp in *Y. pseudotuberculosis* was maintained independent of Hfq at both 37°C and 26°C (Nuss et al., 2015). In addition, the 5′-UTR of the *crp* mRNA in *Y. pseudotuberculosis* is reported to have two transcriptional start sites that produce a 201 and 287 nt long 5′-UTR each, that are predicted to contain riboswitch-like elements (discussed below). It remains to be resolved whether these are legitimate riboswitches and how they function. Additionally, at 37°C the compromised growth phenotype exhibited by a *Y. pestis hfq* mutant can be partially restored by synthesis of Crp, emphasizing that Hfq and Crp are functionally coordinated during *Y. pestis* infection of its mammalian host (Lathem et al., 2014). Because loss of both Crp or Hfq result in attenuation in virulence for both *Y. pestis* and *Y. pseudotuberculosis*, it appears that subtle differences in their regulatory targets may be important in fine tuning virulence in keeping with host nutritional environment. However, the disparate relationship between Crp and Hfq in *Y. pestis* and *Y. pseudotuberculosis* is likely another aspect of the alterations in their post-transcriptional regulatory cascades that drive their distinct disease manifestations. Importantly the *Y. pestis* Crp regulon incorporates the major virulence factor, Pla, and is involved in its direct transcriptional regulation (Zhan et al., 2008; Lathem et al., 2014).

Several direct targets of Crp have been experimentally verified in both *Y. pseudotuberculosis* (Heroven et al., 2012b) and *Y. pestis* (Zhan et al., 2008) including several sRNAs (Nuss et al., 2015). As sRNAs have vital roles in the regulation of carbon metabolism and nutritional stress adaptation, it makes sense that their regulatory function would be closely aligned with that of Crp. The recent study by Nuss et al. (2015) that generated the first single nucleotide resolution transcriptome profile of *Y. pseudotuberculosis* wild type and *crp* mutant strains, uncovered a complex regulatory network composing transcriptional regulators and 53 sRNAs dysregulated by Crp. Northern blotting confirmed expression of 15 of these sRNAs. Further, a newly identified trans-encoded sRNA (Ysr206), as well as two antisense sRNAs (Ysr232 and Ysr114) exclusively expressed in the *crp* mutant were validated as directly targeted with Crp using gel shift assays. Thus key findings regarding Crp regulation of sRNAs, are that Crp acts as a master regulator of many sRNAs; it binds directly or mediates indirect control of sRNAs through other transcriptional regulators, and that Crp regulation of sRNAs occurs especially at stationary growth phase during catabolite repression.

Earlier work by Yan et al. (2013) demonstrated that 10 out of 104 sRNAs identified in the avirulent *Y. pestis* bv. microtus strain, contained Crp-binding sites. Northern analysis verified expression of seven of these sRNAs in the wild type versus the *crp* mutant. The known Crp-regulated CyaR/RyeE sRNA (Johansen et al., 2008; Papenfort et al., 2008) was amongst those identified. Three novel sRNAs were demonstrated to be directly regulated by Crp: sRNAs sR084 encoded in an IG on the pPCP1 plasmid, as well as sR065, are positively regulated, whereas sR066 seem to be negatively regulated by Crp. Deletion mutants of the CyaR and sR084 sRNAs that are expressed in the lungs were not attenuated as was demonstrated by subcutaneous infection in mice. While expression of CyaR at 26°C is Crp-dependent in *Y. pestis,* it is not the case in *Y. pseudotuberculosis* (Nuss et al., 2015) which may be important for adaptation of *Y. pestis* to the flea environment; and likely represents another difference in sRNA regulation that distinguishes the diseases cause by these two pathogens.

Given that for Crp, there appears to be some distinct differences between Crp-regulated targets and its regulation by Hfq between *Y. pestis* vs. *Y. pseudotuberculosis,* a comparative single nucleotide resolution transcriptomic profile of *Y. pestis* similar to that described for *Y. pseudotuberculosis*, is essential. This could provide a more informed understanding of the varied Crp-dependent sRNAs and their targets. Utilization of this technique should be extended to the various sites of infection (e.g., intestines, spleen, lungs, and lymph nodes) of bacteria during systemic dissemination to identify the localized, nutrient-specific disparate regulatory changes in gene expression of the two pathogens that likely fine tunes their adaptation during infection.

Post-Transcriptional Regulation Mediated by mRNA Structural Elements

Long 5′-UTRS of some mRNAs can contain structured *cis*-acting non-coding RNA regulatory elements like riboswitches and RNA thermometers. These elements have a distinctive regulatory mechanism in that they undergo structural rearrangement in response to binding of a small ligand, e.g., metabolites or cofactors, or thermal shifts, i.e., their function is intimately coupled with signal integration.

RNA Thermosensors

Post-transcriptional regulation by RNA thermometers involves the thermally induced unfolding of RNA secondary structures that restricts accessibility to the RBS, and alters translation. Thus far, only one RNA thermometer has been described for the yersiniae. Yet, regulation by such elements seems highly appropriate for *Yersinia* species that exhibit thermo-regulation of gene transcription as the bacteria transitions between the environmental/flea environment into the mammalian host. The thermometer element was found located intercistronically between the *yscW* that encodes a structural component of the T3SS apparatus and the *lcrF* gene that encodes a transcriptional activator of genes encoding effector proteins. The RNA thermometer composes two hairpin mRNA secondary structures. The *lcrF* proximal hairpin contains a sequence of four U's, also called fourU that sequesters the SD sequence, while the secondary structure occludes the start codon. The mammalian host body temperature is, however, sufficient to melt the unstable G-U bonds that hold the SD/fourU sequences together. Once the hairpin structure is destabilized, post-transcriptional controlled initiation of *lcrF* translation ensues leading to differential synthesis of YscW and LcrF proteins and expression of virulence function. Due to the 100% conservation in nucleotide sequence homology in this region, this RNA thermometer mechanism of regulation is likely conserved in all pathogenic *Yersinia* species. Currently the only clearly defined recognition motifs that typify thermometer elements are the fourU and ROSE elements (Krajewski and Narberhaus, 2014), limiting potential of whole genome based *in silico* prediction of such elements.

RNA Riboswitches

Unlike RNA thermometer elements, riboswitches require binding of a ligand, usually a metabolite or cofactor, to induce conformational changes in the RNA secondary that directs premature transcription termination or inhibit initiation of translation. A search for riboswitch elements in *Yersinia* species has been neglected until the recent achievement of a high resolution whole genome transcriptomics of *Y. pseudotuberculosis* (Nuss et al., 2015). In this study, 155 mRNAs composed 5′-UTR regions >200 nucleotides in length which make them prime candidates to encode *cis*-acting regulatory elements like riboswitches. A bioinformatic search for riboswitch-like elements (RLEs) using the RibEx riboswitch explorer (Abreu-Goodger and Merino, 2005) predicted four known and an additional 17 RLEs involved in metabolic and gene expression functions, within these long 5′-UTRs. An example of this is that the mRNA of the *btuB* gene that encodes a Ton-B dependent vitamin B12 receptor, composed of a 315 nucleotide long 5′-UTR, contains a known cobalamin riboswitch element. Unknown RLEs require experimental verification, functional characterization and determination of the metabolites/cofactor ligands that mediate their structural switching.

To date a detailed riboregulatory mechanism of a 5′-UTR Mg^{2+} responsive riboswitch, *mgtA*, which regulates the expression of the downstream encoded Mg^{2+} transporter MgtA, has been described only for *Y. enterocolitica* (Korth and Sigel, 2012). Here, in the 5′-UTR, two stem loops structures form at high Mg^{2+} concentration to inhibit expression of MgtA. At low Mg^{2+} concentration an alternate single stem loop antiterminator structure forms that favors translation of the *mgtA* mRNA and Mg^{2+} uptake. Similar studies are required for the *Y. pestis* and *Y. pseudotuberculosis* to identify disparities in riboswitch mechanisms that may facilitate their distinct host infection phenotypes.

Conclusion

In the last decade, research on sRNA identification and functional analysis has begun to reveal a previously hidden regulatory layer in the already complex gene networks that control cellular function and behavior. As discussed in this review, sRNAs have been shown to act as regulators of *Yersinia* virulence and host adaptation. The underlying emerging theme of the *Yersinia* sRNA studies reviewed here is that sRNAs coordinate metabolic adaptation to enhance the host–pathogen interaction.

Many fundamental questions about sRNA biology remain to be answered, however, before sRNAs can be exploited to disrupt the host–pathogen interaction. For most identified *Yersinia* sRNAs, the exact cellular function and downstream mRNA targets remain to be elucidated. Thus far we have learned that different growth and treatment conditions uncover distinct sRNAs repertoires reflective of that particular set of conditions. So while *in vitro* growth studies representing infection relevant conditions have potential to identify some important host-specific sRNAs, the full complement of these molecules may be overlooked without *in vivo* studies. Only a single study identified *Y. pestis* sRNAs in the biologically relevant context of the lung and spleen, but only one highly induced, and previously annotated sRNA, SsrA, was shown to be important for mouse infection (Yan et al., 2013) similar to a previous study (Okan et al., 2010). Studies identifying, validating and functionally charactering the roles of sRNAs in the various biologically relevant host tissues may prove to be useful in determining the host-specific and niche dependent regulatory mechanisms crucial for virulence of *Yersinia* species. Importantly, comparative analyses of sRNAs and post-transcriptional regulation between *Y. pestis* and *Y. pseudotuberculosis* may provide insight into the evolution of the distinct disease states of these two pathogens.

Because experimental efforts to determine cellular function of sRNAs are time-consuming and labor-intensive, and bioinformatics prediction of target mRNAs remains largely unreliable being confounded by imperfect complementarity between the sRNA and mRNA, researchers are faced with several challenges in the field of sRNA biology. Undoubtedly, however, the mechanisms of sRNA regulation suggests the possibility that inhibition of key sRNA folding or targeted mRNA interactions strategies can be developed as the basis of novel anti-infective strategies, especially in the face of antibiotic resistance. The several instances of obviously different post-transcriptional regulatory control of infection and adaptation between *Y. pestis* and *Y. pseudotuberculosis* discussed in this review, emphasize their important and specific roles in fine tuning adaptation of these pathogens to cause disease.

Author Contributions

LM-C and VV reviewed the literature and contributed to writing and revising this manuscript.

Acknowledgments

This work was supported by an intramural grant from the Washington State University College of Veterinary Medicine grant to VV.

References

Abreu-Goodger, C., and Merino, E. (2005). RibEx: a web server for locating riboswitches and other conserved bacterial regulatory elements. *Nucleic Acids Res.* 33, W690–W692. doi: 10.1093/nar/gki445

Achtman, M., Morelli, G., Zhu, P., Wirth, T., Diehl, I., Kusecek, B., et al. (2004). Microevolution and history of the plague bacillus, *Yersinia pestis. Proc. Natl. Acad. Sci. U.S.A.* 101, 17837–17842. doi: 10.1073/pnas.0408026101

Achtman, M., Zurth, K., Morelli, G., Torrea, G., Guiyoule, A., and Carniel, E. (1999). *Yersinia pestis*, the cause of plague, is a recently emerged clone of *Yersinia pseudotuberculosis. Proc. Natl. Acad. Sci. U.S.A.* 96, 14043–14048. doi: 10.1073/pnas.96.24.14043

Alice, A. F., Naka, H., and Crosa, J. H. (2008). Global gene expression as a function of the iron status of the bacterial cell: influence of differentially expressed genes in the virulence of the human pathogen *Vibrio vulnificus. Infect. Immun.* 76, 4019–4037. doi: 10.1128/IAI.00208-08

Allocati, N., Federici, L., Masulli, M., and Di Ilio, C. (2009). Glutathione transferases in bacteria. *FEBS J.* 276, 58–75. doi: 10.1111/j.1742-4658.2008.06743.x

Argaman, L., Hershberg, R., Vogel, J., Bejerano, G., Wagner, E. G., Margalit, H., et al. (2001). Novel small RNA-encoding genes in the intergenic regions of *Escherichia coli. Curr. Biol.* 11, 941–950. doi: 10.1016/S0960-9822(01)00270-6

Babitzke, P., and Romeo, T. (2007). CsrB sRNA family: sequestration of RNA-binding regulatory proteins. *Curr. Opin. Microbiol.* 10, 156–163. doi: 10.1016/j.mib.2007.03.007

Bai, G., Golubov, A., Smith, E. A., and Mcdonough, K. A. (2010). The importance of the small RNA chaperone Hfq for growth of epidemic *Yersinia pestis*, but not *Yersinia pseudotuberculosis*, with implications for plague biology. *J. Bacteriol.* 192, 4239–4245. doi: 10.1128/JB.00504-10

Barends, S., Karzai, A. W., Sauer, R. T., Wower, J., and Kraal, B. (2001). Simultaneous and functional binding of SmpB and EF-Tu-TP to the alanyl acceptor arm of tmRNA. *J. Mol. Biol.* 314, 9–21. doi: 10.1006/jmbi.2001.5114

Beauregard, A., Smith, E. A., Petrone, B. L., Singh, N., Karch, C., Mcdonough, K. A., et al. (2013). Identification and characterization of small RNAs in *Yersinia pestis. RNA Biol.* 10, 397–405. doi: 10.4161/rna.23590

Bellows, L. E., Koestler, B. J., Karaba, S. M., Waters, C. M., and Lathem, W. W. (2012). Hfq-dependent, co-ordinate control of cyclic diguanylate synthesis and catabolism in the plague pathogen *Yersinia pestis. Mol. Microbiol.* 86, 661–674. doi: 10.1111/mmi.12011

Bobrov, A. G., Kirillina, O., and Perry, R. D. (2005). The phosphodiesterase activity of the HmsP EAL domain is required for negative regulation of biofilm formation in *Yersinia pestis. FEMS Microbiol. Lett.* 247, 123–130. doi: 10.1016/j.femsle.2005.04.036

Bobrov, A. G., Kirillina, O., Ryjenkov, D. A., Waters, C. M., Price, P. A., Fetherston, J. D., et al. (2011). Systematic analysis of cyclic di-GMP signalling enzymes and their role in biofilm formation and virulence in *Yersinia pestis. Mol. Microbiol.* 79, 533–551. doi: 10.1111/j.1365-2958.2010.07470.x

Bobrov, A. G., Kirillina, O., Vadyvaloo, V., Koestler, B. J., Hinz, A. K., Mack, D., et al. (2015). The *Yersinia pestis* HmsCDE regulatory system is essential for blockage of the oriental rat flea (*Xenopsylla cheopis*), a classic plague vector. *Environ. Microbiol.* 17, 947–959. doi: 10.1111/1462-2920.12419

Bollinger, C. J., and Kallio, P. T. (2007). Impact of the small RNA RyhB on growth, physiology and heterologous protein expression in *Escherichia coli. FEMS Microbiol. Lett.* 275, 221–228. doi: 10.1111/j.1574-6968.2007.00880.x

Boughammoura, A., Matzanke, B. F., Bottger, L., Reverchon, S., Lesuisse, E., Expert, D., et al. (2008). Differential role of ferritins in iron metabolism and virulence of the plant-pathogenic bacterium *Erwinia chrysanthemi* 3937. *J. Bacteriol.* 190, 1518–1530. doi: 10.1128/JB.01640-07

Brennan, R. G., and Link, T. M. (2007). Hfq structure, function and ligand binding. *Curr. Opin. Microbiol.* 10, 125–133. doi: 10.1016/j.mib.2007.03.015

Bubeck, S. S., Cantwell, A. M., and Dube, P. H. (2007). Delayed inflammatory response to primary pneumonic plague occurs in both outbred and inbred mice. *Infect. Immun.* 75, 697–705. doi: 10.1128/IAI.00403-06

Bucker, R., Heroven, A. K., Becker, J., Dersch, P., and Wittmann, C. (2014). The pyruvate-tricarboxylic acid cycle node: a focal point of virulence control in the enteric pathogen *Yersinia pseudotuberculosis. J. Biol. Chem.* 289, 30114–30132. doi: 10.1074/jbc.M114.581348

Cathelyn, J. S., Crosby, S. D., Lathem, W. W., Goldman, W. E., and Miller, V. L. (2006). RovA, a global regulator of *Yersinia pestis*, specifically required for bubonic plague. *Proc. Natl. Acad. Sci. U.S.A.* 103, 13514–13519. doi: 10.1073/pnas.0603456103

Chain, P. S., Carniel, E., Larimer, F. W., Lamerdin, J., Stoutland, P. O., Regala, W. M., et al. (2004). Insights into the evolution of *Yersinia pestis* through whole-genome comparison with *Yersinia pseudotuberculosis. Proc. Natl. Acad. Sci. U.S.A.* 101, 13826–13831. doi: 10.1073/pnas.0404012101

Chao, Y., and Vogel, J. (2010). The role of Hfq in bacterial pathogens. *Curr. Opin. Microbiol.* 13, 24–33. doi: 10.1016/j.mib.2010.01.001

Chouikha, I., and Hinnebusch, B. J. (2014). Silencing urease: a key evolutionary step that facilitated the adaptation of *Yersinia pestis* to the flea-borne transmission route. *Proc. Natl. Acad. Sci. U.S.A.* 111, 18709–18714. doi: 10.1073/pnas.1413209111

Cornelis, G. R., Boland, A., Boyd, A. P., Geuijen, C., Iriarte, M., Neyt, C., et al. (1998). The virulence plasmid of *Yersinia*, an antihost genome. *Microbiol. Mol. Biol. Rev.* 62, 1315–1352.

Deng, Z., Liu, Z., Bi, Y., Wang, X., Zhou, D., Yang, R., et al. (2014). Rapid degradation of Hfq-free RyhB in *Yersinia pestis* by PNPase independent of putative ribonucleolytic complexes. *Biomed. Res. Int.* 2014, 798918. doi: 10.1155/2014/798918

Deng, Z., Meng, X., Su, S., Liu, Z., Ji, X., Zhang, Y., et al. (2012). Two sRNA RyhB homologs from *Yersinia pestis* biovar microtus expressed *in vivo* have differential Hfq-dependent stability. *Res. Microbiol.* 163, 413–418. doi: 10.1016/j.resmic.2012.05.006

Dulebohn, D., Choy, J., Sundermeier, T., Okan, N., and Karzai, A. W. (2007). Trans-translation: the tmRNA-mediated surveillance mechanism for ribosome rescue, directed protein degradation, and nonstop mRNA decay. *Biochemistry* 46, 4681–4693. doi: 10.1021/bi6026055

Duss, O., Michel, E., Yulikov, M., Schubert, M., Jeschke, G., and Allain, F. H. T. (2014). Structural basis of the non-coding RNA RsmZ acting as a protein sponge. *Nature* 509, 588–592. doi: 10.1038/Nature13271

Fang, N., Qu, S., Yang, H., Fang, H., Liu, L., Zhang, Y., et al. (2014). HmsB enhances biofilm formation in *Yersinia pestis*. *Front. Microbiol.* 5:685. doi: 10.3389/fmicb.2014.00685

Fortune, D. R., Suyemoto, M., and Altier, C. (2006). Identification of CsrC and characterization of its role in epithelial cell invasion in *Salmonella enterica* serovar Typhimurium. *Infect. Immun.* 74, 331–339. doi: 10.1128/IAI.74.1.331-339.2006

Geng, J., Song, Y., Yang, L., Feng, Y., Qiu, Y., Li, G., et al. (2009). Involvement of the post-transcriptional regulator Hfq in *Yersinia pestis* virulence. *PLoS ONE* 4:e6213. doi: 10.1371/journal.pone.0006213

Goodier, R. I., and Ahmer, B. M. (2001). SirA orthologs affect both motility and virulence. *J. Bacteriol.* 183, 2249–2258. doi: 10.1128/JB.183.7.2249-2258.2001

Gopel, Y., Luttmann, D., Heroven, A. K., Reichenbach, B., Dersch, P., and Gorke, B. (2011). Common and divergent features in transcriptional control of the homologous small RNAs GlmY and GlmZ in Enterobacteriaceae. *Nucleic Acids Res.* 39, 1294–1309. doi: 10.1093/nar/gkq986

Gottesman, S., and Storz, G. (2011). Bacterial small RNA regulators: versatile roles and rapidly evolving variations. *Cold Spring Harb. Perspect. Biol.* 3, a003798. doi: 10.1101/cshperspect.a003798

Guillemet, M. L., and Moreau, P. L. (2008). Fur-dependent detoxification of organic acids by *rpoS* mutants during prolonged incubation under aerobic, phosphate starvation conditions. *J. Bacteriol.* 190, 5567–5575. doi: 10.1128/JB.00577-08

Han, Y., Liu, L., Fang, N., Yang, R., and Zhou, D. (2013). Regulation of pathogenicity by noncoding RNAs in bacteria. *Future Microbiol.* 8, 579–591. doi: 10.2217/fmb.13.20

Heroven, A. K., Bohme, K., and Dersch, P. (2012a). The Csr/Rsm system of *Yersinia* and related pathogens: a post-transcriptional strategy for managing virulence. *RNA Biol.* 9, 379–391. doi: 10.4161/rna.19333

Heroven, A. K., Sest, M., Pisano, F., Scheb-Wetzel, M., Steinmann, R., Bohme, K., et al. (2012b). Crp induces switching of the CsrB and CsrC RNAs in *Yersinia pseudotuberculosis* and links nutritional status to virulence. *Front. Cell. Infect. Microbiol.* 2:158. doi: 10.3389/fcimb.2012.00158

Heroven, A. K., Bohme, K., Rohde, M., and Dersch, P. (2008). A Csr-type regulatory system, including small non-coding RNAs, regulates the global virulence regulator RovA of *Yersinia pseudotuberculosis* through RovM. *Mol. Microbiol.* 68, 1179–1195. doi: 10.1111/j.1365-2958.2008.06218.x

Heroven, A. K., and Dersch, P. (2006). RovM, a novel LysR-type regulator of the virulence activator gene *rovA*, controls cell invasion, virulence and motility of *Yersinia pseudotuberculosis*. *Mol. Microbiol.* 62, 1469–1483. doi: 10.1111/j.1365-2958.2006.05458.x

Heroven, A. K., Nagel, G., Tran, H. J., Parr, S., and Dersch, P. (2004). RovA is autoregulated and antagonizes H-NS-mediated silencing of invasin and *rovA* expression in *Yersinia pseudotuberculosis*. *Mol. Microbiol.* 53, 871–888. doi: 10.1111/j.1365-2958.2004.04162.x

Hindley, J. (1967). Fractionation of 32P-labelled ribonucleic acids on polyacrylamide gels and their characterization by fingerprinting. *J. Mol. Biol.* 30, 125–136. doi: 10.1016/0022-2836(67)90248-3

Hinnebusch, B. J., and Erickson, D. L. (2008). *Yersinia pestis* biofilm in the flea vector and its role in the transmission of plague. *Curr. Top. Microbiol. Immunol.* 322, 229–248.

Hinnebusch, B. J., Perry, R. D., and Schwan, T. G. (1996). Role of the *Yersinia pestis* hemin storage (*hms*) locus in the transmission of plague by fleas. *Science* 273, 367–370. doi: 10.1126/science.273.5273.367

Horler, R. S., and Vanderpool, C. K. (2009). Homologs of the small RNA SgrS are broadly distributed in enteric bacteria but have diverged in size and sequence. *Nucleic Acids Res.* 37, 5465–5476. doi: 10.1093/nar/gkp501

Jarrett, C. O., Deak, E., Isherwood, K. E., Oyston, P. C., Fischer, E. R., Whitney, A. R., et al. (2004). Transmission of *Yersinia pestis* from an infectious biofilm in the flea vector. *J. Infect. Dis.* 190, 783–792. doi: 10.1086/422695

Johansen, J., Eriksen, M., Kallipolitis, B., and Valentin-Hansen, P. (2008). Down-regulation of outer membrane proteins by noncoding RNAs: unraveling the cAMP-CRP- and sigmaE-dependent CyaR-ompX regulatory case. *J. Mol. Biol.* 383, 1–9. doi: 10.1016/j.jmb.2008.06.058

Jousselin, A., Metzinger, L., and Felden, B. (2009). On the facultative requirement of the bacterial RNA chaperone, Hfq. *Trends Microbiol.* 17, 399–405. doi: 10.1016/j.tim.2009.06.003

Kaberdin, V. R., Singh, D., and Lin-Chao, S. (2011). Composition and conservation of the mRNA-degrading machinery in bacteria. *J. Biomed. Sci.* 18, 23. doi: 10.1186/1423-0127-18-23

Karzai, A. W., Roche, E. D., and Sauer, R. T. (2000). The SsrA-SmpB system for protein tagging, directed degradation and ribosome rescue. *Nat. Struct. Biol.* 7, 449–455. doi: 10.1038/75843

Kawamoto, H., Koide, Y., Morita, T., and Aiba, H. (2006). Base-pairing requirement for RNA silencing by a bacterial small RNA and acceleration of duplex formation by Hfq. *Mol. Microbiol.* 61, 1013–1022. doi: 10.1111/j.1365-2958.2006.05288.x

Keiler, K. C. (2008). Biology of trans-translation. *Annu. Rev. Microbiol.* 62, 133–151. doi: 10.1146/annurev.micro.62.081307.162948

Kirillina, O., Fetherston, J. D., Bobrov, A. G., Abney, J., and Perry, R. D. (2004). HmsP, a putative phosphodiesterase, and HmsT, a putative diguanylate cyclase, control Hms-dependent biofilm formation in *Yersinia pestis*. *Mol. Microbiol.* 54, 75–88. doi: 10.1111/j.1365-2958.2004.04253.x

Koo, J. T., Alleyne, T. M., Schiano, C. A., Jafari, N., and Lathem, W. W. (2011). Global discovery of small RNAs in *Yersinia pseudotuberculosis* identifies *Yersinia*-specific small, noncoding RNAs required for virulence. *Proc. Natl. Acad. Sci. U.S.A.* 108, E709–E717. doi: 10.1073/pnas.1101655108

Koo, J. T., and Lathem, W. W. (2012). Global discovery of small noncoding RNAs in pathogenic *Yersinia* species. *Adv. Exp. Med. Biol.* 954, 305–314. doi: 10.1007/978-1-4614-3561-7_38

Korth, M. M., and Sigel, R. K. (2012). Unusually high-affinity Mg(2+) binding at the AU-rich sequence within the antiterminator hairpin of a Mg(2+) riboswitch. *Chem. Biodivers.* 9, 2035–2049. doi: 10.1002/cbdv.201200031

Krajewski, S. S., and Narberhaus, F. (2014). Temperature-driven differential gene expression by RNA thermosensors. *Biochim. Biophys. Acta* 1839, 978–988. doi: 10.1016/j.bbagrm.2014.03.006

Lambrecht, E., Bare, J., Van Damme, I., Bert, W., Sabbe, K., and Houf, K. (2013). Behavior of *Yersinia enterocolitica* in the presence of the bacterivorous *Acanthamoeba castellanii*. *Appl. Environ. Microbiol.* 79, 6407–6413. doi: 10.1128/AEM.01915-13

Lapouge, K., Schubert, M., Allain, F. H., and Haas, D. (2008). Gac/Rsm signal transduction pathway of gamma-proteobacteria: from RNA recognition to regulation of social behaviour. *Mol. Microbiol.* 67, 241–253. doi: 10.1111/j.1365-2958.2007.06042.x

Lathem, W. W., Crosby, S. D., Miller, V. L., and Goldman, W. E. (2005). Progression of primary pneumonic plague: a mouse model of infection, pathology, and bacterial transcriptional activity. *Proc. Natl. Acad. Sci. U.S.A.* 102, 17786–17791. doi: 10.1073/pnas.0506840102

Lathem, W. W., Schroeder, J. A., Bellows, L. E., Ritzert, J. T., Koo, J. T., Price, P. A., et al. (2014). Posttranscriptional regulation of the *Yersinia pestis* cyclic AMP receptor protein Crp and impact on virulence. *MBio* 5, e1038–e1113. doi: 10.1128/mBio.01038-13

Liu, J. M., Livny, J., Lawrence, M. S., Kimball, M. D., Waldor, M. K., and Camilli, A. (2009). Experimental discovery of sRNAs in *Vibrio cholerae* by direct cloning, 5S/tRNA depletion and parallel sequencing. *Nucleic Acids Res.* 37, e46. doi: 10.1093/nar/gkp080

Liu, M. Y., Gui, G., Wei, B., Preston, J. F. III, Oakford, L., Yuksel, U., et al. (1997). The RNA molecule CsrB binds to the global regulatory protein CsrA and antagonizes its activity in *Escherichia coli*. *J. Biol. Chem.* 272, 17502–17510. doi: 10.1074/jbc.272.28.17502

Livny, J., Brencic, A., Lory, S., and Waldor, M. K. (2006). Identification of 17 *Pseudomonas aeruginosa* sRNAs and prediction of sRNA-encoding genes in 10 diverse pathogens using the bioinformatic tool sRNAPredict2. *Nucleic Acids*

Res. 34, 3484–3493. doi: 10.1093/nar/gkl453

Lu, P., Zhang, Y., Hu, Y., Francis, M. S., and Chen, S. (2014). A cis-encoded sRNA controls the expression of *fabH2* in *Yersinia*. *FEBS Lett.* 588, 1961–1966. doi: 10.1016/j.febslet.2014.04.005

Lu, P., Zhang, Y., Li, L., Hu, Y., Huang, L., Li, Y., et al. (2012). Small non-coding RNA SraG regulates the operon YPK_1206-1205 in *Yersinia pseudotuberculosis FEMS Microbiol. Lett.* 331, 37–43. doi: 10.1111/j.1574-6968.2012.02548.x

Masse, E., Escorcia, F. E., and Gottesman, S. (2003). Coupled degradation of a small regulatory RNA and its mRNA targets in *Escherichia coli*. *Genes Dev.* 17, 2374–2383. doi: 10.1101/gad.1127103

Masse, E., and Gottesman, S. (2002). A small RNA regulates the expression of genes involved in iron metabolism in *Escherichia coli*. *Proc. Natl. Acad. Sci. U.S.A.* 99, 4620–4625. doi: 10.1073/pnas.032066599

Masse, E., Vanderpool, C. K., and Gottesman, S. (2005). Effect of RyhB small RNA on global iron use in *Escherichia coli*. *J. Bacteriol.* 187, 6962–6971. doi: 10.1128/JB.187.20.6962-6971.2005

McArthur, S. D., Pulvermacher, S. C., and Stauffer, G. V. (2006). The *Yersinia pestis gcvB* gene encodes two small regulatory RNA molecules. *BMC Microbiol.* 6:52. doi: 10.1186/1471-2180-6-52

Mey, A. R., Craig, S. A., and Payne, S. M. (2005). Characterization of *Vibrio cholerae* RyhB: the RyhB regulon and role of *ryhB* in biofilm formation. *Infect. Immun.* 73, 5706–5719. doi: 10.1128/IAI.73.9.5706-5719.2005

Morelli, G., Song, Y., Mazzoni, C. J., Eppinger, M., Roumagnac, P., Wagner, D. M., et al. (2010). *Yersinia pestis* genome sequencing identifies patterns of global phylogenetic diversity. *Nat. Genet.* 42, 1140–1143. doi: 10.1038/ng.705

Murphy, E. R., and Payne, S. M. (2007). RyhB, an iron-responsive small RNA molecule, regulates *Shigella dysenteriae* virulence. *Infect. Immun.* 75, 3470–3477. doi: 10.1128/IAI.00112-07

Nagel, G., Lahrz, A., and Dersch, P. (2001). Environmental control of invasin expression in *Yersinia pseudotuberculosis* is mediated by regulation of RovA, a transcriptional activator of the SlyA/Hor family. *Mol. Microbiol.* 41, 1249–1269. doi: 10.1046/j.1365-2958.2001.02522.x

Nakao, H., Watanabe, H., Nakayama, S., and Takeda, T. (1995). *yst* gene expression in *Yersinia enterocolitica* is positively regulated by a chromosomal region that is highly homologous to *Escherichia coli* host factor 1 gene (hfq). *Mol. Microbiol.* 18, 859–865. doi: 10.1111/j.1365-2958.1995.18050859.x

Narberhaus, F., and Vogel, J. (2009). Regulatory RNAs in prokaryotes: here, there and everywhere. *Mol. Microbiol.* 74, 261–269. doi: 10.1111/j.1365-2958.2009.06869.x

Nuss, A. M., Heroven, A. K., Waldmann, B., Reinkensmeier, J., Jarek, M., Beckstette, M., et al. (2015). Transcriptomic profiling of *Yersinia pseudotuberculosis* reveals reprogramming of the Crp regulon by temperature and uncovers Crp as a master regulator of small RNAs. *PLoS Genet.* 11:e1005087. doi: 10.1371/journal.pgen.1005087

Nuss, A. M., Schuster, F., Kathrin Heroven, A., Heine, W., Pisano, F., and Dersch, P. (2014). A direct link between the global regulator PhoP and the Csr regulon in *Y. pseudotuberculosis* through the small regulatory RNA CsrC. *RNA Biol.* 11, 580–593. doi: 10.4161/rna.28676

Okan, N. A., Bliska, J. B., and Karzai, A. W. (2006). A Role for the SmpB-SsrA system in *Yersinia pseudotuberculosis* pathogenesis. *PLoS Pathog.* 2:e6. doi: 10.1371/journal.ppat.0020006

Okan, N. A., Mena, P., Benach, J. L., Bliska, J. B., and Karzai, A. W. (2010). The *smpB-ssrA* mutant of *Yersinia pestis* functions as a live attenuated vaccine to protect mice against pulmonary plague infection. *Infect. Immun.* 78, 1284–1293. doi: 10.1128/IAI.00976-09

Oliva, G., Sahr, T., and Buchrieser, C. (2015). Small RNAs, 5' UTR elements and RNA-binding proteins in intracellular bacteria: impact on metabolism and virulence. *FEMS Microbiol. Rev.* 39, 331–349. doi: 10.1093/femsre/fuv022

Otaka, H., Ishikawa, H., Morita, T., and Aiba, H. (2011). PolyU tail of rho-independent terminator of bacterial small RNAs is essential for Hfq action. *Proc. Natl. Acad. Sci. U.S.A.* 108, 13059–13064. doi: 10.1073/pnas.1107050108

Papenfort, K., Pfeiffer, V., Lucchini, S., Sonawane, A., Hinton, J. C., and Vogel, J. (2008). Systematic deletion of *Salmonella* small RNA genes identifies CyaR, a conserved CRP-dependent riboregulator of OmpX synthesis. *Mol. Microbiol.* 68, 890–906. doi: 10.1111/j.1365-2958.2008.06189.x

Papenfort, K., and Vanderpool, C. K. (2015). Target activation by regulatory RNAs in bacteria. *FEMS Microbiol. Rev.* 39, 362–378. doi: 10.1093/femsre/fuv016

Papenfort, K., and Vogel, J. (2010). Regulatory RNA in bacterial pathogens. *Cell Host Microbe* 8, 116–127. doi: 10.1016/j.chom.2010.06.008

Patterson-Fortin, L. M., Vakulskas, C. A., Yakhnin, H., Babitzke, P., and Romeo, T. (2013). Dual posttranscriptional regulation via a cofactor-responsive mRNA leader. *J. Mol. Biol.* 425, 3662–3677. doi: 10.1016/j.jmb.2012.12.010

Perez, N., Trevino, J., Liu, Z., Ho, S. C., Babitzke, P., and Sumby, P. (2009). A genome-wide analysis of small regulatory RNAs in the human pathogen group A Streptococcus. *PLoS ONE* 4:e7668. doi: 10.1371/journal.pone.0007668

Phillips, P., Progulske-Fox, A., Grieshaber, S., and Grieshaber, N. (2014). Expression of *Porphyromonas gingivalis* small RNA in response to hemin availability identified using microarray and RNA-seq analysis. *FEMS Microbiol. Lett.* 351, 202–208. doi: 10.1111/1574-6968.12320

Pouillot, F., Fayolle, C., and Carniel, E. (2008). Characterization of chromosomal regions conserved in *Yersinia pseudotuberculosis* and lost by *Yersinia pestis*. *Infect. Immun.* 76, 4592–4599. doi: 10.1128/IAI.00568-08

Price, P. A., Jin, J., and Goldman, W. E. (2012). Pulmonary infection by *Yersinia pestis* rapidly establishes a permissive environment for microbial proliferation. *Proc. Natl. Acad. Sci. U.S.A.* 109, 3083–3088. doi: 10.1073/pnas.11127 29109

Pujol, C., and Bliska, J. B. (2003). The ability to replicate in macrophages is conserved between *Yersinia pestis* and *Yersinia pseudotuberculosis*. *Infect. Immun.* 71, 5892–5899. doi: 10.1128/IAI.71.10.5892-5899.2003

Qu, S., Zhang, Y., Liu, L., Wang, L., Han, Y., Yang, R., et al. (2013). Cyclic AMP receptor protein is a repressor of adenylyl cyclase gene *cyaA* in *Yersinia pestis*. *Can. J. Microbiol.* 59, 304–310. doi: 10.1139/cjm-2012-0705

Qu, Y., Bi, L., Ji, X., Deng, Z., Zhang, H., Yan, Y., et al. (2012). Identification by cDNA cloning of abundant sRNAs in a human-avirulent *Yersinia pestis* strain grown under five different growth conditions. *Future Microbiol.* 7, 535–547. doi: 10.2217/fmb.12.13

Reichenbach, B., Gopel, Y., and Gorke, B. (2009). Dual control by perfectly overlapping sigma 54- and sigma 70- promoters adjusts small RNA GlmY expression to different environmental signals. *Mol. Microbiol.* 74, 1054–1070. doi: 10.1111/j.1365-2958.2009.06918.x

Reichenbach, B., Maes, A., Kalamorz, F., Hajnsdorf, E., and Gorke, B. (2008). The small RNA GlmY acts upstream of the sRNA GlmZ in the activation of *glmS* expression and is subject to regulation by polyadenylation in *Escherichia coli*. *Nucleic Acids Res.* 36, 2570–2580. doi: 10.1093/nar/gkn091

Rempe, K. A., Hinz, A. K., and Vadyvaloo, V. (2012). Hfq regulates biofilm gut blockage that facilitates flea-borne transmission of *Yersinia pestis*. *J. Bacteriol.* 194, 2036–2040. doi: 10.1128/JB.06568-11

Saier, M. H. Jr. (1998). Multiple mechanisms controlling carbon metabolism in bacteria. *Biotechnol. Bioeng.* 58, 170–174. doi: 10.1002/(SICI)1097-0290(19980420)58:2/3<170::AID-BIT9>3.0.CO;2-I

Santos-Montanez, J., Benavides-Montano, J. A., Hinz, A. K., and Vadyvaloo, V. (2015). *Yersinia pseudotuberculosis* IP32953 survives and replicates in trophozoites and persists in cysts of *Acanthamoeba castellanii*. *FEMS Microbiol. Lett.* 362, fnv091. doi: 10.1093/femsle/fnv091

Schiano, C. A., Bellows, L. E., and Lathem, W. W. (2010). The small RNA chaperone Hfq is required for the virulence of *Yersinia pseudotuberculosis*. *Infect. Immun.* 78, 2034–2044. doi: 10.1128/IAI.01046-09

Schiano, C. A., Koo, J. T., Schipma, M. J., Caulfield, A. J., Jafari, N., and Lathem, W. W. (2014). Genome-wide analysis of small RNAs expressed by *Yersinia pestis* identifies a regulator of the Yop-Ysc type III secretion system. *J. Bacteriol.* 196, 1659–1670. doi: 10.1128/JB.01456-13

Schumacher, M. A., Pearson, R. F., Moller, T., Valentin-Hansen, P., and Brennan, R. G. (2002). Structures of the pleiotropic translational regulator Hfq and an Hfq-RNA complex: a bacterial Sm-like protein. *EMBO J.* 21, 3546–3556. doi: 10.1093/emboj/cdf322

Sebbane, F., Gardner, D., Long, D., Gowen, B. B., and Hinnebusch, B. J. (2005). Kinetics of disease progression and host response in a rat model of bubonic plague. *Am. J. Pathol.* 166, 1427–1439. doi: 10.1016/S0002-9440(10)6 2360-7

Shannon, J. G., Bosio, C. F., and Hinnebusch, B. J. (2015). Dermal neutrophil, macrophage and dendritic cell responses to *Yersinia pestis* transmitted by fleas. *PLoS Pathog.* 11:e1004734. doi: 10.1371/journal.ppat.10 04734

Sharma, C. M., Hoffmann, S., Darfeuille, F., Reignier, J., Findeiss, S., Sittka, A., et al. (2010). The primary transcriptome of the major human pathogen *Helicobacter pylori*. *Nature* 464, 250–255. doi: 10.1038/nature 08756

Sharma, R., Arya, S., Patil, S. D., Sharma, A., Jain, P. K., Navani, N. K., et al. (2014). Identification of novel regulatory small RNAs in *Acinetobacter baumannii*. *PLoS ONE* 9:e93833. doi: 10.1371/journal.pone.0093833

Shepherd, D. P., Li, N., Micheva-Viteva, S. N., Munsky, B., Hong-Geller, E., and Werner, J. H. (2013). Counting small RNA in pathogenic bacteria. *Anal. Chem.* 85, 4938–4943. doi: 10.1021/ac303792p

Silva, I. J., Saramago, M., Dressaire, C., Domingues, S., Viegas, S. C., and Arraiano, C. M. (2011). Importance and key events of prokaryotic RNA decay: the ultimate fate of an RNA molecule. *Wiley Interdiscip. Rev. RNA* 2, 818–836. doi: 10.1002/wrna.94

Simm, R., Morr, M., Kader, A., Nimtz, M., and Romling, U. (2004). GGDEF and EAL domains inversely regulate cyclic di-GMP levels and transition from sessility to motility. *Mol. Microbiol.* 53, 1123–1134. doi: 10.1111/j.1365-2958.2004.04206.x

Sittka, A., Lucchini, S., Papenfort, K., Sharma, C. M., Rolle, K., Binnewies, T. T., et al. (2008). Deep sequencing analysis of small noncoding RNA and mRNA targets of the global post-transcriptional regulator, Hfq. *PLoS Genet.* 4:e1000163. doi: 10.1371/journal.pgen.1000163

Soutourina, O. A., Monot, M., Boudry, P., Saujet, L., Pichon, C., Sismeiro, O., et al. (2013). Genome-wide identification of regulatory RNAs in the human pathogen *Clostridium difficile*. *PLoS Genet.* 9:e1003493. doi: 10.1371/journal.pgen.1003493

Sridhar, J., Kumar, S. S., and Rafi, Z. A. (2009). Small RNA identification in *Enterobacteriaceae* using synteny and genomic backbone retention II. *OMICS J. Integr. Biol.* 13, 261–284. doi: 10.1089/omi.2008.0067

Sun, Y. C., Jarrett, C. O., Bosio, C. F., and Hinnebusch, B. J. (2014). Retracing the evolutionary path that led to flea-borne transmission of *Yersinia pestis*. *Cell Host Microbe* 15, 578–586. doi: 10.1016/j.chom.2014.04.003

Sun, Y. C., Koumoutsi, A., Jarrett, C., Lawrence, K., Gherardini, F. C., Darby, C., et al. (2011). Differential control of *Yersinia pestis* biofilm formation *in vitro* and in the flea vector by two c-di-GMP diguanylate cyclases. *PLoS ONE* 6:e19267. doi: 10.1371/journal.pone.0019267

Swinger, K. K., and Rice, P. A. (2004). IHF and HU: flexible architects of bent DNA. *Curr. Opin. Struct. Biol.* 14, 28–35. doi: 10.1016/j.sbi.2003.12.003

Tran, H. J., Heroven, A. K., Winkler, L., Spreter, T., Beatrix, B., and Dersch, P. (2005). Analysis of RovA, a transcriptional regulator of *Yersinia pseudotuberculosis* virulence that acts through antirepression and direct transcriptional activation. *J. Biol. Chem.* 280, 42423–42432. doi: 10.1074/jbc.M504464200

Urban, J. H., Papenfort, K., Thomsen, J., Schmitz, R. A., and Vogel, J. (2007). A conserved small RNA promotes discoordinate expression of the *glmUS* operon mRNA to activate GlmS synthesis. *J. Mol. Biol.* 373, 521–528. doi: 10.1016/j.jmb.2007.07.035

Urban, J. H., and Vogel, J. (2008). Two seemingly homologous noncoding RNAs act hierarchically to activate *glmS* mRNA translation. *PLoS Biol.* 6:e64. doi: 10.1371/journal.pbio.0060064

Urbanowski, M. L., Stauffer, L. T., and Stauffer, G. V. (2000). The *gcvB* gene encodes a small untranslated RNA involved in expression of the dipeptide and oligopeptide transport systems in *Escherichia coli*. *Mol. Microbiol.* 37, 856–868. doi: 10.1046/j.1365-2958.2000.02051.x

Vadyvaloo, V., Jarrett, C., Sturdevant, D. E., Sebbane, F., and Hinnebusch, B. J. (2010). Transit through the flea vector induces a pretransmission innate immunity resistance phenotype in *Yersinia pestis*. *PLoS Pathog.* 6:e1000783. doi: 10.1371/journal.ppat.1000783

Vanderpool, C. K., and Gottesman, S. (2004). Involvement of a novel transcriptional activator and small RNA in post-transcriptional regulation of the glucose phosphoenolpyruvate phosphotransferase system. *Mol. Microbiol.* 54, 1076–1089. doi: 10.1111/j.1365-2958.2004.04348.x

Wadler, C. S., and Vanderpool, C. K. (2007). A dual function for a bacterial small RNA: SgrS performs base pairing-dependent regulation and encodes a functional polypeptide. *Proc. Natl. Acad. Sci. U.S.A.* 104, 20454–20459. doi: 10.1073/pnas.0708102104

Waters, L. S., and Storz, G. (2009). Regulatory RNAs in bacteria. *Cell* 136, 615–628. doi: 10.1016/j.cell.2009.01.043

Weilbacher, T., Suzuki, K., Dubey, A. K., Wang, X., Gudapaty, S., Morozov, I., et al. (2003). A novel sRNA component of the carbon storage regulatory system of *Escherichia coli*. *Mol. Microbiol.* 48, 657–670. doi: 10.1046/j.1365-2958.2003.03459.x

Yakhnin, A. V., Baker, C. S., Vakulskas, C. A., Yakhnin, H., Berezin, I., Romeo, T., et al. (2013). CsrA activates *flhDC* expression by protecting *flhDC* mRNA from RNase E-mediated cleavage. *Mol. Microbiol.* 87, 851–866. doi: 10.1111/mmi.12136

Yan, Y., Su, S., Meng, X., Ji, X., Qu, Y., Liu, Z., et al. (2013). Determination of sRNA expressions by RNA-seq in *Yersinia pestis* grown *in vitro* and during infection. *PLoS ONE* 8:e74495. doi: 10.1371/journal.pone.0074495

Zhan, L., Han, Y., Yang, L., Geng, J., Li, Y., Gao, H., et al. (2008). The cyclic AMP receptor protein, CRP, is required for both virulence and expression of the minimal CRP regulon in *Yersinia pestis* biovar microtus. *Infect. Immun.* 76, 5028–5037. doi: 10.1128/IAI.00370-08

Zhan, L., Yang, L., Zhou, L., Li, Y., Gao, H., Guo, Z., et al. (2009). Direct and negative regulation of the *sycO-ypkA-ypoJ* operon by cyclic AMP receptor protein (CRP) in *Yersinia pestis*. *BMC Microbiol.* 9:178. doi: 10.1186/1471-2180-9-178

Zhou, D., Tong, Z., Song, Y., Han, Y., Pei, D., Pang, X., et al. (2004). Genetics of metabolic variations between *Yersinia pestis* biovars and the proposal of a new biovar, microtus. *J. Bacteriol.* 186, 5147–5152. doi: 10.1128/JB.186.15.5147-5152.2004

Zhou, W., Russell, C. W., Johnson, K. L., Mortensen, R. D., and Erickson, D. L. (2012). Gene expression analysis of *Xenopsylla cheopis* (Siphonaptera: Pulicidae) suggests a role for reactive oxygen species in response to *Yersinia pestis* infection. *J. Med. Entomol.* 49, 364–370. doi: 10.1603/ME11172

Zimbler, D. L., Schroeder, J. A., Eddy, J. L., and Lathem, W. W. (2015). Early emergence of *Yersinia pestis* as a severe respiratory pathogen. *Nat. Commun.* 6, 7487. doi: 10.1038/ncomms8487

4

The Life Cycle of *L. pneumophila*: Cellular Differentiation is Linked to Virulence and Metabolism

Giulia Oliva[1,2], *Tobias Sahr*[1,2] *and Carmen Buchrieser*[1,2*]

[1] *Institut Pasteur, Biologie des Bactéries Intracellulaires, Paris, France,* [2] *Centre National de la Recherche Scientifique, UMR 3525, Paris, France*

Edited by:
Matthias P. Machner,
National Institutes of Health (NIH),
United States

Reviewed by:
Klaus Heuner,
Robert Koch Institut, Germany
Stephanie Rochelle Shames,
Kansas State University, United States

***Correspondence:**
Carmen Buchrieser
cbuch@pasteur.fr

Legionella pneumophila is a gram-negative bacterium that inhabits freshwater ecosystems, where it is present in biofilm or as planktonic form. *L. pneumophila* is mainly found associated with protozoa, which serve as protection from hostile environments and as replication niche. If inhaled within aerosols, *L. pneumophila* is also able to infect and replicate in human alveolar macrophages, eventually causing the Legionnaires' disease. The transition between intracellular and extracellular environments triggers a differentiation program in which metabolic as well as morphogenetic changes occur. We here describe the current knowledge on how the different developmental states of this bacterium are regulated, with a particular emphasis on the stringent response activated during the transition from the replicative phase to the infectious phase and the metabolic features going in hand. We propose that the cellular differentiation of this intracellular pathogen is closely associated to key metabolic changes in the bacterium and the host cell, which together have a crucial role in the regulation of *L. pneumophila* virulence.

Keywords: *Legionella pneumophila*, regulation, virulence, metabolism, life cycle

ENVIRONMENTAL AND MORPHOGENETIC CHANGES

Whatever strategy microbial pathogens have evolved to successfully infect and replicate in their hosts, they have adapted in the course of evolution to face hostile environments. This adaptation allows them to benefit from the environment of the susceptible host cell and simultaneously to ensure their persistence for another infection cycle. A conspicuous group of bacteria, referred to as facultative and obligate intracellular pathogens, exploits a variety of different hosts to establish a cytosolic or vacuolar niche for replication. Thereby they face and learned to tolerate acidification, starvation, and changes in temperature, oxidative stress and many other host defense mechanisms. Most of these bacteria are also located in the extracellular space between intracellular infection cycles and display thus a dual intracellular/extracellular lifestyle. Among those, the intracellular pathogen *Legionella pneumophila* thrives in fresh water environments, where it either spreads planktonically as free-living microbe or it is associated with biofilm communities (Steinert et al., 2002; Hilbi et al., 2011), but it never has been demonstrated to replicate in these environments. In the environment *Legionella* replicate within eukaryotic phagocytic cells like the environmental amoeba *Acanthamoeba castellanii,* as well as in human monocytes and alveolar macrophages (Horwitz and Silverstein, 1980; Rowbotham, 1986). *L. pneumophila* has successfully adapted to new and challenging environments created by human activities, such as showers, air conditioning systems, water fountains, cooling towers or other artificial water systems facilitating access to humans and human infection, which can result in a severe pneumonia, called Legionnaire's disease or legionellosis (McDade et al., 1977). However, mainly the susceptible population like immunocompromised or elderly persons develop pneumonia caused by *Legionella*, as this bacterium evolved with aquatic protozoa and thus it has not evolved mechanisms to counteract the host defense in healthy humans. In addition to biofilm communities and protozoan predators, *L. pneumophila* has been found to colonize more extreme environmental niches, such as antarctic freshwater lakes at temperature at 0°C as well as extremely acidic habitats and water sources with temperature over 60°C (Hilbi et al., 2011). Therefore, *L. pneumophila* endures in disparate environmental conditions throughout its life cycle with respect to nutrient access and availability, pH, temperature, and host defenses during intracellular replication. The transition between intracellular and extracellular habitats triggers morphogenetic and metabolic changes during the bacterial lifecycle (Molofsky and Swanson, 2004). Accordingly, *L. pneumophila* alternates between different morphogenetic forms including a replicating form (RF), and a transmissive/virulent form that have many distinct features (Molofsky and Swanson, 2004; Brüggemann et al., 2006; Steinert et al., 2007). Starvation and environmental stress induce the transition from the metabolically active, replicating bacteria to motile, stress-resistant virulent bacteria (Molofsky and Swanson, 2004). Moreover, a mature intracellular form (MIF), characterized by bacteria that are highly infectious, motile and cyst-like was described (Garduño et al., 2002; Robertson et al., 2014) as well as viable but non-cultivable (VBNC) forms that develop in response to disparate conditions (Steinert et al., 1997; Al-Bana et al., 2014). The fine-tuned regulation of these different forms ensures the persistence and successful life cycle of this bacterium. Thus, *L. pneumophila* employs a multitude of regulatory elements allowing it to govern its multi-phasic life cycle.

ONE STRATEGY, MULTIPLE HOSTS

In the environment, *L. pneumophila* preferentially establishes a parasitic relationship with protozoa, which provides not only a nutrition source for the persistence, replication and dissemination of *Legionella*, but also functions as shelter offering protection from adverse environmental conditions (Barker et al., 1995; Cunha et al., 2016). Interestingly, bacteria released from protozoa are more infectious, are highly motile and more efficient in surviving and multiplying within human monocytes *in vitro* compared to bacteria grown on agar (Cirillo et al., 1994; Brieland et al., 1997). The protozoan predators (amoebae and ciliates) are the natural hosts of *L. pneumophila*, and humans are accidental hosts as judged by the evidence that only a single and recent case of human-to-human transmission has been reported to date (Correia et al., 2016). Thus, *L. pneumophila* transmission to humans occurs primarily from man-made environmental sources (Hilbi et al., 2010; Newton et al., 2010). The dual host specificity of *Legionella* is thought to be derived from the fact that protozoa are primordial phagocytes and as such they share many similarity at both cellular and molecular level with macrophages. Therefore, the intracellular growth of *L. pneumophila* is very similar in both hosts (Fields et al., 2002; Hilbi et al., 2007) suggesting that the co-evolution of *Legionella* within protozoa had provided the bacterium with an effective strategy to colonize two evolutionally different host cells. Indeed, this co-evolution is reflected in its genome as sequence analyses revealed that *L. pneumophila* as well as *L. longbeachae* have acquired genes coding for proteins with eukaryotic-like properties from its protozoan predators (Cazalet et al., 2004, 2010; de Felipe et al., 2005; Gomez-Valero et al., 2011). These eukaryotic-like proteins were shown to be secreted effectors that mimic the functions of their host counterparts (Cazalet et al., 2004; de Felipe et al., 2005; Nora et al., 2009; Gomez-Valero et al., 2011; Escoll et al., 2016). Their translocation to the host cell is achieved by the Dot/Icm type 4B secretion system (T4BSS), which is indispensible for intracellular replication of this bacterium (Ninio and Roy, 2007; Isberg et al., 2009; Zhu et al., 2011). Thus, this intriguing feature of molecular mimicry is a major virulence strategy developed by this opportunistic bacterium due to a selective pressure from the natural environment (Nora et al., 2009; Escoll et al., 2016).

HUMAN INFECTION

Adaptation of *L. pneumophila* to harsh environmental conditions allowed them to become ubiquitous in human-made aquatic systems where the temperature is higher than the ambient temperature. As consequence, thermally altered aquatic habitats may shift the availability of predators and bacterial preys, eventually promoting *Legionella* replication and the emergence of the disease by inhalation of infected aerosols. Potential sources of *Legionella* transmission include potable water sources, such as fountains, showers and taps, and non-potable sources such as spas, cooling towers and evaporative condensers (Steinert et al., 2007; Newton et al., 2010; Cunha et al., 2016). Upon inhalation of bacteria-contaminated aerosols, *Legionella* reach the lung and are engulfed by alveolar macrophages wherein they can actively replicate, causing a life-threatening pneumonia called Legionnaires' disease (Newton et al., 2010). As *Legionella* is an opportunistic pathogen, persons with chronic lung diseases, elderly, immune-compromised, male gender and smokers are mainly susceptible to contract the disease (Newton et al., 2010; Cunha et al., 2016). Interestingly, not all *Legionella* species seem to be able to cause human disease as among the 58 *Legionella* species currently identified, only about 20 have been associated to human disease. Among those, *L. pneumophila* serogroup 1 is responsible for over 85% of the Legionnaires' disease cases world-wide (Yu et al., 2002; Newton et al., 2010)

LIFE WITHIN A HOST CELL

One of the striking features of *L. pneumophila* is its ability to replicate within a large number of different host cells. The intracellular lifestyle and the adaptation capacity require a series of temporally distinct events leading to the establishment of a successful infection cycle, many of which are provoked by the action of one or more of the over 300 effector proteins known to be secreted by the Dot/Icm secretion system. Following the uptake of *L. pneumophila* by phagocytic cells through conventional or coiling phagocytosis (Bozue and Johnson, 1996; Escoll et al., 2013), this bacterium avoids lysosome-mediated degradation and forms a unique replication-permissive compartment within its host cell (**Figure 1**). This single-membrane *Legionella* containing vacuole (LCV) differs from the cellular compartment containing non-pathogenic bacteria since it does not acidify and it has a distinct membrane identity that is achieved by the recruitment of vesicles rich in lipids and proteins on the cytoplasmic face. Four hours after entry into phagocytes, vesicles derived from rough endoplasmic reticulum (ER) cluster near the nascent LCV. Based on the localization of ER-associated proteins within the LCV, these vesicles, which exit the ER, are able to deliver their content into the vacuoles containing *L. pneumophila* (Robinson and Roy, 2006). In this compartment the bacteria are replicating intravacuolarly, but the LCV was found later in association with the late lysosomal compartment, suggesting that it may also play a role in the replication of the bacteria by providing a nutrient-rich environment (Sturgill-Koszycki and Swanson, 2000). Recent studies demonstrated that *L. pneumophila* evades the host-cell response and interferes also with the host autophagy machinery by modulating the host sphingolipid metabolism or autophagosome formation (Choy et al., 2012; Rolando et al., 2016). The question remains open, whether the manipulation of the host sphingolipd metabolism may not only modulate autophagy, but also provide *L. pneumophila* nutrients for replication. However, additional membrane trafficking events may occur and modulate the intracellular life cycle of this bacterium to manipulate the host response. Following intracellular multiplication, the depletion of nutrients triggers morphological changes and a switch from a replicative form, where bacteria are metabolically active but not infectious, to a transmissive form, which ensures that the bacteria activate the infectious traits for the escape and transmission into a new suitable host or the survival in the environment (Byrne and Swanson, 1998; Garduño et al., 2002; Molofsky and Swanson, 2004; Robertson et al., 2014; **Figure 1**).

THE BIPHASIC LIFE CYCLE OF *L. PNEUMOPHILA*

L. pneumophila Undergoes Morphologic Changes during Its Life Cycle

In a simple model one can describe the *L. pneumophila* life cycle as alternating between two distinct and reciprocal forms: a replicative and a transmissive form that was referred to as microbial differentiation (Molofsky and Swanson, 2004). This term implies physiological, morphogenetic and metabolic changes of the bacterium. Indeed, pronounced morphogenetic changes in the bacterial cell wall, the bacterial shape and in motility as well as the enrichment of the cell in energy-rich polymers are observed (Rowbotham, 1986). Within the LCV, acid-resistant, replicating bacteria exploit the nutrient-rich environment and actively inhibit the phagosome-lysosome fusion to be able to efficiently multiply. Therefore, bacterial density strongly increases whereas nutrient access dramatically decreases over time. Actively replicating bacteria appear as

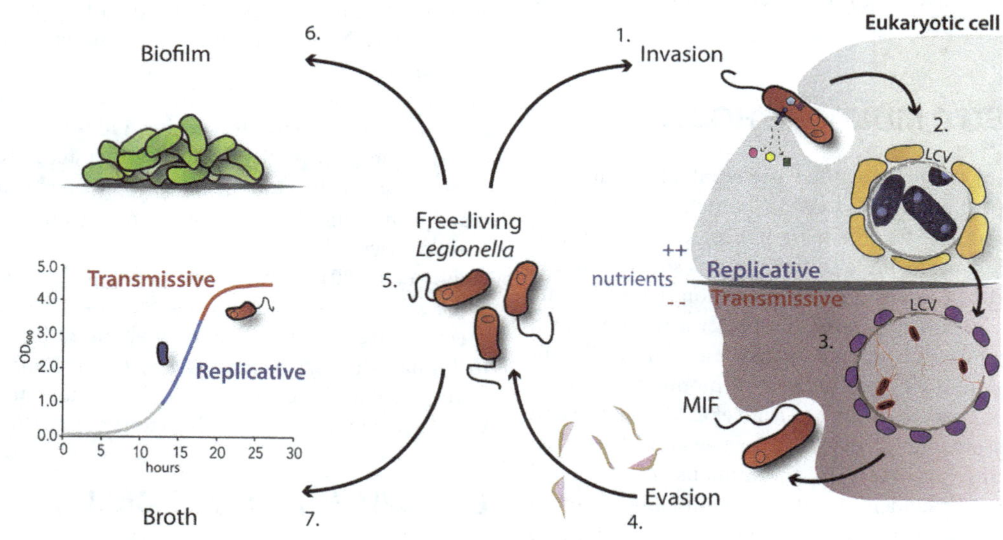

FIGURE 1 | Schematic overview of the *L. pneumophila* morphological states during its growth cycle. 1. Uptake of virulent *L. pneumophila* by the host cell like protozoa or macrophages through convention or coiling phagocytosis. 2. After internalization, the bacteria evade the phagosome-lysosome fusion and start the intracellular multiplication within the LCV, which is surrounded by vesicles (in yellow) rich in lipids and proteins. 3. Nutrient starvation induces the activation of the stringent response and morphological changes. Bacteria express the transmissive traits such as motility (flagella) and become cytotoxic. 4. These infectious bacteria are able to lyse the vacuolar membrane and are released in the extracellular environment. 5. Free-living transmissive bacteria may start a new cycle or persist in the extracellular environment as planktonic form. 6. Alternatively, *L. pneumophila* may be associated within biofilms, either in natural fresh-water habitats or artificial ones. 7. In broth culture, *L. pneumophila* displays also a biphasic life cycle, which closely mimics the replicative and transmissive intracellular forms.

slender rods, are non-motile and display a wavy cell wall (Faulkner and Garduño, 2002). During this metabolically active state, traits related to virulence and transmission such as motility and cytotoxicity are not required thus the replicating bacteria either lack an activator of transmission and/or constitutively express a repressor of transmission traits (Byrne and Swanson, 1998; Molofsky and Swanson, 2003). As local nutrient levels become limiting and disadvantageous conditions are about to be faced, the bacteria convert into the infectious/transmissive variant. Interestingly, virulent bacteria appear as short, stubby rods with blunt ends containing cyst-like inclusions of poly-3-hydroxybutyrate (PHB), and display a smooth thick cell wall (Faulkner and Garduño, 2002). Those phenotypically distinct bacteria coordinately activate the expression of the so-called transmissive traits, which are required for lysosome evasion, escape from the spent host cell, survival in the extracellular environment and the invasion of a new suitable host. After successfully establishing a new intracellular niche, *L. pneumophila* reverts to its replicative form, starting a new cycle (Hammer and Swanson, 1999; Molofsky and Swanson, 2004). To limit costly energy levels, *L. pneumophila* has adopted a strategy employing a reciprocal biphasic expression pathway. Therefore, when conditions are favorable for multiplication, the virulence traits are neither required nor expressed. Conversely, in adverse conditions such as the nutrient deprivation, the bacteria do not replicate (Byrne and Swanson, 1998). Strikingly, the analyses of the global gene expression profiles of *L. pneumophila* in the *in vivo* infection model *A. castellanii* as well as in the *in vitro* broth culture model revealed that the pathogen's life cycle is very similar, as judged by the profound and similar changes in the gene expression program from the replicative/exponential growth phase to the transmissive/post-exponential growth phase of the bacteria (Brüggemann et al., 2006; Faucher et al., 2011). Thus, replicative and transmissive bacteria share a gene expression program similar to that of *in vitro* grown exponential (E) and post-exponential (PE) bacteria, respectively, suggesting that the biphasic life cycle is globally controlled by the bacterial growth phase and by nutrient availability. In addition, intracellular infection of the natural host *A. castellanii* revealed only few strain-specific differences, such as a shorter lag phase of strain *L. pneumophila* Paris and an earlier transition to transmissive form (Brüggemann et al., 2006). Interestingly, the global expression profiles of replicative and transmissive phases of three different *L. pneumophila* strains have been shown to be very similar (Brüggemann et al., 2006). In contrast, comparison of the gene expression program of the E and PE phases of the two major disease-associated species *L. pneumophila* and *L. longbeachae*, revealed clear differences. Particularly, the transition between the replicative and transmissive form is less pronounced in *L. longbeachae*, which seems to regulate this transition mainly by engaging secondary messenger molecules and less transcriptional and post- transcriptional regulators than *L. pneumophila* (Cazalet et al., 2010). Taken together, the transition from the exponential/replicative to the post-exponential/transmissive phase governs a common virulence program engaged of *L. pneumophila* within host cells (Brüggemann et al., 2006; Faucher et al., 2011).

The Key Metabolic Capacity of *L. pneumophila* Is Adapted to Its Biphasic Life Cycle

In response to fluctuating intracellular environmental conditions, *L. pneumophila* certainly requires a well-balanced adaptation of its metabolism. Main questions are (i) what are the essential nutrients required for intracellular proliferation of *L. pneumophila* during infection, (ii) what is the nutrient availability in the LCV and (iii) what are the capacities of the bacterium to catabolize these compounds. The development of a chemical defined liquid medium gave first insights into the nature and physiology of this intracellular pathogen by suggesting that it utilizes only amino acids as energy and carbon sources (Pine et al., 1979; George et al., 1980; Ristroph et al., 1981; Tesh et al., 1983). While formulating this medium, it has been demonstrated that cysteine was an absolute requirement for the bacterial growth and that the addition of soluble ferric pyrophosphate had stimulatory effects. Unlike other microorganisms, *L. pneumophila* has been found to use serine and threonine as a primary supply for energy production rather than any other organic substrate (Pine et al., 1979; George et al., 1980; Ristroph et al., 1981; Fields, 1996). Accordingly, microarray analyses performed during replicative growth of *L. pneumophila* either in broth or upon infection of *A. castellanii*, revealed that while replicating, bacteria express genes indicating that an aerobic metabolism and amino acid catabolism, particularly for serine, threonine, glycine, tyrosine, alanine, and histidine is taking place (Sauer et al., 2005; Brüggemann et al., 2006). However, unexpectedly the up-regulation of genes encoding the Entner-Doudoroff (ED) pathway, as well as of a putative glucokinase, a sugar transporter and the myo-inositol catabolism indicated that *L. pneumophila* may be also able to exploit host carbohydrate-derivatives during the replicative phase of growth within amoebae (Brüggemann et al., 2006). Interestingly, these analyses suggested for the first time that intracellular *L. pneumophila* also may utilize starch or glycogen as judged by the expression of a eukaryotic-like glucoamylase (GamA) during exponential growth (Brüggemann et al., 2006). Indeed, later, it was shown that GamA is responsible for glycogen- and starch-degrading activities of *L. pneumophila* and that it is expressed and active during intracellular replication in *A. castellanii*, suggesting that *L. pneumophila* is degrading glycogen during intracellular replication (Herrmann et al., 2011). Hence, intracellular *L. pneumophila* not only uses amino acids but also diverse carbohydrates as carbon supply (Weiss et al., 1980; Eylert et al., 2010).

However, *L. pneumophila* is auxotrophic for the amino acids Arg, Cys, Ile, Leu, Met, Thr, Val, Ser, Pro, and Phe (Pine et al., 1979; George et al., 1980; Ristroph et al., 1981; Tesh et al., 1983). Moreover, ^{13}C- isotopologue profiling revealed that *L. pneumophila* is able to perform gluconeogenesis and to use the pentose phosphate pathway (PPP), although not all the genes encoding canonical enzymes involved in these pathways are present (Eylert et al., 2010). Based on the presence of a glucose transporter protein and on ^{13}C-tracer experiments, it was reported that glucose, metabolized through the ED and PPP pathways serves as co-substrate for *L. pneumophila,* although the addition of glucose in broth culture does not increase the bacterial growth rate (Tesh et al., 1983; Eylert et al., 2010; Eisenreich and Heuner, 2016; Häuslein et al., 2016). Isotopologue profiling of key metabolites of *L. pneumophila* unveiled a bi-partite metabolism, in which it preferentially uses serine as major carbon, nitrogen and energy supply, whereas glycerol and glucose are shuffled into anabolic processes (Eisenreich and Heuner, 2016; Häuslein et al., 2016; **Figure 2**). In addition, as expected from an intracellular bacterium, which engages a growth phase-dependent program to control its virulence, it was shown that also the carbon and energy sources are metabolized in dependence of the growth phase (Gillmaier et al., 2016; Häuslein et al., 2016). As such, Serine is mainly metabolized during the replicative phase for amino acid (Ser> Ala >Glu> Asp = Gly) and protein biosynthesis (>50 mol%) and for energy production. ^{13}C-labeled serine was found to enter mainly the TCA cycle, generating pyruvate and then acetyl-CoA, and to produce PHB (Eylert et al., 2010; Gillmaier et al., 2016; Häuslein et al., 2016). Conversely, during the post-exponential phase, despite the availability of serine in the medium, serine-dependent protein biosynthesis appears to be reduced, whereas carbon from serine is still used for PHB and fatty acid biosynthesis until the post-exponential phase of growth. Hence, upon entry into the stationary phase and under nutrient starvation, the PHB produced is catabolized by *L. pneumophila*, serving as main carbon and energy storage (James et al., 1999; Eylert et al., 2010; Gillmaier et al., 2016; Häuslein et al., 2016; **Figure 2**).

The other player in this bi-partite metabolism is glucose, which in *L. pneumophila* predominantly enters into the PPP for *the novo* production of histidine and sugars (in particular mannose) and that is also used in lower amounts for the synthesis of other amino acids and PHB. Conversely to serine, glucose is mainly metabolized throughout the late exponential and post-exponential phase of growth (**Figure 2**). As previously mentioned, *L. pneumophila* uses mainly the ED pathway, the gluconeogenesis and the PPP, and to a minor extent the glycolysis to metabolize glucose (Harada et al., 2010; Häuslein et al., 2016). Furthermore, glucose metabolism through the ED pathway is necessary for full fitness of *L. pneumophila* during its biphasic life cycle (Eylert et al., 2010). Earlier studies provided the first indication that glycerol may be used by *L. pneumophila* as carbon source, as deduced from the upregulation of a glycerol-3-phospate dehydrogenase (*glpD*) during intracellular growth in human macrophages (Faucher et al., 2011). Indeed, glycerol is predominantly metabolized during the late and post-exponential phase of growth like the life stage dependent usage of other carbon sources. Similar to glucose, the carbon from glycerol is mainly shuffled into gluconeogenesis and PPP for histidine and mannose production, but only low flux rates of carbon from glycerol into the TCA cycle were reported (Häuslein et al., 2016). In contrast, saturated lipids like palmitate, another carbon source shown to be predominantly used after amino acid depletion,

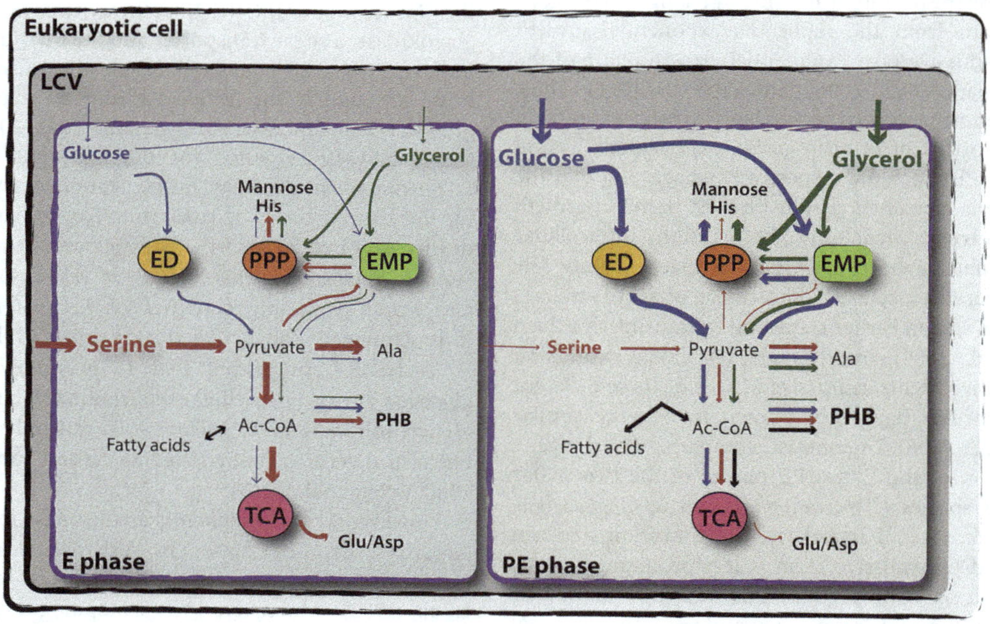

FIGURE 2 | Simplified representation of exponential and post-exponential phase-dependent utilization of serine, glucose and glycerol by *L. pneumophila*. *In vitro* isotopologue labeling experiments using ^{13}C-serine, ^{13}C-glucose and ^{13}C-glycerol revealed a bipartite metabolism in which serine (in red) is mainly used during the exponential phase of growth and enters primarily the TCA cycle, whereas glucose (in blue) and glycerol (in green) are shuffled into anabolic processes during the post-exponential growth phase of *L. pneumophila*. Relative carbon fluxes are depicted by the thickness of the arrows. For more detail, see the text. ED, Entner–Doudoroff pathway; PPP, pentose phosphate pathway; EMP, Embden-Meyerhof-Parnas pathway (glycolysis); TCA, tricarboxylic acid cycle (adapted from Eisenreich and Heuner, 2016).

is used for energy production and PHB synthesis (Häuslein et al., 2017). Taken together, the results from ^{13}C- isotopologue profiling and flux analyses suggested that the life cycle switch of *Legionella* is also reflected by a metabolic shift from amino acids usage during replication to glycerolipids and glucose when entering transmissive phase (Häuslein et al., 2017; **Figure 2**).

The Metabolism of Intracellular *L. pneumophila*

The acquisition of nutrients within host cells is an indispensable prerequisite for *L. pneumophila* multiplication and for a successful infection cycle. The presence and the up-regulation of genes encoding 12 different ABC transporters, amino acid permeases, proteases and phospholipases during intracellular multiplication within host cells suggested how *L. pneumophila* exploits host nutrient to support its intracellular growth (Brüggemann et al., 2006). Indeed, it was shown that intracellular replication of *L. pneumophila* depends on the host cell amino acid transporter SLC1A5 (Wieland et al., 2005) and that it employs the phagosomal transporter A (PhtA) to acquire threonine during growth (Sauer et al., 2005). Furthermore, ^{13}C-Isotopologue compositions of amino acids from bacterial and amoebal proteins showed that *L. pneumophila* takes indeed amino acids up from its host (Schunder et al., 2014). In addition to the above mentioned transport systems, *L. pneumophila* was also reported to employ its effector proteins to generate nutrients for its growth. The *L. pneumophila* effector AnkB (Price et al., 2009; Lomma et al., 2010) is secreted in the host cell where it poly-ubiquitinates its targets leading to their proteolysis by the host proteasome. Price and colleagues suggested that in this way AnkB may generate short peptides and amino acids, which represent nutrients for intracellular bacterial multiplication as these free amino acids may be imported into the LCV *via* different host solute carriers and transporters, such as the glucose (Slc2a1, Slc2a6) and glycerol transporters (Slc37a1) (Price et al., 2011). A recent study reported opposing effects of two Dot/Icm secreted effector families, Lgt and SidE on the master regulator of host amino acid metabolism, the mechanistic target of rapamycin complex 1 (mTORC1). However, these two-effector families work synergistically to inhibit host translation and thereby liberate amino acids for *L. pneumophila* growth (De Leon et al., 2017). Thus, *L. pneumophila* not only employs its own transport systems for the uptake and to use amino acids but also seem to exploit the host proteasome machinery and mTORC1 to generate nutrients. Isoptopologue profiling during the replicative intracellular growth phase showed that *L. pneumophila* uses serine and other amino acids as main carbon and energy sources for protein biosynthesis, amino acid biosynthesis and PHB production (Price et al., 2011). Although less is known about the intracellular metabolism in the late phase of growth, it is likely that *L. pneumophila* gets access to carbohydrate sources such as glycogen, glucose and other polymers from the host upon LCV lysis (Lang and Flieger, 2011). In addition, the transmissive form of *L. pneumophila* contains high amounts of PHB, which serves as energy and carbon storage for the maintenance of the intracellular growth cycle (Gillmaier et al., 2016; Häuslein et al., 2017). Moreover, intracellular *L. pneumophila* metabolizes *myo*-inositol, which was reported to promote infection of *A. castellanii* and macrophages (Manske et al., 2016), and engages the translocated protein MavN, which once integrated in the host-derived LCV membrane, facilitates the acquisition of iron into its vacuole (Isaac et al., 2015). Once nutrients are limited in the LCV, this may be the signal for *L. pneumophila* to activate the infectious traits to escape the spent host. This transition from the replicative to the transmissive/virulent phase is highly regulated by a complex regulatory network, described in the following sections.

Regulation of *L. pneumophila* Differentiation

Triggers of the Stringent Response

After replicating within the LCV to high numbers, nutrients become limited, which triggers complex and coordinated regulation to allow the expression of transmissive traits, which provide *L. pneumophila* with the ability of leaving the host cell, of long-term survival under hostile extracellular conditions and of re-infecting a new host cell. By analogy to *E. coli*, it was proposed that *L. pneumophila,* when starved for amino acids, initiates a stringent response by synthesizing the second messenger guanosine tetraphosphate (p)ppGpp via the synthetase enzyme RelA (Hammer and Swanson, 1999). Indeed, a *L. pneumophila relA* mutant replicates efficiently within either amoebae or macrophages however upon entry into the post-exponential phase of growth, the mutant strain does not produce the second messenger (Zusman et al., 2002). Additionally, virulence traits are poorly expressed when *L. pneumophila* lacks RelA and consequently the alarmone (p)ppGpp (Hammer and Swanson, 1999; Zusman et al., 2002; Dalebroux et al., 2010a). The mild effects displayed by the lack of RelA on the expression of the virulent traits suggested that additional clues and redundant strategies are employed by *L. pneumophila* to govern its virulence (Hammer and Swanson, 1999; Zusman et al., 2002). Indeed *L. pneumophila* is equipped with two ppGpp synthetases, which respond to two distinct metabolic cues. Whereas, RelA synthesizes (p)ppGpp is following fluctuations in amino acid availability, the bifunctional enzyme SpoT leads to the accumulation of the alarmone (p)ppGpp in response to fatty acid depletion. By analogy to *E. coli*, a *L. pneumophila* strain depleted of *relA* and *spoT* lacks (p)ppGpp completely, however whether it results in rRNA transcriptional activation and/or in the synthesis of stable RNA remains unclear (Dalebroux et al., 2009; Dalebroux and Swanson, 2012; Trigui et al., 2015). Thus, the *L. pneumophila* biphasic life cycle requires the fine tuning of the levels of alarmone (p)ppGpp present in the bacteria. When nutrients are abundant, virulent bacteria hydrolyze (p)ppGpp in a SpoT-dependent manner, allowing the bacteria to actively multiply and repress the transmission traits (Molofsky and Swanson, 2003; Dalebroux et al., 2009, 2010a; Trigui et al., 2015). Conversely, as replicating bacteria exhaust the available nutrients within the LCV, (p)ppGpp is produced by RelA and additionally, the equilibrium of SpoT is shifted more toward synthesis instead degradation. This leads to a massive accumulation of the alarmone and triggers the entry into the transmissive state

(Hammer and Swanson, 1999; Molofsky and Swanson, 2004; Dalebroux et al., 2009). SpoT is required throughout the entire infection cycle to mediate (p)ppGpp turnover *via* its hydrolase and synthase activities (Xiao et al., 1991; Potrykus and Cashel, 2008; Dalebroux et al., 2009, 2010a).

Transcriptional Control by Sigma Factors

In *L. pneumophila* the signaling alarmone (p)ppGpp is a key player for the reorganization of the bacterial transcriptome by recruiting sigma factors, allowing the activation of genes necessary for the adaptation to the new condition and the repression of the ones that are no longer required (Dalebroux et al., 2010a). Particularly, the accumulation of (p)ppGpp increases the amount of the alternative sigma factor RpoS ($\sigma^{S/38}$), which results in the regulation of multiple pathways associated with motility and pathogenic functions as well as the activity of transcriptional regulators and Dot/Icm effectors (Hales and Shuman, 1999; Bachman and Swanson, 2004; Trigui et al., 2015). The mechanism that links the accumulation of (p)ppGpp with the expression of RpoS remains to be elucidated, however (p)ppGpp is suggested to destabilize the binding of the vegetative sigma factor $\sigma^{D/70}$ to the core and endorses the recruitment of alternative sigma factors and the expression of their targets, as demonstrated for *E.coli* (Jishage et al., 2002). An additional regulatory element, which acts as cofactor for (p)ppGpp-dependent transcriptional regulation is the RNA polymerase (RNAP) secondary channel interacting protein DksA (Haugen et al., 2008; Potrykus and Cashel, 2008). *L. pneumophila* DksA function may be dependent on bacterial stimuli. In particular, DksA seems to respond to fatty acid stress by inducing bacterial differentiation in a (p)ppGpp-independent manner, as judged by the expression of certain transmissive traits within macrophages (Dalebroux et al., 2010b). However, upon (p)ppGpp accumulation, DksA and (p)ppGpp coordinately regulate the hierarchical cascade for flagellar expression. Therefore, *L. pneumophila* employs both (p)ppGpp and DksA to act independently or cooperatively during bacterial differentiation (Dalebroux et al., 2010b). At the bottom of the hierarchical cascade governing *L. pneumophila* differentiation one can find the flagellar regulon, composed of four different classes of genes, whose coordinated expression is crucial for efficient and maximal virulence of the bacterium (Heuner et al., 1997; Dietrich et al., 2001; Brüggemann et al., 2006; Appelt and Heuner, 2017). Class I genes, which include the genes encoding the flagellar master regulator and the σ^{54} activator protein FleQ together with the alternative sigma factor RpoN (σ^{54}), are required for the expression of the class II genes, leading to the formation of the flagellar basal body, hook and the activation of the regulatory proteins (Jacobi et al., 2004; Steinert et al., 2007; Albert-Weissenberger et al., 2010). Finally, the flagellar sigma factor FliA (σ^{28}) (encoded by a class III gene and regulated by DksA) is directly controlling the flagellar class IV genes such as *flaA* and *fliDS*, encoding the flagellin and the filament cap respectively, leading to the complete formation of the flagellum (Heuner and Steinert, 2003; Jacobi et al., 2004; Brüggemann et al., 2006; Albert-Weissenberger et al., 2007, 2010; Dalebroux et al., 2010b). Interestingly, the flagellar sigma factor FliA is not only implicated in the regulation of the flagellum production but also acts as regulator of virulence genes that are required for the expression of pathways important for cytotoxicity, lysosome evasion, and replication of *L. pneumophila* (Heuner et al., 2002; Molofsky and Swanson, 2004; Brüggemann et al., 2006).

Post-transcriptional Regulation of Transmissive Traits

As in many other bacterial pathogens, *L. pneumophila* post-transcriptional regulation is controlled by two-component systems (TCS), which use protein phosphorylation cascades for signal transduction (Padilla-Vaca et al., 2017). *L. pneumophila* employs at least four distinct TCSs LetA/S, PmrA/B, LsqR/ST, and CpxR/A that govern its differentiation from the replicative to the transmissive state (Gal-Mor and Segal, 2003a; Tiaden et al., 2007; Zusman et al., 2007; Altman and Segal, 2008). Particularly, the TCS LetA/LetS (*Legionella* transmission activator and sensor, respectively) of *L. pneumophila* is an important regulator system for the activation of a large set of virulence phenotypes and the control of the progression into the transmissive state (Hammer et al., 2002; Gal-Mor and Segal, 2003b; Lynch et al., 2003). Probably directly activated by the accumulation of the alarmone (p)ppGpp, LetA is regulating the expression of the small ncRNAs RmsX,Y,Z, which are required to relieve the repression exerted by the global regulator CsrA, an RNA-binding protein, on many virulence genes, thereby ensuring the expression of the transmissive traits (Hovel-Miner et al., 2009; Rasis and Segal, 2009; Sahr et al., 2009; Edwards et al., 2010). The carbon storage regulator protein CsrA was reported to bind more than 450 mRNA targets in *L. pneumophila,* altering their translation, transcription and/or their stability (Sahr et al., 2009, 2017). Among those targets, CsrA affects the expression of the previously mentioned regulators FleQ, RpoS, the quorum sensing regulator LqsR and it also control the expression of over 40 Dot/Icm substrates (Sahr et al., 2017). Moreover, CsrA controls its own expression and the *relA* mRNA in a regulatory feedback loop. This in turn makes the CsrA protein indispensable for *L. pneumophila* thus only conditional or partial mutants could be obtained, which are all however strongly attenuated for intracellular multiplication, underlining its essential role in the life cycle of *L. pneumophila* (Molofsky and Swanson, 2003; Sahr et al., 2017). Another TCS important for virulence gene expression is PmrA/B (Zusman et al., 2007). *L. pneumophila* PmrA/B not only activates the expression of 43 effector-encoding genes but also positively regulates CsrA and consequently post-transcriptional repression of the CsrA-regulated effectors (Zusman et al., 2007; Al-Khodor et al., 2009; Rasis and Segal, 2009) (**Figure 3**). It is likely that a regulatory switch between at least two sets of effectors occurs: one set of effectors, activated by PmrA/B and expressed in the replicative state and the second group of effectors which is regulated by the LetA/S TCS upon entry into the transmissive phase of *L. pneumophila*. Another player in this complex regulatory network, is the TCS LqsRS (*Legionella* quorum sensing), whose role in the regulation of gene expression during the transmissive phase has been extensively studied (Hochstrasser and Hilbi, 2017). Importantly, the production of LqsR is regulated at the post-transcriptional level by the global repressor CsrA (Sahr et al.,

2009, 2017). Finally, the CpxR/A TCS, which acts as dual regulator and thus as an activator and repressor, was shown to control the expression of at least 27 Dot/Icm substrates as well as type II- secreted virulence factors, playing a important role in *L. pneumophila* virulence gene regulation (Gal-Mor and Segal, 2003a; Altman and Segal, 2008).

In addition to TCSs and the RNA-binding protein CsrA, another major player in the regulation of the transition from replicative to transmissive *L. pneumophila* is the RNA binding protein and chaperone Hfq (McNealy et al., 2005; Trigui et al., 2013). This pleiotropic regulatory element is known to modulate gene expression by facilitating the interaction between sRNA and their mRNA targets in diverse bacterial pathogens, controlling pathways related to metabolism, transport, energy production and conversion or membrane proteins (Boudry et al., 2014). In *L. pneumophila*, Hfq expression is influenced by RpoS and LetA regulatory elements as both directly or indirectly turn on *hfq* transcription upon onset of the late post-exponential phase. Furthermore, *L. pneumophila* Hfq regulates its own expression in an auto-regulatory loop (Oliva et al., 2017). Although, only two direct targets *(hfq* mRNA and Anti-hfq sRNA) of *L. pneumophila* Hfq have been identified to date by *in vitro* assays, Hfq was reported to regulate the bacterium's virulence, as judged by the findings that this post-transcriptional regulator promotes motility and is required for efficient multiplication of *L. pneumophila* within *A. castellanii* at environmental temperatures (McNealy et al., 2005; Oliva et al., 2017) (**Figure 3**).

L. pneumophila Engages sRNAs to Control Its Virulence

The complex and hierarchical regulation of the *L. pneumophila* life cycle includes also the recruitment of small RNAs, which ensure a fast and more cost-effective regulation. Previous evidences in *E. coli* showed that the BarA/UvrY TCS (the *L. pneumophila* LetA/S homolog) controls the expression of two sRNAs, named CsrB and CsrC, whose sequences contain GGA motifs, which are the characteristic binding sequences for CsrA (Liu et al., 1997). A first bioinformatics search revealed in *L. pneumophila* the presence of two homologs of CsrB, named RsmY and RsmZ (Kulkarni et al., 2006). Functional analyses confirmed that these sRNAs were the missing regulatory elements linking the LetA/S TCS and the RNA binding protein CsrA in *L. pneumophila* (Rasis and Segal, 2009; Sahr et al., 2009). In detail, LetA binds directly to a conserved consensus sequence upstream the *rsmY/Z* genes, leading to their expression. These sRNAs contain multiple CsrA binding motifs and act as sponge to bind and sequester CsrA from their target mRNAs, leading to the expression of virulence traits. RsmY and RsmZ were the first characterized sRNAs implicated in the regulation of *L. pneumophila* virulence. However, deep RNA sequencing from exponentially (replicative) and post exponentially (virulent) *in vitro* grown *L. pneumophila* have identified more than 700 sRNAs, 60% of which are growth-phase dependently regulated, including a third LetA-dependent sRNA, named RsmX, suggesting that a set of these yet uncharacterized sRNAs, might influence the expression of replication or virulence determinants in *L. pneumophila* (Sahr et al., 2012). Recently, we characterized one of these ncRNAs, a *cis*-encoded sRNA for which we showed that it is implicated in the regulation of the RNA binding protein Hfq (Oliva et al., 2017). This sRNA, named Anti-hfq, is transcribed antisense to the *hfq* transcript and controls the expression of Hfq through a base pairing mechanism during the exponential phase of *L. pneumophila* growth (Oliva et al., 2017). Moreover, it is important to mention that Hfq was reported to influence *L. pneumophila* differentiation by interacting with the major regulatory elements of the cascade. Thus, it is expected that Hfq, acting as RNA chaperone and RNA binder might regulate a number of still unknown sRNAs implicated in bacterial virulence.

Taken together, *L. pneumophila* is equipped with a sophisticated regulatory network, including transcriptional and post-transcriptional regulatory elements, including small non-coding sense and antisense RNAs to control the reciprocal expression of distinct sets of genes under different environmental conditions (**Figure 3**).

CROSS TALK BETWEEN METABOLISM AND THE STRINGENT RESPONSE

Similarly to what has been described in other bacterial pathogens, many regulatory factors implicated in virulence gene expression

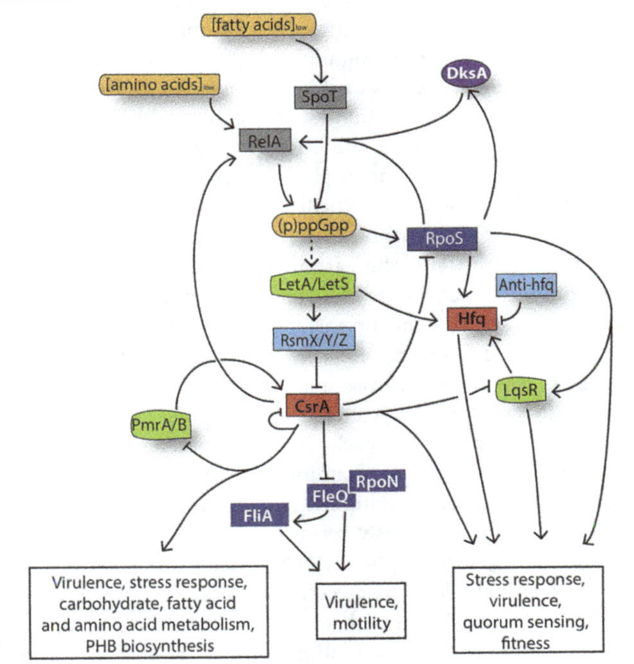

FIGURE 3 | Model the stringent response network governing *L. pneumophila* differentiation. Amino acid and fatty acid starvation triggers RelA and SpoT to produce the alarmone (p)ppGpp. Its accumulation induces the activation of the stress sigma factor RpoS, the LetA/LetS TCS and consequently an increased transcription of the RsmZ, RsmX, and RsmY sRNAs. The three sRNAs act as "sponge," sequestering CsrA and leading thereby to the activation of the infectious traits and to changes in the metabolism. The dashed arrows indicate suggested but not yet confirmed direct interactions.

are also major regulators of metabolic pathways. Indeed, *L. pneumophila* exhibits a bipartite metabolism, which requires a fine-tuned regulation. An intriguing example of a regulator that is important for the expression of virulence and the regulation of metabolic traits is the RNA binding protein CsrA. Interestingly, *L. pneumophila* harbors some of the key genes encoding enzymes of the glycolysis/gluconeogenesis (glyceraldeyde-3-phosphate dehydrogenase or Gap, phosphoglycerate kinase, and pyruvate kinase) and the PPP (transketolase) in one single operon. The combined or individual regulation of these two pathways is under the control of the RNA binding protein CsrA, whose presence ensures the efficient expression of the both parts of this operon (Sahr et al., 2017). When nutrients are abundant CsrA binds within the *gap* transcript, and stabilizes the alternative secondary structure that covers the Rho-dependent transcription termination site. Consequently, this leads to a CsrA-dependent transcription of the glycolysis part of the operon toward gluconeogenesis, which under starvation or stress is not expressed. Another example of how CsrA influences metabolism, is that this regulatory element affects the production of secondary metabolites, in particular thiamine pyrophosphate, ensuring the effecting functioning of central enzymes of the carbohydrate metabolism when required (Sahr et al., 2017).

Indeed, using ^{13}C-isotopologue profiling and carbon-flux analyses of a wild-type and a *csrA* mutant strain confirmed that CsrA plays a major role in regulating the carbon flux between the PPP and the glycolysis (Häuslein et al., 2017). Furthermore, this study highlighted the impact of CsrA on the bipartite metabolism of *L. pneumophila*, as the absence of CsrA induces a reduction of the carbon flux from serine *via* gluconeogenesis into the PPP. By contrast, CsrA has a negative impact on the incorporation and the metabolism of glycerol and glucose. As such, the absence of CsrA results in the increase of the carbon flux from glucose into the PPP and ED pathways and the carbon flux from glycerol into the PPP and the gluconeogenesis (Häuslein et al., 2017). These studies also showed the important influence of CsrA on the production of the storage molecule PHB suggesting that CsrA is a major player in the utilization of the different carbon sources during the biphasic life cycle of *L. pneumophila* (Häuslein et al., 2017). The biphasic life cycle of *L. pneumophila* within the host supports the usage of amino acids as main carbon and energy source during multiplication due to the expression of CsrA that is simultaneously repressing the usage of alternative carbon sources, such as glycerol. Conversely, upon onset of the post-exponential phase of growth, the stress response induces the sRNA RsmX, Y, and Z that sequester CsrA, resulting in an increased utilization of glycerolipids, which along with glucose, mostly trigger the synthesis of lipopolysaccharide sugars through the PPP and in addition, the production of the energy and carbon storage polymer PHB (Häuslein et al., 2017). Hence, CsrA is a major organizer of the biphasic life cycle of *Legionella pneumophila* integrating and coordinating the metabolic carbon switch and the transition between replicative and transmissive traits.

CONCLUDING REMARKS

L. pneumophila is an intracellular opportunistic pathogen, which exploits amoebae and other protozoa as environmental hosts, but that is also able to infect human macrophages, eventually causing Legionnaires' disease, a severe pneumonia that is often fatal when not treated promptly. *L. pneumophila* is ubiquitously found in fresh water habitats, as planktonic form or forming biofilm. In response to diverse and hostile environmental conditions encountered during its life cycle, *L. pneumophila* has evolved sophisticated mechanisms to successfully replicate within different host niches and to also survive in extracellular environments. As such, this intracellular bacterium displays at least two reciprocal stages: a replicative and a transmissive form. The transition between the non-virulent replicative and the virulent non-replicative phase is governed by a complex regulatory network, in which transcriptional and post-transcriptional regulatory elements are engaged to insure an efficient infection cycle. The trigger of this morphological stress response is mainly mediated by metabolic changes and therefore the availability of nutrients in the surroundings. Thus, within the LCV the usage of serine as carbon and energy source supports the multiplication of the bacteria in which the replicating bacteria show a high metabolic activity. Upon amino acid depletion, the stringent response mediates the expression of the virulent traits but in parallel also enables the bacteria to survive for long term under stress and starving condition. This is ensured amongst others by the expression of stress and virulence related genes and an overall metabolic shift leading to the usage of alternative carbon sources like glucose and glycerolipids and an increased production of the storage molecule PHB. Under these conditions, *L. pneumophila* is optimally equipped to escape the spent host, survive for an uncertain period in the extracellular environment and eventually start a new infection cycle.

Taken together, the biphasic life cycle of *L. pneumophila* results in distinct morphological changes and a bipartite carbon metabolism. Thus, during the biphasic life cycle the metabolism influences the transition between replicative and transmissive phase as well as the reciprocal expression of virulence factors and their regulators, in particular CsrA, which is implicated in the regulation of virulence and the metabolism. A comprehensive analysis of *L. pneumophila* adaptation to metabolic cues during the transmissive phase *in vivo* either in amoebae or macrophages is still missing and would provide additional information about the utilization of diverse carbohydrates, and the cross-talk of the regulatory elements which govern *L. pneumophila* virulence. Continuous unrevealing of this complex interplay between metabolism and virulence of *L. pneumophila* may teach us also about host-pathogen interaction in general.

AUTHOR CONTRIBUTIONS

All authors listed have made a substantial, direct and intellectual contribution to the work, and approved it for publication.

ACKNOWLEDGMENTS

Work in the CB laboratory is financed by the Institut Pasteur, the grant n°ANR-10-LABX-62-IBEID, the Fondation pour la Recherche Médicale (FRM) grant N° DEQ20120323697 and the Pasteur-Weizmann consortium ≪ The roles of non-coding RNAs in regulation of microbial life styles and virulence ≫. GO was supported by a stipend from the Pasteur—Paris University (PPU) International PhD Program.

REFERENCES

Al-Bana, B. H., Haddad, M. T., and Garduño, R. A. (2014). Stationary phase and mature infectious forms of *Legionella pneumophila* produce distinct viable but non-culturable cells. *Environ. Microbiol.* 16, 382–395. doi: 10.1111/1462-2920.12219

Albert-Weissenberger, C., Cazalet, C., and Buchrieser, C. (2007). *Legionella pneumophila* - a human pathogen that co-evolved with fresh water protozoa. *Cell Mol. Life Sci.* 64, 432–448. doi: 10.1007/s00018-006-6391-1

Albert-Weissenberger, C., Sahr, T., Sismeiro, O., Hacker, J., Heuner, K., and Buchrieser, C. (2010). Control of flagellar gene regulation in *Legionella pneumophila* and its relation to growth phase. *J. Bacteriol.* 192, 446–455. doi: 10.1128/JB.00610-09

Al-Khodor, S., Kalachikov, S., Morozova, I., Price, C. T., and Abu Kwaik, Y. (2009). The PmrA/PmrB two-component system of *Legionella pneumophila* is a global regulator required for intracellular replication within macrophages and protozoa. *Infect. Immun.* 77, 374–386. doi: 10.1128/IAI.01081-08

Altman, E., and Segal, G. (2008). The response regulator CpxR directly regulates expression of several *Legionella pneumophila icm/dot* components as well as new translocated substrates. *J. Bacteriol.* 90, 1985–1996. doi: 10.1128/JB.01493-07

Appelt, S., and Heuner, K. (2017). The flagellar regulon of *Legionella*-a review. *Front. Cell. Infect. Microbiol.* 7:454. doi: 10.3389/fcimb.2017.00454

Bachman, M. A., and Swanson, M. S. (2004). Genetic evidence that *Legionella pneumophila* RpoS modulates expression of the transmission phenotype in both the exponential phase and the stationary phase. *Infect. Immun.* 72, 2468–2476. doi: 10.1128/IAI.72.5.2468-2476.2004

Barker, J., Scaife, H., and Brown, M. R. (1995). Intraphagocytic growth induces an antibiotic-resistant phenotype of *Legionella pneumophila*. *Antimicrob. Agents Chemother.* 39, 2684–2688. doi: 10.1128/AAC.39.12.2684

Boudry, P., Gracia, C., Monot, M., Caillet, J., Saujet, L., Hajnsdorf, E., et al. (2014). Pleiotropic role of the RNA chaperone protein Hfq in the human pathogen *Clostridium difficile*. *J. Bacteriol.* 196, 3234–3248. doi: 10.1128/JB.01923-14

Bozue, J. A., and Johnson, W. (1996). Interaction of *Legionella pneumophila* with *Acanthamoeba castellanii*: uptake by coiling phagocytosis and inhibition of phagosome-lysosome fusion. *Infect. Immun.* 64, 668–673.

Brieland, J. K, Fantone, J. C., Remick, D. G., LeGendre, M., McClain, M., and Engleberg, N. C. (1997). The role of *Legionella pneumophila*-infected *Hartmannella vermiformis* as an infectious particle in a murine model of Legionnaire's disease. *Infect. Immun.* 65, 5330–5333.

Brüggemann, H., Hagman, A., Jules, M., Sismeiro, O., Dillies, M. A., Gouyette, C., et al. (2006). Virulence strategies for infecting phagocytes deduced from the *in vivo* transcriptional program of *Legionella pneumophila*. *Cell. Microbiol.* 8, 1228–1240. doi: 10.1111/j.1462-5822.2006.00703.x

Byrne, B., and Swanson, M. S. (1998). Expression of *Legionella pneumophila* virulence traits in response to growth conditions. *Infect. Immun.* 66, 3029–3034.

Cazalet, C., Gomez-Valero, L., Rusniok, C., Lomma, M., Dervins-Ravault, D., Newton, H. J., et al. (2010). Analysis of the *Legionella longbeachae* genome and transcriptome uncovers unique strategies to cause Legionnaires' disease. *PLoS Genet.* 6:e1000851. doi: 10.1371/journal.pgen.1000851

Cazalet, C., Rusniok, C., Brüggemann, H., Zidane, N., Magnier, A., Ma, L., et al. (2004). Evidence in the *Legionella pneumophila* genome for exploitation of host cell functions and high genome plasticity. *Nat. Genet.* 36, 1165–1173. doi: 10.1038/ng1447

Choy, A., Dancourt, J., Mugo, B., O'Connor, T. J., Isberg, R. R., Melia, T. J., et al. (2012). The *Legionella* effector RavZ inhibits host autophagy through irreversible Atg8 deconjugation. *Science* 338, 1072–1076. doi: 10.1126/science.1227026

Cirillo, J. D., Falkow, S., and Tompkins, L. S. (1994). Growth of *Legionella pneumophila* in *Acanthamoeba castellanii* enhances invasion. *Infect. Immun.* 62, 3254–3261.

Correia, A. M., Ferreira, J. S., Borges, V., Nunes, A., Gomes, B., Capucho, R., et al. (2016). Probable person-to-person transmission of Legionnaires' disease. *N. Engl. J. Med.* 374, 497–498. doi: 10.1056/NEJMc1505356

Cunha, B. A., Burillo, A., and Bouza, E. (2016). Legionnaires' disease. *Lancet* 387, 376–385. doi: 10.1016/S0140-6736(15)60078-2

Dalebroux, Z. D., Edwards, R. L., and Swanson, M. S. (2009). SpoT governs *Legionella pneumophila* differentiation in host macrophages. *Mol. Microbiol.* 71, 640–658. doi: 10.1111/j.1365-2958.2008.06555.x

Dalebroux, Z. D., Svensson, S. L., Gaynor, E. C., and Swanson, M. S. (2010a). ppGpp conjures bacterial virulence. *Microbiol. Mol. Biol. Rev.* 74, 171–199. doi: 10.1128/MMBR.00046-09

Dalebroux, Z. D., and Swanson, M. S. (2012). ppGpp: magic beyond RNA polymerase. *Nat. Rev. Microbiol.* 10, 203–212. doi: 10.1038/nrmicro2720

Dalebroux, Z. D., Yagi, B. F., Sahr, T., Buchrieser, C., and Swanson, M. S. (2010b). Distinct roles of ppGpp and DksA in *Legionella pneumophila* differentiation. *Mol. Microbiol.* 76, 200–219. doi: 10.1111/j.1365-2958.2010.07094.x

de Felipe, K. S., Pampou, S., Jovanovic, O. S., Pericone, C. D., Ye, S. F., Kalachikov, S., et al. (2005). Evidence for acquisition of *Legionella* type IV secretion substrates via interdomain horizontal gene transfer. *J. Bacteriol.* 187, 7716–7726. doi: 10.1128/JB.187.22.7716-7726.2005

De Leon, J. A., Qiu, J., Nicolai, C. J., Counihan, J. L., Barry, K. C., Xu, L., et al. (2017). Positive and negative regulation of the master metabolic regulator mTORC1 by two families of *Legionella pneumophila* effectors. *Cell Rep.* 21, 2031–2038. doi: 10.1016/j.celrep.2017.10.088

Dietrich, C., Heuner, K., Brand, B. C., Hacker, J., and Steinert, M. (2001). Flagellum of *Legionella pneumophila* positively affects the early phase of infection of eukaryotic host cells. *Infect. Immun.* 69, 2116–2122. doi: 10.1128/IAI.69.4.2116-2122.2001

Edwards, R., Jules, M., Sahr, T., Buchrieser, C., and Swanson, M. (2010). The multi-step design of the LetA/LetS two-component system increases *Legionella pneumophila* versatility. *Infect. Immun.* 78, 2571–2583. doi: 10.1128/IAI.01107-09

Eisenreich, W., and Heuner, K. (2016). The life stage-specific pathometabolism of *Legionella pneumophila*. *FEBS Lett.* 590, 3868–3886. doi: 10.1002/1873-3468.12326

Escoll, P., Mondino, S., Rolando, M., and Buchrieser, C. (2016). Targeting of host organelles by pathogenic bacteria: a sophisticated subversion strategy. *Nat. Rev. Microbiol.* 14, 5–19. doi: 10.1038/nrmicro.2015.1

Escoll, P., Rolando, M., Gomez-Valero, L., and Buchrieser, C. (2013). From amoeba to macrophages: exploring the molecular mechanisms of *Legionella pneumophila* infection in both hosts. *Curr. Top. Microbiol. Immunol.* 376, 1–34. doi: 10.1007/82_2013_351

Eylert, E., Herrmann, V., Jules, M., Gillmaier, N., Lautner, M., Buchrieser, C., et al. (2010). Isotopologue profiling of *Legionella pneumophila*: role of serine and glucose as carbon substrates. *J. Biol. Chem.* 285, 22232–22243. doi: 10.1074/jbc.M110.128678

Faucher, S. P., Mueller, C. A., and Shuman, H. A. (2011). *Legionella pneumophila* transcriptome during intracellular multiplication in human macrophages. *Front. Microbiol.* 2:60. doi: 10.3389/fmicb.2011.00060

Faulkner, G., and Garduño, R. A. (2002). Ultrastructural analysis of differentiation in *Legionella pneumophila*. *J. Bacteriol.* 184, 7025–7041. doi: 10.1128/JB.184.24.7025-7041.2002

Fields, B. S. (1996). The molecular ecology of *Legionellae*. *Trends Microbiol.* 4, 286–290. doi: 10.1016/0966-842X(96)10041-X

Fields, B. S., Benson, R. F., and Besser, R. E. (2002). *Legionella* and Legionnaires' disease: 25 years of investigation. *Clin. Microbiol. Rev.* 15, 506–526. doi: 10.1128/CMR.15.3.506-526.2002

Gal-Mor, O., and Segal, G. (2003a). Identification of CpxR as a positive regulator of icm and dot virulence genes of *Legionella pneumophila*. *J. Bacteriol.* 185, 4908–4919. doi: 10.1128/JB.185.16.4908-4919.2003

Gal-Mor, O., and Segal, , G. (2003b). The *Legionella pneumophila* GacA homolog (LetA) is involved in the regulation of *icm* virulence genes and is required for intracellular multiplication in *Acanthamoeba castellanii*. *Microb. Pathog.* 34, 187–194. doi: 10.1016/S0882-4010(03)00027-5

Garduño, R. A., Garduño, E., Hiltz, M., and Hoffman, P. S. (2002). Intracellular growth of *Legionella pneumophila* gives rise to a differentiated form dissimilar to stationary-phase forms. *Infect. Immun.* 70, 6273–6283. doi: 10.1128/IAI.70.11.6273-6283.2002

George, J. R., Pine, L., Reeves, M. W., and Harrell, W. K. (1980). Amino acid requirements of *Legionella pneumophila*. *J. Clin. Microbiol.* 11, 286–291.

Gillmaier, N., Schunder, E., Kutzner, E., Tlapák, H., Rydzewski, K., Herrmann, V., et al. (2016). Growth-related metabolism of the carbon storage Poly-3-hydroxybutyrate in *Legionella pneumophila*. *J. Biol. Chem.* 291, 6471–6482. doi: 10.1074/jbc.M115.693481

Gomez-Valero, L., Rusniok, C., Cazalet, C., and Buchrieser, C. (2011). Comparative and functional genomics of *Legionella* identified eukaryotic like proteins as key players in host-pathogen interactions. *Front. Microbiol.* 2:208. doi: 10.3389/fmicb.2011.00208

Hales, L. M., and Shuman, H. A. (1999). The *Legionella pneumophila* rpoS gene is required for growth within *Acanthamoeba castellanii*. *J. Bacteriol.* 181, 4879–4889.

Hammer, B. K., and Swanson, M. S. (1999). Co-ordination of *Legionella pneumophila* virulence with entry into stationary phase by ppGpp. *Mol. Microbiol.* 33, 721–731. doi: 10.1046/j.1365-2958.1999.01519.x

Hammer, B. K., Tateda, E. S., and Swanson, M. S. (2002). A two-component regulator induces the transmission phenotype of stationary-phase *Legionella pneumophila*. *Mol. Microbiol.* 44, 107–118. doi: 10.1046/j.1365-2958.2002.02884.x

Harada, E., Iida, K., Shiota, S., Nakayama, H., and Yoshida, S. (2010). Glucose metabolism in *Legionella pneumophila*: dependence on the Entner-Doudoroff pathway and connection with intracellular bacterial growth. *J. Bacteriol.* 192, 2892–2899. doi: 10.1128/JB.01535-09

Haugen, S. P., Ross, W., and Gourse, R. L. (2008). Advances in bacterial promoter recognition and its control by factors that do not bind DNA. *Nat. Rev. Microbiol.* 6, 507–519. doi: 10.1038/nrmicro1912

Häuslein, I., Manske, C., Goebel, W., Eisenreich, W., and Hilbi, H. (2016). Pathway analysis using (13) C-glycerol and other carbon tracers reveals a bipartite metabolism of *Legionella pneumophila*. *Mol. Microbiol.* 100, 229–246. doi: 10.1111/mmi.13313

Häuslein, I., Sahr, T., Escoll, P., Klausner, N., Eisenreich, W., and Buchrieser, C. (2017). *Legionella pneumophila* CsrA regulates a metabolic switch from amino acid to glycerolipid metabolism. *Open Biol* 7:170149. doi: 10.1098/rsob.170149

Herrmann, V., Eidner, A., Rydzewski, K., Blädel, I., Jules, M., Buchrieser, C., et al. (2011). GamA is a eukaryotic-like glucoamylase responsible for glycogen- and starch-degrading activity of *Legionella pneumophila*. *Int. J. Med. Microbiol.* 301, 133–139. doi: 10.1016/j.ijmm.2010.08.016

Heuner, K., Dietrich, C., Skriwan, C., Steinert, M., and Hacker, J. (2002). Influence of the alternative sigma(28) factor on virulence and flagellum expression of *Legionella pneumophila*. *Infect. Immun.* 70, 1604–1608. doi: 10.1128/IAI.70.3.1604-1608.2002

Heuner, K., Hacker, J., and Brand, B. C. (1997). The alternative sigma factor sigma28 of *Legionella pneumophila* restores flagellation and motility to an *Escherichia coli fliA* mutant. *J. Bacteriol.* 179, 17–23. doi: 10.1128/jb.179.1.17-23.1997

Heuner, K., and Steinert, M. (2003). The flagellum of *Legionella pneumophila* and its link to the expression of the virulent phenotype. *Int. J. Med. Microbiol.* 293, 133–143. doi: 10.1078/1438-4221-00259

Hilbi, H., Hoffmann, C., and Harrison, C. F. (2011). *Legionella* spp. outdoors: colonization, communication and persistence. *Environ. Microbiol. Rep.* 3, 286–296. doi: 10.1111/j.1758-2229.2011.00247.x

Hilbi, H., Jarraud, S., Hartland, E., and Buchrieser, C. (2010). Update on Legionnaires' disease: pathogenesis, epidemiology, detection and control. *Mol. Microbiol.* 76, 1–11 doi: 10.1111/j.1365-2958.2010.07086.x

Hilbi, H., Weber, S. S., Ragaz, C., Nyfeler, Y., and Urwyler, S. (2007). Environmental predators as models for bacterial pathogenesis. *Environ. Microbiol.* 9, 563–575. doi: 10.1111/j.1462-2920.2007.01238.x

Hochstrasser, R., and Hilbi, H. (2017). Intra-species and inter-kingdom signaling of *Legionella pneumophila*. *Front. Microbiol.* 8:79. doi: 10.3389/fmicb.2017.00079

Horwitz, M. A., and Silverstein, S. C. (1980). Legionnaires' disease bacterium (*Legionella pneumophila*) multiples intracellularly in human monocytes. *J. Clin. Invest.* 66, 441–450. doi: 10.1172/JCI109874

Hovel-Miner, G., Pampou, S., Faucher, S. P., Clarke, M., Morozova, I., Morozov, P., et al. (2009). SigmaS controls multiple pathways associated with intracellular multiplication of *Legionella pneumophila*. *J. Bacteriol.* 191, 2461–2473. doi: 10.1128/JB.01578-08

Isaac, D. T., Laguna, R. K., Valtz, N., and Isberg, R. R. (2015). MavN is a *Legionella pneumophila* vacuole-associated protein required for efficient iron acquisition during intracellular growth. *Proc. Natl. Acad. Sci. U.S.A.* 112, E5208–E5217. doi: 10.1073/pnas.1511389112

Isberg, R. R., O'Connor, T. J., and Heidtman, M. (2009). The *Legionella pneumophila* replication vacuole: making a cosy niche inside host cells. *Nat. Rev. Microbiol.* 7, 13–24. doi: 10.1038/nrmicro1967

Jacobi, S., Schade, R., and Heuner, K. (2004). Characterization of the alternative sigma factor sigma54 and the transcriptional regulator FleQ of *Legionella pneumophila*, which are both involved in the regulation cascade of flagellar gene expression. *J. Bacteriol.* 186, 2540–2547. doi: 10.1128/JB.186.9.2540-2547.2004

James, B. W., Mauchline, W. S., Dennis, P. J., Keevil, C. W., and Wait, R. (1999). Poly-3-hydroxybutyrate in *Legionella pneumophila*, an energy source for survival in low-nutrient environments. *Appl. Environ. Microbiol.* 65, 822–827.

Jishage, M., Kvint, K., Shingler, V., and Nyström, T. (2002). Regulation of sigma factor competition by the alarmone ppGpp. *Genes Dev.* 16, 1260–1270. doi: 10.1101/gad.227902

Kulkarni, P. R., Cui, X., Williams, J. W., Stevens, A. M., and Kulkarni, R. V. (2006). Prediction of CsrA-regulating small RNAs in bacteria and their experimental verification in *Vibrio fischeri*. *Nucleic Acids Res.* 34, 3361–3369. doi: 10.1093/nar/gkl439

Lang, C., and Flieger, A. (2011). Characterisation of *Legionella pneumophila* phospholipases and their impact on host cells. *Eur. J. Cell Biol.* 90, 903–912. doi: 10.1016/j.ejcb.2010.12.003

Liu, M. Y., Gui, G., Wei, B., Preston, J. F. R., Oakford, L., Yüksel, U., et al. (1997). The RNA molecule CsrB binds to the global regulatory protein CsrA and antagonizes its activity in *Escherichia coli*. *J. Biol. Chem.* 272, 17502–17510. doi: 10.1074/jbc.272.28.17502

Lomma, M., Dervins-Ravault, D., Rolando, M., Nora, T., Newton, H. J., Sansom, F. M., et al. (2010). The *Legionella pneumophila* F-box protein Lpp2082 (AnkB) modulates ubiquitination of the host protein parvin B and promotes intracellular replication. *Cell. Microbiol.* 12, 1272–1291. doi: 10.1111/j.1462-5822.2010.01467.x

Lynch, D., Fieser, N., Glöggler, K., Forsbach-Birk, V., and Marre, R. (2003). The response regulator LetA regulates the stationary-phase stress response in *Legionella pneumophila* and is required for efficient infection of *Acanthamoeba castellanii*. *FEMS Microbiol. Lett.* 219, 241–248. doi: 10.1016/S0378-1097(03)00050-8

Manske, C., Schell, U., and Hilbi, H. (2016). Metabolism of myo-inositol by *Legionella pneumophila* promotes infection of amoeba and macrophages. *Appl. Environ. Microbiol.* 82, 5000–5014. doi: 10.1128/AEM.01018-16

McDade, J. E., Shepard, C. C., Fraser, D. W., Tsai, T. R., Redus, M. A., and Dowdle, W. R. (1977). Legionnaires' disease: isolation of a bacterium and demonstration of its role in other respiratory disease. *N. Engl. J. Med.* 297, 1197–1203. doi: 10.1056/NEJM197712012972202

McNealy, T. L., Forsbach-Birk, V., Shi, C., and Marre, R. (2005). The Hfq homolog in *Legionella pneumophila* demonstrates regulation by LetA and RpoS and interacts with the global regulator CsrA. *J. Bacteriol.* 187, 1527–1532. doi: 10.1128/JB.187.4.1527-1532.2005

Molofsky, A. B., and Swanson, M. S. (2003). *Legionella pneumophila* CsrA is a pivotal repressor of transmission traits and activator of replication. *Mol. Microbiol.* 50, 445–461. doi: 10.1046/j.1365-2958.2003.03706.x

Molofsky, A. B., and Swanson, M. S. (2004). Differentiate to thrive: lessons from the *Legionella pneumophila* life cycle. *Mol. Microbiol.* 53, 29–40. doi: 10.1111/j.1365-2958.2004.04129.x

Newton, H. J., Ang, D. K., van Driel, I. R., and Hartland, E. L. (2010). Molecular pathogenesis of infections caused by *Legionella pneumophila*. *Clin. Microbiol. Rev.* 23, 274–298. doi: 10.1128/CMR.00052-09

Ninio, S., and Roy, C. R. (2007). Effector proteins translocated by *Legionella pneumophila*: strength in numbers. *Trends Microbiol.* 15, 372–380. doi: 10.1016/j.tim.2007.06.006

Nora, T., Lomma, M., Gomez-Valero, L., and Buchrieser, C. (2009). Molecular mimicry: an important virulence strategy employed by *Legionella pneumophila* to subvert host functions. *Future Microbiol.* 4, 691–701. doi: 10.2217/fmb.09.47

Oliva, G., Sahr, T., Rolando, M., Knoth, M., and Buchrieser, C. (2017). A unique cis-encoded small noncoding RNA is regulating *Legionella pneumophila* Hfq expression in a life cycle-dependent manner. *MBio* 8:e02182-16. doi: 10.1128/mBio.02182-16

Padilla-Vaca, F., Mondragón-Jaimes, V., and Franco, B. (2017). General aspects of two-component regulatory circuits in bacteria: domains, signals and roles. *Curr. Protein Pept. Sci.* 18, 990–1004. doi: 10.2174/1389203717666160809154809

Pine, L., George, J. R., Reeves, M. W., and Harrell, W. K. (1979). Development of a chemically defined liquid medium for growth of *Legionella pneumophila*. *J. Clin. Microbiol.* 9, 615–626.

Potrykus, K., and Cashel, M. (2008). (p)ppGpp: still magical? *Annu. Rev. Microbiol.* 62, 35–51. doi: 10.1146/annurev.micro.62.081307.162903

Price, C. T., Al-Khodor, S., Al-Quadan, T., Santic, M., Habyarimana, F., Kalia, A., et al. (2009). Molecular mimicry by an F-box effector of *Legionella pneumophila* hijacks a conserved polyubiquitination machinery within macrophages and protozoa. *PLoS Pathog.* 5:e1000704. doi: 10.1371/journal.ppat.1000704

Price, C. T., Al-Quadan, T., Santic, M., Rosenshine, I., and Abu Kwaik, Y. (2011). Host proteasomal degradation generates amino acids essential for intracellular bacterial growth. *Science* 334, 1553–1557. doi: 10.1126/science.1212868

Rasis, M., and Segal, G. (2009). The LetA-RsmYZ-CsrA regulatory cascade, together with RpoS and PmrA, post-transcriptionally regulates stationary phase activation of *Legionella pneumophila* Icm/Dot effectors. *Mol. Microbiol.* 72, 995–1010. doi: 10.1111/j.1365-2958.2009.06705.x

Ristroph, J. D., Hedlund, K. W., and Gowda, S. (1981). Chemically defined medium for *Legionella pneumophila* growth. *J. Clin. Microbiol* 13, 115–119.

Robertson, P., Abdelhady, H., and Garduño, R. A. (2014). The many forms of a pleomorphic bacterial pathogen-the developmental network of *Legionella pneumophila*. *Front. Microbiol.* 5:670. doi: 10.3389/fmicb.2014.00670

Robinson, C. G., and Roy, C. R. (2006). Attachment and fusion of endoplasmic reticulum with vacuoles containing *Legionella pneumophila*. *Cell. Microbiol.* 8, 793–805. doi: 10.1111/j.1462-5822.2005.00666.x

Rolando, M., Escoll, P., Nora, T., Botti, J., Boitez, V., Bedia, C., et al. (2016). *Legionella pneumophila* S1P-lyase targets host sphingolipid metabolism and restrains autophagy. *Proc. Natl. Acad. Sci. U.S.A.* 113, 1901–1906. doi: 10.1073/pnas.1522067113

Rowbotham, T. J. (1986). Current views on the relationships between *amoebae, legionellae* and man. *Isr. J. Med. Sci.* 22, 678–689.

Sahr, T., Bruggemann, H., Jules, M., Lomma, M., Albert-Weissenberger, C., Cazalet, C., et al. (2009). Two small ncRNAs jointly govern virulence and transmission in *Legionella pneumophila*. *Mol. Microbiol.* 72, 741–762. doi: 10.1111/j.1365-2958.2009.06677.x

Sahr, T., Rusniok, C., Dervins-Ravault, D., Sismeiro, O., Coppee, J. Y., and Buchrieser, C. (2012). Deep sequencing defines the transcriptional map of *L. pneumophila* and identifies growth phase-dependent regulated ncRNAs implicated in virulence. *RNA Biol.* 9, 503–519. doi: 10.4161/rna.20270

Sahr, T., Rusniok, C., Impens, F., Oliva, G., Sismeiro, O., Coppee, J. Y., et al. (2017). The *Legionella pneumophila* genome evolved to accommodate multiple regulatory mechanisms controlled by the CsrA-system. *PLoS Genet.* 13:e1006629. doi: 10.1371/journal.pgen.1006629

Sauer, J. D., Bachman, M. A., and Swanson, M. S. (2005). The phagosomal transporter A couples threonine acquisition to differentiation and replication of *Legionella pneumophila* in macrophages. *Proc. Natl. Acad. Sci. U.S.A.* 102, 9924–9929. doi: 10.1073/pnas.0502767102

Schunder, E., Gillmaier, N., Kutzner, E., Eisenreich, W., Herrmann, V., Lautner, M., et al. (2014). Amino acid uptake and metabolism of *Legionella pneumophila* hosted by *Acanthamoeba castellanii*. *J. Biol. Chem.* 289, 21040–21054. doi: 10.1074/jbc.M114.570085

Steinert, M., Emody, L., Amann, R., and Hacker, J. (1997). Resuscitation of viable but nonculturable *Legionella pneumophila* Philadelphia JR32 by *Acanthamoeba castellanii*. *Appl. Environ. Microbiol.* 63, 2047–2053.

Steinert, M., Hentschel, U., and Hacker, J. (2002). *Legionella pneumophila*: an aquatic microbe goes astray. *FEMS Microbiol. Rev.* 26, 149–162. doi: 10.1111/j.1574-6976.2002.tb00607.x

Steinert, M., Heuner, K., Buchrieser, C., Albert-Weissenberger, C., and Glockner, G. (2007). *Legionella* pathogenicity: genome structure, regulatory networks and the host cell response. *Int. J. Med. Microbiol.* 297, 577–587. doi: 10.1016/j.ijmm.2007.03.009

Sturgill-Koszycki, S., and Swanson, M. S. (2000). *Legionella pneumophila* replication vacuoles mature into acidic, endocytic organelles. *J. Exp. Med.* 192, 1261–1272. doi: 10.1084/jem.192.9.1261

Tesh, M. J., Morse, S. A., and Miller, R. D. (1983). Intermediary metabolism in *Legionella pneumophila*: utilization of amino acids and other compounds as energy sources. *J. Bacteriol.* 154, 1104–1109.

Tiaden, A., Spirig, T., Weber, S. S., Bruggemann, H., Bosshard, R., Buchrieser, C., et al. (2007). The *Legionella pneumophila* response regulator LqsR promotes host cell interactions as an element of the virulence regulatory network controlled by RpoS and LetA. *Cell. Microbiol.* 9, 2903–2920. doi: 10.1111/j.1462-5822.2007.01005.x

Trigui, H., Dudyk, P., Oh, J., Hong, J. I., and Faucher, S. P. (2015). A regulatory feedback loop between RpoS and SpoT supports the survival of *Legionella pneumophila* in water. *Appl. Environ. Microbiol.* 81, 918–928. doi: 10.1128/AEM.03132-14

Trigui, H., Dudyk, P., Sum, J., Shuman, H. A., and Faucher, S. P. (2013). Analysis of the transcriptome of *Legionella pneumophila hfq* mutant reveals a new mobile genetic element. *Microbiology* 159(Pt 8), 1649–1660. doi: 10.1099/mic.0.067983-0

Weiss, E., Peacock, E. M., and Williams, J. C. (1980). Glucose and glutamate metabolism of *Legionella pneumophila*. *Curr. Microbiol.* 4, 1–6. doi: 10.1007/BF02602882

Wieland, H., Ullrich, S., Lang, F., and Neumeister, B. (2005). Intracellular multiplication of *Legionella pneumophila* depends on host cell amino acid transporter SLC1A5. *Mol. Microbiol.* 55, 1528–1537. doi: 10.1111/j.1365-2958.2005.04490.x

Xiao, H., Kalman, M., Ikehara, K., Zemel, S., Glaser, G., and Cashel, M. (1991). Residual guanosine 3',5'-bispyrophosphate synthetic activity of relA null mutants can be eliminated by spoT null mutations. *J. Biol. Chem.* 266, 5980–5990.

Yu, V. L., Plouffe, J. F., Pastoris, M. C., Stout, J. E., Schousboe, M., Widmer, A., et al. (2002). Distribution of *Legionella* species and serogroups isolated by culture in patients with sporadic community-acquired legionellosis: an international collaborative survey. *J. Infect. Dis.* 186, 127–128. doi: 10.1086/341087

Zhu, W., Banga, S., Tan, Y., Zheng, C., Stephenson, R., Gately, J., et al. (2011). Comprehensive identification of protein substrates of the Dot/Icm type iv transporter of *Legionella pneumophila*. *PLoS ONE* 6:e17638. doi: 10.1371/journal.pone.0017638

Zusman, T., Aloni, G., Halperin, E., Kotzer, H., Degtyar, E., Feldman, M., et al. (2007). The response regulator PmrA is a major regulator of the icm/dot type IV secretion system in *Legionella pneumophila* and *Coxiella burnetii*. *Mol. Microbiol.* 63, 1508–1523. doi: 10.1111/j.1365-2958.2007.05604.x

Zusman, T., Gal-Mor, O., and Segal, G. (2002). Characterization of a *Legionella pneumophila relA* insertion mutant and toles of RelA and RpoS in virulence gene expression. *J. Bacteriol.* 184, 67–75. doi: 10.1128/JB.184.1.67-75.2002

5

Legionella RavZ Plays a Role in Preventing Ubiquitin Recruitment to Bacteria-Containing Vacuoles

Tomoko Kubori[1,2], Xuan T. Bui[1], Andree Hubber[1] and Hiroki Nagai[1,2]**

[1] Department of Infectious Disease Control, Research Institute for Microbial Diseases, Osaka University, Suita, Japan,
[2] Department of Microbiology, Graduate School of Medicine, Gifu University, Gifu, Japan

Edited by:
Hayley J. Newton,
University of Melbourne, Australia

Reviewed by:
Hubert Hilbi,
University of Zurich, Switzerland
Thomas Naderer,
Monash University, Australia
Zhao-Qing Luo,
Purdue University, United States

***Correspondence:**
Tomoko Kubori
tkubori@gifu-u.ac.jp
Hiroki Nagai
hnagai@gifu-u.ac.jp

Bacterial pathogens like *Salmonella* and *Legionella* establish intracellular niches in host cells known as bacteria-containing vacuoles. In these vacuoles, bacteria can survive and replicate. Ubiquitin-dependent selective autophagy is a host defense mechanism to counteract infection by invading pathogens. The *Legionella* effector protein RavZ interferes with autophagy by irreversibly deconjugating LC3, an autophagy-related ubiquitin-like protein, from a phosphoglycolipid phosphatidylethanolamine. Using a co-infection system with *Salmonella*, we show here that *Legionella* RavZ interferes with ubiquitin recruitment to the *Salmonella*-containing vacuoles. The inhibitory activity is dependent on the same catalytic residue of RavZ that is involved in LC3 deconjugation. In semi-permeabilized cells infected with *Salmonella*, external addition of purified RavZ protein, but not of its catalytic mutant, induced removal of ubiquitin associated with *Salmonella*-containing vacuoles. The RavZ-mediated restriction of ubiquitin recruitment to *Salmonella*-containing vacuoles took place in the absence of the host system required for LC3 conjugation. These observations suggest the possibility that the targets of RavZ deconjugation activity include not only LC3, but also ubiquitin.

Keywords: *Legionella*, *Salmonella*, ubiquitin, vacuole, effector proteins, autophagy

INTRODUCTION

Legionella pneumophila is a Gram-negative bacterial pathogen that has a wide variety of eukaryotic hosts, ranging from amoebas to humans (Isberg et al., 2009). Host phagocytosis normally mediates entry of *L. pneumophila* into cells. Immediately after entry, *L. pneumophila* remodels the phagosome into a replicative niche, often referred to as a *Legionella*-containing vacuole (LCV). In order to establish the niche, *L. pneumophila* translocates ~300 effector proteins from bacteria into host cells via the specialized Dot/Icm type IV secretion system (T4SS), and by the coordinated function of these effector proteins it modulates a variety of host cellular processes (Ensminger and Isberg, 2009; Isberg et al., 2009; Hubber and Roy, 2010).

Similarly to other bacterial pathogens, *L. pneumophila* utilizes or modulates host ubiquitin pathways for its own benefit (Hubber et al., 2013; Zhou and Zhu, 2015). It was reported that *L. pneumophila* recruits ubiquitin to LCVs by a process depending on the Dot/Icm T4SS and bacterial protein synthesis (Dorer et al., 2006). The ubiquitin associated with LCVs is targeted for host proteasomal degradation, which was suggested to generate amino acids required for bacterial intracellular growth (Price et al., 2011). *L. pneumophila* possesses a large array of effector proteins involved in the modulation of the host ubiquitin system.

Currently, several effector proteins have been found to possess the F-box and U-box domains implicated in E3 ubiquitin ligase activity (Kubori et al., 2008; Hubber et al., 2013). A recent structural analysis revealed that the bacterial protein SidC represents a novel family of E3 ubiquitin ligases (Hsu et al., 2014) containing a unique PI(4)P-binding domain (Luo et al., 2015). SidC associates with LCVs through its high affinity PI(4)P-binding domain (Weber et al., 2006; Ragaz et al., 2008; Dolinsky et al., 2014) and has a direct or indirect role in ubiquitin recruitment to LCVs in early stages of infection (Horenkamp et al., 2014). *L. pneumophila* SidE family proteins exhibit not only deubiquitinase activity (Sheedlo et al., 2015) but also unusual ubiquitin ligase activity independent of the canonical E1 and E2 enzymes (Qiu et al., 2016; Kotewicz et al., 2017).

Autophagy is a conserved eukaryotic catabolic process for breakdown of intracellular components such as proteins or organelles. The target components are sequestered into a membrane-bound compartment called the autophagosome, which eventually fuses with a lysosome, a digestive compartment (Kuballa et al., 2012). Ubiquitin-like conjugation systems including Atg (Autophagy-related genes) proteins are crucially involved in autophagy (He and Klionsky, 2009). Microtubule-associated protein light chain 3 (LC3), an Atg8 homolog in mammalian cells, is a well-known marker of autophagy that localizes in autophagosome membranes (Kabeya et al., 2000). The conjugation of LC3 to phosphatidylethanolamine (PE) on the early autophagosome structure is a critical step of autophagy (Ichimura et al., 2000).

Selective autophagy-mediated clearance of invading microbes is becoming recognized as a potent host immune response (Huang and Brumell, 2014). *Salmonella enterica* serovar Typhimurium has been well-studied as a model intracellular bacterium subjected to autophagy (Narayanan and Edelmann, 2014). Recognition of S. Typhimurium by the host autophagy machinery is thought to be mediated mainly by two signals: ubiquitin, which is associated with cytosolic bacteria or bacteria-containing vacuoles (Birmingham et al., 2006), and a sugar β-galactoside on the luminal surface of bacteria-containing vacuoles that can be exposed by membrane rupture and is sensed by cytosolic receptor galectin-8 (Thurston et al., 2012). Recruitment of autophagic machinery to bacteria or bacterial vacuoles is mediated by LC3-binding adaptor proteins like p62/SQSTM1 and Nuclear dot protein 52 kDa (NDP52) (Thurston et al., 2009; Deretic, 2010; Cemma et al., 2011). In addition to an LC3 binding site, p62 carries a ubiquitin binding site (Zheng et al., 2009), and NDP52 has both ubiquitin and galectin-8 binding sites (Thurston et al., 2012). Another type of adaptor protein, Tecpr-1, which interacts with autophagy proteins Atg5 and WIPI2 (Atg18 homolog in mammalian cells), does not interact with ubiquitin or galectin-8 (Ogawa et al., 2011).

To counteract autophagy, bacteria have developed evasion strategies. The S. Typhimurium effector protein SseL deubiquitinates proteins on *Salmonella*-containing vacuoles (SCVs), preventing recognition by autophagic machinery (Mesquita et al., 2012). The *L. pneumophila* effector RavZ irreversibly deconjugates LC3 from PE (Choy et al., 2012), robustly inhibiting host autophagy. However, even in the absence of RavZ, *L. pneumophila* does not recruit LC3 on the vacuole (Hubber et al., 2017) and is highly resistant to clearance by selective autophagy, suggesting that *L. pneumophila* has alternative mechanisms for autophagy evasion (Choy et al., 2012; Rolando et al., 2016). The fact that *L. pneumophila* rarely recruits LC3 even in the absence of RavZ makes it difficult to identify alternative *L. pneumophila* factors inhibiting LC3 recruitment to LCVs. In an attempt to explore the possible alternative mechanisms, we used SCVs to examine potential roles of *L. pneumophila* effector proteins related to inhibition of selective autophagy. Using co-infection with *L. pneumophila* and S. Typhimurium, we unexpectedly found that RavZ plays a role in interfering with ubiquitin recruitment to the *Salmonella*-containing vacuoles.

MATERIALS AND METHODS

Bacterial Strains, Plasmids, and Culture

S. Typhimurium SR-11 x3181 carrying an mCherry-expressing plasmid (Hubber et al., 2014) was used for infection throughout the study. *Legionella* strains used in this study were derivatives of *L. pneumophila* strains Philadelphia-1 (Lp01) (Berger and Isberg, 1993). *Salmonella* was grown in Luria-Bertani (LB) medium containing 20 μg/ml of chloramphenicol to maintain the mCherry expressing plasmid. *Legionella* strains were grown in liquid N-(2-acetamido)-2-aminoethanesulfonic acid (ACES) buffered yeast extract (AYE) media (Horwitz and Silverstein, 1983). The *ravZ* and *dotA* deletion strains were constructed by allelic exchange as described previously (Zuckman et al., 1999). Eukaryotic expression plasmids for RavZ were constructed by cloning the PCR-amplified wild-type or mutant (C258A) *ravZ* into the p3xFLAG-CMV-10 expression vector (Sigma). Site-directed mutagenesis was conducted using the Quickchange II Site-Directed Mutagenesis Kit according to the manufacturer's recommendations (Agilent Technologies).

Cell Cultures and Transfection

HeLa-FcγRII and HEK293T-FcγRII cells were grown in Dulbecco's modified Eagle medium supplemented with 10% fetal calf serum. Bone marrow derived macrophages (BMDMs) were prepared from A/J mice (Japan SLC, Inc.). The cells were plated onto cover glass in 24-well tissue culture plates 1 day before infection or transfection. Transfection was conducted using Lipofectamine 2000 (Invitrogen) or Fugene HD (Promega) according to the manufacturers' recommendations.

Antibodies

Mouse antibody against mono- and polyubiquitinylated conjugates (FK2) was purchased from Enzo (BML-PW8810). Mouse antibody against ubiquitin (P4D1) was purchased from Cell Signaling (#3936). Mouse antibody against α-tubulin was purchased from Sigma (T6074). Rabbit anti-*Legionella* antibody was purchased from Biodesign (B65051G). Secondary antibodies were purchased; Peroxidase-conjugated goat anti-mouse antibody (62–6520, Invitrogen), Alexa Fluor 488-conjugated goat anti-mouse antibody (A-11029, Thermo Fisher Scientific) and Rhodamine RedX-conjugated goat anti-rabbit antibody (R6394, Thermo Fisher Scientific).

Bacterial Infections

Overnight *Salmonella* culture was diluted 1:30 with fresh LB medium containing 20 μg/ml of Chloramphenicol, grown to an OD_{600} value of about 2.0, and used for infection. Overnight liquid culture of *Legionella* was used for infection. The bacterial cultures were spun down by microfuge, the culture media were replaced with phosphate-buffered saline (PBS), and the suspension was immediately used for infection. HeLa cells were infected with *Salmonella* with multiplicity of infection (MOI) 300 for 1 h. After 20 min of infection, the cell culture medium was replaced with fresh medium containing 50 μg/ml of gentamycin to kill extracellular *Salmonella*. For the co-infection, a pre-mixed solution of *Salmonella* and *Legionella* was prepared and immediately used for infection with MOI 300 of both bacteria for 1 h. The time course study of *Legionella* infection in BMDMs was conducted with MOI 50. After 1 h of infection, the cells were washed three times with PBS to remove extracellular *Legionella* and the culture medium was replaced with fresh medium.

Immunostaining

Infected cells were fixed with 4% (w/v) of paraformaldehyde (PFA) after three washes with PBS and permeabilized with chilled methanol or with 0.1% (w/v) of Triton X-100 in PBS. After blocking with 2% (w/v) of goat serum in PBS for 20 min, the cells on the coverslips were reacted with a primary antibody for 1 h, followed by a secondary antibody for 30 min, at concentrations recommended by the manufacturers.

Protein Purification

BL21(DE3) carrying pET15b-*ravZ* (pNH1678) or pET15b-$ravZ_{C258A}$ (pNH1679) was grown to logarithmic phase at 37°C in 1 l of L-broth. After addition of IPTG to a final concentration of 0.4 mM to induce the production of His-RavZ, cells were cultured overnight at 16°C. *E. coli* cells were recovered by centrifugation and suspended in buffer A (20 mM Tris HCl pH 7.5, 5 mM EDTA) containing complete protease inhibitor cocktail (Roche). After treatment with 250 μg/ml lysozyme for 30 min, cells were disrupted by sonication. Lysate was centrifuged (20,000 g, 20 min) to remove insoluble materials and applied to Q-sepharose Fast Flow (~25 ml bed volume; GE Healthcase) equilibrated with buffer B (20 mM Tris HCl pH 7.5, 10 mM β-mercaptoethanol) in an open column format. After washing with buffer B containing 200 mM NaCl, the fraction containing His-RavZ was eluted with buffer B containing 300 mM NaCl. This fraction was further applied to a HisSelect column (1 ml bed volume; Sigma-Aldridge) equilibrated with buffer C (20 mM Tris HCl pH 7.5, 300 mM NaCl, 10 mM β-mercaptoethanol). After washing with buffer C containing 10 mM imidazole, fractions were eluted with a linear gradient of 10–100 mM imidazole in buffer C. Fractions containing His-RavZ were pooled and concentrated and subjected to a size exclusion column (Superdex 200 16/60 pg; GE healthcare). His-RavZ was eluted in 20 mM Tris HCl pH 7.5, 300 mM NaCl, 1 mM DTT.

Semi-permeabilized Cells Experiment

After 1 h of infection with *Salmonella*, HeLa cells were washed twice with semi-permeabilization buffer [125 mM K(OAc), 2.5 mM $Mg(OAc)_2$, 25 mM Hepes-KOH pH 7.4, 1 mg/ml Glucose, 1 mM DTT]. The cells were treated with 30 μg/ml digitonin in the semi-permeabilization buffer for 4 min at room temperature and washed another three times with the buffer. The permeabilized cells were treated with purified His-tagged wild-type or mutant RavZ proteins with gentle agitation for 1 h at room temperature. After two washes, the cells were fixed with 4% (w/v) PFA and subjected to immunostaining.

Statistical Analysis

Immunofluorescence experiments were conducted with at least 100 bacterial vacuoles counted per single experiment. Values were compared using paired Student's *t*-tests on three independent experiments.

Ethics Statement

All animal experiments were performed in accordance with the institutional guidelines and were approved by the Animal Care and Use Committee of the Research Institute for Microbial Diseases, Osaka University, Japan (Biken-AP-H26-10-0).

RESULTS

LC3 Recruitment to *Salmonella*-Containing Vacuoles Is Inhibited by Co-infecting *L. pneumophila*

As previously described, *Legionella* RavZ is sufficient to inhibit macroautophagy in mammalian cells, but it is not the only determinant preventing LC3 recruitment to LCVs (Choy et al., 2012). In an attempt to identify *L. pneumophila* protein factors responsible for the prevention of LC3 recruitment, we used an S. Typhimurium and *L. pneumophila* co-infection system. Selective autophagy targets a fraction (~25%) of *Salmonella*-containing vacuoles (SCVs) at 1 h post-infection, as evidenced by the LC3-positive SCVs observed in GFP-LC3 expressing HeLa cells challenged with *S.* Typhimurium (**Figure 1** SR-11). When HeLa cells were co-infected with wild-type *L. pneumophila*, LC3 recruitment to SCVs was reduced (**Figure 1** SR-11+Lp01). In contrast, in HeLa cells co-infected with *L. pneumophila* strains lacking functional Dot/Icm T4SS or RavZ, LC3 recruitment to SCVs was not affected (**Figure 1** SR-11+Lp01 *ΔdotA*/*ΔravZ*). These results illustrate that the co-infection experiment can be used to identify a *L. pneumophila* effector responsible for the restriction of LC3 recruitment to bacteria-containing vacuoles. These results also demonstrate that the restriction of LC3 recruitment to SCVs in the experimental condition can be explained solely by the function of the *L. pneumophila* effector RavZ, and any other effectors play only a limited role.

Legionella Co-infection Restricts p62, NDP52, and Ubiquitin Recruitment to SCVs

Adaptor proteins such as p62 and NDP52 carry both LC3- and ubiquitin-binding domains and are proposed to play a critical role in the recognition of invading microorganisms and

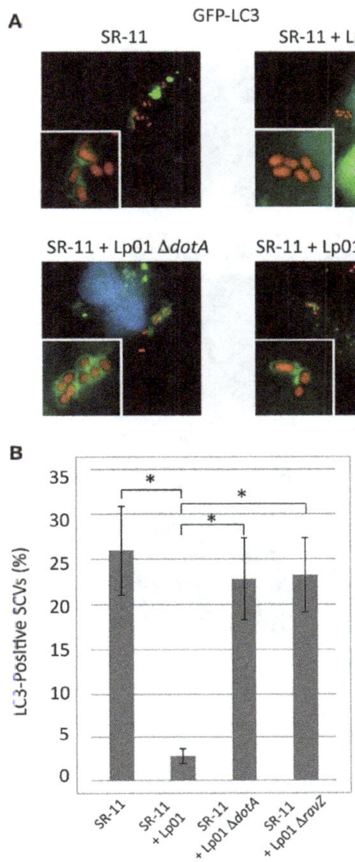

FIGURE 1 | *Legionella* interferes with LC3 recruitment to SCVs in a T4SS- and RavZ-dependent manner. HeLa cells expressing GFP-LC3 were co-infected with *Salmonella* expressing mCherry and *Legionella* for 1 h. **(A)** LC3 recruitment to SCVs was visualized by fluorescence microscopy. **(B)** Average percentage of LC3-positive SCVs in the cells was calculated from three independent examinations. Data present the average ± *SD*; $^*P < 0.02$.

the induction of selective autophagy. We thus examined the effect of co-infecting *L. pneumophila* on the recruitment of these adaptor proteins to SCVs. A fraction of SCVs acquired GFP-p62 or GFP-NDP52 in HeLa cells producing GFP-p62 or GFP-NDP52 respectively at 1 h post-infection (**Figure 2** SR-11). Intriguingly, the recruitment to SCVs of both GFP-p62 and GFP-NDP52 was severely restricted by co-infecting *L. pneumophila,* as in the case of LC3 (**Figures 2A,B,C** SR-11+Lp01). The restriction was dependent on the presence of functional Dot/Icm T4SS in the co-infecting *L. pneumophila* (**Figures 2A,B,C** SR-11+Lp01 *ΔdotA*). RavZ disruption restored the GFP-p62 and GFP-NDP52 recruitment to extents similar to or somewhat less than those observed in HeLa cells infected solely with *S.* Typhimurium (**Figures 2A,B,C** SR-11+*ΔravZ*). We also examined the recruitment to SCVs of Tecpr1-GFP, another autophagy adaptor protein that does not possess ubiquitin-binding activity. In contrast to GFP-p62 and GFP-NDP52, Tecpr1-GFP recruitment to SCVs was not affected by co-infecting *L. pneumophila* (**Figures 2A,D**). These results prompted us to examine ubiquitin recruitment to SCVs, an upstream event before adaptor recruitment, in the co-infection system. Using immunofluorescence microscopy with anti-conjugated-ubiquitin antibody (FK2), we found that the fraction of ubiquitin-positive SCVs was significantly reduced in the presence of co-infecting *L. pneumophila,* (**Figures 3A,B** SR-11+Lp01). As in the case of LC3, p62, and NDP52, the restriction required a functional Dot/Icm T4SS and the effector RavZ of co-infecting *L. pneumophila* (**Figures 3A,B** SR-11+Lp01 *ΔdotA*/*ΔravZ*).

RavZ Restricts Ubiquitin Recruitment to SCVs

The *L. pneumophila* effector RavZ appears to be necessary for the reduction of ubiquitin-positive SCVs in the co-infection condition. To test whether RavZ was sufficient for this reduction, we examined ubiquitin recruitment to SCVs in HeLa cells ectopically expressing RavZ (**Figure 4A**). The fraction of ubiquitin-positive SCVs in 3xFLAG-RavZ producing HeLa cells was lower (~5%) than that in vector-transfected cells (~30%), indicating that RavZ is sufficient for the restriction. RavZ deconjugates LC3, and the $\text{RavZ}_{\text{C258A}}$ mutation disrupts this enzymatic activity. Ectopic expression of 3xFLAG-$\text{RavZ}_{\text{C258A}}$ did not affect ubiquitin recruitment to SCVs, demonstrating that the same enzymatic activity is required for the restriction of ubiquitin recruitment to SCVs (**Figure 4A**).

To further characterize the role of RavZ in decreasing ubiquitin-positive SCVs, we used semi-permeabilized *Salmonella*-infected HeLa cells. The semi-permeabilized cells were washed to remove cytoplasmic proteins and compounds, and purified His-RavZ or His-$\text{RavZ}_{\text{C258A}}$ was added back at various concentrations. LC3 and ubiquitin levels on SCVs were determined by immunofluorescence staining using specific antibodies. Treatment with His-RavZ at a saturating concentration of 25 μg/ml (430 μM) decreased LC3 detectable on SCVs (**Figure 4B**, **Figure S1A**). In the same condition, the fraction of ubiquitin-positive SCVs was reduced two-fold (**Figure 4C**). The added His-RavZ significantly reduced ubiquitin and LC3 levels on SCVs at a concentration as low as 0.008 μg/ml (140 nM), and the reactions were readily saturated as His-RavZ concentration increased (**Figure S1**), indicating the RavZ activity toward ubiquitin had an effective concentration comparable to that of its activity toward LC3. There was a residual fraction of ubiquitin-positive SCVs even at the saturated concentration of RavZ, suggesting that a certain fraction or species of ubiquitin cannot be removed by the action of RavZ (**Figure S1B**). Collectively, these results demonstrated that the activity of RavZ is necessary and sufficient for the reduction of ubiquitin levels on SCVs, presumably by enzymatically removing ubiquitin from unknown substrates on *Salmonella* or SCVs.

RavZ Does Not Affect Ubiquitin Recruitment to LCVs

We examined whether RavZ inhibits ubiquitin recruitment to LCVs in *L. pneumophila*-infected cells. In HeLa cells

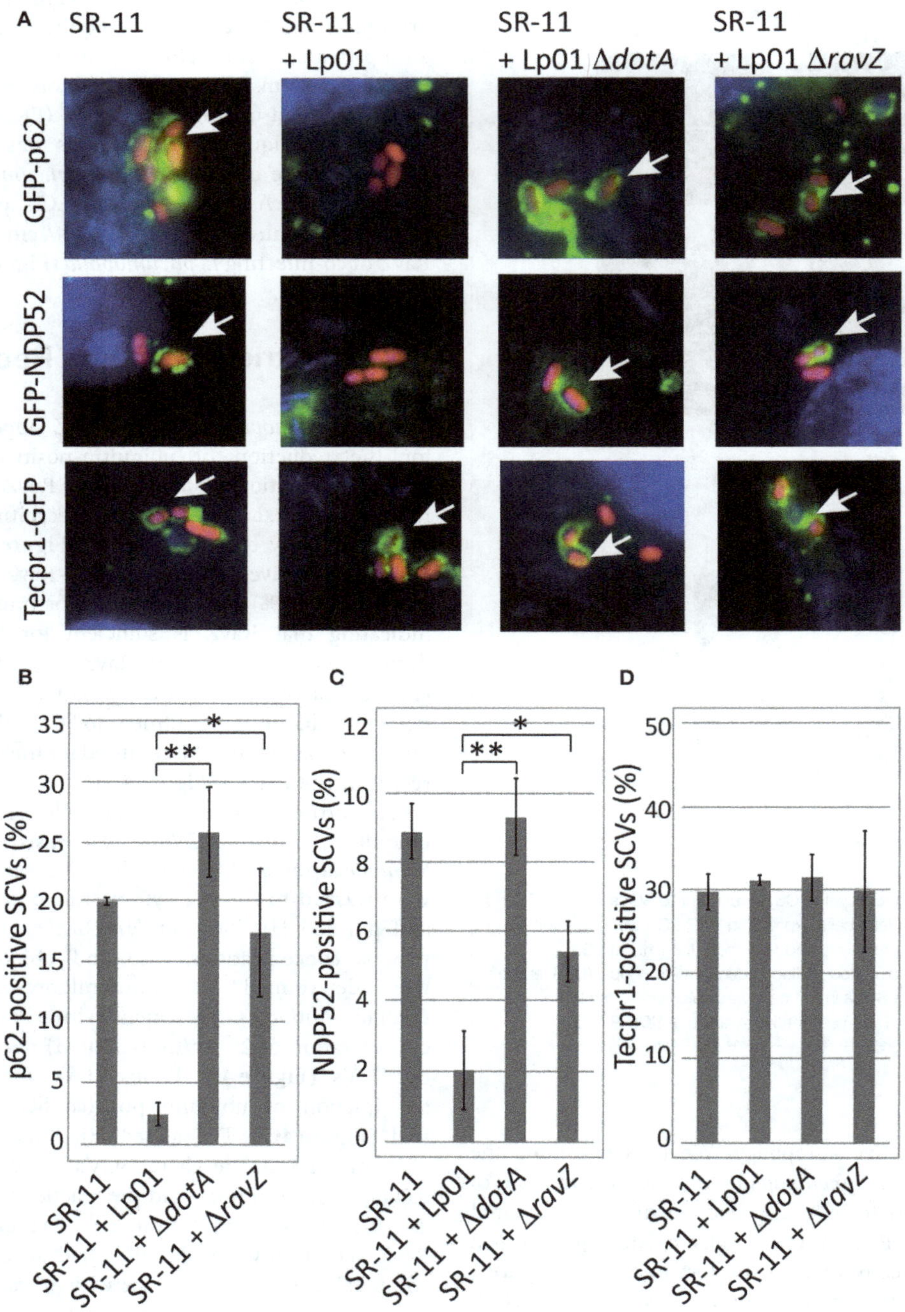

FIGURE 2 | *Legionella* interferes with the recruitment of ubiquitin-dependent adaptor proteins to SCVs in a T4SS- and RavZ-dependent manner. HeLa cells expressing GFP-p62, GFP-NDP52, or Tecpr1-GFP were co-infected with *Salmonella* expressing mCherry and *Legionella* for 1 h. **(A)** Recruitment of the adaptor proteins to SCVs was visualized by fluorescence microscopy. Adaptor-positive SCVs are indicated by arrows. Average percentage of p62 **(B)**, NDP52 **(C)**, or Tecpr1 **(D)**-positive SCVs in transfected cells was calculated from three independent examinations. Data present the average ± *SD*; $^{*}P < 0.02$, $^{**}P < 0.01$.

infected with a *L. pneumophila ΔravZ* strain, the frequency of ubiquitin-positive LCVs was not significantly affected by ectopically expressed 3xFLAG-RavZ or its catalytic mutant (**Figure 4D**). In mouse bone marrow-derived macrophages (BMDMs), infection with *L. pneumophila* resulted in the appearance of LCVs highly decorated with ubiquitin in a T4SS-dependent manner (**Figure 5**). However, up to 10 h post-infection, there was no significant difference in the fractions of ubiquitin-positive LCVs containing wild-type and *ΔravZ* strains. It has been shown that ubiquitin recruitment to LCVs

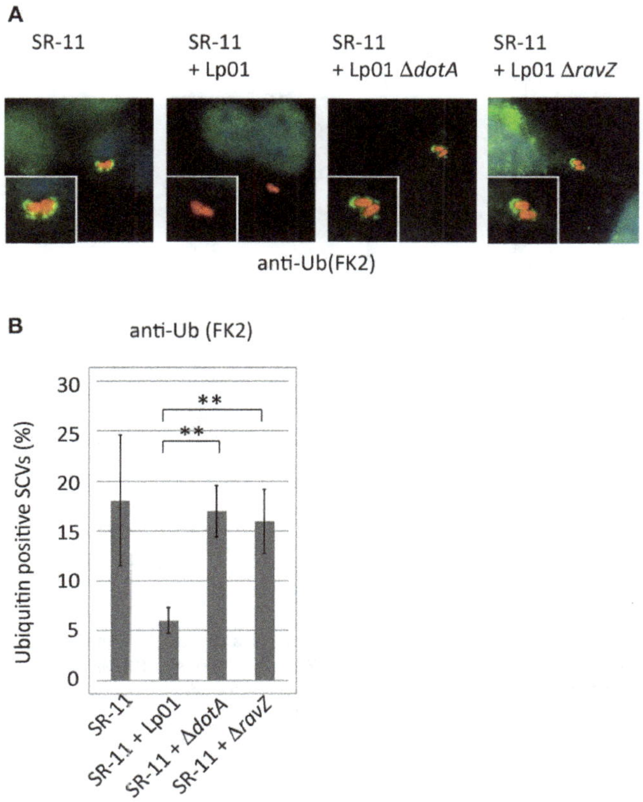

FIGURE 3 | *Legionella* interferes with ubiquitin recruitment to SCVs in a T4SS- and RavZ-dependent manner. HeLa cells were co-infected with *Salmonella* expressing mCherry and *Legionella* for 1 h. **(A)** Recruitment of ubiquitin to SCVs was visualized by immunofluorescence microscopy. Conjugated ubiquitin was stained with the FK2 antibody (green). **(B)** Average percentage of ubiquitin-positive SCVs in the cells calculated from three independent examinations. Data present the average ± *SD*; $^{**}P < 0.01$.

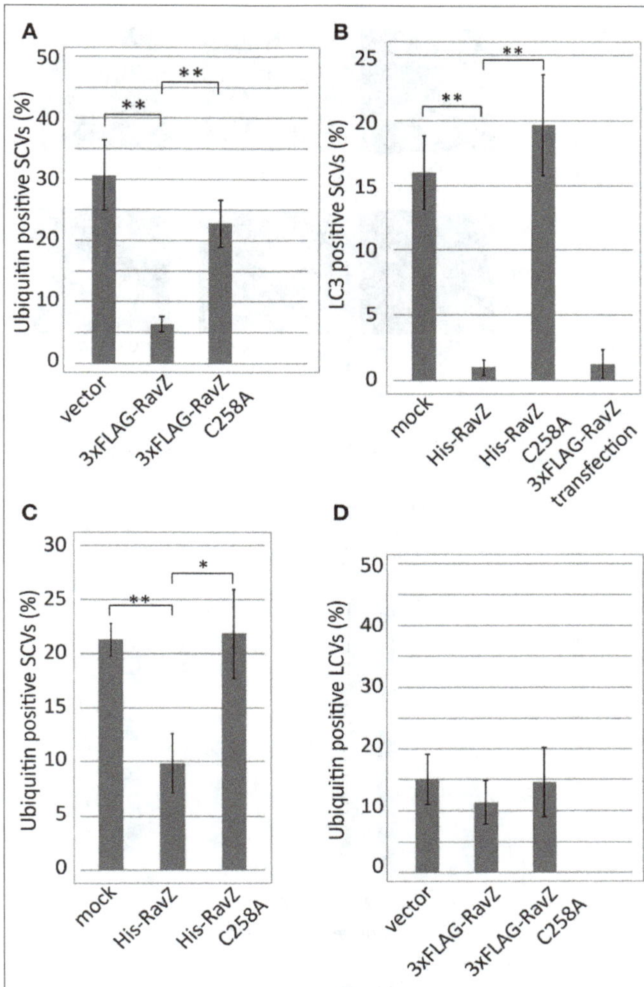

FIGURE 4 | *Legionella* RavZ appears to remove ubiquitin from SCVs in a manner dependent on the catalytic residue responsible for autophagy inhibition. **(A,D)** 3xFLAG-RavZ or its catalytic mutant was expressed in HeLa cells by transfection. The cells were infected with *Salmonella* expressing mCherry **(A)** or *Legionella* **(D)** for 1 h. **(B,C)** The purified His-tagged RavZ or its catalytic mutant protein was supplied into HeLa cells expressing GFP-LC3 **(B)** or HeLa cells **(C)** semi-permeabilized with Digitonin-containing buffer after 1 h of infection with *Salmonella* expressing mCherry. Conjugated ubiquitin was stained with the FK2 antibody. Average percentage of LC3 **(B)** or ubiquitin **(A,C,D)**-positive SCVs or LCVs calculated from three independent examinations. Data present the average ± *SD*; $^{*}P < 0.02$, $^{**}P < 0.01$.

requires bacterial protein synthesis (Dorer et al., 2006). To examine the possibility that the persistent ubiquitin recruitment to LCVs obscures RavZ-dependent ubiquitin reduction, we repeated this experiment in the presence of chloramphenicol, a bacterial translation inhibitor, in order to restrict de novo supply of ubiquitin to LCVs (**Figure S2**). The fraction of ubiquitin-positive LCVs was strongly reduced in the presence of chloramphenicol, while again there was no significant difference in the fractions of ubiquitin-positive LCVs containing wild-type and *ΔravZ* strains. These results do not support a potential contribution of RavZ to removing ubiquitin from LCVs.

The RavZ-Dependent Reduction of Ubiquitin-Positive SCVs Does Not Require a Functional LC3 Conjugation System

Autophagy adaptor proteins p62 and NDP52 can bind both to ubiquitin and to LC3. RavZ possesses the enzymatic activity to irreversibly deconjugate LC3 from PE. Therefore, one of the possible explanations for the reduction of ubiquitin-positive SCVs in the presence of functional RavZ is that RavZ-dependent deconjugation of LC3 leads to destabilization or dissociation of ubiquitinated complexes containing adaptor proteins in the vicinity of SCVs. Alternatively, the apparent decrease in ubiquitin could be attributed to loss of ubiquitinated LC3 from SCVs. These feedback scenarios predict that the reduction of LC3-positive SCVs is a prerequisite for the reduction of ubiquitin-positive SCVs. To test this possibility, we examined the ubiquitin levels on SCVs in Atg7 knockout (KO) MEF cells, in which LC3 conjugation to PE does not take place (Komatsu et al.,

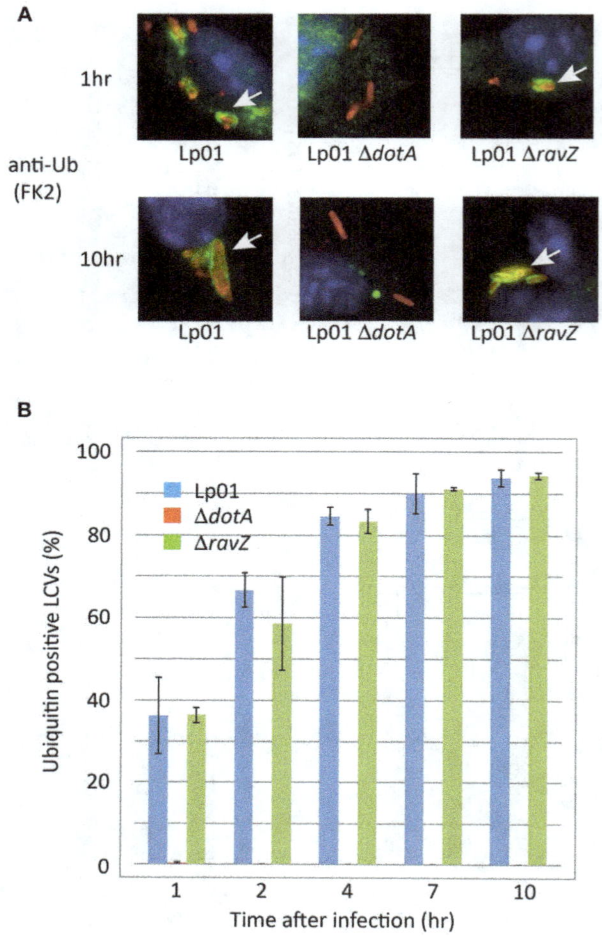

FIGURE 5 | RavZ appears not to affect ubiquitin recruitment to LCVs. Mouse bone marrow-derived macrophages were infected with *Legionella* for the indicated hours. **(A)** Recruitment of ubiquitin to LCVs was visualized by immunofluorescence microscopy. *Legionella* was stained with anti-*Legionella* antibody (red). Conjugated ubiquitin was stained with the FK2 antibody (green). Ubiquitin-positive LCVs are indicated by arrows. **(B)** Percentage of ubiquitin-positive LCVs. Data present the average ± *SD* calculated from three independent experiments.

2005) (**Figure 6**). We found that the RavZ-dependent reduction of ubiquitin-positive SCVs was observed both in wild-type and Atg7 KO MEF cells (**Figure 6**). This strongly suggests that the RavZ-dependent restriction of ubiquitin-positive SCVs does not result from the RavZ-mediated deconjugation of LC3 from PE.

DISCUSSION

Autophagy is carried out by a set of Atg proteins. Among the ~30 Atg proteins, the ubiquitin-like proteins Atg12 and Atg8 have important roles in the process (Geng and Klionsky, 2008). Similarly to the proteins involved in ubiquitin conjugation reactions, E1 and E2-like Atg proteins (Atg7 and Atg3/10 respectively) have been identified (Shintani et al., 1999; Tanida

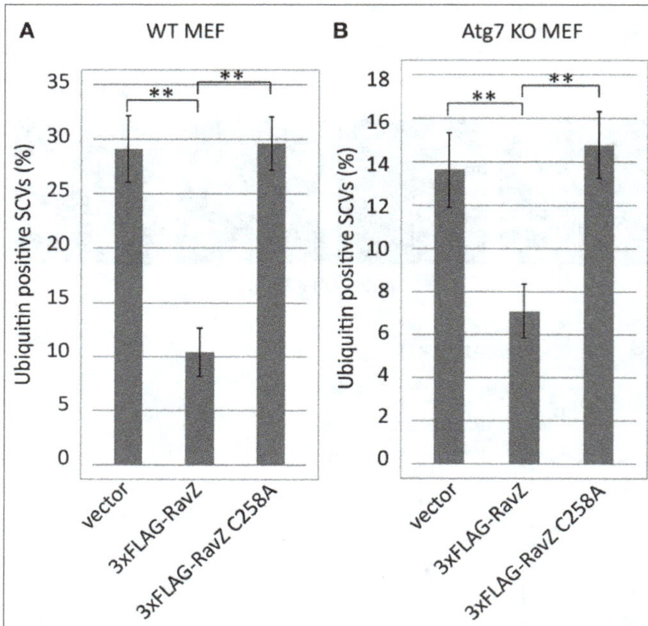

FIGURE 6 | Deficiency for LC3 maturation is not a cause of RavZ-mediated ubiquitin reduction on SCVs. Wild-type **(A)** or Atg7 knockout **(B)** MEF cells expressing RavZ or its catalytic mutant were infected with *Salmonella* expressing mCherry for 1 h. Conjugated ubiquitin was stained with the FK2 antibody. Percentage of ubiquitin-positive SCVs calculated in three independent experiments. Data present the average ± *SD*; $^{**}P < 0.01$.

et al., 1999; Ichimura et al., 2000; Geng and Klionsky, 2008). Atg8 is a yeast counterpart of LC3, and structural studies of Atg8 family proteins including LC3 (Kouno et al., 2005) revealed structural similarity to ubiquitin (Shpilka et al., 2011). Atg8-PE conjugation is a crucial step for autophagosome maturation and is a reversible by the deconjugation enzyme Atg4 (Nakatogawa et al., 2007). RavZ was found to possess Atg4-like activity to deconjugate Atg8 family proteins from PE. There were several critical differences in their enzymatic activities. RavZ cleaves a distinct amide bond of LC3-PE from that targeted by Atg4 (Choy et al., 2012). Therefore, LC3 deconjugation by RavZ is an irreversible process in which deconjugated LC3 is no longer available for reconjugation to PE. Furthermore, RavZ cleaves LC3 conjugated to PE but not to another protein such as YFP, whereas Atg4 can cleave both LC3-PE and LC3-YFP.

The experiment using semi-permeabilized cells demonstrated that pre-existing ubiquitin on SCVs can be targeted by RavZ and that the reduction in ubiquitin-positive SCVs is dependent on the active site of RavZ, which is known to be responsible for LC3 delipidation (**Figure 4**). Furthermore, the effect was observed in the absence of LC3 conjugation to SCVs (**Figure 6**). One possible explanation for these results is that the known amide-bond-cleavage activity of RavZ targets not only LC3-PE but also ubiquitin-conjugated molecules on SCVs. The recently solved crystal structure of RavZ suggests mechanisms for how RavZ preferentially targets the early autophagosomal membrane to cleave lipid-conjugated LC3 (Horenkamp et al., 2015). The structure contained a catalytic domain that showed structural similarity with Ulp family deubiquitinase-like enzymes and a C-terminal domain including a phosphatidylinositol 3-phosphate (PI3P) binding site. These and our results raise the possibility that RavZ has deubiquitinase-like enzymatic activity depending on binding to PI3P present on bacterial vacuoles. A recent report about chemical approaches using semisynthetic LC3 proteins with various modifications described that RavZ activity is strictly dependent on the lipid structure of the substrate (Yang et al., 2017). Together with the fact that RavZ recognizes the LC3 conjugated to a phosphoglycerolipid PE but not LC3-protein conjugates (Choy et al., 2012), it is tempting to speculate that RavZ may have activity toward ubiquitin conjugated to some phosphoglycerolipids containing an amine group, for example PE and phosphatidylserine (PS) in bacteria-containing vacuoles. Cellular polyubiquitin levels are not affected by ectopic expression of RavZ, indicating that RavZ is not likely to be a canonical deubiquitinase (**Figure S3**). Such a ubiquitin-lipid conjugate might play a role as a signal for selective autophagy. Although this is an intriguing hypothesis, the possibility still remains that RavZ might play an indirect role in ubiquitin removal from SCVs, in a manner dependent on the same active site for LC3 deconjugation but not on the removal of LC3. It was reported that a subset of cytosolic *Salmonella* associates with autophagy proteins (Yu et al., 2014). In this context, ubiquitin or LC3 located on not only bacterial vacuoles but also cytosolic bacteria might be targeted by RavZ activity. Future studies to clarify the molecular mechanisms of ubiquitin removal from bacterial-containing vacuoles mediated by RavZ may shed light on the recognition process of selective autophagy, an important host response for restricting invading microorganisms.

AUTHOR CONTRIBUTIONS

TK and HN designed research; TK, XB, AH, and HN performed research; TK, XB, and HN analyzed data; and TK and HN wrote the paper.

ACKNOWLEDGMENTS

We thank Michinaga Ogawa for providing GFP-LC3, GFP-p62, GFP-NDP52, and Tecpr1-GFP expression plasmids. Immortalized Atg7-deficient MEFs were kindly provided by Tatsuya Saitoh on behalf of Masaaki Komatsu. We thank Craig Roy, Hiroki Ando, and Tomoe Kitao for critical discussions, Akiko Okura and Masayo Yamada for technical assistance.

SUPPLEMENTARY MATERIAL

Figure S1 | Concentration-dependency of the effect of RavZ treatment of semi-permeabilized cells on the level of ubiquitin-positive SCVs. After 1 h of infection with Salmonella expressing mCherry, HeLa cells expressing GFP-LC3 were semi-permeabilized with Digitonin-containing buffer and treated with purified His-RavZ at the indicated concentrations. Conjugated ubiquitin was stained with the FK2 antibody. Percentage of LC3 **(A)** or ubiquitin **(B)**-positive SCVs calculated from three independent experiments. Data present the average ± SD.

Figure S2 | RavZ appears not to affect ubiquitin recruitment to LCVs even in the absence of de novo supply of ubiquitin on LCVs. Mouse bone marrow-derived macrophages were infected with Legionella for 7 hs in the absence (dark gray) or presence of 25 μg/ml of chloramphenicol (light gray). Conjugated ubiquitin was stained with the FK2 antibody. Average percentage of ubiquitin-positive LCVs was calculated from three independent examinations. Data present the average ± SD.

Figure S3 | Cellular polyubiquitin levels are not affected by RavZ. HEK293-FcγRII cells were transfected with p3xFLAG-CMV-10 expression vector or derived 3xFLAG-RavZ or RavZ C258A expressing plasmids for 24 h. Cell lysates were applied on 10–20% gradient polyacrylamide gel electrophoresis followed by immunoblotting using anti-ubiquitin antibody (P4D1) for detection of mono and polyubiquitins and anti-α-tubulin for internal loading control.

REFERENCES

Berger, K. H., and Isberg, R. R. (1993). Two distinct defects in intracellular growth complemented by a single genetic locus in *Legionella pneumophila*. *Mol. Microbiol.* 7, 7–19. doi: 10.1111/j.1365-2958.1993.tb01092.x

Birmingham, C. L., Smith, A. C., Bakowski, M. A., Yoshimori, T., and Brumell, J. H. (2006). Autophagy controls Salmonella infection in response to damage to the Salmonella-containing vacuole. *J. Biol. Chem.* 281, 11374–11383. doi: 10.1074/jbc.M509157200

Cemma, M., Kim, P. K., and Brumell, J. H. (2011). The ubiquitin-binding adaptor proteins p62/SQSTM1 and NDP52 are recruited independently to bacteria-associated microdomains to target Salmonella to the autophagy pathway. *Autophagy* 7, 341–345. doi: 10.4161/auto.7.3.14046

Choy, A., Dancourt, J., Mugo, B., O'Connor, T. J., Isberg, R. R., Melia, T. J., et al. (2012). The Legionella effector RavZ inhibits host autophagy through irreversible Atg8 deconjugation. *Science* 338, 1072–1076. doi: 10.1126/science.1227026

Deretic, V. (2010). Autophagy in infection. *Curr. Opin. Cell Biol.* 22, 252–262. doi: 10.1016/j.ceb.2009.12.009

Dolinsky, S., Haneburger, I., Cichy, A., Hannemann, M., Itzen, A., and Hilbi, H. (2014). The *Legionella longbeachae* Icm/Dot substrate SidC selectively binds phosphatidylinositol 4-phosphate with nanomolar affinity and promotes pathogen vacuole-endoplasmic reticulum interactions. *Infect. Immun.* 82, 4021–4033. doi: 10.1128/IAI.01685-14

Dorer, M. S., Kirton, D., Bader, J. S., and Isberg, R. R. (2006). RNA interference analysis of *Legionella* in Drosophila cells: exploitation of early secretory apparatus dynamics. *PLoS Pathog.* 2:e34. doi: 10.1371/journal.ppat.0020034

Ensminger, A. W., and Isberg, R. R. (2009). *Legionella pneumophila* Dot/Icm translocated substrates: a sum of parts. *Curr. Opin. Microbiol.* 12, 67–73. doi: 10.1016/j.mib.2008.12.004

Geng, J., and Klionsky, D. J. (2008). The Atg8 and Atg12 ubiquitin-like conjugation systems in macroautophagy. 'Protein modifications: beyond the usual suspects' review series. *EMBO Rep.* 9, 859–864. doi: 10.1038/embor.2008.163

He, C., and Klionsky, D. J. (2009). Regulation mechanisms and signaling pathways of autophagy. *Annu. Rev. Genet.* 43, 67–93. doi: 10.1146/annurev-genet-102808-114910

Horenkamp, F. A., Kauffman, K. J., Kohler, L. J., Sherwood, R. K., Krueger, K. P., Shteyn, V., et al. (2015). The Legionella anti-autophagy effector RavZ targets the autophagosome via PI3P- and curvature-sensing motifs. *Dev. Cell* 34, 569–576. doi: 10.1016/j.devcel.2015.08.010

Horenkamp, F. A., Mukherjee, S., Alix, E., Schauder, C. M., Hubber, A. M., Roy, C. R., et al. (2014). *Legionella pneumophila* subversion of host vesicular transport by SidC effector proteins. *Traffic* 15, 488–499. doi: 10.1111/tra.12158

Horwitz, M. A., and Silverstein, S. C. (1983). Intracellular multiplication of Legionnaires' disease bacteria (*Legionella pneumophila*) in human monocytes is reversibly inhibited by erythromycin and rifampin. *J. Clin. Invest.* 71, 15–26. doi: 10.1172/JCI110744

Hsu, F., Luo, X., Qiu, J., Teng, Y. B., Jin, J., Smolka, M. B., et al. (2014). The Legionella effector SidC defines a unique family of ubiquitin ligases important for bacterial phagosomal remodeling. *Proc. Natl. Acad. Sci. U.S.A.* 111, 10538–10543. doi: 10.1073/pnas.1402605111

Huang, J., and Brumell, J. H. (2014). Bacteria-autophagy interplay: a battle for survival. *Nat. Rev. Microbiol.* 12, 101–114. doi: 10.1038/nrmicro3160

Hubber, A., Arasaki, K., Nakatsu, F., Hardiman, C., Lambright, D., De Camilli, P., et al. (2014). The machinery at endoplasmic reticulum-plasma membrane contact sites contributes to spatial regulation of multiple Legionella effector proteins. *PLoS Pathog.* 10:e1004222. doi: 10.1371/journal.ppat.1004222

Hubber, A., Kubori, T., Coban, C., Matsuzawa, T., Ogawa, M., Kawabata, T., et al. (2017). Bacterial secretion system skews the fate of Legionella-containing vacuoles towards LC3-associated phagocytosis. *Sci. Rep.* 7:44795. doi: 10.1038/srep44795

Hubber, A., Kubori, T., and Nagai, H. (2013). Modulation of the ubiquitination machinery by Legionella. *Curr. Top. Microbiol. Immunol.* 376, 227–247. doi: 10.1007/82_2013_343

Hubber, A., and Roy, C. R. (2010). Modulation of host cell function by *Legionella pneumophila* type IV effectors. *Annu. Rev. Cell Dev. Biol.* 26, 261–283. doi: 10.1146/annurev-cellbio-100109-104034

Ichimura, Y., Kirisako, T., Takao, T., Satomi, Y., Shimonishi, Y., Ishihara, N., et al. (2000). A ubiquitin-like system mediates protein lipidation. *Nature* 408, 488–492. doi: 10.1038/35044114

Isberg, R. R., O'Connor, T. J., and Heidtman, M. (2009). The *Legionella pneumophila* replication vacuole: making a cosy niche inside host cells. *Nat. Rev. Microbiol.* 7, 13–24. doi: 10.1038/nrmicro1967

Kabeya, Y., Mizushima, N., Ueno, T., Yamamoto, A., Kirisako, T., Noda, T., et al. (2000). LC3, a mammalian homologue of yeast Apg8p, is localized in autophagosome membranes after processing. *EMBO J.* 19, 5720–5728. doi: 10.1093/emboj/19.21.5720

Komatsu, M., Waguri, S., Ueno, T., Iwata, J., Murata, S., Tanida, I., et al. (2005). Impairment of starvation-induced and constitutive autophagy in Atg7-deficient mice. *J. Cell Biol.* 169, 425–434. doi: 10.1083/jcb.200412022

Kotewicz, K. M., Ramabhadran, V., Sjoblom, N., Vogel, J. P., Haenssler, E., Zhang, M., et al. (2017). A single legionella effector catalyzes a multistep ubiquitination pathway to rearrange tubular endoplasmic reticulum for replication. *Cell Host Microbe* 21, 169–181. doi: 10.1016/j.chom.2016.12.007

Kouno, T., Mizuguchi, M., Tanida, I., Ueno, T., Kanematsu, T., Mori, Y., et al. (2005). Solution structure of microtubule-associated protein light chain 3 and identification of its functional subdomains. *J. Biol. Chem.* 280, 24610–24617. doi: 10.1074/jbc.M413565200

Kuballa, P., Nolte, W. M., Castoreno, A. B., and Xavier, R. J. (2012). Autophagy and the immune system. *Annu. Rev. Immunol.* 30, 611–646. doi: 10.1146/annurev-immunol-020711-074948

Kubori, T., Hyakutake, A., and Nagai, H. (2008). Legionella translocates an E3 ubiquitin ligase that has multiple U-boxes with distinct functions. *Mol. Microbiol.* 67, 1307–1319. doi: 10.1111/j.1365-2958.2008.06124.x

Luo, X., Wasilko, D. J., Liu, Y., Sun, J., Wu, X., Luo, Z. Q., et al. (2015). Structure of the legionella virulence factor, sidc reveals a unique PI(4)P-specific binding domain essential for its targeting to the bacterial phagosome. *PLoS Pathog.* 11:e1004965. doi: 10.1145/2818302

Mesquita, F. S., Thomas, M., Sachse, M., Santos, A. J., Figueira, R., and Holden, D. W. (2012). The Salmonella deubiquitinase SseL inhibits selective autophagy of cytosolic aggregates. *PLoS Pathog.* 8:e1002743. doi: 10.1371/journal.ppat.1002743

Nakatogawa, H., Ichimura, Y., and Ohsumi, Y. (2007). Atg8, a ubiquitin-like protein required for autophagosome formation, mediates membrane tethering and hemifusion. *Cell* 130, 165–178. doi: 10.1016/j.cell.2007.05.021

Narayanan, L. A., and Edelmann, M. J. (2014). Ubiquitination as an efficient molecular strategy employed in Salmonella infection. *Front. Immunol.* 5:558. doi: 10.3389/fimmu.2014.00558

Ogawa, M., Yoshikawa, Y., Kobayashi, T., Mimuro, H., Fukumatsu, M., Kiga, K., et al. (2011). A Tecpr1-dependent selective autophagy pathway targets bacterial pathogens. *Cell Host Microbe* 9, 376–389. doi: 10.1016/j.chom.2011.04.010

Price, C. T., Al-Quadan, T., Santic, M., Rosenshine, I., and Abu Kwaik, Y. (2011). Host proteasomal degradation generates amino acids essential for intracellular bacterial growth. *Science* 334, 1553–1557. doi: 10.1126/science.1212868

Qiu, J., Sheedlo, M. J., Yu, K., Tan, Y., Nakayasu, E. S., Das, C., et al. (2016). Ubiquitination independent of E1 and E2 enzymes by bacterial effectors. *Nature* 533, 120–124. doi: 10.1038/nature17657

Ragaz, C., Pietsch, H., Urwyler, S., Tiaden, A., Weber, S. S., and Hilbi, H. (2008). The *Legionella pneumophila* phosphatidylinositol-4 phosphate-binding type IV substrate SidC recruits endoplasmic reticulum vesicles to a replication-permissive vacuole. *Cell. Microbiol.* 10, 2416–2433. doi: 10.1111/j.1462-5822.2008.01219.x

Rolando, M., Escoll, P., Nora, T., Botti, J., Boitez, V., Bedia, C., et al. (2016). *Legionella pneumophila* S1P-lyase targets host sphingolipid metabolism and restrains autophagy. *Proc. Natl. Acad. Sci. U.S.A.* 113, 1901–1906. doi: 10.1073/pnas.1522067113

Sheedlo, M. J., Qiu, J., Tan, Y., Paul, L. N., Luo, Z. Q., and Das, C. (2015). Structural basis of substrate recognition by a bacterial deubiquitinase important for dynamics of phagosome ubiquitination. *Proc. Natl. Acad. Sci. U.S.A.* 112, 15090–15095. doi: 10.1073/pnas.1514568112

Shintani, T., Mizushima, N., Ogawa, Y., Matsuura, A., Noda, T., and Ohsumi, Y. (1999). Apg10p, a novel protein-conjugating enzyme essential for autophagy in yeast. *EMBO J.* 18, 5234–5241. doi: 10.1093/emboj/18.19.5234

Shpilka, T., Weidberg, H., Pietrokovski, S., and Elazar, Z. (2011). Atg8: an autophagy-related ubiquitin-like protein family. *Gen. Biol.* 12, 226. doi: 10.1186/gb-2011-12-7-226

Tanida, I., Mizushima, N., Kiyooka, M., Ohsumi, M., Ueno, T., Ohsumi, Y., et al. (1999). Apg7p/Cvt2p: a novel protein-activating enzyme essential for autophagy. *Mol. Biol. Cell* 10, 1367–1379. doi: 10.1091/mbc.10.5.1367

Thurston, T. L., Ryzhakov, G., Bloor, S., von Muhlinen, N., and Randow, F. (2009). The TBK1 adaptor and autophagy receptor NDP52 restricts the proliferation of ubiquitin-coated bacteria. *Nat. Immunol.* 10, 1215–1221. doi: 10.1038/ni.1800

Thurston, T. L., Wandel, M. P., von Muhlinen, N., Foeglein, A., and Randow, F. (2012). Galectin 8 targets damaged vesicles for autophagy to defend cells against bacterial invasion. *Nature* 482, 414–418. doi: 10.1038/nature10744

Weber, S. S., Ragaz, C., Reus, K., Nyfeler, Y., and Hilbi, H. (2006). Legionella pneumophila exploits PI(4)P to anchor secreted effector proteins to the replicative vacuole. *PLoS Pathog.* 2:e46. doi: 10.1371/journal.ppat.0020046

Yang, A., Pantoom, S., and Wu, Y. W. (2017). Elucidation of the anti-autophagy mechanism of the Legionella effector RavZ using semisynthetic LC3 proteins. *Elife* 6:e23905. doi: 10.7554/eLife.23905

Yu, H. B., Croxen, M. A., Marchiando, A. M., Ferreira, R. B., Cadwell, K., Foster, L. J., et al. (2014). Autophagy facilitates Salmonella replication in HeLa cells. *mBio* 5, e00865–e00814. doi: 10.1128/mBio.00865-14

Zheng, Y. T., Shahnazari, S., Brech, A., Lamark, T., Johansen, T., and Brumell, J. H. (2009). The adaptor protein p62/SQSTM1 targets invading bacteria to the autophagy pathway. *J. Immunol.* 183, 5909–5916. doi: 10.4049/jimmunol.0900441

Zhou, Y., and Zhu, Y. (2015). Diversity of bacterial manipulation of the host ubiquitin pathways. *Cell. Microbiol.* 17, 26–34. doi: 10.1111/cmi.12384

Zuckman, D. M., Hung, J. B., and Roy, C. R. (1999). Pore-forming activity is not sufficient for *Legionella pneumophila* phagosome trafficking and intracellular growth. *Mol. Microbiol.* 32, 990–1001. doi: 10.1046/j.1365-2958.1999.01410.x

6

The Flagellar Regulon of *Legionella*

Sandra Appelt[1] and Klaus Heuner[2*]

[1] Highly Pathogenic Microorganisms, Centre for Biological Threats and Special Pathogens, Robert Koch Institute, Berlin, Germany, [2] Cellular Interactions of Bacterial Pathogens, Centre for Biological Threats and Special Pathogens, Robert Koch Institute, Berlin, Germany

Edited by:
Matthias P. Machner, National Institutes of Health (NIH), United States

Reviewed by:
Ombeline Rossier, Université Paris-Sud, France
Zachary David Dalebroux, University of Oklahoma Health Sciences Center, United States

***Correspondence:**
Klaus Heuner
heunerk@rki.de

The *Legionella* genus comprises more than 60 species. In particular, *Legionella pneumophila* is known to cause severe illnesses in humans. Legionellaceae are ubiquitous inhabitants of aquatic environments. Some Legionellaceae are motile and their motility is important to move around in habitats. Motility can be considered as a potential virulence factor as already shown for various human pathogens. The genes of the flagellar system, regulator and structural genes, are structured in hierarchical levels described as the flagellar regulon. Their expression is modulated by various environmental factors. For *L. pneumophila* it was shown that the expression of genes of the flagellar regulon is modulated by the actual growth phase and temperature. Especially, flagellated *Legionella* are known to express genes during the transmissive phase of growth that are involved in the expression of virulence traits. It has been demonstrated that the alternative sigma-28 factor is part of the link between virulence expression and motility. In the following review, the structure of the flagellar regulon of *L. pneumophila* is discussed and compared to other flagellar systems of different *Legionella* species. Recently, it has been described that *Legionella micdadei* and *Legionella fallonii* contain a second putative partial flagellar system. Hence, the report will focus on flagellated and non-flagellated *Legionella* strains, phylogenetic relationships, the role and function of the alternative sigma factor (FliA) and its anti-sigma-28 factor (FlgM).

Keywords: *Legionella*, flagellar regulon, flagellin, virulence, FliA, FleQ, alternative sigma factor

INTRODUCTION AND OVERVIEW

The Legionellaceae family consists of a single genus: *Legionella* that comprises more than 60 species so far (Gomez-Valero et al., 2009; Bajrai et al., 2016; Khodr et al., 2016). New species are identified continuously (i.e., *Legionella drancourtii, Legionella gresilensis*, and *Legionella beliardensis*), extending the list of known *Legionella* species (Lo Presti et al., 2001; La Scola et al., 2004; Gomez-Valero et al., 2009; Rizzardi et al., 2015; Bajrai et al., 2016; Khodr et al., 2016). More than 20 pathogenic *Legionella* species are known today that differ in their ability to infect hosts and to cause severe to mild diseases in humans (Rizzardi et al., 2015). Human pathogens known to cause the Legionnaires' disease—an atypical pneumonia—are for instance *Legionella pneumophila*, *Legionella micdadei*, and *Legionella longbeachae* (Yu et al., 2002; Whiley and Bentham, 2011). *Legionella* known to cause the Pontiac fever—a mild flu-like disease—are for instance *Legionella feelei, L. micdadei* and *Legionella anisa*, but also *L. pneumophila*. Often *Legionella* strains of the same species and same serogroup cause one of the mentioned diseases (Swanson and Hammer, 2000; Fields et al., 2002; Wang et al., 2015).

Nevertheless, it was assumed that humans are accidental hosts of *Legionella* species within which the bacterium replicates (Horwitz and Silverstein, 1980; Cianciotto et al., 1989; Horwitz, 1992; Fields, 1996; Neumeister et al., 1997; Newton et al., 2010). Known natural hosts are protozoa, especially free-living amoebae: *Acanthamoeba* spp., *Naegleria* spp., or *Hartmanella vermiformis* (Barbaree et al., 1986; Rowbotham, 1986; Fields, 1996; Atlas, 1999; Fields et al., 2002; Greub and Raoult, 2004; Abdel-Nour et al., 2013; Richards et al., 2013; Cateau et al., 2014). Accordingly, in general, *Legionella* species are prevalent inhabitants of soil, mud and above all of aquatic environments (Fliermans et al., 1981; Fields, 1996; Atlas, 1999; Gomez-Valero et al., 2009; Declerck, 2010; Schalk et al., 2014; Currie and Beattie, 2015). The ability of *L. pneumophila* to grow within biofilms made by *Klebsiella pneumoniae* or *Pseudomonas aeruginosa* in aquatic or wet environments raised questions about their host-free persistence (Stewart et al., 2012). In connection with favorable aquatic habitats and potential protozoa hosts, especially flagella-driven motility of some *Legionella* spp. is an important feature needed to move around, to find new hosts and to form maybe even biofilms (Kirov et al., 2004; Danhorn and Fuqua, 2007; Heuner and Albert-Weissenberger, 2008). For *Legionella* infecting humans, motility may also be crucial for spreading within lungs of patients, as flagellated forms of *L. pneumophila* were detected in alveolar spaces (Chandler et al., 1980; Jager et al., 2014). Recently, it was published that *L. feelei* strains that cause Legionnaires' disease are flagellated while *L. feelei* strains that cause the Pontiac fever are non-flagellated (Wang et al., 2015). The majority of *Legionella* species are flagellated (Elliott and Johnson, 1981; Bornstein et al., 1991; Bangsborg et al., 1995; Heuner et al., 1995), but not all pathogenic *Legionella* have a complete flagellar regulon (i.e., *L. longbeachae* and *Legionella oakridgensis*, see below) (Orrison et al., 1983; Heuner et al., 1995; Cazalet et al., 2010; Kozak et al., 2010; Brzuszkiewicz et al., 2013).

The assumption that the expression of flagella and virulence are linked was already made at an early stage (Rowbotham, 1986) and later on confirmed. It was shown that there is a regulatory link between the expression of a virulent phenotype and the flagellum (Pruckler et al., 1995; Byrne and Swanson, 1998; Hammer et al., 2002; Gal-Mor and Segal, 2003; Molofsky et al., 2005; Heuner and Albert-Weissenberger, 2008; Albert-Weissenberger et al., 2010; Schulz et al., 2012). The expression of flagellar genes is regulated on the flagellar regulon, extensively investigated in *L. pneumophila* due to its biphasic intracellular life cycle during which the bacterium undergoes a shape change. Within the host, after replication inside of *Legionella*-containing vacuoles (LCVs), when nutrients become limited, *L. pneumophila* differentiates into a flagellated, non-replicating form. The flagellated, transmissible, mature form is stress-resistant, virulent and metabolically resting as well as infectious (abbr. MIF) (Rowbotham, 1986; Byrne and Swanson, 1998; Heuner et al., 1999; Swanson and Hammer, 2000; Faulkner and Garduno, 2002; Garduno et al., 2002; Hammer et al., 2002; Molofsky and Swanson, 2004; Fonseca and Swanson, 2014; Eisenreich and Heuner, 2016). The actual release process of mature forms is still under discussion: either the bacteria are released from the LCV into the environment by lysis of the host or the bacteria are released first into the cytosol of the host and then after an additional putative round of replication into the environment (Rowbotham, 1986; Molmeret et al., 2004). The latter hypothesis implies that the flagellum is produced inside the cytosol of the host and not in LCVs as proposed earlier. Furthermore, there is also a possibility that the bacteria are released by the host *via* a non-lytic mechanism (Chen et al., 2004; Bouyer et al., 2007; Berk et al., 2008). However, the released form is well-prepared to reinfect new hosts or to differentiate into a viable-but-nonculturable form (VBNC) meant to enable a long-term survival of the bacteria (Rowbotham, 1986; Steinert et al., 1997; Ohno et al., 2003; Molmeret et al., 2010; Al-Bana et al., 2014). VBNC forms can be resuscitated when they are taken up by amoebae (Steinert et al., 1997; Ohno et al., 2003; Al-Bana et al., 2014). Further, different morphological forms of *L. pneumophila* have been recently discussed (Robertson et al., 2014). Next to the biphasic intracellular life cycle it was shown that *L. pneumophila* exhibits also a life stage-specific bipartite metabolism, an area for further investigations (Schunder et al., 2014; Eisenreich and Heuner, 2016; Gillmaier et al., 2016; Hauslein et al., 2016).

MOTILITY OF *LEGIONELLA*

Different forms of bacterial motility are known including swarming, twitching and sliding. Notably the flagellum—next to pili—allows bacteria to move. Bacterial motility is often related to chemotactic behavior that enables a bacterium to locate special environmental conditions and to get closer to higher concentrations of attractants (Szurmant and Ordal, 2004; Hazelbauer et al., 2008; Micali and Endres, 2016). Some *Legionella* have a chemotaxis system (*L. longbeachae, Legionella parisiensis,* and *Legionella bozemanii*) but most *Legionella* do not have the corresponding genes (e.g., *L. pneumophila*, *L. micdadei*, and *L. oakridgensis*). Moreover, the ability of *Legionella* to swarm and to show a chemotaxis behavior has not yet been reported.

Twitching motility is based on a functional type IV pilus. The ability to move forward by twitching has been reported for *L. pneumophila* (Coil and Anne, 2009; Hoppe et al., 2017). In addition, sliding motility, a surfactant-mediated motility, has been described for *L. pneumophila* (Stewart et al., 2009).

THE FLAGELLUM AND THE FLAGELLAR REGULON

The Structure of the Flagellum and Flagellar Systems

Most *Legionella* species are motile due to a single polar flagellum (**Figure 1**) (Chandler et al., 1980; Elliott and Johnson, 1982; Heuner et al., 1995). More than 50 genes are involved in the expression of functional flagella, and due to high metabolic costs, a tight regulation is essential (Chilcott and Hughes, 2000; McCarter, 2006; Osterman et al., 2015). The flagellum of *Legionella* consists of a basal body, a hook structure and a filament (**Figure 2**; Heuner and Steinert, 2003; Heuner and Albert-Weissenberger, 2008). For the assembly of the flagellum, needed proteins (hook, rod and the filament forming proteins) are exported out of the cell by a flagellum-specific export apparatus, a type III-like secretion system (T3SS) (Heuner and Albert-Weissenberger, 2008; Altegoer and Bange, 2015).

The basal body consists of a rod, three rings ["membrane/supramembrane" (MS), "peptidoglycan" (P), and "lipopolysaccharide" (L)] and a motor switch complex (MotAB). The MotAB is providing the energy for the rotation of the flagellum (Minamino and Imada, 2015). For the formation of the slightly curved hook structure, FlgE and FlgD are essential. The hook cap protein FlgD assists when FlgE is incorporated into the hook structure (Altegoer and Bange, 2015).

Interestingly, an uncommonly straight hook has been reported for *L. pneumophila* mutant strains (Δ*fliA*, Δ*fliD*, and Δ*flaA*, i.e., a flagellin mutant) (**Figure 1**) (Schulz et al., 2012). The findings give credit to the assumption that *L. pneumophila* might have a straight hook that is hard to detect in wildtype strains. The filament consists mainly out of a single protein, the flagellin (FlaA or FliC) (**Figure 2**). The cap protein FliD is essential for the assembly of flagellin subunits into the filament. To assemble the filament, flagellin is exported through the filament structure by a flagellum-specific export apparatus (T3SS) and assembled at the tip of the filament (Heuner and Albert-Weissenberger, 2008; Altegoer and Bange, 2015).

More details about the flagellum structure can be found in dedicated review articles (Aldridge and Hughes, 2002; Heuner and Steinert, 2003; Macnab, 2003; Pallen et al., 2005; Heuner and Albert-Weissenberger, 2008; Altegoer and Bange, 2015).

Next to the regular flagellar system, a second putative flagellar system was suspected for two *Legionella* species: *Legionella fallonii* and *L. micdadei* (Gomez-Valero et al., 2014). Comparative genome analysis led to the suspicion that the strains do have homologs to flagellar genes of *L. pneumophila*. The identified genetic region is comprised of genes that encode a putative basal body, a secretion system, as well as a putative hook structure. No homologs to *flaA* or *fliD* were found in the predicted genomic region. Yet, further investigations are needed to find out if a T3SS or a putative second flagellum is encoded. Additionally, *in silico* investigation, performed on the draft genome sequence of *Legionella israelensis*, identified a similar operon. A BLAST search using the operon (10,041 bp, ctg_064, *L. israeliensis* draft-genome; Burstein et al., 2016) as query identified similar genes in *L. drancourtii*, *L. fallonii*, *L. worsleiensis*, *L. quateirensis*, *L. birminghamensis*, and *L. drozanskii* (Heuner, unpublished results). Initial findings show that the operon is present in two out of three *Legionella* clades (**Figure 3**), leaving a margin for additional studies.

The Flagellar Regulon

The expression of flagellar genes of *L. pneumophila* is regulated in a hierarchical cascade (**Figure 4**) (Heuner et al., 1995, 2006; Heuner and Steinert, 2003; Jacobi et al., 2004; Albert-Weissenberger et al., 2010; Schulz et al., 2012). Their expression depends on growth phase, temperature, medium viscosity and nutrient availability (e.g., amino acids and fatty acids) (Ott et al., 1991; Byrne and Swanson, 1998; Heuner et al., 1999, 2006; Heuner and Albert-Weissenberger, 2008).

In short, intracellular alarmone accumulation—ppGpp, a signal molecule, produced when environmental conditions are unfavorable (e.g., limited nutrient supply)—is regulated by RelA and SpoT. RelA senses the amount of available intracellular amino acids and SpoT senses the amount of intracellular fatty acids (Hammer and Swanson, 1999; Dalebroux et al., 2009, 2010).

Alarmone accumulation triggers the activation of an alternative stationary-phase sigma factor (RpoS) and of the two component system LetA/S (Byrne and Swanson, 1998; Hammer et al., 2002; Zusman et al., 2002; Molofsky and Swanson, 2004; Dalebroux et al., 2009, 2010; Edwards et al., 2009; Rasis and Segal, 2009; Sahr et al., 2009) (**Figure 4**). RpoS and LetA/S promote the transcription of small regulatory RNAs (RsmX, RsmY, RsmZ). RsmX plays a role in the virulence of *L. pneumophila* (Sahr et al., 2012). The transcription of *rsmZ/rmsY* is also influenced by a quorum sensing system regulator, LqsR (Tiaden et al., 2007; Schell et al., 2016). The two regulatory RNAs are able to bind a number of carbon storage regulator molecules (CsrA) at once (Sahr et al., 2017). CsrA is a negative regulator and through the binding on RsmY or RsmZ, other targets of the regulatory RNAs can be expressed. The expression of transmissive traits starts and main activator proteins (e.g., FleQ) are produced and flagellar genes are expressed (Zusman et al., 2002; Molofsky and Swanson, 2003; Rasis and Segal, 2009; Sahr et al., 2009, 2017; Albert-Weissenberger et al., 2010). Notably, some parts of the function of the negative regulator CsrA (*flaA* expression and motility of *L. pneumophila*) can be "complemented" by ectopically expressed *csrT*, a CsrA-like regulatory gene associated with integrative conjugative elements (Abbott et al., 2015).

CrsA is also controlling a major regulator involved in the expression of flagella, FleQ (Sahr et al., 2017). FleQ is responsible for the expression of early flagellar genes belonging to class II and III genes in an RpoN-dependent and RpoN-independent pathway (Jacobi et al., 2004; Albert-Weissenberger et al., 2010; Schulz et al., 2012). RpoN is an enhancer-binding protein encoding an alternative sigma factor that initiates transcription when activator proteins like FleQ, FleR, and PilR (Jacobi et al., 2004) are present. When class II and III genes are expressed, the activity of FliA—the alternative sigma factor—leads to the expression of class IV genes and the assembly of the flagellum (**Figure 4**). RpoN and FleR seem to be responsible for a negative feedback loop on flagellar genes (Albert-Weissenberger et al., 2010). It was found that RpoS and FlaR (transcriptional regulator FlaR, LysR family member) are also involved in the expression of the flagellin gene (Heuner et al., 2000; Bachman and Swanson, 2001, 2004; Rasis and Segal, 2009; Sahr et al., 2009). The production of FlaA is also regulated by cyclic di-GMP, shown by the analysis of a gene (*cdgS13*) coding for a protein with diguanylate cyclase activity (Levi et al., 2011). The influence of cyclic di-GMP on flagellum-based motility has been shown for other bacteria than *Legionella* species (Wolfe and Visick, 2008).

THE FLAGELLUM AND VIRULENCE

Already early on, it has been hypothesized that virulence and flagellum expression are genetically linked with each other.

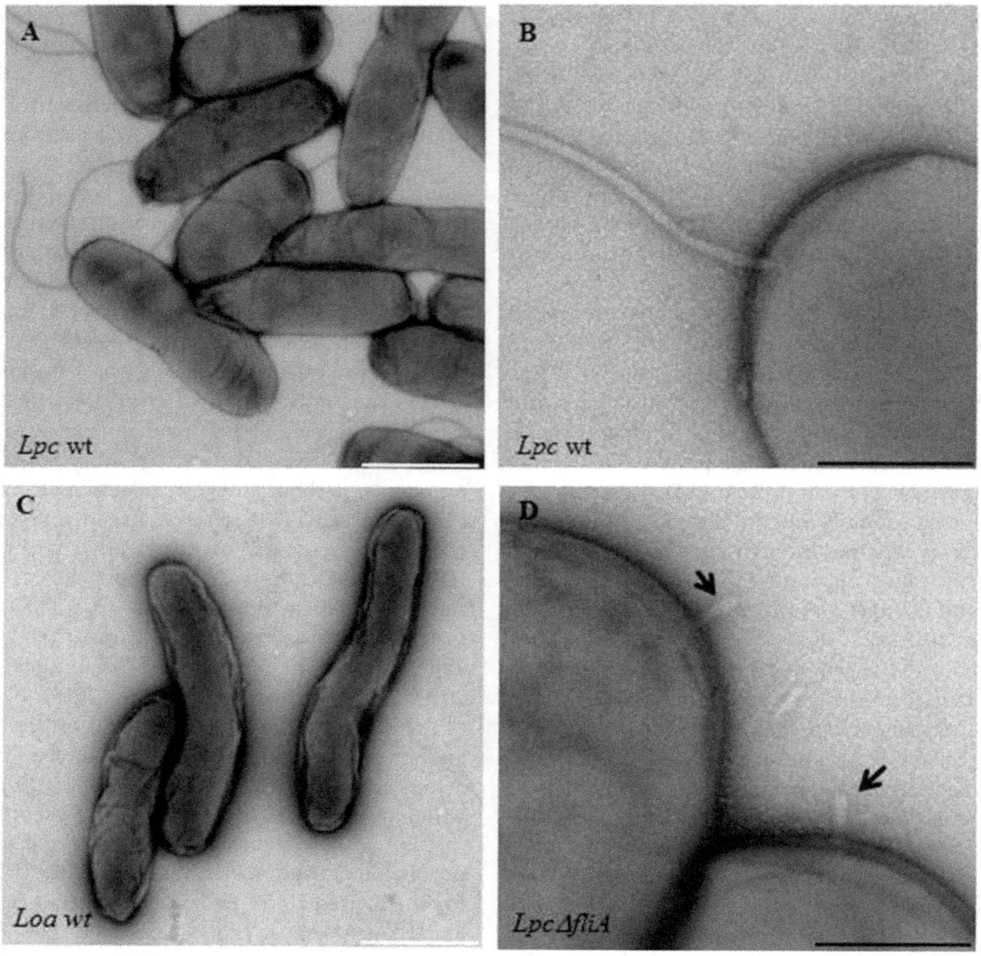

FIGURE 1 | Electron microscopic images of flagellated and non-flagellated *Legionella* wild-type strains as a flagella mutant strain. Shown are **(A,B)** *Legionella pneumophila* Corby (Lpc wt) and **(C)** *Legionella oakridgensis* (Loa wt) wild-type strains that are either representative of flagellated or non-flagellated *Legionella*. An electron microscopic image of a *fliA L. pneumophila* mutant strain (Lpc Δ*fliA*), is also shown **(D)**. The straight hook structure of the mutant strain is indicated by black arrows **(D)**. Bacteria were grown in AYE medium at 30°C and the samples were stained with 0.5% uranyl acetate. White scale bars: 1 μm; black scale bars: 200 nm.

It has been shown that motile, transmissive *Legionella* were more infectious for amoebae than non-motile replicative phase *Legionella* (Rowbotham, 1986; Pruckler et al., 1995; Bosshardt et al., 1997; Byrne and Swanson, 1998; Hammer et al., 2002; Heuner et al., 2002; Molofsky et al., 2005; Heuner and Albert-Weissenberger, 2008). Different experiments could show that the motility but not the flagellin promotes the contact with host cells. Motility increases the infectivity and the fitness. Furthermore, it turned out that the flagellum is not necessary for intracellular replication (Pruckler et al., 1995; Dietrich et al., 2001; Polesky et al., 2001; Heuner et al., 2002; Jacobi et al., 2004; Molofsky et al., 2005; Schulz et al., 2012).

The four major regulators of the flagellar regulon (RpoN, FleQ, FleSR, FliA) seems to be involved in the invasion process of *L. pneumophila* into hosts. These findings point out the proposed link between virulence traits and flagellum expression (Dietrich et al., 2001; Hammer et al., 2002; Molofsky et al., 2005; Heuner and Albert-Weissenberger, 2008; Albert-Weissenberger et al., 2010; Schulz et al., 2012). Especially the FliA regulon plays an important role (please, see the section: FliA and its implication in virulence below). *FliA*, but not the flagellin (*flaA*), is involved into the ability of *L. pneumophila* to form biofilm that allow bacteria to survive whenever environmental conditions are not favorable. (Mampel et al., 2006). Notably unwanted biofilms are a health issue causing a significant amount of nosocomial infections (Bryers, 2008). *Legionella* are known to survive within biofilm of other bacteria (e.g., *Klebsiella pneumophila* and *Pseudomonas aeruginosa*) (Molofsky et al., 2005; Stewart et al., 2012). More recently, findings about *L. pneumophila's* ability to form biofilms by itself in natural environments and on medical devices have attracted attention (Lau and Ashbolt, 2009; Abu Khweek et al., 2013). Biofilm formation is regulated by temperature, surface material and intracellular growth (Konishi et al., 2006; Piao et al., 2006; Bigot et al., 2013) and biofilm-derived *L. pneumophila* do

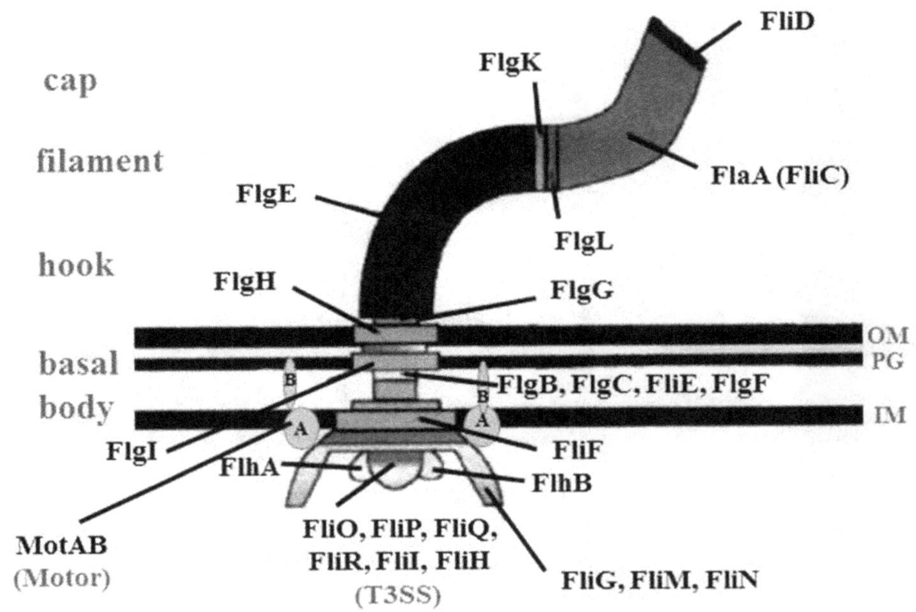

FIGURE 2 | Overview of the structure of a flagellum. Schematic drawing of a flagellum consisting of basal body, hook structure and filament. Cap, FliD protein that is essential for the assembly of the filament, Motor, MotAB proteins that provide energy for the rotation; T3SS, type III-like secretion system; FlgD (not shown), the hook cap protein needed to incorporate FlgE; IM, inner membrane; OM, outer membrane; PG, peptidoglycan layer; (Heuner and Albert-Weissenberger (2008), modified).

not express flagellin (Abu Khweek et al., 2013). More information about biofilms and *L. pneumophila* can be found in a recent review (Abdel-Nour et al., 2013).

The flagellum also affects the resistance of hosts to Legionnaires' disease and when *Legionella* do not produce flagellin they can evade the innate immune response in macrophages (Hawn et al., 2003; Molofsky et al., 2006; Ren et al., 2006; Abu Khweek et al., 2013). Resistance is mediated by the Naip5/Ipaf-dependent recognition of flagellin, which induces a protective immunity in non-A/J mouse models (Ricci et al., 2005). Detailed information about addressed points can be found in dedicated reviews (Fontana and Vance, 2011; Schell et al., 2016; Mascarenhas and Zamboni, 2017).

THE ALTERNATIVE SIGMA FACTOR 28

FliA Expression

One of the major regulators involved in the expression of the flagellum is FliA and an increased alarmone level leads to accumulation of functional FliA (Bruggemann et al., 2006; Heuner et al., 2006; Dalebroux et al., 2010). The alternative sigma factor (σ^{28}) is directly involved in the regulation and expression of the flagellin gene (*flaA*) and others (**Figure 4**). A *fliA* mutant of *L. pneumophila* does not produce flagellin and is consequently non-flagellated. Moreover, a Δ*fliA* mutant of *Escherichia coli* can be completed with a *fliA* gene of *L. pneumophila* (Heuner et al., 1995, 1997, 2002; Bruggemann et al., 2006; Albert-Weissenberger et al., 2010; Schulz et al., 2012).

The expression of flagellar class III and IV genes is induced in a FleQ-dependent manner. The FliA-regulated class IV genes are involved in the assembly of the filament and flagella motility (*flgL, fliD, flaA, motY*). Both lead to the complete synthesis of the flagellum (Jacobi et al., 2004; Albert-Weissenberger et al., 2010). The *fliA* gene itself is expressed in a FleQ-dependent but RpoN-independent manner (Albert-Weissenberger et al., 2010). Nevertheless, FleQ and RpoN are not necessary for a basal expression of *fliA*. For a basal expression, *fliA* is transcribed from a putative sigma-70 promoter element and later, during the exponential phase, the expression of *fliA* is induced in a FleQ-dependent manner (Schulz et al., 2012). Accordingly, it was hypothesized that during the exponential phase the basal *fliA* promotor activity may be mediated by DksA independent of the ppGpp concentration, whereas during the post-exponential phase DksA cooperates with ppGpp to activate *fliA* (Dalebroux et al., 2010). The identification of the transcription start point of *fliA* corroborates the presence of a putative DksA binding site, an A/T rich discriminator site (Schulz et al., 2012).

FliA and Its Implication in Virulence

FliA is a regulator that is also involved in the expression of putative virulence genes (Bruggemann et al., 2006; Albert-Weissenberger et al., 2010; Tlapak et al., 2017).

Several investigations performed on a *fliA* mutant strain of *L. pneumophila* pointed out that the mutant (at low MOI) is not replicating in host cells anymore (*Dictyostelium discoideum*). The *fliA* mutant seems to be less infectious for macrophages and non-cytotoxic to bone marrow-derived macrophages. Moreover, the mutant has a reduced fitness

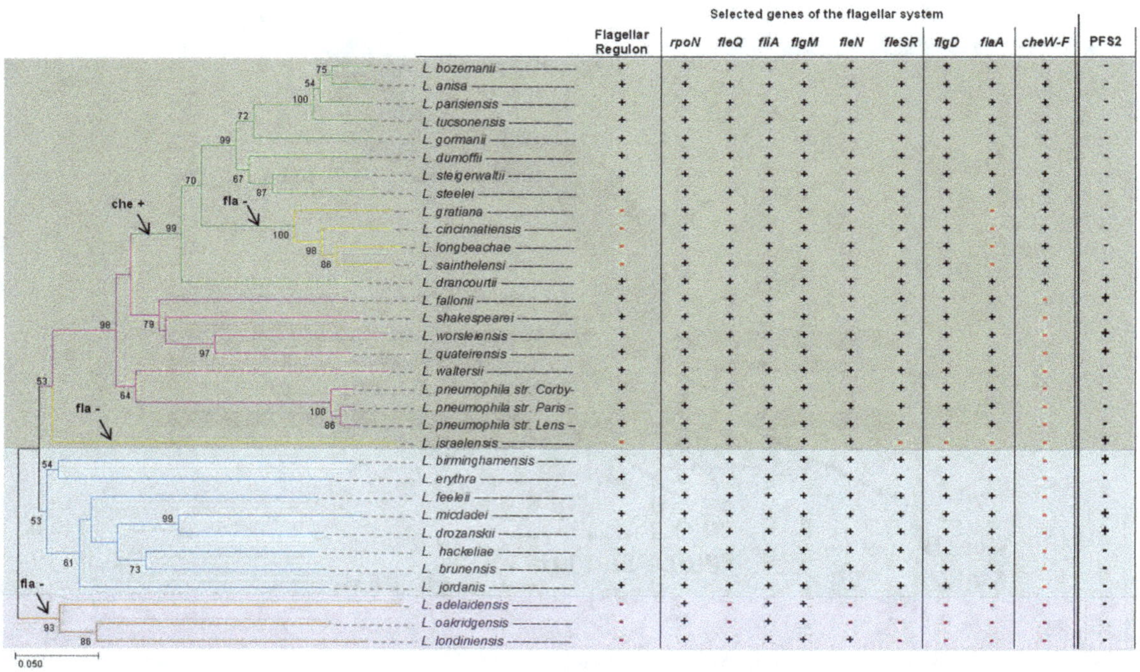

FIGURE 3 | Evolutionary relationships of *Legionella* species and distribution of flagellar regulon and chemotaxis genes. Shown is a phylogenetic tree of *Legionella* spp. based on the *fliA* gene (left side) and also the distribution of flagellar regulon and chemotaxis genes (right side). The phylogenetic tree was reconstructed on the basis of the *fliA* gene using the Neighbor-Joining method (Saitou and Nei, 1987) and a bootstrap of 1,000 (Felsenstein, 1985). The bootstrap >50 is reported next to the branches. The phylogenetic distances were computed using the Tamura-Kumar method (Tamura and Kumar, 2002) and refers to the units of the number of base substitutions per site. For the evolutional analysis, performed in MEGA7 (Kumar et al., 2016), position containing gaps and/or missing data were eliminated. To determine whether selected genes belonging to the flagellar regulon and chemotaxis genes are present among investigated *Legionella*, an *in silico* BLAST search was performed using *Legionella* draft genomes recently published by Burstein et al. (2016). The complete structure of genes of the flagellar regulon of *L. pneumophila* strain Corby, *L. longbeachae*, *L. israeliensis*, and *L. oakridgensis* are given in **Figure 5**. Data for *L. pneumophila* strains were taken from Cazalet et al. (2010), Glockner et al. (2008), and Albert-Weissenberger et al. (2010). *Legionella* clade I is highlighted in green, clade II in blue and clade III in purple. On the left, in the phylogenetic tree, possible time points when the flagellin gene (*flaA*) and the chemotaxis (*che*) genes were lost (–) or acquired (+) are indicated with arrows. On the right, the presence (+) and/or absence (–) of selected genes belonging to the flagellar regulon, the chemotaxis operon or an operon encoding a putative second flagellum or a putative T3SS (PFS2, *based on unpublished data*) are indicated, respectively.

potential in amoebae (Dietrich et al., 2001; Hammer et al., 2002; Heuner et al., 2002; Jacobi et al., 2004; Molofsky et al., 2005; Heuner and Albert-Weissenberger, 2008; Schulz et al., 2012). Likewise, a *fliA* mutant strain of *L. oakridgensis* showed a reduced fitness in its host (*Acanthamoeba lenticulata*) (Tlapak et al., 2017). In addition, another *L. pneumophila fliA* mutant strain exhibited a reduced ability to form biofilms (Mampel et al., 2006).

FliA is obviously a virulence factor, and target genes of *fliA* were investigated to understand its implication for virulence (Bruggemann et al., 2006; Albert-Weissenberger et al., 2010; Tlapak et al., 2017). Target genes of *fliA* in *L. pneumophila* strains include genes of the flagellar regulon (e.g., *flaA* and *flgM*) and others (e.g., *enhA* and *lvrA*), illustrated in **Figure 4** and listed in **Table 1** (Bruggemann et al., 2006; Albert-Weissenberger et al., 2010; Schulz et al., 2012). Other identified FliA-dependent genes encode for putative virulence factors corroborating the involvement of FliA in the establishment of *Legionella* infections. Identified putative virulence factors are: *lpp0952, lpp1290*, and *lpp0972.* The first one, *lpp0952,* is coding for a GGDEF/EAL and PAS/PAC domain protein (Bruggemann et al., 2006; Albert-Weissenberger et al., 2010). The *two* remaining genes are homologs of the enhanced entry proteins EnhA. Respective homologs were also found in *L. longbeachae* which is non-flagellated, putatively associated with the flagellar system (Kozak et al., 2010).

Recently, especially *L. oakridgensis* simplified the identification of *fliA* targets potentially involved in virulence (Tlapak et al., 2017). *L. oakridgensis* strains are non-flagellated, the entire flagellar regulon is missing and only homologs of FliA and FlgM are present (**Figure 5**). *L. oakridgensis* is less infectious then *L. pneumophila*, but still causes Legionnaires' disease. In addition, *L. oakridgensis* replicates in guinea pigs, in human cell lines, in *Acanthamoeba lenticulata* and for growth in media no additional cysteine is needed. (Orrison et al., 1983; Fields et al., 1986; O'Connell et al., 1996; Neumeister et al., 1997; Lo Presti et al., 2001; Brzuszkiewicz et al., 2013). Nevertheless, *L. oakridgensis* exhibits a functional T4SS, homologs of known virulence factors, as well as newly identified virulence factors (Brzuszkiewicz et al., 2013). *L. oakridgensis* is used to investigate

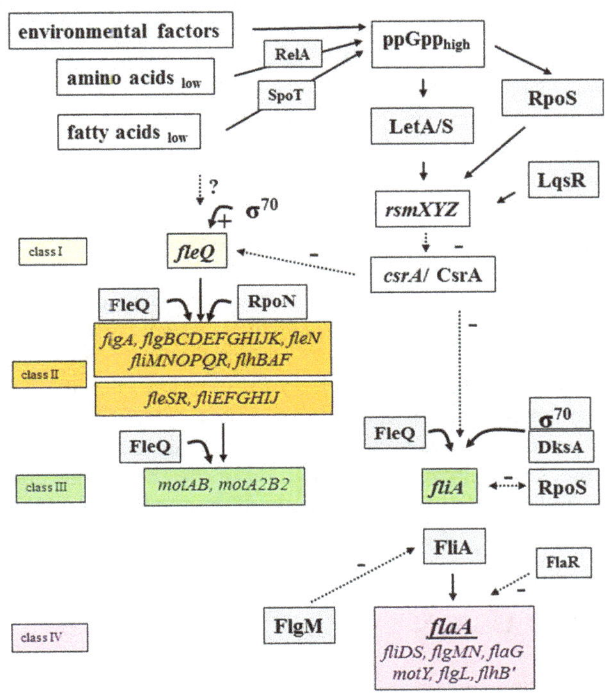

FIGURE 4 | The flagellar regulon cascade. The regulatory cascade which leads to the expression of proteins that are needed to construct a functional flagellum is shown (Heuner and Albert-Weissenberger, 2008; Sahr et al., 2017, modified). Specific environmental factors lead to the activation of the pathway through alarmone (ppGpp) accumulation. Some main players of the flagellar regulon cascade are: the activator protein FleQ, the alternative sigma factor 28 (FliA), the two-component system (LetA/S) and the carbon storage regulator (CsrA) as examples. FleQ is the master regulator of the flagellar genes (Class II and class III). FliA is the regulator of flagellar genes of class IV. Black lined, continuous arrows refer to positive regulation events. The broken lines in black refer to negative regulatory effects of respective factors. Question marks (?) indicate pathway points with putative mode of actions that need to be specified by further investigations. RsmXYZ, regulatory RNAs; FlaA, flagellin; FlaR, transcriptional regulator (LysR family); FleQ and FleR, sigma factor 54 activator protein; FlgM, anti-sigma 28 factor; RpoS, alternative sigma 38 factor.

FliA since a *fliA L. oakridgensis* knockout will not cause the inactivation of the entire flagellar regulon and target genes of FliA can still be identified as well as genes involved in the expression of virulence traits. However, mutant strain analyses aimed at identifying target genes of FliA in *L. oakridgensis* yielded no results for putative FliA-dependent virulence genes yet (**Table 1**; Tlapak et al., 2017). Additional investigations are needed including phenotypic characterizations and deletion analyses of further target genes of FliA.

FliA-FlgM Interaction in *Legionella oakridgensis*

Flagellated bacteria regulate the FliA activity often post-transcriptionally. For *Salmonella, Escherichia*, and *Vibrio* species it is known that an anti-sigma-28 factor (FlgM) binds FliA, preventing the binding of FliA to FliA-dependent promoter sites, and FliA-dependent genes are consequently not translated. After assembling of the hook-basal body structure, FlgM is exported and FliA is not repressed anymore (Gillen and Hughes, 1991; Ohnishi et al., 1992; Chilcott and Hughes, 2000; Aldridge et al., 2006). In *Helicobacter pylori,* the FlgM protein is inactivated instead of being exported out of the cells (Rust et al., 2009). FliA-FlgM interactions in *Legionella* are still unknown. Recent findings suggest that FliA-FlgM interaction might be different than in other flagellated bacteria, at least for *L. oakridgensis*. Respective species do not have a flagellum (**Figure 1C**), a flagellar regulon (**Figure 5**) and a basal body although *flgM* and *fliA* homologs are present which encode for functional FlgM and FliA proteins (Brzuszkiewicz et al., 2013; Tlapak et al., 2017); consequently, the mechanism controlling FliA-FlgM interactions must be different.

For *L. pneumophila* as well as for *L. oakridgensis* it was found that the expression of FlgM or homologs is sigma-28-dependent (Albert-Weissenberger et al., 2010; Tlapak et al., 2017). Although *L. oakridgensis* has no flagellar system, the expression of *fliA*-dependent genes is growth phase- and temperature-dependent (Heuner et al., 1999; Tlapak et al., 2017). Moreover, for *L. oakridgensis* it was demonstrated that FlgM is a negative regulator of FliA-dependent genes and the protein itself seems to be degraded in a growth phase- and temperature-dependent manner (Tlapak et al., 2017). Thus, it seems likely that, as described for *H. pylori*, FlgM in *L. oakridgensis* is degraded by protease activity instead of being secreted. However, investigations are needed to show if FlgM in *L. pneumophila* is effectively secreted in a basal body-dependent manner.

DISTRIBUTION OF THE FLAGELLAR SYSTEM AMONG *LEGIONELLA* SPECIES

Phylogenetic reconstruction—based on concatenated amino acid alignment of 78 orthologous ORFs—divided the *Legionella* species into three major clades (clade I and clade II, clade III) (Burstein et al., 2016). Clade I is comprised of most *Legionella* species including *L. pneumophila, L. parisiensis, L. bozemanii*, and *L. longbeachae* (Burstein et al., 2016). Clade II comprises among others *L. feelei* and *L. micdadei* and clade III, a deep-branching clade, includes three members: *Legionella adelaidensis, L. oakridgensis* and *Legionella londiniensis* (Burstein et al., 2016). Phylogenetic reconstructions performed herein, yielded similae results that are given in **Figure 3**. As opposed to former investigations, the phylogenetic tree was reconstructed on the basis of the *fliA* gene.

The flagellar system can be found in *Legionella* species classified as clade I or II (Cazalet et al., 2004, 2010; Chen et al., 2004; Bruggemann et al., 2006; Glockner et al., 2008; Kozak et al., 2010; Brzuszkiewicz et al., 2013; Gomez-Valero et al., 2014; Burstein et al., 2016), but not in clade III *Legionella*. Clade III *Legionella* do not have a functional flagellar system and do not have most of the flagellar regulon genes, except for *fliA* and its anti-sigma factor *flgM* (*L. oakridgensis, L. adelaidensis,* and *L. londiniensis*) (Cazalet et al., 2010; Brzuszkiewicz et al., 2013; Tlapak et al., 2017) and two additional genes: *fleQ* and *fleN* (*L. londiniensis*) (**Figures 3, 5**). Also, some clade I *Legionella* species do not have a functional flagellar system

TABLE 1 | Genes belonging to the FliA regulon (FliA target genes) of *L. pneumophila* Paris (Lpp)*.

Gene name	Annotation	FC
§ lpp1294, flaA	Flagelline	0.003
lpp1293, flaG	Unknown	0.007
§ lpp0972	Similar to enhanced entry protein EnhA	0.010
#§ lpp2282	Unknown	0.024
lpp1746, fliA	Sigma factor 28	0.042
#§ lpp2998	Similar to conserved hypothetical protein	0.045
§ lpp1292, fliD	Flagellar capping protein	0.045
lpp0197	Similar to adenine specific DNA methylase	0.046
lpp1291, fliS	Similar to flagellar protein FliS	0.048
lpp1745, motA	Flagellar motor protein MotA	0.052
§ lpp1290	Similar to enhanced entry protein EnhA	0.053
§ lpp1841	Unknown	0.059
lpp0968, flgN	Hook-associated protein	0.068
#lpp0969, flgM	Anti-sigma-28 factor	0.081
lpp0198	Similar to Type III RM enzyme- helicase subunit	0.097
lpp1050	Unknown	0.114
§ lpp3034, motY	Similar to sodium-type flagellar protein MotY	0.116
lpp1743	Similar to hypothetical poteins	0.122
§ lpp0952	Regulatory protein (GGDEF and EAL domains)	0.122
lpp2281	Similar to membrane-associated metalloprotease proteins	0.133
lpp2376	Similar to *Legionella vir* region protein LvrA	0.165
lpp0763	Weakly similar to *L. pneumophila* IcmL protein	0.187
lpp1941	Unknown	0.202
lpp2634	Similar to hypothetical proteins	0.228
lpp1568, plaB	Phospholipase	0.232
§ lpp2260	Unknown	0.262
lpp2635, flhB′	Similar to FlhB protein	0.295
lpp0010	Similar to GTP-binding protein HflX	0.296
lpp0009	Similar to host factor-1 protein	0.366
plpp0131	Similar to alanyl tRNA synthetase	0.370
lpp1742, prfB	Highly similar to peptide chain release factor 2	0.371
lpp2386	Unknown	0.374
lpp1234, flgL	Flagellar hook-associated protein FlgL	0.379

From Table S7 (Albert-Weissenberger et al., 2010), modified; #Homolog gene belonging to the FliA-regulon of L. oakridgensis (data from Tlapak et al., 2017); §Belonging to the FliA regulon of L. pneumophila Paris replicating in A. castellanii (data from Bruggemann et al., 2006); FC, fold-change values.

(clade I: *L. longbeachae*, **Figure 5**, *Legionella gratiana*, *Legionella cincinnatiensis*, *Legionella sainthelensi*, and *L. israelensis*). It has been hypothesized that the loss of flagellar genes has not happened recently (Kozak et al., 2010). This is corroborated by the finding that *L. longbeachae* and all subclade members are negative for the flagellar regulon but positive for genes coding for the sigma factor FliA, the regulator FleN and the two component system comprising of FleR and FleS, as well as FlgD (**Figure 3**) (Cazalet et al., 2010; Kozak et al., 2010). Also *L. israelensis* is negative for *flaA* (Heuner et al., 1995) and most flagellar regulon genes, except: *fleQ, fliA, fleN*, and *flgM* (**Figure 5**). The finding allows to assume that the flagellar system may have been lost at different time points during the evolution of *Legionella* species (**Figure 3**). In addition, some genes that have regulatory functions outside of the flagellar system are still present (Albert-Weissenberger et al., 2010; Cazalet et al., 2010; Kozak et al., 2010; Tlapak et al., 2017). The same applies to *flgD* which is involved in the assembly of the hook structure of the flagellum with unassigned hypothetical alternative functions.

The investigated *Legionella* genomes were also screened for the presence/absence of genes of major regulators of the flagellar system as well as of the chemotaxis operon (**Figure 3**). It was found that the genes of the chemotaxis operon are only found in a subclade of the clade I *Legionella*. *L. longbeachae* is the first *Legionella* species described to exhibit chemotaxis genes (Cazalet et al., 2010; Kozak et al., 2010) that do not have flagellar genes. It seemed paradoxical that *L. pneumophila* is flagella positive but chemotaxis negative and *L. longbeachae* is flagella negative but chemotaxis positive. The distribution of the chemotaxis operon may indicate that the chemotaxis operon was acquired by a common 'ancestor' of this sub-tree clade (**Figure 3**).

CONCLUSION

The review aimed to summarize knowledge gained about flagella and the flagellar regulon of different *Legionella* species. The majority of *Legionella* species exhibit genes encoding for a functional flagellum and they are flagellated (Elliott and Johnson, 1981; Bornstein et al., 1991; Bangsborg et al., 1995; Heuner et al., 1995). Motility increases infectivity and fitness, helping the bacteria to reach new hosts after successful replication within protozoan host cells and release into aquatic environments.

Some *Legionella*—including some pathogenic species (e.g., *L. longbeachae*, and *L. oakridgensis*)—are not flagellated and most flagellar regulon genes are absent (Orrison et al., 1983; Heuner et al., 1995; Cazalet et al., 2010; Kozak et al., 2010; Brzuszkiewicz et al., 2013). With the advance in molecular techniques and the ability to produce and to process metagenomics datasets, it was found that some of the non-flagellated *Legionella* have still parts of the flagellar regulon, mainly genes with regulator functions.

However, in *Legionella* flagellum synthesis is associated with the expression of a virulence phenotype; and motility can be seen as a virulence and a fitness factor in *Legionella* and other bacteria. The alternative sigma factor FliA is also involved in the expression of virulence traits. FliA-dependent putative virulence genes were already identified by initial investigations that need to be extended. Also, additional investigations are needed to determine the role of FliA and molecular mechanisms of FliA-FlgM interactions in *Legionellae*. FlgM and FliA, two main players involved in the expression of the flagellum genes, are still present in non-flagellated *Legionella*, a promising take-off for future investigations. Nevertheless, the flagellum is not necessarily needed for an intracellular replication within host cells. Moreover, in some hosts the Naip5/Ipaf-dependent recognition of flagellin can cause an innate immune response leading to resistance against *Legionella* infections (Molofsky et al.,

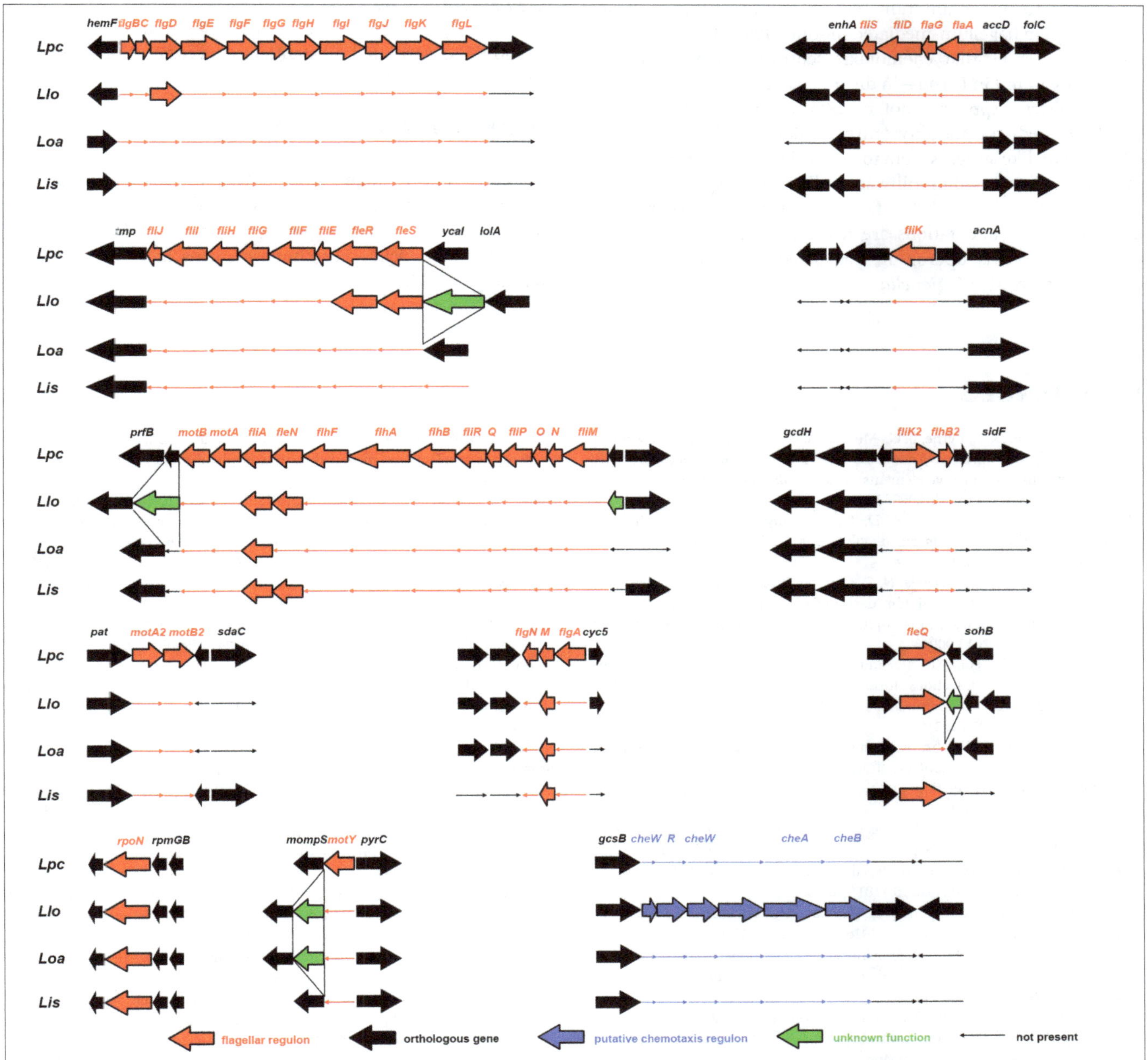

FIGURE 5 | Structure of the operons of the flagellar regulon of *L. pneumophila* Corby (Lpc), *L. longbeachae* (Llo), *L. israeliensis* (Lis), and *L. oakridgensis* (Loa). Genes are indicated by arrows and gene names are given above. Flagellar and chemotaxis genes are indicated in red and blue, respectively; other genes in black. Genes not present in a strain are indicated by thin arrows (Brzuszkiewicz et al., 2013, modified).

2006; Ren et al., 2006). For example, it was reported that biofilm-derived *L. pneumophila* without flagellin expression evade the innate immune response in macrophages (Abu Khweek et al., 2013), as it was suggested for the non-flagellated *L. longbeachae* (Cazalet et al., 2010; Kozak et al., 2010). It seems that under certain conditions, the loss of the flagellum may increase the fitness of bacteria. For instance, *L. pneumophila* which can be found mainly in aquatic environments, is still flagellated whereas *L. longbeachae* which can be found predominantly in soil, is non-flagellated (Kozak et al., 2010). Nevertheless, if the loss of the flagellar system from *Legionella* species depends on the habitat or environmental conditions remains unanswered.

As outlined, flagellated and non-flagellated *Legionella* are positive for genes belonging to the chemotaxis operon. The ability of *Legionella* to swarm and to show off a chemotaxis behavior has not yet been reported. Interestingly, some chemotaxis-positive and flagellar operon-negative *Legionella* (e.g., *L. longbeachae*) give credit to the assumption that chemotaxis genes may not be involved in flagellum-mediated motility. Recent investigations could even show that chemotaxis sensory systems—different from those found in *E. coli*—in distinct bacteria (e.g., *Myxococcus* spp., *Geobacter* spp.) are not necessarily involved in bacterial flagellum-mediated motility (Kirby, 2009; Kozak et al., 2010). Chemotaxis-like systems seem to be involved among other things in type IV pilus-based motility and cell to cell interaction and/or social motility (Kirby, 2009; Kozak et al., 2010). Accordingly, additional experimentations are needed to investigate the role of the chemotaxis operon in flagellated and non-flagellated, chemotaxis-positive *Legionella*.

AUTHOR CONTRIBUTIONS

SA and KH contributed substantial to the conception and design of the work. SA and KH wrote the paper.

ACKNOWLEDGMENTS

We would like to thank Roland Grunow for assistance with different aspects of the presented work and for copy-editing. This work received financial support by grant HE2854/5-1+2/9-1 from the Deutsche Forschungsgemeinschaft DFG (Bonn, Germany) to KH.

REFERENCES

Abbott, Z. D., Flynn, K. J., Byrne, B. G., Mukherjee, S., Kearns, D. B., and Swanson, M. S. (2015). csrT represents a new class of csrA-like regulatory genes associated with integrative conjugative elements of *Legionella pneumophila*. *J. Bacteriol.* 198, 553–564. doi: 10.1128/JB.00732-15

Abdel-Nour, M., Duncan, C., Low, D. E., and Guyard, C. (2013). Biofilms: the stronghold of *Legionella pneumophila*. *Int. J. Mol. Sci.* 14, 21660–21675. doi: 10.3390/ijms141121660

Abu Khweek, A., Fernandez Davila, N. S., Caution, K., Akhter, A., Abdulrahman, B. A., Tazi, M., et al. (2013). Biofilm-derived *Legionella pneumophila* evades the innate immune response in macrophages. *Front. Cell. Infect. Microbiol.* 3:18. doi: 10.3389/fcimb.2013.00018

Al-Bana, B. H., Haddad, M. T., and Garduno, R. A. (2014). Stationary phase and mature infectious forms of *Legionella pneumophila* produce distinct viable but non-culturable cells. *Environ. Microbiol.* 16, 382–395. doi: 10.1111/1462-2920.12219

Albert-Weissenberger, C., Sahr, T., Sismeiro, O., Hacker, J., Heuner, K., and Buchrieser, C. (2010). Control of flagellar gene regulation in *Legionella pneumophila* and its relation to growth phase. *J. Bacteriol.* 192, 446–455. doi: 10.1128/JB.00610-09

Aldridge, P. D., Karlinsey, J. E., Aldridge, C., Birchall, C., Thompson, D., Yagasaki, J., et al. (2006). The flagellar-specific transcription factor, sigma28, is the Type III secretion chaperone for the flagellar-specific anti-sigma28 factor FlgM. *Genes Dev.* 20, 2315–2326. doi: 10.1101/gad.380406

Aldridge, P., and Hughes, K. T. (2002). Regulation of flagellar assembly. *Curr. Opin. Microbiol.* 5, 160–165. doi: 10.1016/S1369-5274(02)00302-8

Altegoer, F., and Bange, G. (2015). Undiscovered regions on the molecular landscape of flagellar assembly. *Curr. Opin. Microbiol.* 28, 98–105. doi: 10.1016/j.mib.2015.08.011

Atlas, R. M. (1999). *Legionella*: from environmental habitats to disease pathology, detection and control. *Environ. Microbiol.* 1, 283–293. doi: 10.1046/j.1462-2920.1999.00046.x

Bachman, M. A., and Swanson, M. S. (2001). RpoS co-operates with other factors to induce *Legionella pneumophila* virulence in the stationary phase. *Mol. Microbiol.* 40, 1201–1214. doi: 10.1046/j.1365-2958.2001.02465.x

Bachman, M. A., and Swanson, M. S. (2004). Genetic evidence that *Legionella pneumophila* RpoS modulates expression of the transmission phenotype in both the exponential phase and the stationary phase. *Infect. Immun.* 72, 2468–2476. doi: 10.1128/IAI.72.5.2468-2476.2004

Bajrai, L. H., Azhar, E. I., Yasir, M., Jardot, P., Barrassi, L., Raoult, D., et al. (2016). *Legionella saoudiensis* sp. nov., isolated from a sewage water sample. *Int. J. Syst. Evol. Microbiol.* 66, 4367–4371. doi: 10.1099/ijsem.0.001357

Bangsborg, J. M., Hindersson, P., Shand, G., and Hoiby, N. (1995). The *Legionella micdadei* flagellin: expression in *Escherichia coli* K 12 and DNA sequence of the gene. *APMIS* 103, 869–877. doi: 10.1111/j.1699-0463.1995.tb01446.x

Barbaree, J. M., Fields, B. S., Feeley, J. C., Gorman, G. W., and Martin, W. T. (1986). Isolation of protozoa from water associated with a legionellosis outbreak and demonstration of intracellular multiplication of *Legionella pneumophila*. *Appl. Environ. Microbiol.* 51, 422–424.

Berk, S. G., Faulkner, G., Garduno, E., Joy, M. C., Ortiz-Jimenez, M. A., and Garduno, R. A. (2008). Packaging of live *Legionella pneumophila* into pellets expelled by *Tetrahymena* spp. does not require bacterial replication and depends on a Dot/Icm-mediated survival mechanism. *Appl. Environ. Microbiol.* 74, 2187–2199. doi: 10.1128/AEM.01214-07

Bigot, R., Bertaux, J., Frere, J., and Berjeaud, J. M. (2013). Intra-amoeba multiplication induces chemotaxis and biofilm colonization and formation for *Legionella*. *PLoS ONE* 8:e77875. doi: 10.1371/journal.pone.0077875

Bornstein, N., Marmet, D., Dumaine, M. H., Surgot, M., and Fleurette, J. (1991). Detection of flagella in 278 *Legionella* strains by latex reagent sensitized with antiflagellum immunoglobulins. *J. Clin. Microbiol.* 29, 953–956.

Bosshardt, S. C., Benson, R. F., and Fields, B. S. (1997). Flagella are a positive predictor for virulence in *Legionella*. *Microb. Pathog.* 23, 107–112. doi: 10.1006/mpat.1997.0134

Bouyer, S., Imbert, C., Rodier, M. H., and Hechard, Y. (2007). Long-term survival of *Legionella pneumophila* associated with *Acanthamoeba castellanii* vesicles. *Environ. Microbiol.* 9, 1341–1344. doi: 10.1111/j.1462-2920.2006.01229.x

Bruggemann, H., Hagman, A., Jules, M., Sismeiro, O., Dillies, M. A., Gouyette, C., et al. (2006). Virulence strategies for infecting phagocytes deduced from the *in vivo* transcriptional program of *Legionella pneumophila*. *Cell. Microbiol.* 8, 1228–1240. doi: 10.1111/j.1462-5822.2006.00703.x

Bryers, J. D. (2008). Medical biofilms. *Biotechnol. Bioeng.* 100, 1–18. doi: 10.1002/bit.21838

Brzuszkiewicz, E., Schulz, T., Rydzewski, K., Daniel, R., Gillmaier, N., Dittmann, C., et al. (2013). *Legionella oakridgensis* ATCC 33761 genome sequence and phenotypic characterization reveals its replication capacity in amoebae. *Int. J. Med. Microbiol.* 303, 514–528. doi: 10.1016/j.ijmm.2013.07.003

Burstein, D., Amaro, F., Zusman, T., Lifshitz, Z., Cohen, O., Gilbert, J. A., et al. (2016). Genomic analysis of 38 *Legionella* species identifies large and diverse effector repertoires. *Nat. Genet.* 48, 167–175. doi: 10.1038/ng.3481

Byrne, B., and Swanson, M. S. (1998). Expression of *Legionella pneumophila* virulence traits in response to growth conditions. *Infect. Immun.* 66, 3029–3034.

Cateau, E., Delafont, V., Hechard, Y., and Rodier, M. H. (2014). Free-living amoebae: what part do they play in healthcare-associated infections? *J. Hosp. Infect.* 87, 131–140. doi: 10.1016/j.jhin.2014.05.001

Cazalet, C., Gomez-Valero, L., Rusniok, C., Lomma, M., Dervins-Ravault, D., Newton, H. J., et al. (2010). Analysis of the *Legionella longbeachae* genome and transcriptome uncovers unique strategies to cause Legionnaires' disease. *PLoS Genet.* 6:e1000851. doi: 10.1371/journal.pgen.1000851

Cazalet, C., Rusniok, C., Bruggemann, H., Zidane, N., Magnier, A., Ma, L., et al. (2004). Evidence in the *Legionella pneumophila* genome for exploitation of host cell functions and high genome plasticity. *Nat. Genet.* 36, 1165–1173. doi: 10.1038/ng1447

Chandler, F. W., Thomason, B. M., and Hebert, G. A. (1980). Flagella on Legionnaires' disease bacteria in the human lung. *Ann. Intern. Med.* 93, 715–716. doi: 10.7326/0003-4819-93-5-715

Chen, J., de Felipe, K. S., Clarke, M., Lu, H., Anderson, O. R., Segal, G., et al. (2004). *Legionella* effectors that promote nonlytic release from protozoa. *Science* 303, 1358–1361. doi: 10.1126/science.1094226

Chilcott, G. S., and Hughes, K. T. (2000). Coupling of flagellar gene expression to flagellar assembly in *Salmonella enterica* serovar *typhimurium* and *Escherichia coli*. *Microbiol. Mol. Biol. Rev.* 64, 694–708. doi: 10.1128/MMBR.64.4.694-708.2000

Cianciotto, N., Eisenstein, B. I., Engleberg, N. C., and Shuman, H. (1989). Genetics and molecular pathogenesis of *Legionella pneumophila*, an intracellular parasite of macrophages. *Mol. Biol. Med.* 6, 409–424.

Coil, D. A., and Anne, J. (2009). Twitching motility in *Legionella pneumophila*. *FEMS Microbiol. Lett.* 293, 271–277. doi: 10.1111/j.1574-6968.2009.01532.x

Currie, S. L., and Beattie, T. K. (2015). Compost and *Legionella longbeachae*: an emerging infection? *Perspect. Public Health* 135, 309–315. doi: 10.1177/1757913915611162

Dalebroux, Z. D., Edwards, R. L., and Swanson, M. S. (2009). SpoT governs *Legionella pneumophila* differentiation in host macrophages. *Mol. Microbiol.* 71, 640–658. doi: 10.1111/j.1365-2958.2008.06555.x

Dalebroux, Z. D., Yagi, B. F., Sahr, T., Buchrieser, C., and Swanson, M. S. (2010). Distinct roles of ppGpp and DksA in *Legionella pneumophila* differentiation. *Mol. Microbiol.* 76, 200–219. doi: 10.1111/j.1365-2958.2010.07094.x

Danhorn, T., and Fuqua, C. (2007). Biofilm formation by plant-associated bacteria. *Annu. Rev. Microbiol.* 61, 401–422. doi: 10.1146/annurev.micro.61.080706.093316

Declerck, P. (2010). Biofilms: the environmental playground of *Legionella pneumophila*. *Environ. Microbiol.* 12, 557–566. doi: 10.1111/j.1462-2920.2009.02025.x

Dietrich, C., Heuner, K., Brand, B. C., Hacker, J., and Steinert, M. (2001). Flagellum of *Legionella pneumophila* positively affects the early phase of infection of eukaryotic host cells. *Infect. Immun.* 69, 2116–2122. doi: 10.1128/IAI.69.4.2116-2122.2001

Edwards, R. L., Dalebroux, Z. D., and Swanson, M. S. (2009). *Legionella pneumophila* couples fatty acid flux to microbial differentiation and virulence. *Mol. Microbiol.* 71, 1190–1204. doi: 10.1111/j.1365-2958.2009.06593.x

Eisenreich, W., and Heuner, K. (2016). The life stage-specific pathometabolism of *Legionella pneumophila*. *FEBS Lett.* 590, 3868–3886. doi: 10.1002/1873-3468.12326

Elliott, J. A., and Johnson, W. (1981). Immunological and biochemical relationships among flagella isolated from *Legionella pneumophila* serogroups 1, 2, and 3. *Infect. Immun.* 33, 602–610.

Elliott, J. A., and Johnson, W. (1982). Virulence conversion of *Legionella pneumophila* serogroup 1 by passage in guinea pigs and embryonated eggs. *Infect. Immun.* 35, 943–946.

Faulkner, G., and Garduno, R. A. (2002). Ultrastructural analysis of differentiation in *Legionella pneumophila*. *J. Bacteriol.* 184, 7025–7041. doi: 10.1128/JB.184.24.7025-7041.2002

Felsenstein, J. (1985). Confidence limits on phylogenies: an approach using the bootstrap. *Evolution* 39, 783–791. doi: 10.1111/j.1558-5646.1985.tb00420.x

Fields, B. S. (1996). The molecular ecology of legionellae. *Trends Microbiol.* 4, 286–290. doi: 10.1016/0966-842X(96)10041-X

Fields, B. S., Barbaree, J. M., Shotts, E. B. Jr., Feeley, J. C., Morrill, W. E., Sanden, G. N., et al. (1986). Comparison of guinea pig and protozoan models for determining virulence of *Legionella* species. *Infect. Immun.* 53, 553–559.

Fields, B. S., Benson, R. F., and Besser, R. E. (2002). *Legionella* and Legionnaires' disease: 25 years of investigation. *Clin. Microbiol. Rev.* 15, 506–526. doi: 10.1128/CMR.15.3.506-526.2002

Fliermans, C. B., Cherry, W. B., Orrison, L. H., Smith, S. J., Tison, D. L., and Pope, D. H. (1981). Ecological distribution of *Legionella pneumophila*. *Appl. Environ. Microbiol.* 41, 9–16.

Fonseca, M. V., and Swanson, M. S. (2014). Nutrient salvaging and metabolism by the intracellular pathogen *Legionella pneumophila*. *Front. Cell. Infect. Microbiol.* 4:12. doi: 10.3389/fcimb.2014.00012

Fontana, M. F., and Vance, R. E. (2011). Two signal models in innate immunity. *Immunol. Rev.* 243, 26–39. doi: 10.1111/j.1600-065X.2011.01037.x

Gal-Mor, O., and Segal, G. (2003). The *Legionella pneumophila* GacA homolog (LetA) is involved in the regulation of icm virulence genes and is required for intracellular multiplication in *Acanthamoeba castellanii*. *Microb. Pathog.* 34, 187–194. doi: 10.1016/S0882-4010(03)00027-5

Garduno, R. A., Garduno, E., Hiltz, M., and Hoffman, P. S. (2002). Intracellular growth of *Legionella pneumophila* gives rise to a differentiated form dissimilar to stationary-phase forms. *Infect. Immun.* 70, 6273–6283. doi: 10.1128/IAI.70.11.6273-6283.2002

Gillen, K. L., and Hughes, K. T. (1991). Molecular characterization of flgM, a gene encoding a negative regulator of flagellin synthesis in *Salmonella typhimurium*. *J. Bacteriol.* 173, 6453–6459. doi: 10.1128/jb.173.20.6453-6459.1991

Gillmaier, N., Schunder, E., Kutzner, E., Tlapak, H., Rydzewski, K., Herrmann, V., et al. (2016). Growth-related metabolism of the carbon storage poly-3-hydroxybutyrate in *Legionella pneumophila*. *J. Biol. Chem.* 291, 6471–6482. doi: 10.1074/jbc.M115.693481

Glockner, G., Albert-Weissenberger, C., Weinmann, E., Jacobi, S., Schunder, E., Steinert, M., et al. (2008). Identification and characterization of a new conjugation/type IVA secretion system (trb/tra) of *Legionella pneumophila* Corby localized on two mobile genomic islands. *Int. J. Med. Microbiol.* 298, 411–428. doi: 10.1016/j.ijmm.2007.07.012

Gomez-Valero, L., Rusniok, C., and Buchrieser, C. (2009). *Legionella pneumophila*: population genetics, phylogeny and genomics. *Infect. Genet. Evol.* 9, 727–739. doi: 10.1016/j.meegid.2009.05.004

Gomez-Valero, L., Rusniok, C., Rolando, M., Neou, M., Dervins-Ravault, D., Demirtas, J., et al. (2014). Comparative analyses of *Legionella* species identifies genetic features of strains causing Legionnaires' disease. *Genome Biol.* 15, 505. doi: 10.1186/PREACCEPT-1086350395137407

Greub, G., and Raoult, D. (2004). Microorganisms resistant to free-living amoebae. *Clin. Microbiol. Rev.* 17, 413–433. doi: 10.1128/CMR.17.2.413-433.2004

Hammer, B. K., and Swanson, M. S. (1999). Co-ordination of *Legionella pneumophila* virulence with entry into stationary phase by ppGpp. *Mol. Microbiol.* 33, 721–731. doi: 10.1046/j.1365-2958.1999.01519.x

Hammer, B. K., Tateda, E. S., and Swanson, M. S. (2002). A two-component regulator induces the transmission phenotype of stationary-phase *Legionella pneumophila*. *Mol. Microbiol.* 44, 107–118. doi: 10.1046/j.1365-2958.2002.02884.x

Hauslein, I., Manske, C., Goebel, W., Eisenreich, W., and Hilbi, H. (2016). Pathway analysis using (13) C-glycerol and other carbon tracers reveals a bipartite metabolism of *Legionella pneumophila*. *Mol. Microbiol.* 100, 229–246. doi: 10.1111/mmi.13313

Hawn, T. R., Verbon, A., Lettinga, K. D., Zhao, L. P., Li, S. S., Laws, R. J., et al. (2003). A common dominant TLR5 stop codon polymorphism abolishes flagellin signaling and is associated with susceptibility to legionnaires' disease. *J. Exp. Med.* 198, 1563–1572. doi: 10.1084/jem.20031220

Hazelbauer, G. L., Falke, J. J., and Parkinson, J. S. (2008). Bacterial chemoreceptors: high-performance signaling in networked arrays. *Trends Biochem. Sci.* 33, 9–19. doi: 10.1016/j.tibs.2007.09.014

Heuner, K., and Albert-Weissenberger, C. (2008). "The flagellar regulon of *Legionella pneumophila* and the expression of virulence traits," in *Legionella: Molecular Microbiology*, eds K. Heuner and M. S. Swanson (Norfolk, UK: Horizon Scientific Press), 249.

Heuner, K., Bender-Beck, L., Brand, B. C., Luck, P. C., Mann, K. H., Marre, R., et al. (1995). Cloning and genetic characterization of the flagellum subunit gene (*flaA*) of *Legionella pneumophila* serogroup 1. *Infect. Immun.* 63, 2499–2507.

Heuner, K., Brand, B. C., and Hacker, J. (1999). The expression of the flagellum of *Legionella pneumophila* is modulated by different environmental factors. *FEMS Microbiol. Lett.* 175, 69–77. doi: 10.1111/j.1574-6968.1999.tb13603.x

Heuner, K., Dietrich, C., Skriwan, C., Steinert, M., and Hacker, J. (2002). Influence of the alternative σ^{28} factor on virulence and flagellum expression of *Legionella pneumophila*. *Infect. Immun.* 70, 1604–1608. doi: 10.1128/IAI.70.3.1604-1608.2002

Heuner, K., Dietrich, C., Steinert, M., Gobel, U. B., and Hacker, J. (2000). Cloning and characterization of a *Legionella pneumophila*-specific gene encoding a member of the LysR family of transcriptional regulators. *Mol. Gen. Genet.* 264, 204–211. doi: 10.1007/s004380000310

Heuner, K., Hacker, J., and Brand, B. C. (1997). The alternative sigma factor sigma28 of *Legionella pneumophila* restores flagellation and motility to an *Escherichia coli fliA* mutant. *J. Bacteriol.* 179, 17–23. doi: 10.1128/jb.179.1.17-23.1997

Heuner, K., Jacobi, S., Albert, C., Steinert, M., Brüggemann, H., and Buchrieser, C. (2006). "Gene expression and virulence in *Legionella*: the flagellar regulon," in *Legionella: State of the Art 30 Years After its Recognition*, eds N. Cianciotto, Abu, Y. Kwaik, Edelstein, P. Fields, B. Geary, D. Harrison, T. Joseph, C. Ratcliff, R. J. Stout, and Swanson (Washington, DC: ASM), 327–332.

Heuner, K., and Steinert, M. (2003). The flagellum of *Legionella pneumophila* and its link to the expression of the virulent phenotype. *Int. J. Med. Microbiol.* 293, 133–143. doi: 10.1078/1438-4221-00259

Hoppe, J., Unal, C. M., Thiem, S., Grimpe, L., Goldmann, T., Gassler, N., et al. (2017). PilY1 Promotes *Legionella pneumophila* infection of human lung tissue explants and contributes to bacterial adhesion, host cell invasion, and twitching motility. *Front. Cell. Infect. Microbiol.* 7:63. doi: 10.3389/fcimb.2017.00063

Horwitz, M. A. (1992). Interactions between macrophages and *Legionella pneumophila*. *Curr. Top. Microbiol. Immunol.* 181, 265–282. doi: 10.1007/978-3-642-77377-8_10

Horwitz, M. A., and Silverstein, S. C. (1980). Legionnaires' disease bacterium (*Legionella pneumophila*) multiples intracellularly in human monocytes. *J. Clin. Invest.* 66, 441–450. doi: 10.1172/JCI109874

Jacobi, S., Schade, R., and Heuner, K. (2004). Characterization of the alternative sigma factor sigma54 and the transcriptional regulator FleQ of *Legionella pneumophila*, which are both involved in the regulation cascade of flagellar gene expression. *J. Bacteriol.* 186, 2540–2547. doi: 10.1128/JB.186.9.2540-2547.2004

Jager, J., Marwitz, S., Tiefenau, J., Rasch, J., Shevchuk, O., Kugler, C., et al. (2014). Human lung tissue explants reveal novel interactions during *Legionella pneumophila* infections. *Infect. Immun.* 82, 275–285. doi: 10.1128/IAI.00703-13

Khodr, A., Kay, E., Gomez-Valero, L., Ginevra, C., Doublet, P., Buchrieser, C., et al. (2016). Molecular epidemiology, phylogeny and evolution of *Legionella. Infect. Genet. Evol.* 43, 108–122. doi: 10.1016/j.meegid.2016.04.033

Kirby, J. R. (2009). Chemotaxis-like regulatory systems: unique roles in diverse bacteria. *Annu. Rev. Microbiol.* 63, 45–59. doi: 10.1146/annurev.micro.091208.073221

Kirov, S. M., Castrisios, M., and Shaw, J. G. (2004). Aeromonas flagella (polar and lateral) are enterocyte adhesins that contribute to biofilm formation on surfaces. *Infect. Immun.* 72, 1939–1945. doi: 10.1128/IAI.72.4.1939-1945.2004

Konishi, T., Yamashiro, T., Koide, M., and Nishizono, A. (2006). Influence of temperature on growth of *Legionella pneumophila* biofilm determined by precise temperature gradient incubator. *J. Biosci. Bioeng.* 101, 478–484. doi: 10.1263/jbb.101.478

Kozak, N. A., Buss, M., Lucas, C. E., Frace, M., Govil, D., Travis, T., et al. (2010). Virulence factors encoded by *Legionella longbeachae* identified on the basis of the genome sequence analysis of clinical isolate D-4968. *J. Bacteriol.* 192, 1030–1044. doi: 10.1128/JB.01272-09

Kumar, S., Stecher, G., and Tamura, K. (2016). MEGA7: Molecular Evolutionary Genetics Analysis Version 7.0 for bigger datasets. *Mol. Biol. Evol.* 33, 1870–1874. doi: 10.1093/molbev/msw054

La Scola, B., Birtles, R. J., Greub, G., Harrison, T. J., Ratcliff, R. M., and Raoult, D. (2004). *Legionella drancourtii* sp. nov., a strictly intracellular amoebal pathogen. *Int. J. Syst. Evol. Microbiol.* 54(Pt 3), 699–703. doi: 10.1099/ijs.0.02455-0

Lau, H. Y., and Ashbolt, N. J. (2009). The role of biofilms and protozoa in *Legionella pathogenesis*: implications for drinking water. *J. Appl. Microbiol.* 107, 368–378. doi: 10.1111/j.1365-2672.2009.04208.x

Levi, A., Folcher, M., Jenal, U., and Shuman, H. A. (2011). Cyclic diguanylate signaling proteins control intracellular growth of *Legionella pneumophila. MBio* 2:e00316–10. doi: 10.1128/mBio.00316-10

Lo Presti, F., Riffard, S., Meugnier, H., Reyrolle, M., Lasne, Y., Grimont, P. A., et al. (2001). *Legionella gresilensis* sp. nov. and *Legionella beliardensis* sp. nov., isolated from water in France. *Int. J. Syst. Evol. Microbiol.* 51(Pt 6), 1949–1957. doi: 10.1099/00207713-51-6-1949

Macnab, R. M. (2003). How bacteria assemble flagella. *Annu. Rev. Microbiol.* 57, 77–100. doi: 10.1146/annurev.micro.57.030502.090832

Mampel, J., Spirig, T., Weber, S. S., Haagensen, J. A., Molin, S., and Hilbi, H. (2006). Planktonic replication is essential for biofilm formation by *Legionella pneumophila* in a complex medium under static and dynamic flow conditions. *Appl. Environ. Microbiol.* 72, 2885–2895. doi: 10.1128/AEM.72.4.2885-2895.2006

Mascarenhas, D. P., and Zamboni, D. S. (2017). Inflammasome biology taught by *Legionella pneumophila*. *J. Leukoc. Biol.* 101, 841–849. doi: 10.1189/jlb.3MR0916-380R

McCarter, L. L. (2006). Regulation of flagella. *Curr. Opin. Microbiol.* 9, 180–186. doi: 10.1016/j.mib.2006.02.001

Micali, G., and Endres, R. G. (2016). Bacterial chemotaxis: information processing, thermodynamics, and behavior. *Curr. Opin. Microbiol.* 30, 8–15. doi: 10.1016/j.mib.2015.12.001

Minamino, T., and Imada, K. (2015). The bacterial flagellar motor and its structural diversity. *Trends Microbiol.* 23, 267–274. doi: 10.1016/j.tim.2014.12.011

Molmeret, M., Bitar, D. M., Han, L., and Kwaik, Y. A. (2004). Disruption of the phagosomal membrane and egress of *Legionella pneumophila* into the cytoplasm during the last stages of intracellular infection of macrophages and *Acanthamoeba polyphaga*. *Infect. Immun.* 72, 4040–4051. doi: 10.1128/IAI.72.7.4040-4051.2004

Molmeret, M., Jones, S., Santic, M., Habyarimana, F., Esteban, M. T., and Kwaik, Y. A. (2010). Temporal and spatial trigger of post-exponential virulence-associated regulatory cascades by *Legionella pneumophila* after bacterial escape into the host cell cytosol. *Environ. Microbiol.* 12, 704–715. doi: 10.1111/j.1462-2920.2009.02114.x

Molofsky, A. B., Byrne, B. G., Whitfield, N. N., Madigan, C. A., Fuse, E. T., Tateda, K., et al. (2006). Cytosolic recognition of flagellin by mouse macrophages restricts *Legionella pneumophila* infection. *J. Exp. Med.* 203, 1093–1104. doi: 10.1084/jem.20051659

Molofsky, A. B., Shetron-Rama, L. M., and Swanson, M. S. (2005). Components of the *Legionella pneumophila* flagellar regulon contribute to multiple virulence traits, including lysosome avoidance and macrophage death. *Infect. Immun.* 73, 5720–5734. doi: 10.1128/IAI.73.9.5720-5734.2005

Molofsky, A. B., and Swanson, M. S. (2003). *Legionella pneumophila* CsrA is a pivotal repressor of transmission traits and activator of replication. *Mol. Microbiol.* 50, 445–461. doi: 10.1046/j.1365-2958.2003.03706.x

Molofsky, A. B., and Swanson, M. S. (2004). Differentiate to thrive: lessons from the *Legionella pneumophila* life cycle. *Mol. Microbiol.* 53, 29–40. doi: 10.1111/j.1365-2958.2004.04129.x

Neumeister, B., Schoniger, S., Faigle, M., Eichner, M., and Dietz, K. (1997). Multiplication of different *Legionella* species in Mono Mac 6 cells and in *Acanthamoeba castellanii*. *Appl. Environ. Microbiol.* 63, 1219–1224.

Newton, H. J., Ang, D. K., van Driel, I. R., and Hartland, E. L. (2010). Molecular pathogenesis of infections caused by *Legionella pneumophila*. *Clin. Microbiol. Rev.* 23, 274–298. doi: 10.1128/CMR.00052-09

O'Connell, W. A., Dhand, L., and Cianciotto, N. P. (1996). Infection of macrophage-like cells by *Legionella* species that have not been associated with disease. *Infect. Immun.* 64, 4381–4384.

Ohnishi, K., Kutsukake, K., Suzuki, H., and Lino, T. (1992). A novel transcriptional regulation mechanism in the flagellar regulon of *Salmonella typhimurium*: an antisigma factor inhibits the activity of the flagellum-specific sigma factor, sigma F. *Mol. Microbiol.* 6, 3149–3157. doi: 10.1111/j.1365-2958.1992.tb01771.x

Ohno, A., Kato, N., Yamada, K., and Yamaguchi, K. (2003). Factors influencing survival of *Legionella pneumophila* serotype 1 in hot spring water and tap water. *Appl. Environ. Microbiol.* 69, 2540–2547. doi: 10.1128/AEM.69.5.2540-2547.2003

Orrison, L. H., Cherry, W. B., Tyndall, R. L., Fliermans, C. B., Gough, S. B., Lambert, M. A., et al. (1983). *Legionella oakridgensis*: unusual new species isolated from cooling tower water. *Appl. Environ. Microbiol.* 45, 536–545.

Osterman, I. A., Dikhtyar, Y. Y., Bogdanov, A. A., Dontsova, O. A., and Sergiev, P. V. (2015). Regulation of flagellar gene expression in bacteria. *Biochem. Mosc.* 80, 1447–1456. doi: 10.1134/S000629791511005X

Ott, M., Messner, P., Heesemann, J., Marre, R., and Hacker, J. (1991). Temperature-dependent expression of flagella in *Legionella*. *J. Gen. Microbiol.* 137, 1955–1961. doi: 10.1099/00221287-137-8-1955

Pallen, M. J., Penn, C. W., and Chaudhuri, R. R. (2005). Bacterial flagellar diversity in the post-genomic era. *Trends Microbiol.* 13, 143–149. doi: 10.1016/j.tim.2005.02.008

Piao, Z., Sze, C. C., Barysheva, O., Iida, K., and Yoshida, S. (2006). Temperature-regulated formation of mycelial mat-like biofilms by *Legionella pneumophila*. *Appl. Environ. Microbiol.* 72, 1613–1622. doi: 10.1128/AEM.72.2.1613-1622.2006

Polesky, A. H., Ross, J. T., Falkow, S., and Tompkins, L. S. (2001). Identification of *Legionella pneumophila* genes important for infection of amoebas by signature-tagged mutagenesis. *Infect. Immun.* 69, 977–987. doi: 10.1128/IAI.69.2.977-987.2001

Pruckler, J. M., Benson, R. F., Moyenuddin, M., Martin, W. T., and Fields, B. S. (1995). Association of flagellum expression and intracellular growth of *Legionella pneumophila*. *Infect. Immun.* 63, 4928–4932.

Rasis, M., and Segal, G. (2009). The LetA-RsmYZ-CsrA regulatory cascade, together with RpoS and PmrA, post-transcriptionally regulates stationary phase activation of *Legionella pneumophila* Icm/Dot effectors. *Mol. Microbiol.* 72, 995–1010. doi: 10.1111/j.1365-2958.2009.06705.x

Ren, T., Zamboni, D. S., Roy, C. R., Dietrich, W. F., and Vance, R. E. (2006). Flagellin-deficient *Legionella* mutants evade caspase-1- and Naip5-mediated macrophage immunity. *PLoS Pathog.* 2:e18. doi: 10.1371/journal.ppat.0020018

Ricci, M. L., Torosantucci, A., Scaturro, M., Chiani, P., Baldassarri, L., and Pastoris, M. C. (2005). Induction of protective immunity by *Legionella pneumophila* flagellum in an A/J mouse model. *Vaccine* 23, 4811–4820. doi: 10.1016/j.vaccine.2005.05.013

Richards, A. M., Von Dwingelo, J. E., Price, C. T., and Abu Kwaik, Y. (2013). Cellular microbiology and molecular ecology of *Legionella*-amoeba interaction. *Virulence* 4, 307–314. doi: 10.4161/viru.24290

Rizzardi, K., Winiecka-Krusnell, J., Ramliden, M., Alm, E., Andersson, S., and Byfors, S. (2015). *Legionella norrlandica* sp. nov., isolated from the biopurification systems of wood processing plants. *Int. J. Syst. Evol. Microbiol.* 65(Pt 2), 598–603. doi: 10.1099/ijs.0.068940-0

Robertson, P., Abdelhady, H., and Garduno, R. A. (2014). The many forms of a pleomorphic bacterial pathogen-the developmental network of *Legionella pneumophila*. *Front. Microbiol.* 5:670. doi: 10.3389/fmicb.2014.00670

Rowbotham, T. J. (1986). Current views on the relationships between amoebae, legionellae and man. *Isr. J. Med. Sci.* 22, 678–689.

Rust, M., Borchert, S., Niehus, E., Kuehne, S. A., Gripp, E., Bajceta, A., et al. (2009). The *Helicobacter pylori* anti-sigma factor FlgM is predominantly cytoplasmic and cooperates with the flagellar basal body protein FlhA. *J. Bacteriol.* 191, 4824–4834. doi: 10.1128/JB.00018-09

Sahr, T., Bruggemann, H., Jules, M., Lomma, M., Albert-Weissenberger, C., Cazalet, C., et al. (2009). Two small ncRNAs jointly govern virulence and transmission in *Legionella pneumophila*. *Mol. Microbiol.* 72, 741–762. doi: 10.1111/j.1365-2958.2009.06677.x

Sahr, T., Rusniok, C., Dervins-Ravault, D., Sismeiro, O., Coppee, J. Y., and Buchrieser, C. (2012). Deep sequencing defines the transcriptional map of *L. pneumophila* and identifies growth phase-dependent regulated ncRNAs implicated in virulence. *RNA Biol.* 9, 503–519. doi: 10.4161/rna.20270

Sahr, T., Rusniok, C., Impens, F., Oliva, G., Sismeiro, O., Coppee, J. Y., et al. (2017). The *Legionella pneumophila* genome evolved to accommodate multiple regulatory mechanisms controlled by the CsrA-system. *PLoS Genet.* 13:e1006629. doi: 10.1371/journal.pgen.1006629

Saitou, N., and Nei, M. (1987). The neighbor-joining method: a new method for reconstructing phylogenetic trees. *Mol. Biol. Evol.* 4, 406–425.

Schalk, J. A., Euser, S. M., van Heijnsbergen, E., Bruin, J. P., den Boer, J. W., and de Roda Husman, A. M. (2014). Soil as a source of *Legionella pneumophila* sequence type 47. *Int. J. Infect. Dis.* 27, 18–19. doi: 10.1016/j.ijid.2014.05.009

Schell, U., Simon, S., and Hilbi, H. (2016). Inflammasome recognition and regulation of the *Legionella* flagellum. *Curr. Top. Microbiol. Immunol.* 397, 161–181. doi: 10.1007/978-3-319-41171-2_8

Schulz, T., Rydzewski, K., Schunder, E., Holland, G., Bannert, N., and Heuner, K. (2012). FliA expression analysis and influence of the regulatory proteins RpoN, FleQ and FliA on virulence and *in vivo* fitness in *Legionella pneumophila*. *Arch. Microbiol.* 194, 977–989. doi: 10.1007/s00203-012-0833-y

Schunder, E., Gillmaier, N., Kutzner, E., Eisenreich, W., Herrmann, V., Lautner, M., et al. (2014). Amino acid uptake and metabolism of *Legionella pneumophila* hosted by *Acanthamoeba castellanii*. *J. Biol. Chem.* 289, 21040–21054. doi: 10.1074/jbc.M114.570085

Steinert, M., Emody, L., Amann, R., and Hacker, J. (1997). Resuscitation of viable but non-culturable *Legionella pneumophila* philadelphia JR32 by Acanthamoeba castellanii. *Appl. Environ. Microbiol.* 63, 2047–2053.

Stewart, C. R., Muthye, V., and Cianciotto, N. P. (2012). *Legionella pneumophila* persists within biofilms formed by *Klebsiella pneumoniae*, *Flavobacterium* sp., and *Pseudomonas fluorescens* under dynamic flow conditions. *PLoS ONE* 7:e50560. doi: 10.1371/journal.pone.0050560

Stewart, C. R., Rossier, O., and Cianciotto, N. P. (2009). Surface translocation by *Legionella pneumophila*: a form of sliding motility that is dependent upon type II protein secretion. *J. Bacteriol.* 191, 1537–1546. doi: 10.1128/JB.01531-08

Swanson, M. S., and Hammer, B. K. (2000). *Legionella pneumophila* pathogesesis: a fateful journey from amoebae to macrophages. *Annu. Rev. Microbiol.* 54, 567–613. doi: 10.1146/annurev.micro.54.1.567

Szurmant, H., and Ordal, G. W. (2004). Diversity in chemotaxis mechanisms among the bacteria and archaea. *Microbiol. Mol. Biol. Rev.* 68, 301–319. doi: 10.1128/MMBR.68.2.301-319.2004

Tamura, K., and Kumar, S. (2002). Evolutionary distance estimation under heterogeneous substitution pattern among lineages. *Mol. Biol. Evol.* 19, 1727–1736. doi: 10.1093/oxfordjournals.molbev.a003995

Tiaden, A., Spirig, T., Weber, S. S., Bruggemann, H., Bosshard, R., Buchrieser, C., et al. (2007). The *Legionella pneumophila* response regulator LqsR promotes host cell interactions as an element of the virulence regulatory network controlled by RpoS and LetA. *Cell. Microbiol.* 9, 2903–2920. doi: 10.1111/j.1462-5822.2007.01005.x

Tlapak, H., Rydzewski, K., Schulz, T., Weschka, D., Schunder, E., and Heuner, K. (2017). Functional analysis of the alternative sigma-28 factor FliA and Its Anti-sigma factor FlgM of the non-flagellated *Legionella* species, *L. oakridgensis*. *J. Bacteriol.* 199:e00018-17. doi: 10.1128/JB.00018-17

Wang, C., Saito, M., Tanaka, T., Amako, K., and Yoshida, S. (2015). Comparative analysis of virulence traits between a *Legionella* feeleii strain implicated in Pontiac fever and a strain that caused Legionnaires' disease. *Microb. Pathog.* 89, 79–86. doi: 10.1016/j.micpath.2015.09.004

Whiley, H., and Bentham, R. (2011). *Legionella longbeachae* and legionellosis. *Emerg. Infect. Dis.* 17, 579–583. doi: 10.3201/eid1704.100446

Wolfe, A. J., and Visick, K. L. (2008). Get the message out: cyclic-Di-GMP regulates multiple levels of flagellum-based motility. *J. Bacteriol.* 190, 463–475. doi: 10.1128/JB.01418-07

Yu, V. L., Plouffe, J. F., Pastoris, M. C., Stout, J. E., Schousboe, M., Widmer, A., et al. (2002). Distribution of *Legionella* species and serogroups isolated by culture in patients with sporadic community-acquired legionellosis: an international collaborative survey. *J. Infect. Dis.* 186, 127–128. doi: 10.1086/341087

Zusman, T., Gal-Mor, O., and Segal, G. (2002). Characterization of a *Legionella pneumophila* relA insertion mutant and toles of RelA and RpoS in virulence gene expression. *J. Bacteriol.* 184, 67–75. doi: 10.1128/JB.184.1.67-75.2002

7

Hijacking of the Host Ubiquitin Network by *Legionella pneumophila*

Jiazhang Qiu[1,2*] *and Zhao-Qing Luo*[1,2,3*]

[1] *Center of Infection and Immunity, First Hospital, Jilin University, Changchun, China,* [2] *Key Laboratory of Zoonosis, Ministry of Education, College of Veterinary Medicine, Jilin University, Changchun, China,* [3] *Department of Biological Sciences, Purdue Institute for Inflammation, Immunology and Infectious Diseases, Purdue University, West Lafayette, IN, United States*

Edited by:
Elizabeth L. Hartland,
Hudson Institute of Medical Research, Australia

Reviewed by:
Shivani Pasricha,
University of Melbourne, Australia
Hubert Hilbi,
University of Zurich, Switzerland

****Correspondence:***
Jiazhang Qiu
qiujiazhang1983@163.com
Zhao-Qing Luo
luoz@purdue.edu

Protein ubiquitination is critical for regulation of numerous eukaryotic cellular processes such as protein homeostasis, cell cycle progression, immune response, DNA repair, and vesicular trafficking. Ubiquitination often leads to the alteration of protein stability, subcellular localization, or interaction with other proteins. Given the importance of ubiquitination in the regulation of host immunity, it is not surprising that many infectious agents have evolved strategies to interfere with the ubiquitination network with sophisticated mechanisms such as functional mimicry. The facultative intracellular pathogen *Legionella pneumophila* is the causative agent of Legionnaires' disease. *L. pneumophila* is phagocytosed by macrophages and is able to replicate within a niche called Legionella-containing vacuole (LCV). The biogenesis of LCV is dependent upon the Dot/Icm type IV secretion system which delivers more than 330 effector proteins into host cytosol. The optimal intracellular replication of *L. pneumophila* requires the host ubiquitin-proteasome system. Furthermore, membranes of the bacterial phagosome are enriched with ubiquitinated proteins in a way that requires its Dot/Icm type IV secretion system, suggesting the involvement of effectors in the manipulation of the host ubiquitination machinery. Here we summarize recent advances in our understanding of mechanisms exploited by *L. pneumophila* effector proteins to hijack the host ubiquitination pathway.

Keywords: type IV secretion, effectors, posttranslational modification, bacterial virulence, cell signaling

INTRODUCTION

Post-translational modification (PTM) is a biochemical mechanism in which amino-acid residues in a protein are covalently modified by specific enzymes. PTMs regulate the function of most proteins, thereby allowing the modulation of a wide range of cellular processes, which permits cells to respond to endogenous developmental signals or external stimuli imposed by environmental changes. More than 200 types of PTM have been described, including ubiquitination which is among one of best studied (Deribe et al., 2010).

The Eukaryotic Ubiquitination Network

Ubiquitination is a central signaling system that is conserved among all eukaryotic organisms (Hershko and Ciechanover, 1998; **Figure 1A**). Ubiquitination is defined as the covalent conjugation of one or several ubiquitin moieties to residues (mostly lysines) of target proteins. The conventional conjugation of proteins with ubiquitin occurs through the universally conserved three-enzyme cascade (Hershko and Ciechanover, 1998). Free ubiquitin is first activated by E1

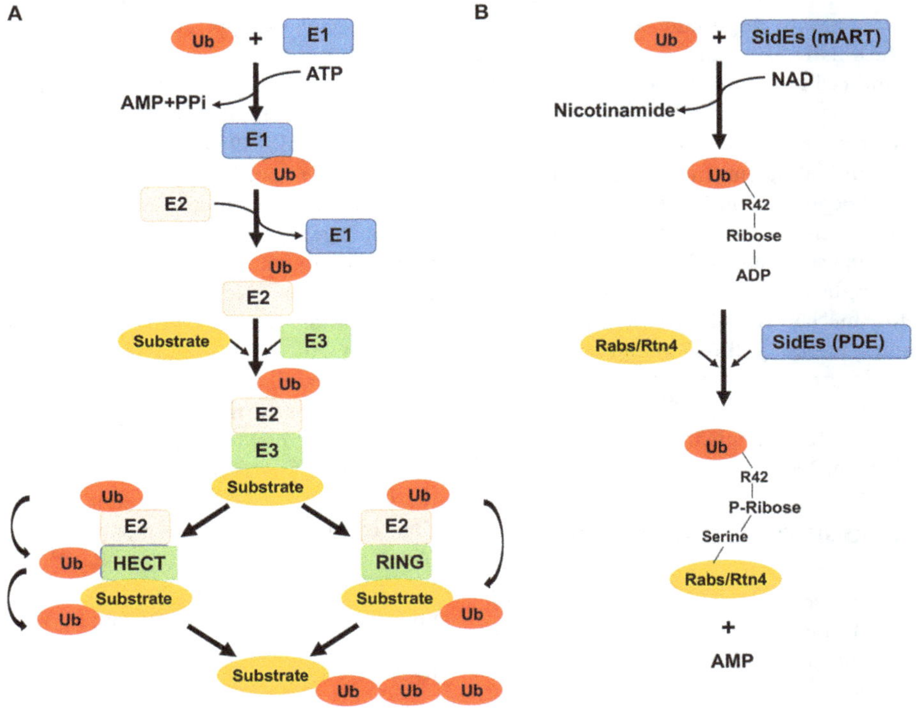

FIGURE 1 | Enzymes and chemical reactions involved in ubiquitination catalyzed by the canonical mechanism and by members of the SidE family proteins. **(A)** In the canonical mechanism, a ubiquitin molecule is activated by E1 at the expense of an ATP. The activated ubiquitin is first linked to E1 via a labile thioester bond prior to being transferred to the E2 conjugation enzyme, also linked by a thioester bond. The final step of the reaction differs greatly among different groups of E3 enzymes which dictate substrate specificity. For members of the HECT family of E3 enzymes (left), a reaction intermediate is formed, again by the formation of a thiolester bond between ubiquitin and E3, from where it is finally linked to lysine residues of the substrate. For other groups of E3 enzymes such as the RING family, the ubiquitin moiety is directly transferred to the substrate without the formation of an intermediate. **(B)** The reaction catalyzed by the SidE family begins with ubiquitin activation by ADP-ribosylation at Arg_{42} to produce the reaction intermediate ADP-ribosylated ubiquitin (ADPR-Ub), a nicotinamide moiety is released in this step of the reaction. In the second reaction, ADPR-Ub is cleaved by a phosphodiesterase (PDE) activity also embedded in these proteins, resulting in the attachment of phosphoribosylated ubiquitin to serine residues of the substrate and the release of AMP. How ubiquitin is recognized by the mART motif is unknown, nor is the mechanism of substrate recognition presumably by the PDE domain.

(ubiquitin-activating enzyme) at the expense of ATP to form a ubiquitin-AMP intermediate that is used to modify E1 by a thiol-ester linkage formed between the carboxyl-terminus of ubiquitin and a cysteine residue on E1. The E1-linked ubiquitin is then transferred via a transthiolation reaction to a cysteine residue on E2 (ubiquitin-conjugating enzyme). Finally, E3 (ubiquitin-protein ligase) catalyzes the covalent attachment of ubiquitin to substrate via a isopeptide bond formed between the C-terminal end of ubiquitin to the ε-amino group, mostly on a lysine residue (Hershko and Ciechanover, 1998). In eukaryotic cells, there are two genes that encode E1 enzymes and dozens of genes encode E2 enzymes (Ye and Rape, 2009). Since E3 enzymes play an important role in determining substrate specificity, there are a large number of genes (over 1,000 in human genome in estimation) encoding E3 enzymes (Rytkönen and Holden, 2007). The large number of E3 ligases are classified into three major types according to the presence of different catalytic motifs and the mechanisms of catalysis. Members of the HECT (Homologous to the E6AP C-terminus) domain family E3 ligases require the formation of a thiol-ester intermediate with ubiquitin on the active cysteine residue prior to being transferred to substrates (Metzger et al., 2012). Members of the RING (really interesting new gene) family E3 ligases function as adaptors that bind to both E2 and the substrate, thereby facilitating the direct transfer of ubiquitin molecule from E2 to the substrate (Metzger et al., 2012). The RING–IBR (In-Between-RINGs)–RING (RBR) type of E3s catalyze ubiquitination through a RING-HECT hybrid mechanism (Wenzel et al., 2011; Metzger et al., 2014).

The effect of ubiquitination to a large extent depends on the length and linkage type of the ubiquitin chain attached to the protein. Based on the length of the ubiquitin chain, ubiquitination can be divided into mono-ubiquitination, multi-monoubiquitination, and polyubiquitination. Monoubiquitination and multi-monoubiquitination have been shown to regulate subcellular protein localization, endocytosis, and the recruitment of ubiquitin-binding proteins (Haglund and Dikic, 2005). The formation of polyubiquitin chains can occur on one of the seven lysine residues (K6, K11, K27, K29, K33, K48, and K63) and the amino terminal methionine (M1) (Komander and Rape, 2012). Polyubiquitin chains linked via K48 render the modified proteins to be recognized by the proteasome for destruction. In contrast, polyubiquitin chains conjugated via K63 controls a wide range of important cellular signaling involved in processes such as DNA repair, endocytosis, vesicle trafficking, immunity, and cell cycle progression (Haglund and Dikic, 2005).

Ubiquitination is a reversible process catalyzed by a group of proteins known as deubiquitinating enzymes or deubiquitinases (DUBs) that cleave the isopeptide bond between ubiquitin and the modified protein. Therefore, DUBs act to recycle ubiquitin and restore the ubiquitinated substrate back to its original form. The human genome is predicted to encode nearly 100 DUBs, which according to the mechanism of action, are classified into five different families: the ubiquitin-C-terminal hydrolases (UCHs), ubiquitin-specific proteases (USPs), Machado-Joseph domain (MJD) DUBs, ovarian-tumor (OTU) domain DUBs, and the Jab1/Mov34/Mpr1 Pad1 N-terminal+ (MPN+) (JAMM) domain proteases (Wilkinson, 2009).

Subversion of Ubiquitination by Bacterial Pathogens

Prokaryotic cells do not possess genes coding for ubiquitin, therefore the prototypical ubiquitination pathway is absent in bacteria. Proteins modified by ubiquitination are critical regulators in virtually every eukaryotic cellular process. Thus, effective hijacking of the host ubiquitination system is essential for the success of many pathogens in their evasion of host immunity or their exploitation of host resources. For symbiotic and pathogenic bacteria, such exploitation is achieved by virulence factors that are either secreted into the extracellular milieu (which enter the host cells via various mechanisms) or are directly translocated into the cytosol of host cells via specialized secretion systems (e.g., Type III and Type IV secretion systems; Ashida et al., 2014). Some examples are species of *Shigella*, enteropathogenic *Escherichia coli* (EPEC), as well as pathogenic species of *Salmonella*, *Legionella* and *Chlamydia*. Accumulating evidence has shown that bacterial effector proteins exploit the host ubiquitination machinery by diverse strategies (Ashida et al., 2014; Zhou and Zhu, 2015).

The study of how bacteria co-opt the host ubiquitination machinery is a rapidly growing research field, and great progress has been made in past decades (Ashida et al., 2014). In this review, we will focus on discussing the current knowledge of effectors utilized by *Legionella pneumophila* to interfere with host ubiquitination signaling pathways. Strategies used by other human or plant bacterial pathogens will not be covered, and readers are referred to other excellent reviews (Rytkönen and Holden, 2007; Ashida et al., 2014; Zhou and Zhu, 2015; Lin and Machner, 2017).

Intracellular Replication of *Legionella pneumophila*

L. pneumophila is an opportunistic human pathogen that causes Legionnaires' disease, a form of potentially fatal pneumonia (Rowbotham, 1980). The genus *Legionella* was originally described in 1979 after the bacterium was identified following an outbreak of lethal pneumonia that affected participants of the 1976 American Legion Convention in Philadelphia (Fraser et al., 1977). *Legionella* spp. are ubiquitous environmental bacteria, found in freshwater niches and soil where they exist as parasites of unicellular eukaryotes such as amoebae, which are considered their natural hosts and the major source of evolutionary pressure (Moliner et al., 2010). Inhalation of aerosols contaminated by *Legionella* spp. by susceptible individuals can lead to lung infection due to robust intracellular replication in alveolar macrophages (Newton et al., 2010). The majority of human infections are caused by serogroup 1 of *L. pneumophila* and *L. longbeachae* (Newton et al., 2010). We will focus our discussion on *L. pneumophila*, the best-studied species of this pathogen.

The intracellular life cycle of *L. pneumophila* in human cells is similar to that in amoebae, which is characterized by quick establishment and maturation of the *Legionella*-containing vacuole (LCV) into a compartment with features typical for the rough endoplasmic reticulum (Swanson and Isberg, 1995). The maturation of this compartment is accompanied by sequential intimate interactions with organelles such as the ER, mitochondria, and ribosomes. The early LCV undergoes phosphoinositide conversion from PI(3)P to PI(4)P (Weber et al., 2014), the acquisition of ER resident proteins (Swanson and Isberg, 1995; Lu and Clarke, 2005) and expansion due to ER remodeling probably in part driven by the large GTPase Atlastin (Steiner et al., 2017). It is believed that this conversion allows the LCV to evade fusion with the lysosomal network (Isberg et al., 2009).

Intracellular replication of *L. pneumophila* depends completely on the Dot/Icm type IV secretion system (T4SS), which translocates more than 330 protein substrates into host cells (Finsel and Hilbi, 2015; Ensminger, 2016). These effectors comprise more than 10% of the genes predicted to code for proteins, which represent the largest arsenal of effectors among characterized bacterial pathogens. Considerable progress has been made in biochemical and cell biological studies of these effectors in the past decade, which revealed the manipulation of diverse host processes by sophisticated and novel mechanisms (Qiu and Luo, 2017).

The importance of the ubiquitin network in *L. pneumophila* virulence was first observed in a study aiming at identifying host factors important for its LCV formation and intracellular replication (Dorer et al., 2006). One of the targets found was Cdc48/p97 (Dorer et al., 2006), an AAA-ATPase that is critical for many ubiquitin-dependent processes including ER-associated degradation (ERAD) (Jarosch et al., 2002). Cdc48/p97 also recognizes ubiquitinated proteins, and often acts as a chaperone to facilitate the delivery of ubiquitinated proteins to the proteasome (Gallagher et al., 2014). This study also found that the LCV is decorated with ubiquitinated proteins shortly after its formation and such decoration requires the Dot/Icm transporter (Dorer et al., 2006), which suggests the co-option of host ubiquitination by Dot/Icm effectors. Here we will highlight the subversion of the host ubiquitination machinery by Dot/Icm effectors that function by mimicking known mechanisms or by unprecedented modes of action.

L. PNEUMOPHILA EFFECTORS THAT FUNCTION AS E3 UBIQUITIN LIGASES

Pathogen-mediated ubiquitination is mostly catalyzed by virulence factors that mimic the function of E3 ligases (Maculins et al., 2016). In the case of *L. pneumophila*, a large cohort of effector proteins are known to be involved in ubiquitination, either by mimicking classic E3 ligase families or by completely novel mechanisms (**Table 1**).

U-Box and F-Box E3 Ligases

The RING type E3 ligases which contain a conserved RING domain constitute the large majority of known E3s in eukaryotic cells (Metzger et al., 2014). The RING domain consists of 40–60 residues and coordinates two Zn^{2+} ions in a cross-braced arrangement to form a platform for binding to E2s. RING-type domains can either exist as single-subunit proteins which tend to form homodimers and heterodimers, or as multi-subunit assemblies, including Cullin RING E3 ligase complexes (CRLs). Each CRL subfamily consists of a Cullin protein serving as scaffold, a small RING protein (in most cases Rbx1/Roc1/Hrt1), an adaptor protein and a protein for substrate binding. The best-studied CRLs are the SCF (Skp1-Cul1-F-box protein) family, which contains the RING-domain protein Rbx1, Cullin 1, SKP1 (S-phase-kinase associated protein 1), and an F-box domain-containing protein that directly binds SKP1 (Schulman et al., 2000). In addition, F-box-containing proteins are capable of recognizing specific substrates via leucine-rich repeat (LRR) or WD40 protein-binding domains. The U box is a motif of 70 amino acids that is present in proteins from yeast to humans; it is capable of assembling poly-ubiquitin chains (Hatakeyama et al., 2001). Due to the structural similarity between U-box and RING domain, U-box-containing E3 ligases are classified as RING-type E3s. The U-box E3s use intramolecular interactions other than zinc chelation to maintain the RING finger motif due to the absence of canonical cysteine residues for Zn^{2+} coordination (Hatakeyama et al., 2001). A wide range of host signaling pathways are controlled by ubiquitination catalyzed by the RING-type family E3 ligase, and this mechanism is often targeted by bacterial pathogens for their own advantage.

A study aiming at screening for genes that encode proteins with features typical for eukaryotic proteins in the *L. pneumophila* genome identified proteins that harbor domain structures known to be involved in ubiquitin manipulation. These include two proteins that harbor an F-box domain and one gene product that harbors a U-box domain (de Felipe et al., 2005). Currently, seven F-box-containing proteins (LegU1, LicA, Lpg1975/Lpp1959, AnkB/LegAU13), PpgA/Lpg2224 Lpg2525, and Lpp2486 (Only in strain Paris) and two U-box-containing proteins (LubX/LegU2 and GobX) have been identified in *L. pneumophila* strain Philadelphia 1 (Hubber et al., 2014). Without exception, these proteins are translocated to the host cytosol via the Dot/Icm machinery during infection. Although, the exact number varies, proteins that contain these domains exist in predicted effectors among all sequenced Legionella species (Burstein et al., 2016). Four of them LegU1, AnkB, LubX, and GobX have been proven to possess E3 ligase activity through biochemical studies (Kubori et al., 2008; Ensminger and Isberg, 2010; Ensminger, 2016).

Effector proteins are generally thought to target host proteins. However, LubX is capable of binding and ubiquitinating SidH, another *L. pneumophila* effector protein, leading to its degradation by the proteasome (Kubori et al., 2010). LubX is thus designated as a "metaeffector," an effector that regulates the activity of one or more other effectors. Expression and thus the translocation of LubX only occur several hours after bacterial uptake by host cells, and peaks at 10 h post infection. This delayed translocation of LubX to the host cytosol results in the shutdown of SidH within the host cells at later stages of infection, suggesting a temporal regulation of SidH activity by LubX (Kubori et al., 2010; **Figure 2A**). These results suggest that SidH is only beneficial for bacterial infection in the first several hours after uptake. Indeed, disruption of *lubX* led to the persistence of intracellular SidH accompanied by a hyper-lethal phenotype of *L. pneumophila* in a fly infection model (Kubori et al., 2010).

LubX contains two domains that have a remarkable similarity to the eukaryotic U-box. LubX has ubiquitin ligase activity with a preference for the UbcH5a or UbcH5c E2 enzymes (Kubori et al., 2008). Structural studies have provided more detailed insights in the molecular mechanism adopted by U-box domains of LubX (Quaile et al., 2015). The two U-box domains are structurally similar and both have adopted the typical fold of their eukaryotic counterparts. The structure of LubX in complex with E2 enzyme UBE2D2 highlighted the remarkable differences in recognizing E2 enzymes between the U-box domains within LubX. Although, the U-box folds are highly conserved, there are significant variations of residues in U-box 2 that are critical for the formation of canonical E2 binding site in most U-boxes, preclude E2 association by U-box 2 (Quaile et al., 2015). Additionally, among the surface-exposed residues in LubX, Arg_{121} was the only residue identified to be critical for the interaction between LubX and SidH (One of the *L. pneumophila* effector proteins). However, Arg_{121} localizes to the alpha C helix connecting the two U-box domains, and thus is not part of the U-box fold (Quaile et al., 2015). Notably, the U-box 2 domain employed by Kubori et al. is required for the association of SidH and also contain the alpha C helix, and thus the Arg_{121} is included (Kubori et al., 2010). Interaction of LubX and SidH might occur over a large area, and require the contribution of a number of residues for interaction. Therefore, single point mutation likely fails to disrupt the interaction surface. Instead, Arg_{121} may be important for the stabilization of the LubX structure (Quaile et al., 2015).

LubX also binds to the host factor Cdc2-like kinase 1 (Clk1) and directs its polyubiquitination *in vitro*. The N-terminal U-box domain (U-box 1) of LubX is essential for ubiquitin ligation, and serves as the E2 binding site, while the C-terminal U-box (U-box 2) is dispensable for interaction with Clk1 (Kubori et al., 2008). Thus, LubX has a non-canonical U-box domain that functions to mediate substrate recognition rather than E2 binding, which is a function not previously reported for eukaryotic U-box domains. The reason that LubX adopts a U-box domain for substrate binding is unclear. Clk kinases have been

TABLE 1 | *L. pneumophila* Dot/Icm effectors involved in ubiquitination.

Effectors (gene number)	Aliases	Interactor/Substrate	Enzymatic activity	Function	References
lpg0171	legU1	SKP1, Cullin 1, BAT3	F-Box protein, E3 ubiquitin ligase	Unknown	Ensminger and Isberg, 2010
lpg1408	licA	SKP1	F-Box protein	Unknown	Ensminger and Isberg, 2010
lpg2144/lpp2082	legAU13/ankB	SKP1, Cullin 1, Parvin B	F-Box protein, E3 ubiquitin ligase	Recruitment of polyubiquitinated species to LCV; Generation of amino acids for *L. pneumophila* replication	Price et al., 2009, 2011; Ensminger and Isberg, 2010; Lomma et al., 2010
lpg2224	PpgA	Unknown	F-Box protein	Unknown	Ensminger and Isberg, 2010
lpg2525	–	Unknown	F-Box protein	Unknown	Ensminger and Isberg, 2010
lpp2486	–	Unknown	F-Box protein	Unknown	
lpg2455	GobX	Unknown	U-Box protein, E3 ubiquitin ligase	Unknown	Lin et al., 2015
lpg2830	LegU2/LubX	Clk1, SidH	U-Box protein, E3 ubiquitin ligase	SidH degradation	Kubori et al., 2008, 2010
lpg2510 lpg2511	SdcA and SidC	Unknown	E3 ubiquitin ligase	Recruitment of ER vesicles and polyubiquitinated species to LCV	Hsu et al., 2014
lpg0234	SidE	Rab1, Rab6a, Rab30, Rab33b, Rtn4	All-in-one ubiquitin conjugation enzyme; Deubiquitinase	Intracellular replication; regulation of ubiquitin dynamics on the LCV; Recruitment of ER markers to the LCV; ER tubule Rearrangement.	Sheedlo et al., 2015; Bhogaraju et al., 2016; Qiu et al., 2016; Kotewicz et al., 2017
lpg2153	SdeC				
lpg2156	SdeB				
lpg2157	SdeA				
lpg1148	LupA	Unknown	Deubiquitinase	Unknown	Urbanus et al., 2016
lpg2155	SidJ	Rab1, Rab6a, Rab30, Rab33b, Rtn4	Phosphodiesterase, Deubiquitinase	Recruitment of ER markers to the LCV; Regulation of SidEs-mediated substrates modification	Liu and Luo, 2007; Qiu et al., 2017

shown to interact with, and phosphorylate, serine- and arginine-rich (SR) proteins, which in turn regulate mRNA splicing (Prasad et al., 1999). Inhibition of Clk kinases interferes with intracellular growth of *L. pneumophila*, suggesting that these enzymes regulate pathways important for the development of the bacterial vacuole (Kubori et al., 2008). However, deletion of *lubX* did not cause any growth defect in mouse macrophages or in protozoan cells; in addition, over-expression of LubX in cells only ubiquitinates a small amount of Clk1 (Kubori et al., 2008). Therefore, the mechanism used by Clk1 to modulate *L. pneumophila* growth and whether Clk1 is ubiquitinated by LubX under infection conditions as well as the consequences of modification remain to be clarified.

Another Dot/Icm effector protein GobX possesses a central domain that has a secondary structure remotely similar to U-box motif. GobX exhibits E3 ubiquitin ligase activity in reactions with the E2 enzymes UbcH5a, -5b, -5c, or UbcH6 (Lin et al., 2015). GobX exhibit limited homology at the primary sequence level to other U-box domains, however, the conserved hydrophobic/aromatic residues involved in E2 interaction used by other U-box proteins are also present in the secondary structure of GobX, as mutations in Ile-58 or Trp-87 strongly attenuated its ubiquitination activity (Lin et al., 2015). In addition, the hydrophobic lipid palmitate is covalently attached to Cys_{175} of GobX, which allows the protein to specifically localize to the Golgi apparatus (Lin et al., 2015). Therefore, GobX exploits host cell S-palmitoylation to gain accurate host subcellular targeting. Similar to most Dot/Icm effector proteins, GobX is dispensable for intracellular survival and proliferation of *L. pneumophila* within host cells, which again highlights the potential functional redundancy within the effector repertoire (Lin et al., 2015). Host substrates of GobX are currently unknown, which limits our understanding of how its ubiquitination activity benefits intracellular bacterial growth.

To date, all sequenced *L. pneumophila* strains encode genes with predicted F-box domains. For example, strain Philadelphia-1 harbors five F-box motif-containing proteins (Ensminger and Isberg, 2010). The F-box is a motif that is best known for its role in interaction with other proteins such as SKP1, a core component of the SCF complex (Skaar et al., 2013). Since bacteria do not produce any SKP1, CUL1, or RBX1, *L. pneumophila* F-box proteins require SCF components provided by host cells to be functional. All the F-box motif-containing proteins are delivered into host cells during infection through the Dot/Icm apparatus (Ensminger and Isberg, 2010). Three of the *L. pneumophila* F-box proteins, LegU1, AnkB, and LicA were able to interact with SKP1 in mammalian cells, indicating the presence of a functional F-box domain within these proteins (Ensminger and Isberg, 2010). In contrast, no interaction of SKP1 was detected in cells expressing either PpgA/Lpg2224 or Lpg2525, suggesting that these proteins may not function as canonical F-box proteins (Ensminger and Isberg, 2010). Additionally, LegU1 and AnkB also interact with Cullin 1 and integrated into functional SCF complexes which may confer E3 ligase activity. Although, LicA binds to SKP1, it fails to interact with Cullin 1, suggesting that this effector is unable to form a functional E3 ligase. Alternatively, LicA may

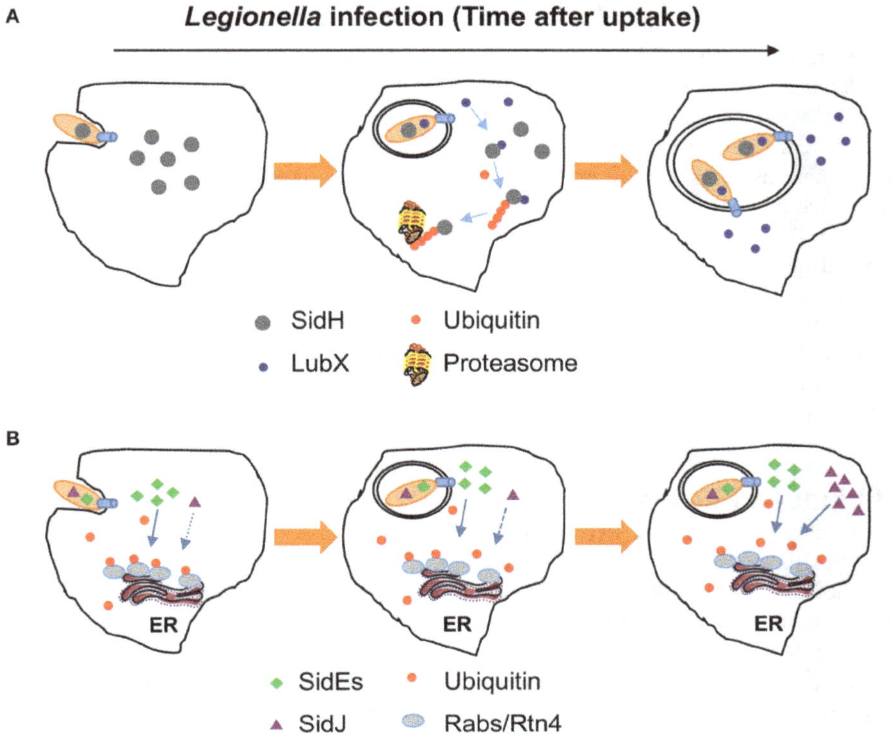

FIGURE 2 | Temporal regulation of effector activity by effectors. **(A)** Regulation of SidH by LubX. The expression of *lubX* does not become apparent until after several hours postinfection. This ubiquitin E3 ligase functions with the host machinery to ubiquitinate SidH, resulting its degradation by the proteasome. **(B)** Regulation of SidEs by SidJ. SidEs catalyze the ubiquitination of RTN4 or ER-associated Rab small GTPases such as Rab33b whereas SidJ reverses such modification by its phosphodiesterase activity. In the early phase of infection, the ratio between translocated SidEs and SidJ favors ubiquitination of relevant substrates, which is beneficial for the biogenesis of the LCV. Several hours after bacterial uptake, the activity of SidJ becomes dominant due to higher amount of translocated protein, which reverses the ubiquitination imposed by the SidEs.

require a different set of eukaryotic proteins to form an active E3 complex (Ensminger and Isberg, 2010). In addition, due to the presence of another predicted choline kinase domain, it is also possible that LicA functions to modulate SKP1 activity (Ensminger and Isberg, 2010).

Consistent with the predicted function of F-box proteins, SCF complexes formed by LegU1 or AnkB exhibit E3 ligase activity (Ensminger and Isberg, 2010). The E2 proteins UBCH5a and UBCH5c stimulate robust formation of self-ubiquitinated LegU1 and LegAU13. LegU1 binds the host protein HLA-B-associated transcript 3 (BAT3) and specifically directs its polyubiquitination (Ensminger and Isberg, 2010). BAT3 is an abundant and highly conserved protein in higher eukaryotes; it participates in the regulation of a wide range of host processes including apoptosis, the response to ER stress, p53-regulated gene expression, and Hsp70 stability (Desmots et al., 2008; Sasaki et al., 2008). However, the biological role of LegU1-catalyzed polyubiquitination of BAT3 in *L. pneumophila* pathogenesis remains unknown. LegU1 also interacts with another effector protein Lpg2160 via the BAT3-LegU1 complex. Yet, LegU1 does not detectably ubiquitinate Lpg2160 (Ensminger and Isberg, 2010). Since both LegU1 and Lpg2160 interact with BAT3, they might have overlapping functions during *L. pneumophila* infection. The host proteins targeted by AnkB for ubiquitination are still mysterious. Lomma et al. reported that Lpp2082, the ortholog of AnkB in *Legionella* strain Paris, interacts with the host protein Parvin β/ParvB, an endogenously ubiquitinated protein (Lomma et al., 2010). Surprisingly, expression of Lpp2082 in cells led to a decrease of ubiquitinated ParvB (Lomma et al., 2010). Lpp2082 might modulate ParvB ubiquitination by competing with the interaction sites normally used by eukaryotic E3 ligases. ParvB functions as a pro-apoptotic protein (Fukuda et al., 2003; Zhang et al., 2004), and thus reduction of ParvB ubiquitination by Lpp2082 will compromise its pro-apoptotic effects. Indeed, Lpp2082 is implicated in apoptotic signaling, as supported by the evidence that infection of cells with a Lpp2082 deficient mutant strain led to a significant reduction of caspase-3 activity (Lomma et al., 2010).

It is possible that some of the F-box and U-box proteins are responsible for the enrichment of ubiquitin species on the LCV (Dorer et al., 2006). These effectors may function to facilitate the degradation of proteins by proteasome that are detrimental to intracellular bacterial growth. This notion is consistent with the fact that inhibition of proteasome activity leads to arrest in the development of the LCV (Dorer et al., 2006). In *L. pneumophila* strains AA100 and Paris, significant defects in recruitment of ubiquitinated species to LCV were observed in AA100 strain with insertion mutation of *ankB* gene or Paris strain with an in-frame deletion of *Lpp2082* (Price et al., 2009; Lomma et al., 2010). In addition, these strains also displayed striking intracellular growth defect in mouse macrophages and *Acanthamoeba castellanii* (Al-Khodor et al., 2008; Lomma et al., 2010). However, the importance of AnkB in *L. pneumophila* virulence differs greatly among different strains. In strain Philadelphia-1, single mutants lacking one of the F-box and U-box proteins which has established E3 ligase activity, including LegU1, AnkB, LegU2, did not show significant growth defect within any examined host cells (Ivanov and Roy, 2009). Strikingly, recruitment of ubiquitinated proteins to the LCV was also not affected either by single mutants or a quadruple mutant strain lacking *legU1, ankB, licA,* and *legU2* (Ivanov and Roy, 2009). The genetic background of different strains and/or subtle differences in mutant strain construction may be responsible for such discrepancy.

New Type of E3 Ligases (SidC and SdcA)

The mature LCV is characterized by an enrichment of a particular phosphoinositide lipid, PI(4)P. Upon being delivered by the Dot/Icm machinery into the host cytosol, SidC is highly enriched on the LCV membrane via its PI4P binding domain located at the C-terminal end of the protein (Luo and Isberg, 2004; Ragaz et al., 2008). SidC and its paralog SdcA were shown to play a key role in recruiting ER-derived vesicles and ubiquitinated proteins onto the LCV, which requires the N-terminal domain of SidC and SdcA (Ragaz et al., 2008). Therefore, SidC and SdcA have since been considered as tethering factors for host proteins. Several groups determined the crystal structure of N-terminal domain of SidC, which showed a novel fold without resemblance to any characterized proteins (Gazdag et al., 2014; Horenkamp et al., 2014; Hsu et al., 2014). Detailed sequence homology analysis revealed a canonical Cys-His-Asp (C46, H444, and D446) catalytic triad located at the surface of SidC, a motif usually found in cysteine proteases and deubiquitinase (Hsu et al., 2014). Further study revealed that instead of acting as hydrolytic enzymes, SidC and SdcA exhibit E3 ligase activity in a mechanism that requires the C_{46}-H_{444}-D_{446} catalytic triad (Hsu et al., 2014). Among the several E2 enzymes tested, SidC functions most efficiently with UbcH7 and preferentially catalyzes the formation of K11 and K33-linked polyubiquitin chains (Hsu et al., 2014). SdcA shares 72% sequence identity with SidC; Yet, it prefers UbcH5 for efficient poly-ubiquitin chain assembly. The molecular mechanism of differential preference for E2 enzymes by SidC and SdcA remains to be studied. The E3 activity is essential for the recruitment of ER proteins and ubiquitinated proteins to the LCV by SidC and SdcA, as a C46A mutation abolishes this activity (Hsu et al., 2014). Further structural analysis of a larger portion of SidC that encompasses the E3 ligase domain and the PI4P binding domain (Luo et al., 2015) suggests that the PI4P binding domain masked the active site of the E3 ligase domain (Luo et al., 2015). Indeed, the activity of full-length SidC is lower than its truncation mutants lacking the PI4P-binding domain (Luo et al., 2015), suggesting that PI4P association leads to an "open" conformation where the catalytic sites of the SidC E3 ligase domain are exposed (Luo et al., 2015). Taken together, these findings suggest an intramolecular regulation model for SidC. Further, binding to PI4P may not only compartmentalize the activity of SidC and SdcA to the LCV but also maximize their activity. Such regulation would reduce non-specific protein ubiquitination and exert less unintended interference of host processes.

The observation that SidC and SdcA play important roles for the recruitment of ER-derived vesicles to the LCV suggests these E3 ligases manipulate the function of host proteins involved in vesicle trafficking. Indeed, the small GTPase Rab1, a key regulator of ER to the *cis*-Golgi trafficking, was mono-ubiquitinated during *L. pneumophila* infection in an manner that requires SidC and SdcA (Horenkamp et al., 2014). However, mono-ubiquitination of Rab1 was not detected in cells coexpressing these two proteins or in reactions containing all of the components required for the activity of SidC and SdcA (Hsu et al., 2014). Future studies need to focus on the identification of the substrates modified by SidC and SdcA, which will definitely shed light on how these E3 ligases benefit intracellular bacterial replication.

All-in-One Ubiquitin E3 Ligases

The three-enzyme cascade is the fundamental principle of all described ubiquitination events, in which E1 and E2 enzymes are indispensible for the reaction to occur (Qiu et al., 2016; **Figure 1A**). However, recent studies of the *L. pneumophila* SidE effector family (SidEs) rewrote this strict rule of ubiquitination (Qiu et al., 2016). SidEs distinguish themselves from most of the *L. pneumophila* Dot/Icm effectors by their importance in intracellular bacterial growth in the protozoan host *Dictyostelium discoideum* (Luo and Isberg, 2004; Bardill et al., 2005). Bioinformatics analysis identified a putative mono-ADP-ribosyltransferase (mART) motif located in the middle of all SidE family proteins such as SdeA that is essential for their toxicity to yeast and for the ability to complement a mutant lacking this effector family (Qiu et al., 2016). Proteins containing an mART motif usually catalyze mono-ADP-ribosylation of arginine residues in target proteins with nicotinamide adenine dinucleotide (NAD) as the substrate (Simon et al., 2014). However, no ADP-ribosylation activity was detected in reactions containing recombinant SdeA. Surprisingly, when expressed in mammalian cells, SidEs were found to induce ubiquitination of several ER-associated Rab small GTPases including Rab33b and Rab1, in a manner that requires the mART motif (Qiu et al., 2016). Further analysis revealed that SidEs catalyze ubiquitination by a mechanism that is fundamentally different from the classical three-enzyme cascade (Qiu et al., 2016). First, the reaction is independent of the host ubiquitination machinery and does not require E1 and E2 enzymes; second, instead of ATP, it utilizes NAD as the energy source; Third, SidEs activate ubiquitin via ADP-ribosylation of Arg_{42} of the modifier molecule to produce the reaction intermediate ADP-ribosylated ubiquitin (ADPR-Ub). Consistent with this observation, the two glycine residues in the carboxyl end of ubiquitin essential for the canonical reaction are not required for the new reaction. It also suggests that the ubiquitin is linked to the substrate via a covalent bond that differs from the isopeptide bond used by most canonical reactions (**Figure 1B**). This discovery represents the first example of an ubiquitin-specific mART, as well as the first documentation of E1/E2 independent ubiquitination (Bhogaraju and Dikic, 2016).

Two subsequent studies revealed that ADPR-Ub produced by the mART motif is utilized by a phosphodiesterase (PDE) activity also embedded in SidEs to modify target proteins (Bhogaraju et al., 2016; Kotewicz et al., 2017). In this reaction, the phosphodiester bond between the two phosphate groups in ADPR-Ub is cleaved by a phosphodiesterase activity conferred by the PDE domain, leading to the release of AMP and attachment of phosphoribosylated ubiquitin (PR-Ub) to serine residues in the presence of target proteins or the production of the free PR-Ub when water is the acceptor molecule (Bhogaraju et al., 2016; Kotewicz et al., 2017; **Figure 1B**).

Ubiquitination of Rab33b by SdeA detectably affects its activity in GTP loading and hydrolysis but did not detectably affect its stability (Qiu et al., 2016). However, how ubiquitination of the Rabs by SidEs contributes to *L. pneumophila* virulence remains to be studied. Notably, SidEs appear to have multiple structurally diverse substrates in host cells. These ligases ubiquitinate reticulon 4 (Rtn4), a protein that regulates the dynamics of the tubular ER. Ubiquitination of Rtn4 causes a rearrangement in tubule ER and its enrichment on the LCV (Kotewicz et al., 2017). It is anticipated that SidEs likely attack additional host proteins. Interestingly, both ADPR-Ub and PR-Ub produced by the activity of SidEs potently impair the conventional ubiquitination reaction by blocking the activation of E1 and E2 enzymes, leading to the interference of a wide range of ubiquitination-dependent cellular events including mitophagy and TNF signaling (Bhogaraju et al., 2016). Because a SdeC mutant defective in the PDE activity was unable to restore the virulence of the *L. pneumophila* mutant lacking the SidE effector family, ubiquitination of substrates but not the interference of host normal ubiquitination events is responsible for the role of the SidEs in bacterial virulence (Kotewicz et al., 2017). Nevertheless, the strong inhibitory effects of ADPR-Ub and PR-Ub suggest that eukaryotic cells may regulate ubiquitin signaling by producing these molecules from endogenous enzymes (Bhogaraju and Dikic, 2016; Bhogaraju et al., 2016).

Although, the biochemical mechanism of SidEs-mediated ubiquitination has been largely elucidated, several questions remain. First, how do SidEs recognize ubiquitin and substrates? Second, how do the mART and PDE motifs coordinate their activity? Are these two activities channeled or do they function independently of each other? Third, how does the activity of SidEs contribute to virulence? Future structural and cell biological studies will continue to provide exciting insights into these questions.

Pathogenic bacteria, especially intracellular pathogens often acquire toxins or effector proteins by horizontal gene transfer during their coevolution with host cells. It is therefore likely that eukaryotic cells utilize mechanisms similar to that by SidEs for ubiquitination. It is possible that proteins harboring domains capable of producing and utilizing ADPR-Ub form complexes to modify their substrates. The identification of such enzymes will surely lead to better appreciation of the cellular processes regulated by ubiquitin.

L. PNEUMOPHILA DEUBIQUITINASES (DUBS)

Ubiquitination is a reversible process and the removal of ubiquitin from modified proteins is carried out by the action of a large family of proteases known as deubiquitinase (DUBs). DUBs specifically catalyze the cleavage of isopeptide linkage between ubiquitin and substrate or within poly-ubiquitin chains, resulting in the release of ubiquitin as well as the termination or alteration of biological events of the substrate proteins (Wilkinson, 2009). DUBs have been found to be employed by several bacterial pathogens to effectively modulate the host signaling pathway regulated by ubiquitin (Zhou and Zhu, 2015). Examples include SseL of *Salmonella enterica* Typhimurium, ChlaDub1, and ChlaDub2 by *Chlamydia trachomatis*, and ElaD by *E. coli* (Misaghi et al., 2006; Catic et al., 2007; Rytkönen and Holden, 2007; Rytkönen et al., 2007). Not surprisingly, recent studies revealed that *L. pneumophila* contains effector proteins with DUB activity that play an important role in remodeling the bacterial phagosome (Sheedlo et al., 2015).

DUBs That Cleave Isopeptide Bonds

In addition to the mART and PDE domains mentioned above, the SidE family proteins harbor a DUB domain located at its amino terminal end, characterized by the presence of the Cys_{118}-His_{64}-Asp_{80} catalytic triad found in many proteases (Sheedlo et al., 2015). The DUB activity of SidEs exhibits a preference for K63-linked poly-ubiquitin chains (Sheedlo et al., 2015). These DUBs are also active against Neddylation, indicating substrate promiscuity (Sheedlo et al., 2015). The DUB activity of SidEs plays a role in the enrichment of polyubiquitin to the LCV, as more vacuoles harboring the SidEs deletion mutant are positive in the association with polyubiquitinated species (Sheedlo et al., 2015). Yet, the DUB activity of SidEs is not required for maximal intracellular bacterial replication (Sheedlo et al., 2015); This DUB may function to release ubiquitin from modified proteins to provide a reaction precursor for the ligase activity conferred by the mART and PDE domains.

Another *L. pneumophila* effector protein LupA (Lpg1148) also harbors a DUB domain. LupA catalyzes the removal of ubiquitin from target proteins, a reaction that requires the predicted C-H-D catalytic triad (Urbanus et al., 2016). LupA rescues the yeast growth defect caused by the *L. pneumophila* effector protein LegC3; In addition, LupA removes ubiquitin modification from LegC3 when co-expressed in mammalian cells (Urbanus et al., 2016). Collectively, these observations suggest that the activity of LegC3 may depend upon on ubiquitination by one or more host E3 ligases, and LupA functions to inactivate it via specific deubiquitination (Urbanus et al., 2016). The biological role of LupA during bacterial infection requires further investigation.

A DUB That Cleaves the Linkage Induced by Members of the SidE Family Effectors

One unique feature of *L. pneumophila* effectors is the regulation of one effector activity by another effector, with the latter being designated as metaeffector (Kubori et al., 2010). Such regulation is achieved by affecting protein stability exemplified by LubX and SidH and more commonly by effectors with opposite biochemical activities such as SidD and Lem3 that antagonize the activity of SidM and AnkX, respectively (Tan and Luo, 2011; Tan et al., 2011). Although, the biological significance is not known, regulation by direct protein-protein interactions between effectors has also been suggested (Urbanus et al., 2016). The activity of SidEs is regulated by another effector SidJ, which itself is also required for maximal bacterial intracellular growth (Liu and Luo, 2007). SidJ suppressed yeast toxicity of members of the SidE family (Havey and Roy, 2015; Jeong et al., 2015), suggesting that it may reverse the modification imposed by the ligases. Indeed, recombinant SidJ effectively removes ubiquitin from modified substrates such as Rab33b by cleaving the phosphodiester bond between ubiquitin and substrate protein (Qiu et al., 2017; **Figure 3A**). These results establish SidJ as a phosphodiesterase (PDE). Although, it has been reported that some DUBs cleave non-isopeptide bond like oxyester and thio-ester linkages, this is the first DUB known to cleave a phosphodiester linkage (Ronau and Hochstrasser, 2017). Substitution mutants failed to rescue SidEs-induced toxicity against eukaryotic cells also failed to complement the Δ*sidJ* mutant in infection. Without exception, these mutants have almost completely lost the DUB activity (Qiu et al., 2017). Thus, the DUB activity is responsible for the role of SidJ in *L. pneumophila* infection.

Surprisingly, SidJ also displays activity against isopeptide bond, and is able to deubiquitinate from proteins modified by the canonical pathway (**Figure 3B**). SidJ hydrolyzes K11, K33, K48, and K63-linked diubiquitin, but with a preference for the K63 linkage. Intriguingly, although it is sensitive to N-ethylmaleimide, a commonly used inhibitor that reacts with and blocks active thiol groups, mutations in any of the three cysteine residues did not detectably affect the activity of SidJ (Qiu et al., 2017), suggesting that SidJ is not a member of the thiol protease family. SidJ may represent a unique DUB which uses a novel catalytic mechanism to cleave both isopeptide and phosphoribosyl linkage. Whether these two activities are conferred by a single or two catalytic motifs are unknown. It is worth noting that SidJ purified from *L. pneumophila* but not from *E. coli* showed the DUB activity (Qiu et al., 2017), suggesting that SidJ needs co-factor(s) unique to *L. pneumophila* to function. The exact catalytic mechanism of SidJ and the nature of such co-factor(s) await further investigation.

Although, SidJ is constitutively expressed in broth-grown *L. pneumophila* (Liu and Luo, 2007), the quantity of protein translocated into host cells significantly increase as infection proceeds (Qiu et al., 2017; **Figure 2B**). The increasing ratios between SidJ and SidEs in infected cells render the DUB activity to become dominant at later infection phases, thus allowing temporal regulation of the activity of SidEs. Indeed, the amount of ubiquitinated Rab33b begins to decrease several hours after infection with wild type bacteria, but such decrease was delayed in infections using the Δ*sidJ* mutant (Qiu et al., 2017). Based on its ability to make SidEs undetectable from the LCV by immunostaining, it has been suggested that SidJ also spatially regulates the activity of SidEs (Jeong et al., 2015). Yet, the mechanism of such regulation, even if it exists, is unknown. The fact that the ligase activity of SdeA (likely other members of the SidE family, too) does not affect its cellular localization suggests that self-ubiquitination is not important for the association of SdeA with specific organelles. As a result, the DUB activity of SidJ unlikely plays a role in altering the cellular localization of SdeA.

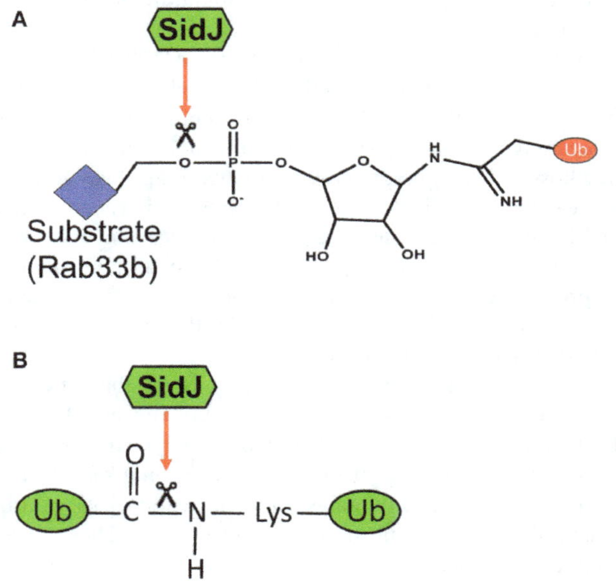

FIGURE 3 | A diagram of the dual biochemical activities of SidJ. **(A)** Cleavage of phosphodiester bond by SidJ. SidJ specifically recognizes the phophosdiester bond that links phosphoribosylated ubiquitin to serine residue on the substrate. This activity allows the substrate to return to its unmodified form; it also produces phosphoribosylated ubiquitin, which may be further hydrolyzed by enzymes from either the host or the bacterium or both. **(B)** Cleavage of isopeptide bond by SidJ. SidJ recognizes and cleaves the isopeptide bond that links ubiquitin to lysine residue of the substrate or another ubiquitin moiety. This activity produces free ubiquitin. Since mutants that retain only one of these two activities have not been isolated, the importance of the classical DUB activity of SidJ is not clear.

Intriguingly, the amount of ubiquitinated substrates eventually decreases in cells infected with the Δ*sidJ* mutant, suggesting the existence of additional bacterial proteins or host enzymes capable of reversing ubiquitination induced by SidEs. Such enzymes from host cells may function with the putative endogenous ligases that catalyze NAD-dependent ubiquitination to regulate certain cellular processes.

CONCLUDING REMARKS

L. pneumophila encodes a large cohort of effectors to modulate the host ubiquitination system for its benefit, which emphasizes the importance of the ubiquitin network in the virulence of this pathogen. The diverse strategies ranging from functional mimicry of canonical E3 ligases or DUBs to mechanisms of completely different chemistry employed by this pathogen have deepened our understanding in not only bacterial pathogenesis but also in cell biology of the host cell. Despite the progress in biochemical characterization of these ubiquitin-editing effectors, our understanding of their role in the biogenesis of the LCV remains limited. A major challenge is that we know very little about the host proteins specifically targeted by these enzymes, let alone the biological significance of the modification imposed by these effectors. Given the complexity of the regulation of the ubiquitin network, it is anticipated that more Dot/Icm effectors involved in hijacking this signaling mechanism will be uncovered. A detailed understanding of their biochemical activities and the coordination of these effectors during bacterial infection will provide insights into both *L. pneumophila* pathogenesis and signaling in eukaryotic cells.

AUTHOR CONTRIBUTIONS

JQ: prepared the first draft of the manuscript; Z-QL: initated the points to dicuss, revised and finalize the manuscript and prepared **Figure 3**.

ACKNOWLEDGMENTS

We thank members of our laboratory for helpful discussion. Our work is supported by NIH-NIAID grants R21AI105714 and R01AI127465. The content is solely the responsibility of the authors and does not necessarily represent the official views of the National Institutes of Health. JQ is a recipient of the "Young 1000 Talents" Award of the Chinese government.

REFERENCES

Al-Khodor, S., Price, C. T., Habyarimana, F., Kalia, A., and Abu Kwaik, Y. (2008). A Dot/Icm-translocated ankyrin protein of *Legionella pneumophila* is required for intracellular proliferation within human macrophages and protozoa. *Mol. Microbiol.* 70, 908–923. doi: 10.1111/j.1365-2958.2008.06453.x

Ashida, H., Kim, M., and Sasakawa, C. (2014). Exploitation of the host ubiquitin system by human bacterial pathogens. *Nat. Rev. Microbiol.* 12, 399–413. doi: 10.1038/nrmicro3259

Bardill, J. P., Miller, J. L., and Vogel, J. P. (2005). IcmS-dependent translocation of SdeA into macrophages by the *Legionella pneumophila* type IV secretion system. *Mol. Microbiol.* 56, 90–103. doi: 10.1111/j.1365-2958.2005.04539.x

Bhogaraju, S., and Dikic, I. (2016). Cell biology: ubiquitination without E1 and E2 enzymes. *Nature* 533, 43–44. doi: 10.1038/nature17888

Bhogaraju, S., Kalayil, S., Liu, Y. B., Bonn, F., Colby, T., Matic, I., et al. (2016). Phosphoribosylation of ubiquitin promotes serine ubiquitination and impairs conventional ubiquitination. *Cell* 167, 1636.e13–1649.e13. doi: 10.1016/j.cell.2016.11.019

Burstein, D., Amaro, F., Zusman, T., Lifshitz, Z., Cohen, O., Gilbert, J. A., et al. (2016). Genomic analysis of 38 Legionella species identifies large and diverse effector repertoires. *Nat. Genet.* 48, 167–175. doi: 10.1038/ng.3481

Catic, A., Misaghi, S., Korbel, G. A., and Ploegh, H. L. (2007). ElaD, a Deubiquitinating protease expressed by E. *PLoS ONE* 2:e381. doi: 10.1371/journal.pone.0000381

de Felipe, K. S., Pampou, S., Jovanovic, O. S., Pericone, C. D., Ye, S. F., Kalachikov, S., et al. (2005). Evidence for acquisition of Legionella type IV secretion substrates via interdomain horizontal gene transfer. *J. Bacteriol.* 187, 7716–7726. doi: 10.1128/JB.187.22.7716-7726.2005

Deribe, Y. L., Pawson, T., and Dikic, I. (2010). Post-translational modifications in signal integration. *Nat. Struct. Mol. Biol.* 17, 666–672. doi: 10.1038/nsmb.1842

Desmots, F., Russell, H. R., Michel, D., and McKinnon, P. J. (2008). Scythe regulates apoptosis-inducing factor stability during endoplasmic reticulum stress-induced apoptosis. *J. Biol. Chem.* 283, 3264–3271. doi: 10.1074/jbc.M706419200

Dorer, M. S., Kirton, D., Bader, J. S., and Isberg, R. R. (2006). RNA interference analysis of Legionella in *Drosophila* cells: exploitation of early secretory apparatus dynamics. *PLoS Pathog.* 2:e34. doi: 10.1371/journal.ppat.0020034

Ensminger, A. W. (2016). *Legionella pneumophila*, armed to the hilt: justifying the largest arsenal of effectors in the bacterial world. *Curr. Opin. Microbiol.* 29, 74–80. doi: 10.1016/j.mib.2015.11.002

Ensminger, A. W., and Isberg, R. R. (2010). E3 ubiquitin ligase activity and targeting of BAT3 by multiple *Legionella pneumophila* translocated substrates. *Infect. Immun.* 78, 3905–3919. doi: 10.1128/IAI.00344-10

Finsel, I., and Hilbi, H. (2015). Formation of a pathogen vacuole according to *Legionella pneumophila*: how to kill one bird with many stones. *Cell. Microbiol.* 17, 935–950. doi: 10.1111/cmi.12450

Fraser, D. W., Tsai, T. R., Orenstein, W., Parkin, W. E., Beecham, H. J., Sharrar, R. G., et al. (1977). Legionnaires' disease: description of an epidemic of pneumonia. *N. Engl. J. Med.* 297, 1189–1197. doi: 10.1056/NEJM197712012972201

Fukuda, T., Chen, K., Shi, X., and Wu, C. (2003). PINCH-1 is an obligate partner of integrin-linked kinase (ILK) functioning in cell shape modulation, motility, and survival. *J. Biol. Chem.* 278, 51324–51333. doi: 10.1074/jbc.M309122200

Gallagher, P. S., Clowes Candadai, S. V., and Gardner, R. G. (2014). The requirement for Cdc48/p97 in nuclear protein quality control degradation depends on the substrate and correlates with substrate insolubility. *J. Cell Sci.* 127, 1980–1991. doi: 10.1242/jcs.141838

Gazdag, E. M., Schobel, S., Shkumatov, A. V., Goody, R. S., and Itzen, A. (2014). The structure of the N-terminal domain of the Legionella protein SidC. *J. Struct. Biol.* 186, 188–194. doi: 10.1016/j.jsb.2014.02.003

Haglund, K., and Dikic, I. (2005). Ubiquitylation and cell signaling. *EMBO J.* 24, 3353–3359. doi: 10.1038/sj.emboj.7600808

Hatakeyama, S., Yada, M., Matsumoto, M., Ishida, N., and Nakayama, K. I. (2001). U box proteins as a new family of ubiquitin-protein ligases. *J. Biol. Chem.* 276, 33111–33120. doi: 10.1074/jbc.M102755200

Havey, J. C., and Roy, C. R. (2015). Toxicity and SidJ-mediated suppression of toxicity require distinct regions in the SidE family of *Legionella pneumophila* effectors. *Infect. Immun.* 83, 3506–3514. doi: 10.1128/IAI.00497-15

Hershko, A., and Ciechanover, A. (1998). The ubiquitin system. *Annu. Rev. Biochem.* 67, 425–479. doi: 10.1146/annurev.biochem.67.1.425

Horenkamp, F. A., Mukherjee, S., Alix, E., Schauder, C. M., Hubber, A. M., Roy, C. R., et al. (2014). *Legionella pneumophila* subversion of host vesicular transport by SidC effector proteins. *Traffic* 15, 488–499. doi: 10.1111/tra.12158

Hsu, F., Luo, X., Qiu, J., Teng, Y. B., Jin, J., Smolka, M. B., et al. (2014). The Legionella effector SidC defines a unique family of ubiquitin ligases important for bacterial phagosomal remodeling. *Proc. Natl. Acad. Sci. U.S.A.* 111, 10538–10543. doi: 10.1073/pnas.1402605111

Hubber, A., Kubori, T., and Nagai, H. (2014). Modulation of the ubiquitination machinery by Legionella. *Curr. Top. Microbiol. Immunol.* 376, 227–247. doi: 10.1007/82_2013_343

Isberg, R. R., O'Connor, T. J., and Heidtman, M. (2009). The *Legionella pneumophila* replication vacuole: making a cosy niche inside host cells. *Nat. Rev. Microbiol.* 7, 13–24. doi: 10.1038/nrmicro1967

Ivanov, S. S., and Roy, C. R. (2009). Modulation of ubiquitin dynamics and suppression of DALIS formation by the *Legionella pneumophila* Dot/Icm system. *Cell. Microbiol.* 11, 261–278. doi: 10.1111/j.1462-5822.2008.01251.x

Jarosch, E., Taxis, C., Volkwein, C., Bordallo, J., Finley, D., Wolf, D. H., et al. (2002). Protein dislocation from the ER requires polyubiquitination and the AAA-ATPase Cdc48. *Nat. Cell Biol.* 4, 134–139. doi: 10.1038/ncb746

Jeong, K. C., Sexton, J. A., and Vogel, J. P. (2015). Spatiotemporal regulation of a *Legionella pneumophila* T4SS substrate by the metaeffector SidJ. *PLoS Pathog.* 11:e1004695. doi: 10.1371/journal.ppat.1004695

Komander, D., and Rape, M. (2012). The ubiquitin code. *Annu. Rev. Biochem.* 81, 203–229. doi: 10.1146/annurev-biochem-060310-170328

Kotewicz, K. M., Ramabhadran, V., Sjoblom, N., Vogel, J. P., Haenssler, E., Zhang, M. Y., et al. (2017). A single legionella effector catalyzes a multistep ubiquitination pathway to rearrange tubular endoplasmic reticulum for replication. *Cell Host Microbe* 21, 169–181. doi: 10.1016/j.chom.2016.12.007

Kubori, T., Hyakutake, A., and Nagai, H. (2008). Legionella translocates an E3 ubiquitin ligase that has multiple U-boxes with distinct functions. *Mol. Microbiol.* 67, 1307–1319. doi: 10.1111/j.1365-2958.2008.06124.x

Kubori, T., Shinzawa, N., Kanuka, H., and Nagai, H. (2010). Legionella metaeffector exploits host proteasome to temporally regulate cognate effector. *PLoS Pathog.* 6:e1001216. doi: 10.1371/journal.ppat.1001216

Lin, Y. H., Doms, A. G., Cheng, E., Kim, B., Evans, T. R., and Machner, M. P. (2015). Host cell-catalyzed S-palmitoylation mediates golgi targeting of the Legionella ubiquitin ligase Gobx. *J. Biol. Chem.* 290, 25766–25781. doi: 10.1074/jbc.M115.637397

Lin, Y. H., and Machner, M. P. (2017). Exploitation of the host cell ubiquitin machinery by microbial effector proteins. *J. Cell Sci.* 130, 1985–1996. doi: 10.1242/jcs.188482

Liu, Y., and Luo, Z. Q. (2007). The *Legionella pneumophila* effector SidJ is required for efficient recruitment of endoplasmic reticulum proteins to the bacterial phagosome. *Infect. Immun.* 75, 592–603. doi: 10.1128/IAI.01278-06

Lomma, M., Dervins-Ravault, D., Rolando, M., Nora, T., Newton, H. J., Sansom, F. M., et al. (2010). The *Legionella pneumophila* F-box protein Lpp2082 (AnkB) modulates ubiquitination of the host protein parvin B and promotes intracellular replication. *Cell. Microbiol.* 12, 1272–1291. doi: 10.1111/j.1462-5822.2010.01467.x

Lu, H., and Clarke, M. (2005). Dynamic properties of Legionella-containing phagosomes in Dictyostelium amoebae. *Cell. Microbiol.* 7, 995–1007. doi: 10.1111/j.1462-5822.2005.00528.x

Luo, X., Wasilko, D. J., Liu, Y., Sun, J., Wu, X., Luo, Z. Q., et al. (2015). Structure of the Legionella virulence factor, SidC reveals a unique PI(4)P-specific binding domain essential for its targeting to the bacterial phagosome. *PLoS Pathog.* 11:e1004965. doi: 10.1371/journal.ppat.1004965

Luo, Z. Q., and Isberg, R. R. (2004). Multiple substrates of the Legionella pneumophila Dot/Icm system identified by interbacterial protein transfer. *Proc. Natl. Acad. Sci. U.S.A.* 101, 841–846. doi: 10.1073/pnas.0304916101

Maculins, T., Fiskin, E., Bhogaraju, S., and Dikic, I. (2016). Bacteria-host relationship: ubiquitin ligases as weapons of invasion. *Cell Res.* 26, 499–510. doi: 10.1038/cr.2016.30

Metzger, M. B., Hristova, V. A., and Weissman, A. M. (2012). HECT and RING finger families of E3 ubiquitin ligases at a glance. *J. Cell Sci.* 125, 531–537. doi: 10.1242/jcs.091777

Metzger, M. B., Pruneda, J. N., Klevit, R. E., and Weissman, A. M. (2014). RING-type E3 ligases: master manipulators of E2 ubiquitin-conjugating enzymes and ubiquitination. *Biochim. Biophys. Acta* 1843, 47–60. doi: 10.1016/j.bbamcr.2013.05.026

Misaghi, S., Balsara, Z. R., Catic, A., Spooner, E., Ploegh, H. L., and Starnbach, M. N. (2006). *Chlamydia trachomatis*-derived deubiquitinating enzymes in mammalian cells during infection. *Mol. Microbiol.* 61, 142–150. doi: 10.1111/j.1365-2958.2006.05199.x

Moliner, C., Fournier, P. E., and Raoult, D. (2010). Genome analysis of microorganisms living in amoebae reveals a melting pot of evolution. *FEMS Microbiol. Rev.* 34, 281–294. doi: 10.1111/j.1574-6976.2009.00209.x

Newton, H. J., Ang, D. K., van Driel, I. R., and Hartland, E. L. (2010). Molecular pathogenesis of infections caused by *Legionella pneumophila*. *Clin. Microbiol. Rev.* 23, 274–298. doi: 10.1128/CMR.00052-09

Prasad, J., Colwill, K., Pawson, T., and Manley, J. L. (1999). The protein kinase Clk/Sty directly modulates SR protein activity: both hyper- and hypophosphorylation inhibit splicing. *Mol. Cell. Biol.* 19, 6991–7000. doi: 10.1128/MCB.19.10.6991

Price, C. T., Al-Khodor, S., Al-Quadan, T., Santic, M., Habyarimana, F., Kalia, A., et al. (2009). Molecular mimicry by an F-box effector of *Legionella pneumophila* hijacks a conserved polyubiquitination machinery within macrophages and protozoa. *PLoS Pathog.* 5:e1000704. doi: 10.1371/journal.ppat.1000704

Price, C. T. D., Al-Quadan, T., Santic, M., Rosenshine, I., and Abu Kwaik, Y. (2011). Host proteasomal degradation generates amino acids essential for intracellular bacterial growth. *Science* 334, 1553–1557. doi: 10.1126/science.1212868

Qiu, J., and Luo, Z.-Q. (2017). *Legionella* and *Coxiella* effectors: strength in diversity and activity. *Nat. Rev. Microbiol.* 15, 591–605. doi: 10.1038/nrmicro.2017.67

Qiu, J., Yu, K., Fei, X., Liu, Y., Nakayasu, E. S., Piehowski, P. D., et al. (2017). A unique deubiquitinase that deconjugates phosphoribosyl-linked protein ubiquitination. *Cell Res.* 27, 865–881. doi: 10.1038/cr.2017.66

Qiu, J., Sheedlo, M. J., Yu, K., Tan, Y., Nakayasu, E. S., Das, C., et al. (2016). Ubiquitination independent of E1 and E2 enzymes by bacterial effectors. *Nature* 533, 120–124. doi: 10.1038/nature17657

Quaile, A. T., Urbanus, M. L., Stogios, P. J., Nocek, B., Skarina, T., Ensminger, A. W., et al. (2015). Molecular characterization of LubX: functional divergence of the U-box fold by *Legionella pneumophila*. *Structure* 23, 1459–1469. doi: 10.1016/j.str.2015.05.020

Ragaz, C., Pietsch, H., Urwyler, S., Tiaden, A., Weber, S. S., and Hilbi, H. (2008). The *Legionella pneumophila* phosphatidylinositol-4 phosphate-binding type IV substrate SidC recruits endoplasmic reticulum vesicles to a replication-permissive vacuole. *Cell. Microbiol.* 10, 2416–2433. doi: 10.1111/j.1462-5822.2008.01219.x

Ronau, J. A., and Hochstrasser, M. (2017). The DUB blade goes snicker-snack: novel ubiquitin cleavage by a Legionella effector protein. *Cell Res.* 27, 845–846. doi: 10.1038/cr.2017.80

Rowbotham, T. J. (1980). Preliminary report on the pathogenicity of *Legionella pneumophila* for freshwater and soil amoebae. *J. Clin. Pathol.* 33, 1179–1183. doi: 10.1136/jcp.33.12.1179

Rytkönen, A., and Holden, D. W. (2007). Bacterial interference of ubiquitination and deubiquitination. *Cell Host Microbe* 1, 13–22. doi: 10.1016/j.chom.2007.02.003

Rytkönen, A., Poh, J., Garmendia, J., Boyle, C., Thompson, A., Liu, M., et al. (2007). SseL, a Salmonella deubiquitinase required for macrophage killing and virulence. *Proc. Natl. Acad. Sci. U.S.A.* 104, 3502–3507. doi: 10.1073/pnas.0610095104

Sasaki, T., Marcon, E., McQuire, T., Arai, Y., Moens, P. B., and Okada, H. (2008). Bat3 deficiency accelerates the degradation of Hsp70-2/HspA2 during spermatogenesis. *J. Cell Biol.* 182, 449–458. doi: 10.1083/jcb.200802113

Schulman, B. A., Carrano, A. C., Jeffrey, P. D., Bowen, Z., Kinnucan, E. R. E., Finnin, M. S., et al. (2000). Insights into SCF ubiquitin ligases from the structure of the Skp1-Skp2 complex. *Nature* 408, 381–386. doi: 10.1038/35042620

Sheedlo, M. J., Qiu, J. Z., Tan, Y. H., Paul, L. N., Luo, Z. Q., and Das, C. (2015). Structural basis of substrate recognition by a bacterial deubiquitinase important for dynamics of phagosome ubiquitination. *Proc. Natl. Acad. Sci. U.S.A.* 112, 15090–15095. doi: 10.1073/pnas.1514568112

Simon, N. C., Aktories, K., and Barbieri, J. T. (2014). Novel bacterial ADP-ribosylating toxins: structure and function. *Nat. Rev. Microbiol.* 12, 599–611. doi: 10.1038/nrmicro3310

Skaar, J. R., Pagan, J. K., and Pagano, M. (2013). Mechanisms and function of substrate recruitment by F-box proteins. *Nat. Rev. Mol. Cell Biol.* 14, 369–381. doi: 10.1038/nrm3582

Steiner, B., Swart, A. L., Welin, A., Weber, S., Personnic, N., Kaech, A., et al. (2017). ER remodeling by the large GTPase atlastin promotes vacuolar growth of *Legionella pneumophila*. *EMBO Rep.* 18, 1817–1836. doi: 10.15252/embr.201743903

Swanson, M. S., and Isberg, R. R. (1995). Association of *Legionella pneumophila* with the macrophage endoplasmic reticulum. *Infect. Immun.* 63, 3609–3620.

Tan, Y., Arnold, R. J., and Luo, Z. Q. (2011). *Legionella pneumophila* regulates the small GTPase Rab1 activity by reversible phosphorylcholination. *Proc. Natl. Acad. Sci. U.S.A* 108, 21212–21217. doi: 10.1073/pnas.1114023109

Tan, Y., and Luo, Z. Q. (2011). *Legionella pneumophila* SidD is a deAMPylase that modifies Rab1. *Nature* 475, 506–509. doi: 10.1038/nature10307

Urbanus, M. L., Quaile, A. T., Stogios, P. J., Morar, M., Rao, C., Di Leo, R., et al. (2016). Diverse mechanisms of metaeffector activity in an intracellular bacterial pathogen, *Legionella pneumophila*. *Mol. Syst. Biol.* 12, 893. doi: 10.15252/msb.20167381

Weber, S., Wagner, M., and Hilbi, H. (2014). Live-cell imaging of phosphoinositide dynamics and membrane architecture during Legionella infection. *MBio* 5, e00839–e00813. doi: 10.1128/mBio.00839-13

Wenzel, D. M., Lissounov, A., Brzovic, P. S., and Klevit, R. E. (2011). UBCH7 reactivity profile reveals parkin and HHARI to be RING/HECT hybrids. *Nature* 474, 105–108. doi: 10.1038/nature09966

Wilkinson, K. D. (2009). DUBs at a glance. *J. Cell Sci.* 122, 2325–2329. doi: 10.1242/jcs.041046

Ye, Y., and Rape, M. (2009). Building ubiquitin chains: E2 enzymes at work. *Nat. Rev. Mol. Cell Biol.* 10, 755–764. doi: 10.1038/nrm2780

Zhang, Y. J., Chen, K., Tu, Y. Z., and Wu, C. Y. (2004). Distinct roles of two structurally closely related focal adhesion proteins, alpha-parvins and beta-parvins, in regulation of cell morphology and survival. *J. Biol. Chem.* 279, 41695–41705. doi: 10.1074/jbc.M401563200

Zhou, Y., and Zhu, Y. (2015). Diversity of bacterial manipulation of the host ubiquitin pathways. *Cell. Microbiol.* 17, 26–34. doi: 10.1111/cmi.12384

8

A Virulence Associated Siderophore Importer Reduces Antimicrobial Susceptibility of *Klebsiella pneumoniae*

Robeena Farzand[1,2], Kumar Rajakumar[1], Michael R. Barer[1], Primrose P. E. Freestone[1], Galina V. Mukamolova[1], Marco R. Oggioni[3] and Helen M. O'Hare[1,4*]

[1] Leicester Microbial Sciences and Infectious Diseases Network, Department of Respiratory Sciences, University of Leicester, Leicester, United Kingdom, [2] Department of Microbiology, Kohat University of Science and Technology, Kohat, Pakistan, [3] Department of Genetics and Genome Biology, University of Leicester, Leicester, United Kingdom, [4] Department of Molecular and Cell Biology, University of Leicester, Leicester, United Kingdom

Edited by:
Xian-Zhi Li,
Health Canada, Canada

Reviewed by:
Catherine Llanes,
University of Franche-Comté, France
Laura A. Mike,
University of Michigan, United States
Francesca Short,
Macquarie University, Australia

****Correspondence:***
Helen M. O'Hare
hmo7@le.ac.uk;
hmo7@leicester.ac.uk

The accessory genomes of many pathogenic bacteria include ABC transporters that scavenge metal by siderophore uptake and ABC transporters that contribute to antimicrobial resistance by multidrug efflux. There are mechanistic and recently recognized structural similarities between siderophore importer proteins and efflux pumps. Here we investigated the influence of siderophore importer YbtPQ on antimicrobial resistance of *Klebsiella pneumoniae*. YbtPQ is encoded in the yersiniabactin cluster in a prevalent mobile genetic element ICEKp, and is also common in pathogenicity islands of *Escherichia coli* and *Yersinia* species, where yersiniabactin enhances virulence. Deletion of ICEKp increased the susceptibility of *K. pneumoniae* to all antimicrobials tested. The mechanism was dependent on the yersiniabactin importer YbtPQ and may involve antimicrobial efflux, since it was affected by the inhibitor reserpine. The element ICEKp is naturally highly mobile, indeed the accessory genome of *K. pneumoniae* is recognized as a reservoir of genes for the emergence of hospital outbreak strains and for transfer to other Gram-negative pathogens. Introduction of ICEKp, or a plasmid encoding YbtPQ, to *E. coli* decreased its susceptibility to a broad range of antimicrobials. Thus a confirmed siderophore importer, on a rapidly evolving and highly mobile element capable of interspecies transfer, may have a secondary function exporting antimicrobials.

Keywords: *Klebsiella pneumoniae*, yersiniabactin, ABC transporter, antimicrobial efflux, siderophore, mobile genetic element, integrative conjugative element

INTRODUCTION

ABC transporters in the accessory genomes of bacterial pathogens significantly influence both virulence and antimicrobial resistance. Siderophore importers scavenge metals from the host and efflux pumps export antimicrobials, and the presence of such transporters on mobile genetic elements is associated with both disease severity and treatment failure.

The nature of this transport is specific and unidirectional, due to the specific interactions between substrate and binding cavity, and the asymmetry of ATP-powered conformational changes through inward facing, closed and outward-facing forms. Broad-specificity multidrug efflux pumps are an apparent exception, and these have binding cavities with multiple sites that can interact with diverse antimicrobials (Du et al., 2018). Other rare examples of bispecific or multispecific unidirectional transport have been reported in bacteria and eukaryotes, such as siderophore export by multidrug efflux pumps, antibiotic entry through an asparagine importer, and chloroquine transport by a plasmodium peptide transporter (Hannauer et al., 2012; Smith et al., 2019; Shafik et al., 2020). By contrast, bidirectional ABC transporters that import one substrate and export another, are compatible with the mechanistic models but are unknown.

Siderophore importers are plausible candidates for such bidirectional transport since they have a spacious substrate binding cavity that might accommodate other molecules, and the structural organization is exporter-like (Arnold et al., 2020; Wang et al., 2020). Furthermore, in the context of hospital outbreak strains, a secondary function in antimicrobial export could provide a selective advantage.

The yersiniabactin siderophore cluster is prevalent and spreading in *Klebsiella pneumoniae* (Lam et al., 2018a,b) and was found in integrative and conjugative elements known as ICEKp1 or ICEKp3 in approximately half the *K. pneumoniae* clinical isolates tested in a recent United Kingdom and global study (Farzand et al., 2019). This cluster is also common in pathogenicity islands in *Escherichia coli* and *Yersinia* where it enhances virulence (Schubert et al., 2004; Lawlor et al., 2007; Perry and Fetherston, 2011; Koh et al., 2017).

The cargo genes of ICEKp1 vary, and the example studied here, referred to as ICEKp from a clinical isolate of *K. pneumoniae*, the yersiniabactin cluster is the only cargo (Lam et al., 2018a,b; Farzand et al., 2019). Annotation of ICEKp and the yersiniabactin cluster is provided (**Figures 1, 2**). Yersiniabactin bound to iron is imported by a heterodimeric ABC transporter YbtPQ encoded within the yersiniabactin cluster, as shown by biochemical and phenotypic assays (Bearden et al., 1997; Fetherston et al., 1999; Lawlor et al., 2005; Koh et al., 2017; Wang et al., 2020). Another putative transporter encoded in the cluster is, YbtX, a permease of the MFS (major facilitator) superfamily, whose function is unknown, as its knockout in *Y. pestis* affected neither secretion nor utilization of yersiniabactin (Perry and Fetherston, 2011).

Antimicrobial resistance of *K. pneumoniae* arises from the combination of RND (resistance-nodulation-cell division family) efflux pumps and antibiotic resistance genes in the core genome [1] plus resistance determinants encoded on mobile genetic elements [17, 18]. Systematic deletion of mobile genetic elements from a clinical isolate has proven their contribution to antimicrobial resistance and generated a strain that is tractable for study by safe genetic modification [17, 18] and has been used to study gene mobilization by conjugation [9]. An ICEKp deletion mutant of this strain [9], apparently showed increased susceptibility to antimicrobials (reported below) despite the lack of obvious efflux pumps or antibiotic resistance genes

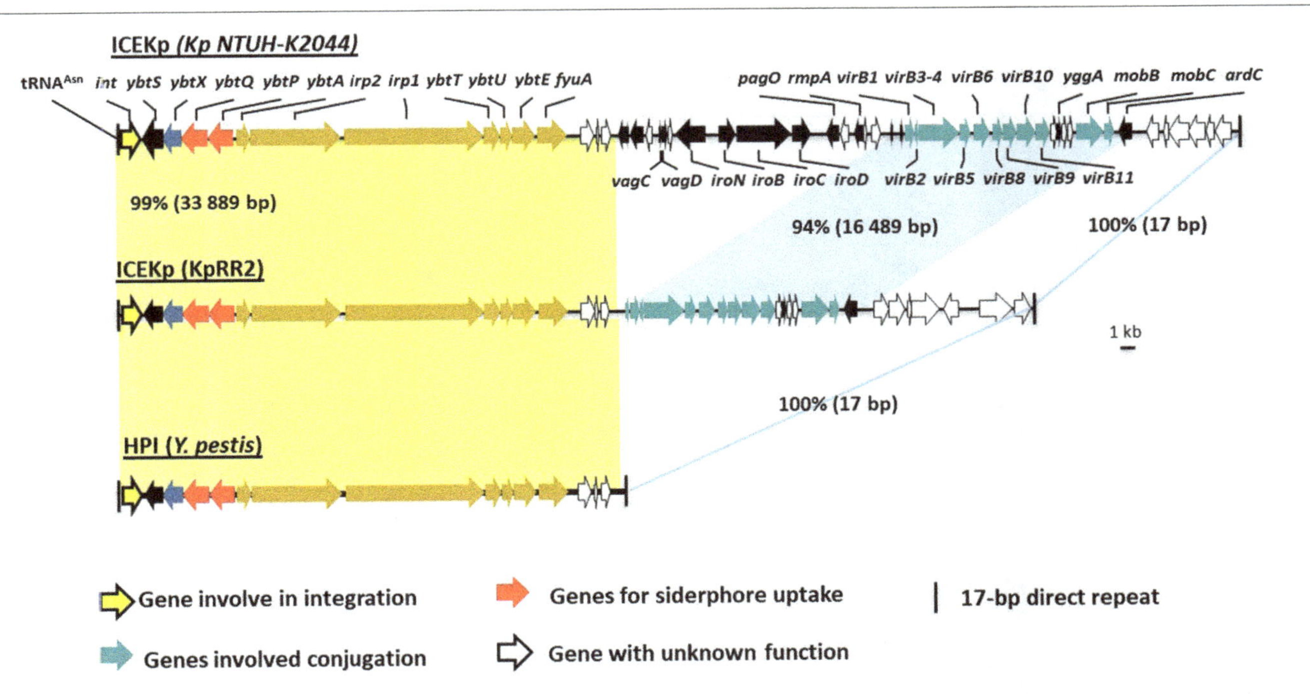

FIGURE 1 | Comparison of ICEKp in the study strain KpRR2 with the first characterized ICEKp in strain NTUH-K2044, and with the yersiniabactin gene cluster in *Yersinia pestis*. The first characterization of ICEKp in *K. pneumoniae* was in strain Kp NTUH-K2044 (Lin et al., 2008). Yellow shading indicates the synteny and 99% DNA identity between the yersiniabactin gene cluster in the study strain KpRR2 and that in *Yersinia pestis* (AF091251) High Pathogenicity Island (HPI). The genes cluster responsible for conjugation are indicated by blue shading region, these share 94% identity.

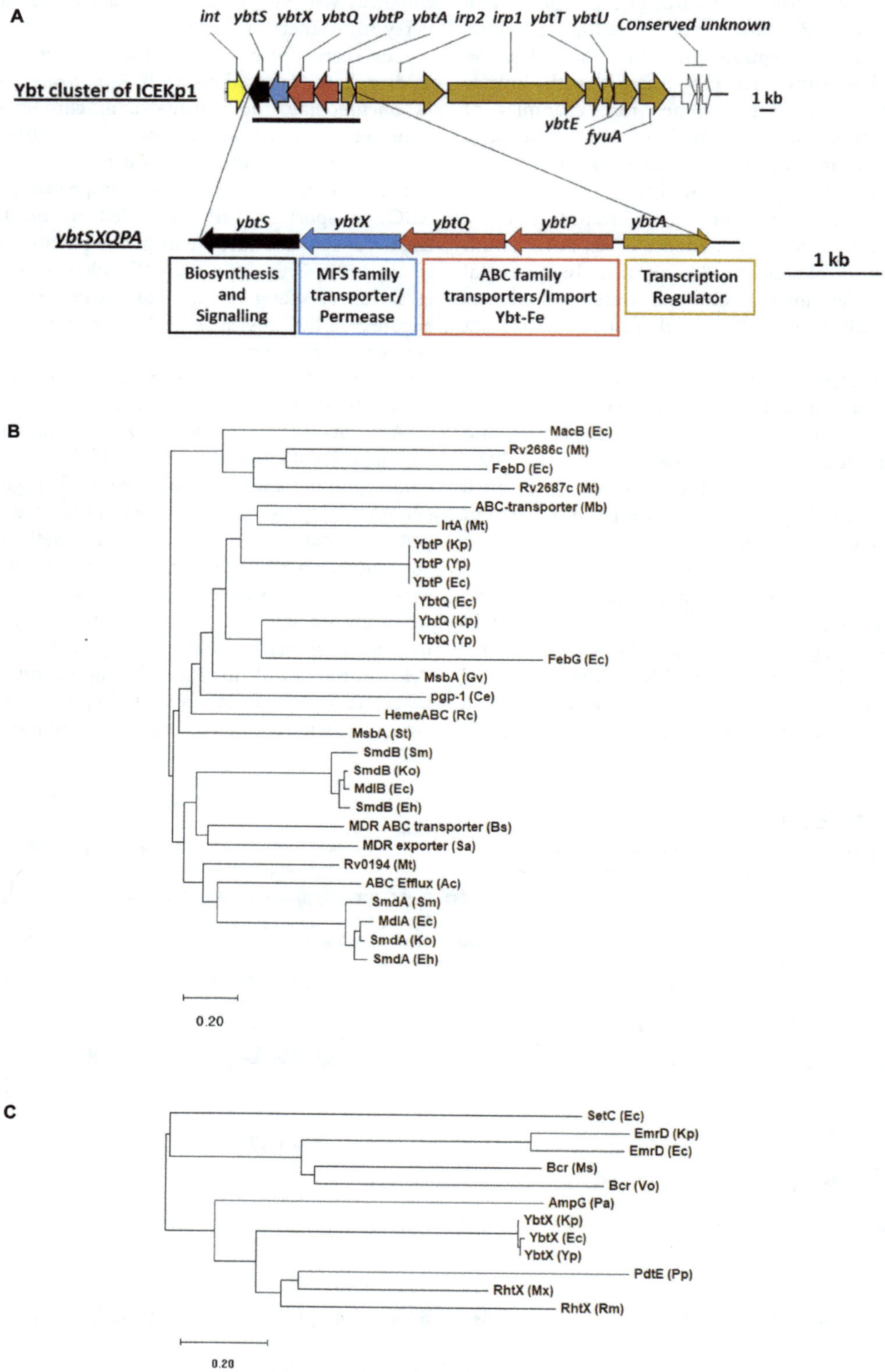

FIGURE 2 | The cargo genes of ICEKp from *K. pneumoniae* KpRR2 are an iron siderophore biosynthesis cluster (yersiniabactin) that includes an ABC transporter and an MFS family permease. **(A)** The entire cargo genes of ICEKp are shown, with annotation. The only transmembrane transporters amongst the cargo genes are in the four-gene operon: *ybtP-ybtS*. This operon is unique to yersiniabactin clusters and conserved with >80% amino acid identity in all organisms. The previously reported deletion strain ΔICE lacks the entire ICEKp (and therefore lacks the entire yersiniabactin cluster). **(B)** YbtP and YbtQ share 34% amino acid identity with each

(Continued)

FIGURE 2 | Continued
other, and form a heterodimeric transporter YbtPQ, homologous to ABC transporters of drugs, heme, lipid A and enterobactin. ABC transporters are recognized by their nucleotide binding domain: the C-terminal domain of YbtP and YbtQ (residues 342–556 and 344–558) share 29% and 31% identity with the nucleotide binding domain of MacB. **(C)** YbtX belongs to the major facilitator superfamily (MFS) and is a homolog of aerobactin and rhizobactin transporters, multidrug efflux transporters and peptidoglycan recycling transporters. Ce, *Caenorhabditis elegans;* Ec, *E. coli*; Mt, *Mycobacterium tuberculosis;* Gv, *Gloeobacter violaceus;* Ko, *Klebsiella oxytoca;* Kp, *K. pneumoniae*; Mb, *Methanobrevibacter*; Ms, *Mycobacterium smegmatis;* Mx, *Myxococcus xanthus;* Pa, *Pseudomonas aeruginosa;* Pp, *Pseudomonas putida;* Sa, *Staphylococcus aureus*; Rc, *Rhodobacter capsulatus;* Rm, *Rhizobium meliloti*; St, *Salmonella typhimurium;* Vo, *Vibrio orientalis;* Yp, *Yersinia pestis*. Functional annotations of proteins: AmpG (muropeptide MFS transporter), Bcr (MDR efflux), EmrD (MDR efflux), FebDG (ferric enterobactin transporter), IrtA (iron transporter), MacAB (MDR efflux), MdlAB (MDR efflux), MsbA (lipid ABC transporter), PdtE (inner membrane permease) Pgp-1 (multidrug resistance protein), SetC (sugar efflux system), SmdAb (MDR efflux), RhtX (rhizobactin transporter), Rv0194 (MDR ABC transporter), Rv2886c/Rv2887c (MDR efflux). Phylogenetic tree of amino acids sequence of ybt genes was inferred by maximum-likelihood method using MEGA10 (Kumar et al., 2018). Bootstrap values were calculated with 1000 replications. The bar shows changes observed between the two sequences; 0.20 means 20% amino acid changes.

in ICEKp, leading us to select this system to investigate, by genetic manipulation, whether YbtPQ influences antimicrobial susceptibility of *K. pneumoniae* and *E. coli.* We found that YbtPQ was necessary and sufficient to confer a modest but significant reduction in susceptibility (increase in minimum inhibitory concentration) for a broad range of antimicrobials (all antimicrobials tested) on both organisms, even in the absence of the rest of the yersiniabactin cluster or external yersiniabactin. The effect was likely due to antimicrobial efflux, since it was blocked by the efflux pump inhibitor reserpine.

RESULTS

ICEKp and Its Transporter Gene Cluster Influence the Antimicrobial Susceptibility of *K. pneumoniae*

The influence of ICEKp and its cargo genes on antimicrobial susceptibility was investigated by measuring growth inhibition of *K. pneumoniae* by antimicrobials on agar using Estrips (minimum inhibitory concentration assay) and disks (zone of inhibition assay). We used the available ICEKp deletion mutant ΔICE that lacks the entire ICEKp element (Farzand et al., 2019) and had been constructed in strain KpRR2, which was derived from clinical isolate HS11286 (Liu et al., 2012; Bi et al., 2015). We initially tested a frontline treatment, the β-lactam cephalosporin ceftazidime, plus antimicrobials from four additional classes to target cell wall biosynthesis, protein synthesis and transcription. The minimum inhibitory concentrations of ΔICE were significantly lower than that of KpRR2 for all five antimicrobials, i.e., ceftazidime, erythromycin, streptomycin, rifampicin, tetracyclin ($p < 0.05$, **Table 1**). A plasmid pSXPQA was constructed to reintroduce the yersiniabactin transporter gene cluster (five genes: *ybtS, ybtX, ybtP, ybtQ, ybtA*), and this plasmid restored the MIC to parental levels for all antimicrobials (**Table 1**).

Some proteins that transport antimicrobials are specific for a particular class of antimicrobial, while others are polyspecific (Du et al., 2018). To study larger numbers of antimicrobials, we used agar disk-diffusion assay, which is the official susceptibility test used in many clinical laboratories due to its simplicity, low cost, and suitability for high throughput (Balouiri et al., 2016). We verified the change in susceptibility to rifampicin and tested an additional frontline treatment (the β-lactam carbapenem imipenem), an additional example each of a tetracycline (doxycycline), and an aminoglycoside (gentamicin), two DNA gyrase inhibitors (ofloxacin and novobiocin) and a treatment of last resort (cationic peptide polymyxin B or colistin). For all eight antimicrobials, ΔICE was more sensitive than KpRR2 (significantly larger zone of inhibition, $p < 0.05$, **Figure 3**) whereas the plasmid-complemented mutant was not significantly different from the parent strain.

ICEKp and the Transporter Gene Cluster Reduce Antimicrobial Susceptibility by Enhancing Antimicrobial Efflux

ABC transporters that efflux antimicrobials can be blocked by the inhibitor reserpine (Schmitz et al., 1998; Dhanarani et al., 2017). To determine whether ICEKp affects antimicrobial susceptibility by causing efflux of antimicrobials, we repeated the disk diffusion assay using reserpine (**Figure 4**). Reserpine significantly enhanced the susceptibility of the parental strain and the complemented strain ($p < 0.05$) but not the mutant strain to tetracycline and trimethoprim (**Figure 4**) and other antimicrobials (**Supplementary Figure 1**), which would be consistent with the effects of ICEKp and the transporter plasmid acting through an efflux mechanism.

The ABC Transporter YbtPQ Alone Is Sufficient to Reduce Antimicrobial Susceptibility of *K. pneumoniae* ΔICE

Separate plasmids were constructed to determine which component(s) of the yersiniabactin transporter cluster influence antimicrobial susceptibility. *ybtPQ* and *ybtX* were chosen since these encode transmembrane proteins, and *ybtS* was chosen since it may be cotranscribed with *ybtPQ* and *ybtX*. Reintroducing the YbtPQ transporter using plasmid pPQ was necessary and sufficient to significantly reduce the zone of inhibition of ΔICE, such that it matched the parent strain for trimethoprim (**Figure 5**) and other antimicrobials (**Supplementary Figure 2**), whereas plasmids encoding YbtX or YbtS caused no significant change. Importantly, when ΔICE carrying plasmid pPQ was tested in the presence and absence of reserpine, we observed that the reduction in antimicrobial susceptibility conferred by *ybtPQ* was reversed by reserpine (**Figure 5B** and **Table 2**). Since YbtPQ reversed the changes in antimicrobial susceptibility caused by ICEKp deletion, we did not investigate the expression level or

TABLE 1 | Minimum inhibitory concentrations (MIC) for *K. pneumoniae* KpRR2, the mutant ΔICE and complemented strain ΔICE + pSXPQA for five antimicrobials.

Strain	MIC for antibiotic μg/ml*: Mean (Range)				
	Tetracycline	**Streptomycin**	**Erythromycin**	**Ceftazidime**	**Rifampicin**
KpRR2	3.2 (2–4)	2.1 (1–4)	35.6 (22–46)	1.9 (1.5–2)	11.2 (8–12)
ΔICE	1.8 (1.5–2)	0.37 (0.2–0.5)	8.7 (6–12)	0.6 (0.5–0.75)	2.3 (1.5–3)
ΔICE + pSXPQA	3 (2.5–4)	1.9 (1–4)	32.2 (21–38)	2 (1.5–2.5)	11.8 (11–12)
Fold change KpRR2 vs. ΔICE	**1.8**	**5.7**	**4.1**	**3.2**	**4.9**

**range tested using E-test strips (5 replicates): tetracycline, erythromycin, ceftazidime 0.016–256 μg/ml, streptomycin 0.064–1024 μg/ml, rifampicin 0.002–32 μg/ml.*

putative functions of YbtX and the function(s) of this putative transporter remain unknown.

Transfer of the Yersiniabactin Importer YbtPQ to *E. coli* Reduced Antimicrobial Susceptibility by an Efflux Mechanism

ICEKp transfers efficiently from *K. pneumoniae* to *E. coli* by conjugation (Farzand et al., 2019). Transconjugant *E. coli* were produced and were significantly less sensitive than parental *E. coli* HB101 to tetracycline and trimethoprim (**Figure 6A**). Mirroring the results in *K. pneumoniae*, the efflux inhibitor reserpine abrogated the effect of ICEKp on antimicrobial susceptibility in *E. coli* (**Figure 6A**), and the plasmid encoding YbtPQ was sufficient to significantly reduce antimicrobial susceptibility (**Figure 6B**). The effect of YbtPQ was broad, since antimicrobials from diverse classes were chosen, and the of zone of inhibition was significantly reduced for all tested (**Figure 6B**). Reserpine significantly reduced the effect of YbtPQ on antimicrobial susceptibility (**Figures 6C,D**).

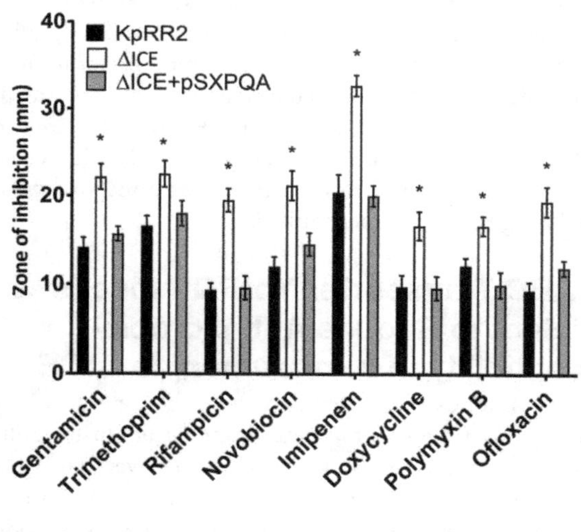

FIGURE 3 | Deletion of the integrative and conjugative element ICEKp from *K. pneumoniae* rendered it more sensitive to multiple classes of antimicrobial, and the effect could be reversed by reintroduction of the yersiniabactin transporter gene cluster. A deletion mutant lacking the entire ICEKp element, ΔICE, had a significantly larger zone of inhibition than the parent strain KpRR2 for all antimicrobials tested. Reintroduction of the yersiniabactin transporter gene cluster on plasmid pSXPQA restored the zone size to that of the parental strain. Data show the mean and standard deviation of three replicates. The quantity of antimicrobial per disk was: gentamicin 10 μg, trimethoprim 5 μg, rifampicin 5 μg, novobiocin 30 μg, imipenem 10 μg, doxycycline 30 μg, polymyxin B 300 μg, ofloxacin 5 μg. Each strain was compared with the parent strain by Student's *t* test, and significant differences in zone diameter were indicated with an asterisk. *$p < 0.05$.

DISCUSSION

While there is considerable progress in conjugating siderophores with antimicrobials for 'Trojan horse' delivery of antimicrobials via siderophore uptake systems (Klahn and Bronstrup, 2017), the promiscuous ability of siderophore transporters to transport unmodified antimicrobials has yet to be determined. Here we demonstrate that the yersiniabactin cluster of *K. pneumoniae* affects antimicrobial susceptibility via the transporter YbtPQ, and this trait can be transmitted by conjugation to *E. coli* and therefore potentially other pathogenic *Enterobacteriaceae*. This suggests that two selective pressures, iron acquisition and antimicrobial resistance, might drive the acquisition, spread and evolution of elements carrying the yersiniabactin cluster. Notably, the same selective advantage may also apply to other siderophore gene clusters.

Direct efflux of antimicrobials by YbtPQ is the simplest explanation of the effects of YbtPQ on antimicrobial susceptibility of *K. pneumoniae* and *E. coli*. However, this is the first data indicating that any ABC transporter might be a bifunctional importer and exporter and will require further investigation into the potential activity and the mechanism. The structure of the inward-open conformation of the YbtPQ transporter is consistent with antimicrobial access to the cavity, which could lead to transport by the measured basal ATPase activity, or by antimicrobial-enhancement of ATP binding and hydrolysis (Wang et al., 2020). YbtPQ is thought to span the inner membrane, so any antimicrobial efflux from the periplasm would necessarily involve direct or indirect coupling to an outer membrane protein. Apart from the putative efflux activity of YbtPQ, alternative explanations for its effect on antimicrobial susceptibility could include YbtPQ-induced changes in gene expression or YbtPQ-catalyzed transport of other molecules. KpRR2 ΔICE and *E. coli* HB101 make only one siderophore, enterobactin, which is structurally and chemically dissimilar from yersiniabactin.

The reduction in antimicrobial susceptibility conferred by YbtPQ was broad: susceptibility was reduced for 13 out of

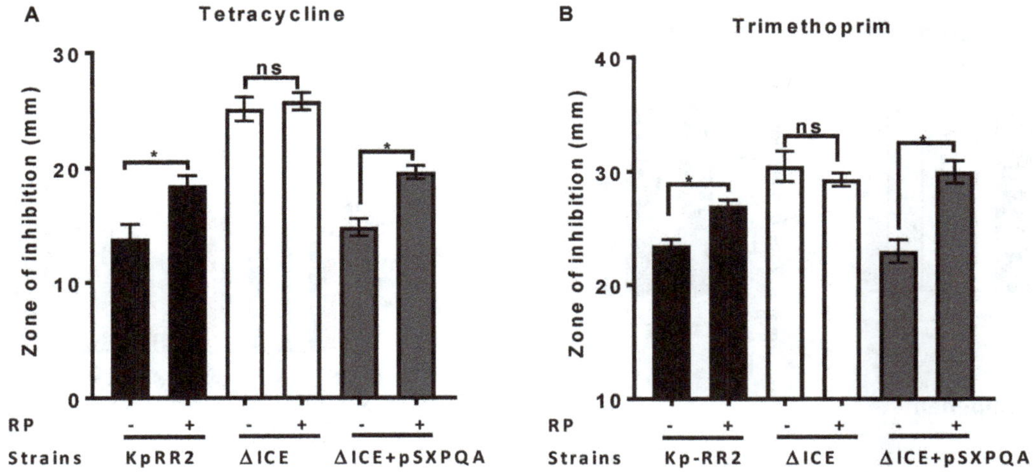

FIGURE 4 | Efflux pump inhibitor reserpine increased the susceptibility of *K. pneumoniae*, but not the ICEKp knockout, to antimicrobials. Addition of reserpine (RP, 50 µg/ml) significantly increased the diameter of the zone of inhibition of study strain KpRR2 and the plasmid-complemented mutant ΔICE + pSXPQA for trimethoprim **(A)** and tetracycline **(B)**. By contrast, the presence (+) or absence (-) of reserpine had no significant effect on the zones of inhibition of the knockout strain ΔICE. Trimethoprim was used at 5 µg per disk and tetracycline at 25 µg per disk. Data are the mean and standard deviation of three replicates. *indicates $p < 0.05$ using Student's *t* test. "ns" indicates $p > 0.05$. The assay was also performed with other antimicrobials giving similar results (**Supplementary Figure 1**).

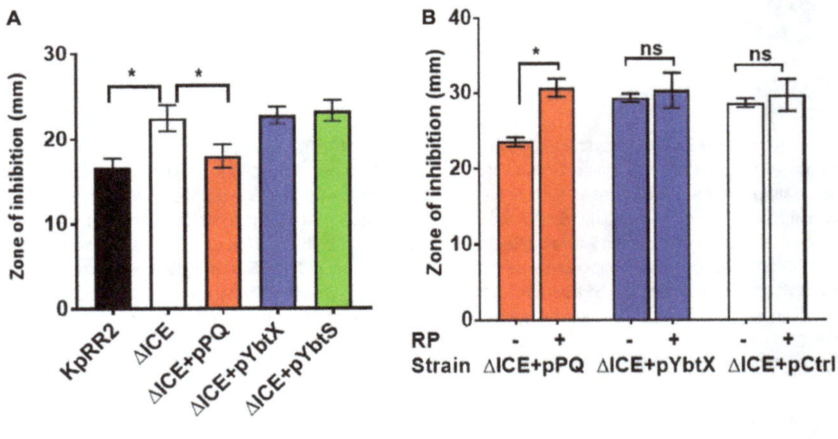

FIGURE 5 | The yersiniabactin importer YbtPQ reduced antimicrobial susceptibility of *K. pneumoniae* ICEKp mutant and the effect of YbtPQ was reversed by reserpine. **A** The enhanced trimethoprim susceptibility (larger zone of inhibition) of mutant ΔICE was fully complemented by plasmid pPQ, encoding transporter YbtPQ. Plasmids pYbtX and pYbtS did not complement the defect of ΔICE (no significant change in zone of inhibition compared to ΔICE). Trimethoprim was used at 5 µg per disk. **B** The enhanced imipenem susceptibility of mutant ΔICE was fully complemented by plasmid pPQ, encoding transporter YbtPQ, but not by the vector control, pCtrl. The effect of plasmid pPQ on imipenem susceptibility was reversed by the addition of reserpine (RP) 50 µg/disk). Data are the mean and standard deviation of at least three replicates. *indicates $p < 0.05$ using Student's *t* test. The effect of plasmid YbtPQ on antimicrobial susceptibility was also tested with additional antimicrobials (**Supplementary Figure 2**) and the effect of YbtPQ on antimicrobial susceptibility in the presence/absence of reserpine was additionally tested by MIC assay (**Table 2**).

TABLE 2 | Minimum inhibitory concentrations (MIC) for *K. pneumoniae* mutant ΔICE containing control plasmid or plasmids encoding YbtPQ or YbtX in the presence or absence of reserpine (RP).

Strain	MIC for antibiotic µg/ml*: Mean (Range)			
	Tetracycline No RP	**Tetracycline + RP**	**Streptomycin No RP**	**Streptomycin + RP**
ΔICE + pCtrl	2.1 (1.5–2.8)	1.6 (1.5–2)	0.51 (0.3–0.75)	0.43 (0.3–0.5)
ΔICE + pPQ	4.2 (3–5.5)	2.1 (1.5–3)	1.9 (1.8–2)	0.67 (0.62–0.7)
ΔICE + pYbtX	1.8 (1.7–2)	1.6 (1.5–1.8)	0.66 (0.5–0.75)	0.38 (0.3–0.5)

range tested using E-test strips (3 replicates): tetracycline 0.016–256 µg/ml, streptomycin 0.064–1024 µg/ml.

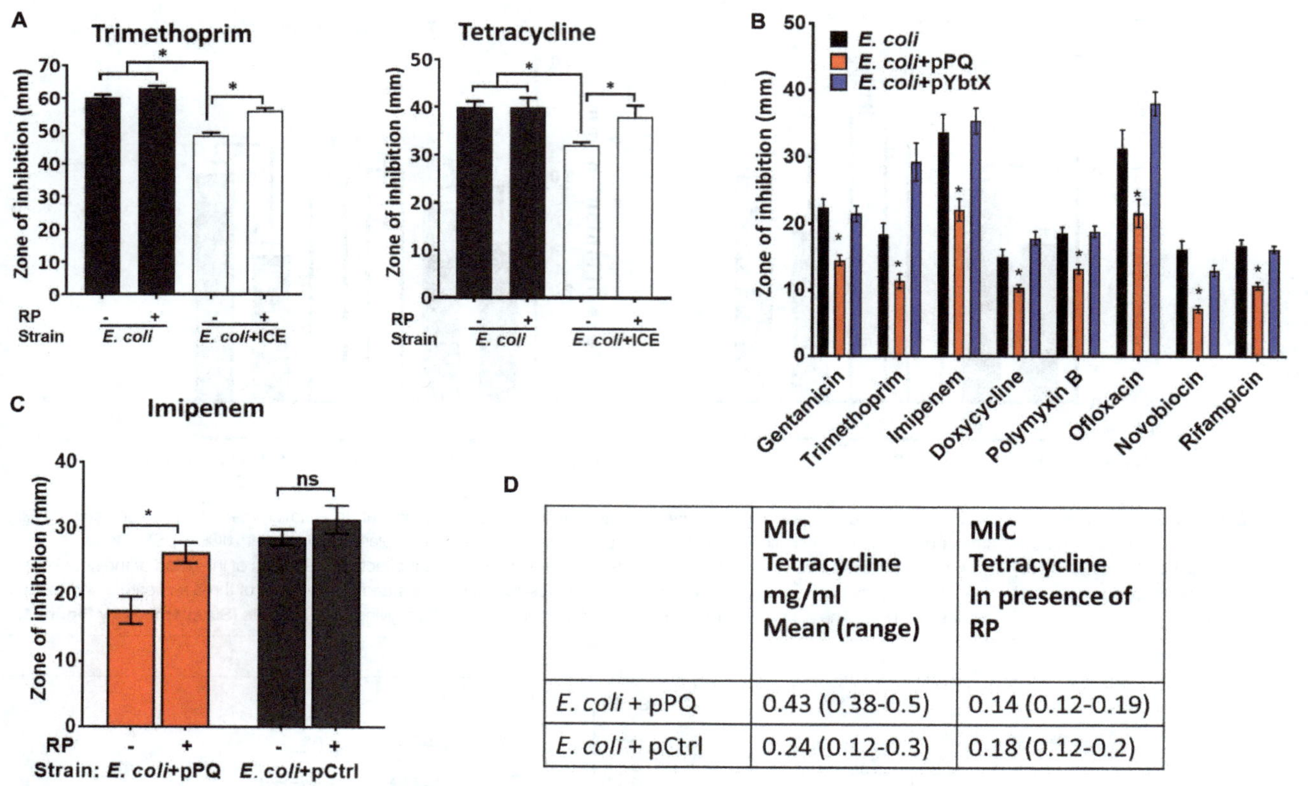

	MIC Tetracycline mg/ml Mean (range)	MIC Tetracycline In presence of RP
E. coli + pPQ	0.43 (0.38-0.5)	0.14 (0.12-0.19)
E. coli + pCtrl	0.24 (0.12-0.3)	0.18 (0.12-0.2)

FIGURE 6 | Antimicrobial susceptibility of *E. coli* was reduced by ICEKp or by YbtPQ alone. **(A)** Transconjugant *E. coli* + ICE carrying the ICEKp was significantly less sensitive than *E. coli* to trimethoprim and tetracycline in a disk diffusion assay (25 and 10 μg per disk, respectively). Inhibition of efflux using reserpine (50 μg/disk) significantly increased the susceptibility of the transconjugant to each antimicrobial. Reserpine (RP) had no significant effect on the susceptibility of parental *E. coli* to either antimicrobial. **(B)** *E. coli* carrying plasmid pPQ, encoding YbtPQ, were significantly less susceptible to all antimicrobials tested, whereas control plasmid pYbtX had no significant effect on antimicrobial susceptibility. From left to right, disks contained 10 μg, 5 μg, 10 μg, 30 μg, 300 units, 5 μg, 30 μg, or 5 μg of antimicrobial. **(C)** The effect of YbtPQ on susceptibility to imipenem was reduced by reserpine. pCtrl indicates a control plasmid without *ybtPQ* (vehicle only control). Data are the mean and standard deviation of six replicates. Strains and conditions were compared by Student's *t* test. * indicates $p < 0.05$. **(D)** The effect of YbtPQ on the minimum inhibitory concentrations (MIC) of *E. coli* for tetracycline was determined using E-test assay in the presence or absence of reserpine (RP 50 μg/ml) in triplicate (range tested was 0.016–256 μg/ml).

13 tested antimicrobials, and for some of antimicrobials like imipenem, the deletion of YbtPQ changed *K. pneumoniae* from resistant to sensitive according to the breakpoint in EUCAST (European Committee on Antimicrobial Susceptibility Testing 2020) (Kronvall et al., 2011). Broad specificity is a feature of some antimicrobial drug transporters, for example AcrB, an RND efflux pump (Yu et al., 2003), which reduces susceptibility to all antimicrobial classes tested in this study, and 10 out of 13 of the specific antimicrobials. As mentioned above, introduction of YbtPQ could affect antimicrobial susceptibility by multiple mechanisms including changes in expression of porins and transporters, therefore the broad reduction in antimicrobial susceptibility conferred by YbtPQ could reflect specificity of YbtPQ itself or other proteins and processes that YbtPQ affects. The inhibitor reserpine, which was used to test whether reduced antimicrobial susceptibility could be due to antimicrobial efflux, inhibits ABC, RND, and MFS transporters (Shaheen et al., 2019).

The yersiniabactin cluster is one of the key virulence associated factors reported in surveillance studies of outbreaks and spread of *K. pneumoniae* and here we demonstrate a potential selective advantage for this cluster in the presence of antimicrobials. Given that expression of the cluster is upregulated in infection models when iron availability is limited, this advantage could be significant in patients infected with *K. pneumoniae* who are receiving antibiotic treatment.

Other siderophore transporters from related or unrelated clusters might similarly influence antimicrobial resistance. The virulence enhancing piscibactin plasmids, carrying a YbtPQ homolog, are transmissible between species and enhance virulence of the economically fish aquaculture pathogen, *Photobacterium damselae* subsp *piscidia* (Osorio et al., 2015; Thode et al., 2018).

The facile movement of a virulence determinant from a multidrug resistant clinical isolate to another Gram-negative bacterium, with associated reduction in antimicrobial susceptibility, highlights the complex threat posed by the evolution and spread of drug resistance loci and drug resistant pathogens.

MATERIALS AND METHODS

Bacterial Strains and Plasmids

Klebsiella pneumoniae and *E. coli* were cultivated in accordance with biosafety and institutional safety procedures using LB (Luria-Bertani or Lysogeny broth) or Luria agar. When needed for selection or maintenance of plasmids, chloramphenicol was added at 30 μg/ml. Reserpine was purchased from Fisher, dissolved in DMSO (dimethyl sulfoxide), filter sterilized, and added to agar at 50 μg/ml when indicated (final 0.1% DMSO).

K. pneumoniae KpRR2 was derived from clinical isolate HS11286 in a previous study (Bi et al., 2015). The GenBank Accession number for the HS11286 chromosome is CP003200. The chromosome sequence includes the *asn* tRNA-associated mobile genetic element ICEKp. ICEKp is also deposited separately as ICE*Kpn*HS11286-1 in the ICEberg database, ID 180 (Bi et al., 2012). Strain ΔICE was derived from KpRR2 previously by deletion of the entire ICEKp (Farzand et al., 2019).

Filter Mating for Introduction of ICEKp to *E. coli* by Conjugation

A plasmid pOriT containing the chloramphenicol resistance cassette and origin of transfer (Farzand et al., 2019) was introduced to KpRR2 to act as a selectable marker for conjugation. Filter mating was used as described previously to transfer ICEKp with pOriT from KpRR2 to *E. coli* HB1010. Transconjugants were selected on LA with streptomycin 50 μg/ml and chloramphenicol 30 μg/ml (Farzand et al., 2019). PCR was used to confirm the presence of ICEKp and to verify the species using an *E. coli*-specific primer pair. Three transconjugants, termed *E. coli* + ICE, had equivalent phenotypes and were used in parallel for all experiments. Strain genotypes and primer sequences are listed in **Supplementary Tables 1, 2**.

Construction of Complementation Plasmids

Plasmids pSXPQA, pPQ, pYbtX, and pYbtS were constructed by PCR amplification of the named gene(s) and cloning using the HD infusion cloning method (Raman and Martin, 2014) into plasmid pACYC184 (Chang and Cohen, 1978; Rose, 1988). Plasmid pSXPQA contained the five genes KPHS_34610-KPHS_34650 (*ybtS-ybtA* respectively) with their native promoter. The other three plasmids pPQ, pYbtX, and pYbtS, used the *crp* promoter. A control vector pCtrl was produced from pPQ by replacing *ybtPQ* with non-coding sequence of matching length taken from pOriT. In all cases, inserts replaced the tetracycline marker gene. Plasmid construction was verified by sequencing. Plasmids were introduced to *K. pneumoniae* and *E. coli* by electroporation (0.2 cm cuvette, 25 kV/cm, 25 μF, 200 Ω). The sequences of genes *ybtS* to *ybtA* are available in GenBank Accession number NC_016845.1 (Gene ID: 11848490, 11848491, 11848492, 11848493, and 11848494).

Antibiotic Susceptibilities by Disk Diffusion and E-test Method

Colonies from overnight LA plate cultures were picked and suspended in sterile 0.9% NaCl. The turbidity of the suspension was adjusted to match McFarland 0.5 standard (~0.8 at OD_{600}). The suspension was spread evenly on Mueller Hinton agar within 15 min of preparation using a sterile cotton swab to create a semi-confluent growth. The inoculum was allowed to dry for 10 min before applying the antibiotic disks or E-test strips then the plates were incubated at 37°C for 24 h. Disks and E-test strips containing antimicrobial were purchased from Oxoid and BioMerieux respectively. Any plates with uneven growth were discarded. The zone of growth inhibition around the disks was measured in mm and results are reported as the mean and standard deviation of three or more independent experiments, for which each experiment had three technical replicates. Representative images are provided in **Supplementary Figure 3**. For E-test, the MIC (μg/ml) was read from the scale on the E-test strip where the symmetrical inhibition ellipse edge intercepts the strip and results are reported as the mean of five independent replicates.

AUTHOR CONTRIBUTIONS

RF designed the study, performed the experiments, analyzed the data, and wrote the manuscript. KR designed the study. MB, PF, GM, and MO analyzed the data. HO'H analyzed the data and wrote the manuscript. All authors contributed to the article, reviewed the manuscript and approved the submitted version.

REFERENCES

Arnold, F. M., Weber, M. S., Gonda, I., Gallenito, M. J., Adenau, S., Egloff, P., et al. (2020). The ABC exporter IrtAB imports and reduces mycobacterial siderophores. *Nature* 580, 413–417. doi: 10.1038/s41586-020-2136-9

Balouiri, M., Sadiki, M., and Ibnsouda, S. K. (2016). Methods for in vitro evaluating antimicrobial activity: a review. *J. Pharm. Anal.* 6, 71–79. doi: 10.1016/j.jpha.2015.11.005

Bearden, S. W., Fetherston, J. D., and Perry, R. D. (1997). Genetic organization of the yersiniabactin biosynthetic region and construction of avirulent mutants in *Yersinia pestis*. *Infect. Immun.* 65, 1659–1668. doi: 10.1128/iai.65.5.1659-1668.1997

Bi, D., Jiang, X., Sheng, Z. K., Ngmenterebo, D., Tai, C., Wang, M., et al. (2015). Mapping the resistance-associated mobilome of a carbapenem-resistant *Klebsiella pneumoniae* strain reveals insights into factors shaping these regions and facilitates generation of a 'resistance-disarmed' model organism. *J. Antimicrob. Chemother.* 70, 2770–2774. doi: 10.1093/jac/dkv204

Bi, D., Xu, Z., Harrison, E. M., Tai, C., Wei, Y., He, X., et al. (2012). ICEberg: a web-based resource for integrative and conjugative elements found in Bacteria. *Nucleic Acids Res.* 40, D621–D626.

Chang, A. C., and Cohen, S. N. (1978). Construction and characterization of amplifiable multicopy DNA cloning vehicles derived from the P15A cryptic miniplasmid. *J. Bacteriol.* 134, 1141–1156. doi: 10.1128/jb.134.3.1141-1156.1978

Dhanarani, S., Congeevaram, S., Piruthiviraj, P., Park, J. H., and Kaliannan, T. (2017). Inhibitory effects of reserpine against efflux pump activity of antibiotic resistance bacteria. *Chem. Biol. Lett.* 4, 69–72.

Du, D., Wang-Kan, X., Neuberger, A., Van Veen, H. W., Pos, K. M., Piddock, L. J., et al. (2018). Multidrug efflux pumps: structure, function and regulation. *Nat. Rev. Microbiol.* 16, 523–539.

Farzand, R., Rajakumar, K., Zamudio, R., Oggioni, M. R., Barer, M. R., and O'hare, H. M. (2019). ICEKp2: description of an integrative and conjugative element in *Klebsiella pneumoniae*, co-occurring and interacting with ICEKp1. *Sci. Rep.* 9:13892.

Fetherston, J. D., Bertolino, V. J., and Perry, R. D. (1999). YbtP and YbtQ: two ABC transporters required for iron uptake in *Yersinia pestis. Mol. Microbiol.* 32, 289–299. doi: 10.1046/j.1365-2958.1999.01348.x

Hannauer, M., Braud, A., Hoegy, F., Ronot, P., Boos, A., and Schalk, I. J. (2012). The PvdRT-OpmQ efflux pump controls the metal selectivity of the iron uptake pathway mediated by the siderophore pyoverdine in *Pseudomonas aeruginosa Environ. Microbiol.* 14, 1696–1708. doi: 10.1111/j.1462-2920.2011.02674.x

Klahn, P., and Bronstrup, M. (2017). Bifunctional antimicrobial conjugates and hybrid antimicrobials. *Nat. Prod. Rep.* 34, 832–885. doi: 10.1039/c7np00006e

Koh, E. I., Robinson, A. E., Bandara, N., Rogers, B. E., and Henderson, J. P. (2017). Copper import in *Escherichia coli* by the yersiniabactin metallophore system. *Nat. Chem. Biol.* 13, 1016–1021. doi: 10.1038/nchembio.2441

Kronvall, G., Giske, C. G., and Kahlmeter, G. (2011). Setting interpretive breakpoints for antimicrobial susceptibility testing using disk diffusion. *Int. J. Antimicrob. Agents* 38, 281–290. doi: 10.1016/j.ijantimicag.2011.04.006

Kumar, S., Stecher, G., Li, M., Knyaz, C., and Tamura, K. (2018). MEGA X: molecular evolutionary genetics analysis across computing platforms. *Mol. Biol. Evol.* 35, 1547–1549. doi: 10.1093/molbev/msy096

Lam, M. M. C., Wick, R. R., Wyres, K. L., Gorrie, C. L., Judd, L. M., Jenney, A. W. J., et al. (2018a). Genetic diversity, mobilisation and spread of the yersiniabactin-encoding mobile element ICEKp in *Klebsiella pneumoniae* populations. *Microb. Genom.* 4:e000196.

Lam, M. M. C., Wyres, K. L., Judd, L. M., Wick, R. R., Jenney, A., Brisse, S., et al. (2018b). Tracking key virulence loci encoding aerobactin and salmochelin siderophore synthesis in *Klebsiella pneumoniae. Genome Med.* 10:77.

Lawlor, M. S., Hsu, J., Rick, P. D., and Miller, V. L. (2005). Identification of *Klebsiella pneumoniae* virulence determinants using an intranasal infection model. *Mol. Microbiol.* 58, 1054–1073. doi: 10.1111/j.1365-2958.2005.04918.x

Lawlor, M. S., O'connor, C., and Miller, V. L. (2007). Yersiniabactin is a virulence factor for *Klebsiella pneumoniae* during pulmonary infection. *Infect. Immun.* 75, 1463–1472. doi: 10.1128/iai.00372-06

Lin, T.-L., Lee, C.-Z., Hsieh, P.-F., Tsai, S.-F., and Wang, J.-T. (2008). Characterization of integrative and conjugative element ICEKp1-associated genomic heterogeneity in a *Klebsiella pneumoniae* strain isolated from a primary liver abscess. *J. Bacteriol.* 190, 515–526. doi: 10.1128/jb.01219-07

Liu, P., Li, P., Jiang, X., Bi, D., Xie, Y., Tai, C., et al. (2012). Complete genome sequence of *Klebsiella pneumoniae subsp. pneumoniae* HS11286, a multidrug-resistant strain isolated from human sputum. *J. Bacteriol.* 194, 1841–1842. doi: 10.1128/jb.00043-12

Osorio, C. R., Rivas, A. J., Balado, M., Fuentes-Monteverde, J. C., Rodriguez, J., Jimenez, C., et al. (2015). A transmissible plasmid-borne pathogenicity island confers piscibactin biosynthesis in the fish pathogen *Photobacterium damselae subsp. piscicida. Appl. Environ. Microbiol.* 81, 5867–5879. doi: 10.1128/aem.01580-15

Perry, R. D., and Fetherston, J. D. (2011). Yersiniabactin iron uptake: mechanisms and role in *Yersinia pestis* pathogenesis. *Microbes Infect.* 13, 808–817. doi: 10.1016/j.micinf.2011.04.008

Raman, M., and Martin, K. (2014). One solution for cloning and mutagenesis: in-fusion® HD cloning plus. *Nat. Methods* 11:972.

Rose, R. E. (1988). The nucleotide sequence of pACYC184. *Nucleic Acids Res.* 16:355. doi: 10.1093/nar/16.1.355

Schmitz, F.-J., Fluit, A., Lückefahr, M., Engler, B., Hofmann, B., Verhoef, J., et al. (1998). The effect of reserpine, an inhibitor of multidrug efflux pumps, on the in-vitro activities of ciprofloxacin, sparfloxacin and moxifloxacin against clinical isolates of *Staphylococcus aureus. J. Antimicrob. Chemother.* 42, 807–810. doi: 10.1093/jac/42.6.807

Schubert, S., Rakin, A., and Heesemann, J. (2004). The *Yersinia* high-pathogenicity island (HPI): evolutionary and functional aspects. *Int. J. Med. Microbiol.* 294, 83–94. doi: 10.1016/j.ijmm.2004.06.026

Shafik, S. H., Cobbold, S. A., Barkat, K., Richards, S. N., Lancaster, N. S., Llinas, M., et al. (2020). The natural function of the malaria parasite's chloroquine resistance transporter. *Nat. Commun.* 11:3922.

Shaheen, A., Afridi, W. A., Mahboob, S., Sana, M., Zeeshan, N., Ismat, F., et al. (2019). Reserpine is the new addition into the repertoire of AcrB efflux pump inhibitors. *Mol. Biol. (Mosk)* 53, 674–684.

Smith, D. D. N., Williams, A. N., Verrett, J. N., Bergbusch, N. T., Manning, V., Trippe, K., et al. (2019). Resistance to two vinylglycine antibiotic analogs is conferred by inactivation of two separate amino acid transporters in *Erwinia amylovora. J. Bacteriol.* 201:e00658-18.

Thode, S. K., Rojek, E., Kozlowski, M., Ahmad, R., and Haugen, P. (2018). Distribution of siderophore gene systems on a *Vibrionaceae* phylogeny: database searches, phylogenetic analyses and evolutionary perspectives. *PLoS One* 13:e0191860. doi: 10.1371/journal.pone.0191860

Wang, Z., Hu, W., and Zheng, H. (2020). Pathogenic siderophore ABC importer YbtPQ adopts a surprising fold of exporter. *Sci. Adv.* 6:eaay7997. doi: 10.1126/sciadv.aay7997

Yu, E. W., Aires, J. R., and Nikaido, H. (2003). AcrB multidrug efflux pump of *Escherichia coli*: composite substrate-binding cavity of exceptional flexibility generates its extremely wide substrate specificity. *J. Bacteriol.* 185, 5657–5664. doi: 10.1128/jb.185.19.5657-5664.2003

9

Identification of Conserved ABC Importers Necessary for Intracellular Survival of *Legionella pneumophila* in Multiple Hosts

*Amrita Lama, Samuel L. Drennan, Rudd C. Johnson, Grace L. Rubenstein and Eric D. Cambronne**

Department of Molecular Microbiology and Immunology, Oregon Health and Science University, Portland, OR, United States

Edited by:
Elizabeth L. Hartland, Hudson Institute of Medical Research, Australia

Reviewed by:
Julia Walochnik, Medical University of Vienna, Austria
Sunny Shin, University of Pennsylvania, United States

***Correspondence:**
Eric D. Cambronne
cambronn@ohsu.edu

It is established that the human pathogen *Legionella pneumophila* becomes significantly augmented for infection of macrophages after intracellular growth in amoebae when compared to like-strains cultivated in laboratory media. Based on this observation, we reasoned that the most critical virulence determinants of *L.p.* are expressed by responding to stimuli generated by the protozoan host specifically; a process we term "protozoan-priming." We sought to identify *L.p.* virulence factors that were required for replication in amoebae in order to highlight the genes necessary for production of the most infectious form of the bacterium. Using a transposon mutagenesis screen, we successfully identified 12 insertions that produced bacteria severely attenuated for growth in amoebae, while retaining a functional Dot/Icm type IVb secretion system. Seven of these insertion mutants were found dispensable for growth in macrophages, revealing attractive therapeutic targets that reside upstream of the pathogen-human interface. Two candidates identified, *lpg0730* and *lpg0122* were required for survival and replication in amoebae and macrophage host cells. Both genes are conserved among numerous important human pathogenic bacteria that can persist or replicate in amoebae. Each gene encodes a component of an ATP binding cassette (ABC) transport complex of unknown function. We demonstrate the *lpg0730* ortholog in *Francisella tularensis* subsp. *novicida* to be essential for colonization of both protozoan and mammalian host cells, highlighting conserved survival mechanisms employed by bacteria that utilize protozoa as an environmental reservoir for replication.

Keywords: *Legionella*, pathogenesis, transposon mutagenesis, *Acanthamoeba castellanii*, macrophage, Dot/Icm T4b secretion system, ABC transporter, *Francisella*

INTRODUCTION

Legionella pneumophila (*L.p.*) is a Gram-negative bacterium predominantly associated with freshwater environments. Free-living bacteria persist in water, yet replication is restricted to the confines of host protozoan cells (amoebae), where the bacterium is a facultative intracellular parasite (Fields, 1996; Abu Kwaik et al., 1998). It is an opportunistic human pathogen, where egress from host amoebae generates the most infectious form of the bacterium (Cirillo et al., 1994, 1999; Brieland et al., 1996). Aerosolization of contaminated water sources provides an invariant route of transmission, providing access to resident lung alveolar macrophages. Intracellular colonization and replication in host macrophages occurs in a similar fashion to the bacterial life-cycle in protozoa, manifesting a spectrum of pathologies collectively termed "legionellosis." The range of presentation, which is largely owed to host immune-competence, includes self-limited flu-like illness (Pontiac fever) to a severe and often fatal pneumonia called Legionnaires' disease (Burillo et al., 2017).

In order to survive and replicate in eukaryotic cells, *L.p.* requires a specialized type IVb secretion system termed Dot/Icm (Marra et al., 1992; Berger and Isberg, 1993; Segal and Shuman, 1997). This multi-protein secretion apparatus functions to deliver up to 300 effector proteins into the host-cell (Zhu et al., 2011). Effectors collectively function in the subversion of host-cellular processes to facilitate establishment of a replication-permissive compartment, often termed *Legionella*-containing vacuole (LCV) (Roy, 2002). Strains defective for Dot/Icm-mediated transport of effectors are rendered largely avirulent (Roy et al., 1998). *L.p.* developed these intracellular survival strategies through close associations with protozoa in the environment. Strategies adopted for survival in protozoa therefore directly translated to efficient replication in the hostile environment encountered within the macrophage of the lung.

Importantly, evidence dating back over 20 years demonstrated that *L.p.* cultivated in amoebae prior to infection of macrophage cell lines or murine hosts were both hyper-invasive and hyper-virulent; upwards of 100-fold when compared to like-strains cultured in laboratory media; a phenomenon we term "protozoan-priming" (**Figure 1**) (Cirillo et al., 1994, 1999; Brieland et al., 1996, 1997a,b; Drennan et al., 2013). This effect has been experimentally ascribed to a transition from an intracellular replicative phase to a non-dividing transmissive form (Molofsky and Swanson, 2004). Similarly "transmissive" bacteria have been modeled *in vitro* by culturing *L.p.* to early stationary phase. Based on these early observations however, the discrepancy between the virulence properties associated with protozoan-primed vs. *in vitro* cultured bacteria could be exploited to reveal key determinants for *L.p.* dissemination.

We speculated that *L.p.* must respond to environmental stimuli provided exclusively by the protozoan host-cell in order to activate the expression of genes that contribute to the observed augmented infection phenotypes. With this notion in mind, we first sought to comprehensively identify genes necessary for intracellular survival in amoebae specifically, to potentially uncover therapeutic targets that would prevent generation of the highly virulent and most biologically relevant form of *L.p.* in the context of human infection. Additionally, these targets would have potential to reside upstream of the bacterial-human interface.

To this end, we conducted a genome-wide transposon mutagenesis screen in *L.p.* using intracellular survival and replication in the model protozoan *Acanthamoeba castellanii* (*A.c.*) as a primary evaluation criterion (Holden et al., 1984). Mutants that were attenuated or failed to replicate in *A.c.* (as determined using fluorescence microscopy) were next subjected to additional rounds of screening. Potential *dot/icm* insertions were selected-against based on sensitivity to growth on artificial media containing sodium chloride, and confirmed for a functional Dot/Icm transporter using an adenylate-cyclase reporter assay (Vogel et al., 1996; Cambronne and Roy, 2007). Host-cell specificity among several candidate insertions was demonstrated by examining intracellular replication in additional host cell types including murine and human macrophages.

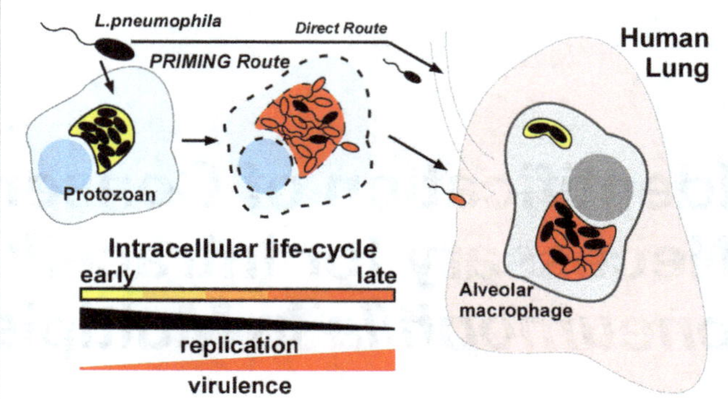

FIGURE 1 | Protozoan-primed *Legionella pneumophila* are augmented for dissemination to human macrophages. *L.p.* strains that have completed an intracellular life-cycle in natural-host protozoa are up to 100x more infectious (*Priming route*) than like-strains cultivated in laboratory medium or persisting in freshwater (*Direct route*). Response to specific host-cellular cues must trigger expression of key virulence determinants for alveolar macrophage infection in the human lung.

Of particular interest were insertions in *lpg0730* and *lpg0122*, each encoding a structural component of a distinct ATP binding cassette (ABC) transport complex (Theodoulou and Kerr, 2015). The assignment of substrate to each transport complex is unresolved. However, we found both *loci* conserved among several bacterial pathogens that can utilize protozoa as an intermediate reservoir for proliferation and transmission in the environment. Further, we demonstrated that disruption of the *lpg0730* ortholog in *Francisella tularensis* subsp. *novicida* was essential for colonization of both protozoan and mammalian host-cells. Our data suggest that Lpg0730-containing ABC transport complexes therefore represent a conserved intracellular survival determinant that represents an attractive target for inhibiting proliferation in environmental host cells.

RESULTS

Construction and Screening of *L. pneumophila* Mutant Library

We first generated a fluorescently-tractable isogenic *L.p.* strain harboring a single copy of *gfp*$_{mut3}$ on the chromosome that would serve as a wild-type representation for mutagenesis (JR32::*gfp*) (Cormack et al., 1996). GFP production was driven by dual promoters in tandem. The isopropyl β-D-1-thiogalactopyranoside (IPTG)-inducible *tac* promoter was located immediately 5′ to the *icmR* promoter; active in early stationary phase (Neild and Roy, 2003). The *gfp* construct was inserted 3′ to the stop codon of the *wipA* effector locus (**Figure 2A**). This location was chosen due to the large 710 bp stretch of non-coding sequence 3′ to the monocistronic *wipA*. WipA, which was recently reported to harbor tyrosine phosphatase activity had been previously determined dispensable for intracellular survival of *L.p.* in both macrophage and amoebae (Ninio et al., 2005; Pinotsis and Waksman, 2017). We additionally constructed an isogenic Δ*dotA::gfp* strain for use as a negative control for intracellular replication (Roy et al., 1998). Both JR32::*gfp* and JR32Δ*dotA::gfp* produced GFP when cultured *in vitro* in the presence of IPTG (**Figure 2B**/*insets*). However, WT was the only strain capable of supporting intracellular replication, which could be detected using fluorescence microscopy (**Figure 2B**).

In order to provide unbiased coverage of the non-essential *L.p.* genome, we constructed a library of individual insertion mutants of *L.p.* strain JR32::*gfp* using a modified minimariner transposon (mini-minimariner) mutagenic strategy (Murata et al., 2006). For library construction, we used pNH3503 plasmid carrying the mini-minimariner transposon. This transposon targets TA di-nucleotides on the chromosome, integrating in a random fashion (**Figure 2C**) (Murata et al., 2006). Greater than 4,000 insertions were isolated using 20 individual rounds of mutagenesis, where 200–250 isolates were collected per round. PCR analysis of randomly selected mutants demonstrated the presence of transposon in every isolate examined (not shown). Individual mutants were cataloged in 96-well format for preservation (**Figure 2C**).

To screen the library, we first examined whether each mutant could replicate intracellularly in the model host protozoan *A.c.* (Holden et al., 1984). Individual mutants were used to infect *A.c.* in 96-well plate format for 18 h. Infected wells were examined via fluorescence microscopy. Mutants that were attenuated or unable to replicate intracellularly, as judged by reduction or absence of mature fluorescent vacuoles, were selected for further rounds of screening (**Figure 2C**).

A secondary screen of candidate mutants was performed in 24-well format under controlled multiplicity of infection. Successful candidates that were validated as attenuated or defective in intracellular replication were next tested for their capacity to grow under conditions of elevated sodium chloride concentration. WT *L.p.* cannot grow on media containing (150–200 mM) NaCl, a consequence of a functional Dot/Icm type IVb secretion system (Vogel et al., 1996). Conversely, *L.p.* mutations in several *loci* encoding structural components of the Dot/Icm transporter were shown to render *L.p.* permissive for growth on elevated [NaCl] (**Figure 2C**). This important selection criterion therefore allowed for identification of isolates that failed to divide in host cells yet presumably retained a functional Dot/Icm transporter. Candidates that satisfied the screening criteria were next used to infect a panel of mammalian host-cell types to evaluate specificity. Thirty-eight insertions were selected for further characterization, where 18 failed to replicate in *A.c.* and an additional 20 were attenuated by 50% or greater when compared to WT (as measured subjectively by total mature replication vacuoles per microscopic field).

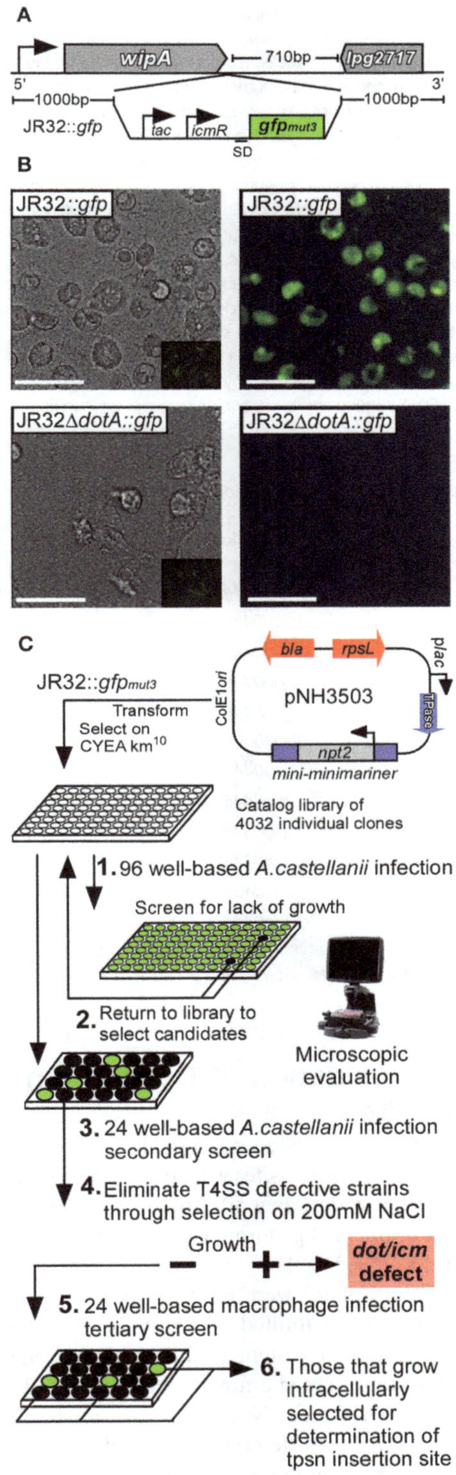

FIGURE 2 | Schematic representation of genome-wide transposon mutagenesis of *L. pneumophila* and survival screen in multiple host-cell types. **(A)** gfp_{mut3} with tandem promoters (*tac*, *icmR*) was inserted on the *L.p.* JR32 chromosome 3′ to the *wipA* (*lpg2718*) locus using homologous recombination. **(B)** The JR32::*gfp* strain was used to generate a T4SS defective mutant

(Continued)

FIGURE 2 | through in-frame deletion of *dotA*, resulting in JR32Δ*dotA*::*gfp*. Both strains were cultured to post-exponential phase in the presence of IPTG (insets) and used to infect *A.c.* cells. Eighteen hours post-infection, light (left column) and fluorescence (right column) micrographs were captured to visualize replication-permissive vacuoles. Scale bar = 50 μm. **(C)** JR32::*gfp* was subjected to transposon mutagenesis using the plasmid pNH3503, where kanamycin resistant clones were transferred to 96-well culture plates. Each clone was cultured to post-exponential phase and used to infect *A.c.* cultures for 18 h (1). Individual wells were examined microscopically to evaluate abundance of fluorescent vacuoles. Mutants that lacked fluorescence were targeted for secondary infections of *A.c.* using controlled MOI (2–3). Mutants that passed secondary screen were subjected to cultivation on media supplemented with 150 mM NaCl in order to select against Dot/Icm-defective strains (4). Salt-sensitive mutants were subsequently used to infect CHO FcγRII-transgenic, J774.A1, and THP-1 monolayers to evaluate survival capacity using fluorescence microscopy (5). All remaining mutants were subjected to DNA sequencing to determine transposon insertion sites (6).

TABLE 1 | NaCl-sensitive *dot/icm* insertions.

Isolate	Insertion	Annotation
B4-3	*lpg2674*	*dotD*
A3-4	*lpg0453*	*dotE/icmC*
E9-4	*lpg0454*	*dotP/icmD*
E2-9	*lpg0446*	*dotL/icmO*
E2-16	*lpg0446*	*dotL/icmO*
H12-32	*lpg0446*	*dotL/icmO*
C11-13	*lpg0445*	*dotM/icmP*
G9-32	*lpg0445*	*dotM/icmP*
F5-40	*lpg0445*	*dotM/icmP*
B7-29	*lpg0445*	*dotM/icmP*
E7-27	*lpg0455*	*dotN/icmJ*
A2-28	*lpg0444*	*icmQ*

Exhibited zero growth in A.c., CHO-FcγRII, J774.A1, THP-1.

Sequence Validation of Transposon Insertion Sites

Of the 38 isolates, 24 were successfully sequence validated, where transposon insertion sites were determined either by using the transposon insertion as a site for priming and subsequent amplification of purified genomic DNA or with "arbitrary" PCR (O'Toole and Kolter, 1998).

Twelve of the isolates were found in *dot/icm*-encoding *loci*, where insertions were limited to eight of the 27 genes that comprise the transport system (**Table 1**). These eight *dot/icm* genes therefore represented a functionally distinct class, as each failed to grow on elevated [NaCl]. An additional 12 insertions were located at the sites described in (**Table 2**). Each of these mutants were severely attenuated or failed to replicate in one of the four eukaryotic hosts examined. The remaining 14 mutants, displaying a range of intermediate phenotypes in *A.c.* infection were not sequenced but were cataloged. Infection phenotypes for these isolates are described in (Table S1).

In addition to measuring survival and replication in *A.c.*, each mutant was used to infect FcγRII transgenic Chinese Hamster Ovarian (CHO) cells (Nagai et al., 2005). Here bacteria were opsonized with polyclonal antisera directed against heat-killed WT *L.p.* prior to infection, which effectively stimulated phagocytosis and allowed for measure of survival in an artificial host system. Mutants were also used to infect the murine macrophage cell line J774.A1, or phorbol myristate acetate (PMA)-differentiated human THP-1 macrophages. Survival values indicated in **Table 2** represent approximates of the total number of fluorescent vacuoles per microscopic field compared as a percentage to the total number of observed vacuoles in an infection using WT *L.p.* over the same time-course of infection.

Overall, the 24 mutants identified using the selection criteria could be divided into two classes: (I) *dot/icm* machinery mutants that retained salt-sensitivity, or (II) mutants that were attenuated for intracellular survival in one or multiple host cell types. We found some degree of host cell-specificity associated with survival among seven of the 12 class 2 insertions (**Table 2**). Additionally, 3 of 12 class II insertions were located in intergenic regions.

Of the insertions identified as a result of limited or complete failure to produce mature replication vacuoles in *A.c.*, three interrupted *dot/icm* effector encoding genes. The product of *sdhA* has been implicated in vacuolar integrity and is necessary for survival in macrophages (Creasey and Isberg, 2012). The promoter regions of effector-encoding *lem25* (*lpg2422*) and *rvfA* (*lpg1797*) were also interrupted (Huang et al., 2011). Complementation studies using plasmid-borne copies of *lpg2422* or *lpg2423* (encoded opposite direction from transposon insertion) in the H2-15 isolate failed to restore intracellular survival. Further the *lpg1797* ORF alone, or in the context of native promoter failed to complement the A9-20 isolate. Therefore, the contribution of these transposon insertions to observed phenotypes remains unresolved.

Two additional insertions were found to interrupt genes encoding enzymes involved in amino acid metabolism. *lpg2276* encodes a Glu/Leu/Phe/Val- family dehydrogenase, a NAD^+ or $NADP^+$-dependent enzyme that de-aminates the amino acid to a keto-acid form, which can be assimilated into the Kreb's cycle. The *lpg1811* locus encodes aspartokinase-diaminopimelate decarboxylase, an enzyme important in lysine synthesis. The requirement for this gene was strictly limited to intracellular survival in the protozoan host.

Additional sequenced insertions included the *proQm* activator of ProP osmoprotectant transporter, which functions on ProP at a post-translational level. ProP responds to osmotic stress functioning as a zwitterion/proton symporter (Chaulk et al., 2011). Loss of *proQm* was found more detrimental for survival in the protozoan host. The D12-34 insertion was located between *lspG* and *lspF*, structural components of the type II secretion system in *L.p.* The Lsp secretion system has been previously implicated in intracellular survival of *L.p.* through secretion of multiple enzymes to the extracellular confines of the replication-permissive vacuolar compartment (Cianciotto, 2005). We also found *hsp90* to be important for optimal survival in the protozoan cell, remaining dispensable in each of the metazoan cell types.

TABLE 2 | Sequence-validated gene insertions identified in screen.

Isolate	Insertion	Annotation	Function	Intracellular survival (% of WT infection)			
				A.c.[a]	CHO[b]	J774	THP-1
C4-1	***lpg0730***	***perM***	**Membrane permease**	**0**	**0**	**100**[c]	**0**[c]
H2-16	*lpg0730*	*perM*	Membrane permease	0	0	100	0
C9-34	*lpg0730*	*perM*	Membrane permease	0	0	100	0
C6-28	*lpg2276*		Glu/Leu/Phe/Val dehydrogenase	0	0	100	100
A9-13	*lpg1811*	*lysC*	Aspartokinase-diaminopimelate decarboxylase	0	100	100	100
H2-15	lpg2422-23	***lem25**-hyp*	Dot/Icm effector Lem25 - hypothetical	0	0	50	50
D1-37	***lpg0122***		**ABC transporter-ATP binding protein**	**10**	**10**	**100**	**10**[c]
C1-1	*lpg0376*	***sdhA***	Dot/Icm effector protein SdhA	10	10	10	0
G12-15	*lpg0133*	*proQm*	Activator of ProP osmoprotectant transporter	10	50	50	100
D12-34	*lpg1362-63*	*gspG-gspF*	Type II secretion	10	100	100	0
A12-18	*lpg1369*	*htpG*	Heat shock protein Hsp90	50	100	100	100
A9-20	*lpg1796-97*	*lysR-**rvfA***	LysR family transcriptional regulator – Dot/Icm effector protein RvfA	50	100	100	100

Red indicates genes targeted in this study. Blue indicates Dot/Icm effector-encoding genes.
If two genes are indicated for an isolate, insertion was in intergenic region.
[a]*Acanthamoeba castellanii.*
[b]*Opsonized bacteria used to infect CHO FcγRII monolayers.*
[c]*As measured over 72 h using CFU counts.*

Identification of Putative ABC Import Complexes

Of particular interest, was that four of the insertions interrupted putative ATP-binding cassette (ABC) transporter components, *lpg0730* and *lpg0122*. (**Table 2**) (Theodoulou and Kerr, 2015). The *lpg0730* locus is the final gene in a transcriptional unit that contains as many as 9 ORFs. It resides 3′ to the gene encoding phosphatidylglycerophosphate phosphatase (*pgpA*) involved in lipid metabolism and membrane homeostasis (**Figure 3A**) (Funk et al., 1992). Three unique transposon insertion sites were identified in the screen, highlighting the importance of this locus for survival in *A.c* (**Figure 3B**). The 1,053 bp *lpg0730* ORF encodes a 350 aa (38.7 kDa) protein with 9 putative transmembrane segments. It is annotated as a membrane permease of the UPF0118 superfamily, which encompasses many substrate-binding transporters. *E. coli* YdiK is the archetype of this family of proteins, whose function is not determined but it is a member of the *purR* regulon, responsive to purine concentration (Cho et al., 2011). These permeases share strong homology with a class of transmembrane domain (TMD) proteins found in ABC transport systems.

The *lpg0122* locus is the second gene in a bi-cistronic transcriptional unit (**Figure 3A**). The D1-37 isolate was interrupted by the transposon at nucleotide position 543 of the 1,299 bp *lpg0122* ORF (**Figure 3B**). The gene encodes a 432 aa (48.4 kDa) protein annotated as a nucleotide binding domain (NBD) component of an ABC transport system. Its sequence falls under the NtrD/SsuB transporter family that encompasses numerous ATP-binding cassette domains that are involved in the import of nitrates and sulfonates. Lpg0122 shares its N-terminal 252 aa with TauB, involved in taurine import (Pereira et al., 2015). The C-terminal region is annotated as an AAA-ATPase associated region. The product of *lpg0121* is annotated as an ABC membrane permease (TMD).

We surmised *lpg0730* and *lpg0122* individually represented components of distinct ABC *"importer"* complexes, which could be extrapolated based on amino acid sequence and their arrangement on the *L.p.* genome. For ABC import complexes, the TMD and NBD components are generally distinct proteins, whereas in ABC exporters the TMD and NBD are each sub-domains of a single fusion protein (**Figure 3C**) (Theodoulou and Kerr, 2015). It is more likely that Lpg0122 (NBD) associates with the co-expressed Lpg0121 (TMD) than with Lpg0730. Furthermore, transposon interruptions of *lpg0730* or *lpg0122* exhibited different requirements for intracellular survival based on host cell type. Both Lpg0730 and Lpg0122 were determined to be highly conserved in Gram-negative and some Gram-positive bacteria when translated protein was used as template for homology search. This included several notable human pathogens that can use freshwater amoebae as an environmental intermediate. Multiple sequence alignment and phylogenetic assignment of Lpg0730, Lpg0122, and corresponding orthologous proteins are depicted in **Figures S1, S2**, respectively.

Intracellular survival relative to WT was grossly estimated visually for *lpg0730::Tn* and *lpg0122::Tn* as part of the screening process. In addition to *A.c.*, where survival rates approximated zero and 10 percent respectively, polyclonal anti-*legionella*-opsonized strains were used to infect FcγRII-transgenic CHO monolayers to measure survival independent from internalization (Nagai et al., 2005). Here, both insertions generated identical survival phenotypes to those observed in *A.c.* When used to infect murine J774.A1 macrophages, *lpg0730* and *lpg0122* were found dispensable for intracellular survival, and

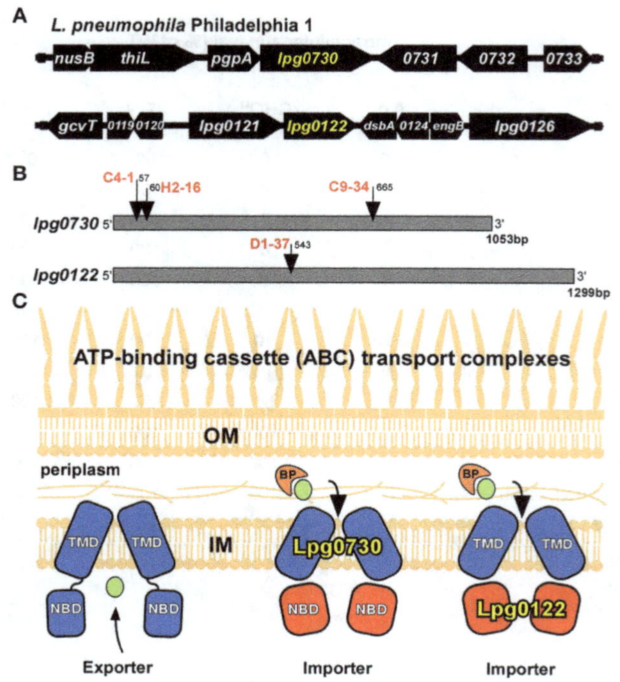

FIGURE 3 | Transposon insertions found in ABC import complex components. **(A)** Location on the *L.p.* genome of *lpg0730* and *lpg0122*. **(B)** Transposon insertion sites (nucleotide relative to +1) on the *lpg0730* and *lpg0122* ORF's for candidates identified in screen with clone designations in red. **(C)** Proposed location and function of Lpg0730 and Lpg0122 in the *L.p.* inner membrane. BP, binding protein (orange); TMD, transmembrane domain (blue); NBD, nucleotide binding domain (red); IM, inner membrane; OM, outer membrane, transported substrate depicted in green.

performed identical to WT. These results suggested that each of these *loci* were critical for survival in particular host cell types (*A.c.*, CHO FcγRII) while remaining completely dispensable in another (J774A.1). Curiously, although *lpg0122::Tn* also performed similar to WT during infection of human THP-1 macrophages, each of the three *lpg0730::Tn* mutants failed to replicate, suggesting a gradient of substrate concentration or availability among the macrophage hosts (**Table 2**).

Host Cell-Specific Activation of Candidate Genes

Some degree of host cell-specificity was observed for seven of 12 of the non-*dot/icm* insertion mutants. Differential genetic requirements for intracellular survival were most pronounced in *lpg0730*, *lpg1811*, and *lpg2276* insertion mutants (**Table 2**). We sought to determine the expression profile of these *loci* in WT *L.p.* in the context of intracellular growth in multiple established host cell types. Two closely related *Acanthamoeba* species (*A. castellanii*, *A. polyphaga*), a more distantly related amoebae species (*Hartmanella vermiformis*), and J774.A1 macrophages were selected for the analyses. Total RNA was isolated from bacteria either immediately after exposure to host cells or 18 h post-infection (prior to host cell egress). Quantitative PCR was performed using primer sets to generate ~150 bp amplicons from *lpg0730*, *lpg1811*, *lpg2276*, and DNA gyrase B (*gyrB*), which was used for normalization. As depicted in **Figure 4** each of the genes examined were highly activated in all three amoebae species while remaining inactivated or slightly repressed in J774.A1 over the time-course of infection. These results suggest that the micro-environments encountered by *L.p.* in various host phagosomes must be chemically diverse.

Characterization of *lpg0730* and *lpg0122* Mutant Strains

Our mutagenesis strategy selected only for non-essential genes. We therefore examined growth kinetics of the C4-1 (*lpg0730*) and D1-37 (*lpg0122*) insertion mutants through direct comparison to WT cultured in ACES-buffered yeast extract broth (AYE). As shown in **Figures 5A,B** the growth kinetics of all three strains were indistinguishable. Similar results were obtained through culture in defined synthetic media (not shown) (Warren and Miller, 1979). These results indicate that both *lpg0730* and *lpg0122* are dispensable for *L. p.* growth *in vitro*.

It is possible that the severe intracellular growth attenuation observed for the C4-1 and D1-37 insertion mutants in *A.c.* was a result of detrimental effects on the type IV secretion pathway, which is essential for pathogenesis. We sought to determine whether interruptions in *lpg0730* or *lpg0122* generated strains that were defective for *dot/icm*-dependent translocation of effector proteins. Reporter proteins were utilized that consisted of the catalytic domain of the calmodulin-dependent adenylate cyclase (Cya) from *Bordetella pertussis* fused to the amino terminus of the established effectors RalF or SidG (Cambronne and Roy, 2007). The production of cyclic adenosine-monophosphate (cAMP) resulting from the translocation of a Cya-effector hybrid into CHO-FcγRII cells was used to measure productive translocation. As indicated in **Figure 5C**, both the Cya-RalF and Cya-SidG hybrids were translocated to the host cytosol with higher fidelity in the C4-1 (*lpg0730*) background than in the parent JR32 strain. Translocation of Cya-RalF in the D1-37 (*lpg0122*) strain was slightly attenuated when compared with JR32. However, there was no significant difference in translocation efficiency of Cya-SidG, an *icmSW*-dependent effector protein. Cya-SidG translocation was previously reported to be attenuated with deletions in either *icmS* or *icmW*, which form an adaptor complex in the *L.p.* cytoplasm that promotes delivery of a class of effectors to the substrate receptor complex (**Figure 5C**) (Cambronne and Roy, 2007). Even though SidG translocation is reduced in *icmS* or *icmW* mutants, both strains can form replication-permissive vacuoles with reduced kinetics in *A.c* (Coers et al., 2000; Tilney et al., 2001). It is likely that a mutation in *lpg0122* does not have a direct effect on type IV secretion, but may indirectly affect kinetics of vacuole maturation.

Both *lpg0730* and *lpg0122* were next targeted for deletion using homologous recombination in a tri-parental mating scheme. We successfully generated a marker-less deletion of the *lpg0122* locus (Δ*lpg0122*) on the JR32::*gfp* genetic background. Multiple attempts to generate a similar deletion of the *lpg0730* locus,

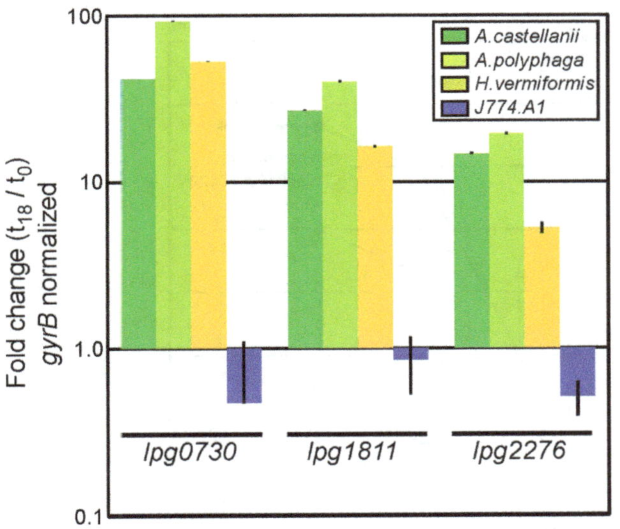

FIGURE 4 | Genes identified in screen are exclusively activated during infection in protozoa. qPCR analysis using total bacterial RNA isolated immediately after exposure to, or 18 h post-infection of indicated host cells. Bars indicate fold-change of indicated transcripts (t_{18}/t_0) normalized to DNA gyrase B (*gyrB*). Infections were performed in triplicate (three biological replicates) with standard deviations indicated.

including production of variant recombination vectors, failed to generate the desired strain. We therefore continued our studies using the C4-1 transposon insertion.

We next performed genetic complementation *in trans* by cloning *lpg0730* or *lpg0122* into a low-copy plasmid, with expression guided by an IPTG-inducible *tac* promoter. Sequence validated plasmids were transformed either into the C4-1 (*lpg0730*) or Δ*lpg0122* strains, in parallel to empty vector (pJB1806). Parent JR32, C4-1, and Δ*lpg0122* harboring appropriate plasmids were cultured to early stationary phase and used to infect *A.c.* After 18 h of infection *A.c.* cultures were imaged using light and fluorescence microscopy. A productive infection with WT JR32::*gfp* can be visualized in **Figure 5D**. No vacuoles were visualized with the C4-1 (*lpg0730*) strain, and the *lpg0122* deletion phenocopied the D1-37 isolate, with sparse vacuoles and low fluorescence. Production of Lpg0730 or Lpg0122 *in trans* fully restored *A.c.* infection fidelity to WT levels when evaluated microscopically (**Figure 5D**).

To quantitatively evaluate contribution of *lpg0730* and *lpg0122* to intracellular survival of *L.p.*, *lpg0730::Tn* (C4-1) or Δ*lpg0122* were used to infect either *A.c.* or THP-1 macrophages over a 72 h time course. Parallel infections were performed with WT (JR32::*gfp*), *dotA* deletion (JR32::*gfp*, Δ*dotA*), or complement strains. The *lpg0730::Tn* strain failed to replicate in either *A.c.* (**Figure 6A**) or THP-1 (**Figure 6C**), and was less persistent than Δ*dotA* in both hosts. Production of Lpg0730 via low-copy plasmid restored survival and replication of *lpg0730::Tn* to WT levels in *A.c.* and THP-1 (**Figures 6A,C**). The *lpg0122* deletion strain (Δ*lpg0122*) was replication competent in both *A.c.* and THP-1 with significantly reduced kinetics. In *A.c.*, maximal CFU

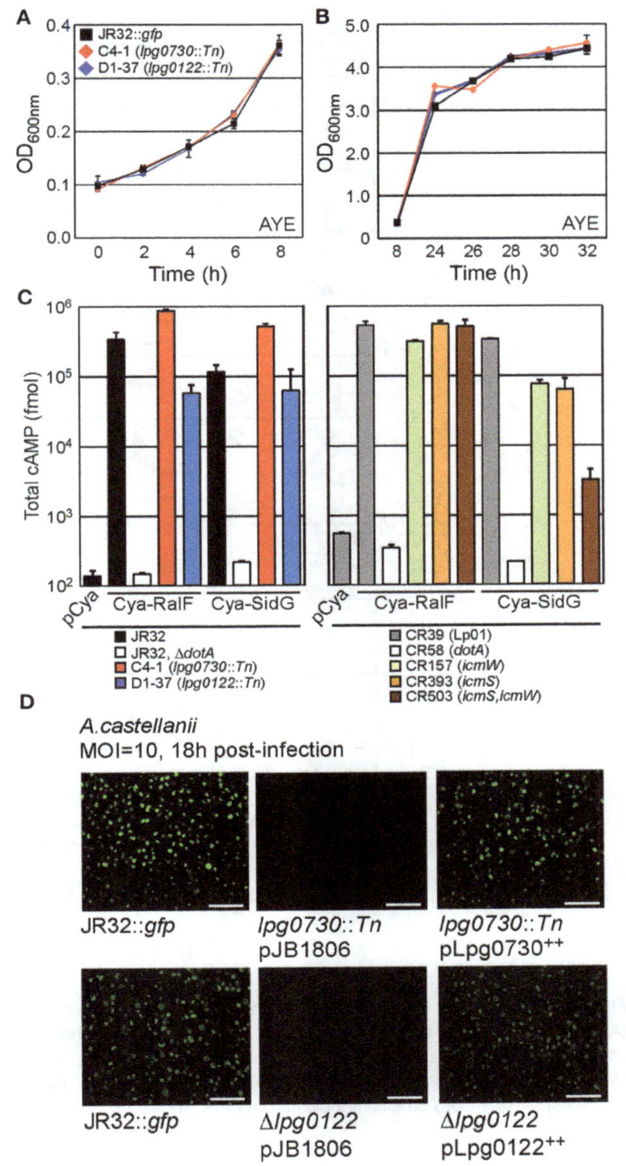

FIGURE 5 | *lpg0730* and *lpg0122* are dispensable for replication *in vitro* and do not interfere with Dot/Icm-mediated translocation of effectors. **(A)** Indicated strains were suspended in AYE broth at an $OD_{600\ nm} = 0.1$ and cultivated at 37°C on shaker. Spectrophotometric measurements were taken at indicated times spanning lag to exponential phase transition. **(B)** As in **(A)** with measurements taken to cover late exponential to early stationary phase transition. **(C)** *L.p.* strains (indicated with colored bars) harboring plasmids producing adenylate cyclase (Cya) fusions to the *L.p.* effectors RalF or SidG were used to infect CHO FcγRII monolayers for 1 h. Cells were lysed after which total cAMP was extracted and quantified using immunosorbent assay. Bars indicate total cAMP produced per well (fmol). Experiments were performed in triplicate (three biological replicates) with standard deviations indicated. **(D)** *lpg0730::Tn* and Δ*lpg0122* were transformed with either empty vector (pJB1806) or plasmids that supported inducible expression of Lpg0730 or Lpg0122. Resulting strains were cultured to post-exponential phase in parallel with JR32::*gfp* and used to infect *A.c.* cultures for 18 h (MOI = 10). Representative fluorescence micrographs are shown for infected *A.c.* cells with each indicated bacterial strain. Fluorescent *foci* represent individual replication vacuoles. Scale bar = 100 μm.

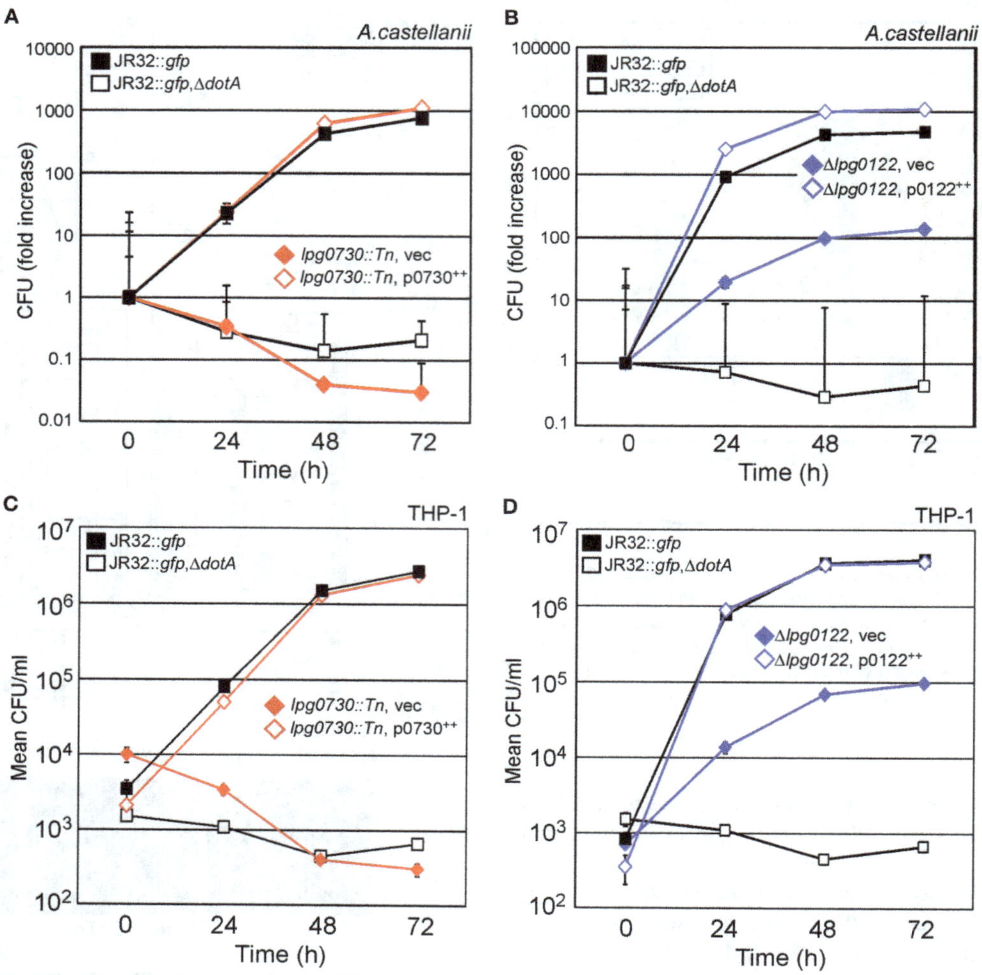

FIGURE 6 | Intracellular survival and replication of *lpg0730::Tn* and Δ*lpg0122* in *A.c.* and THP-1 human macrophages is fully restored via complementation *in trans*. **(A,B)** Indicated strains were used to infect *A.c.* cultures (MOI = 1), where host cells were lysed at indicated times and total CFU were determined by colony count after serial dilution and cultivation on CYEA. Results are shown as fold increase over 72 h. Experiments were performed in triplicate (three biological replicates) with standard deviations indicated. **(C,D)** As in **(A,B)** with 48 h PMA-differentiated THP-1 monocytes substituted for *A.c.* Graphs indicate mean CFU/ml (three biological replicates) with standard deviations indicated.

counts were achieved 72 h post-infection and were nearly 100-fold lower than CFU's recovered in the WT infection (**Figure 6B**). Similar attenuation was observed over the infection time course in THP-1 (**Figure 6D**). Similar to Lpg0730, plasmid-derived Lpg0122 restored intracellular survival and replication to WT levels in both *A.c.* and THP-1 (**Figures 6B,D**). Overall both expression constructs complemented the growth attenuation observed in the two mutant strains.

Conservation of *lpg*0730 and *lpg*0122 in Pathogens that Colonize Protozoa

In addition to *Legionella* species, many notable human pathogens are capable of colonizing protozoan cells. *Mycobacterium*, *Salmonella*, *Staphylococcus*, and many other Gram-negative and Gram-positive species can use amoebae as a nutrient rich environmental intermediate for replication and dissemination. We used the amino acid sequences of Lpg0730 and Lpg0122 to search for homologous and orthologous proteins in a subset of pathogenic species known to colonize amoebae use the BLASTP algorithm. To our surprise, both proteins were conserved in nearly every bacterial species examined. We next performed multiple sequence alignment and phylogenetic analysis using Clustal Omega. Of the species examined, Lpg0730, Lpg0122, and their respective homologs/orthologs separated into three clades from a common ancestral protein. The phylogenetic arrangement and multiple sequence alignments for Lpg0730 and Lpg0122 are shown in **Figures S1, S2** respectively.

We sought to determine whether Lpg0730 homologs/orthologs were necessary for intracellular survival. We first constructed a C-terminal FLAG epitope-tagged version of Lpg0730 by cloning this construct into the same vector used

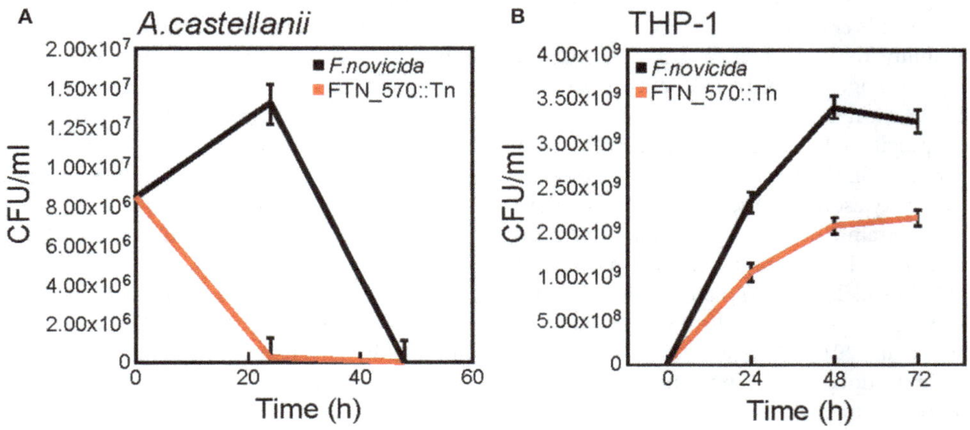

FIGURE 7 | The *lpg0730* ortholog *FTN0570* from *Francisella tularensis* subsp. *novicida* is required for persistence/survival in *A.c.* and human THP-1 macrophages. **(A)** *F.novicida* WT (U112) or FTN0570 mutant (FTN0570::Tn) were used to infect *A.c.* cultures (MOI = 10), where host cells were lysed at indicated times and CFU/ml was determined by colony count after serial dilution and cultivation on supplemented TSA. **(B)** As in **(A)** except strains were used to infect differentiated THP-1 macrophages for indicated times prior to host cell lysis. Graphs indicate mean CFU/ml (three biological replicates) with standard deviations indicated.

for complementation. Using the same *A.c.* infection strategy employed in **Figure 5D**, WT (JR32::*gfp*), and *dotA* deletion (JR32::*gfp*, Δ*dotA*), and *lpg0730::Tn* (C4-1) pLpg0730$_{FLAG}$ were evaluated microscopically 18 h post-infection. Similar to expression of Lpg0730 alone appending the C-terminal FLAG epitope had no detrimental effect on intracellular survival and the fidelity of infection was restored to WT levels (**Figure S3**).

Lpg0730 is assigned by sequence to the PerM family of membrane permeases, which is exemplified by the YdiK protein in *E. coli*. Alignment of Lpg0730 and YdiK is shown in **Figure S1**. We sought to determine whether an Lpg0730 homolog from *L. longbeachae* (76% identity) or PerM orthologs from *Salmonella enterica* subsp. Typhimurium (25.2% identity) or *Francisella tularemia* subsp. *novicida* (22.6% identity) could complement the intracellular survival defect of the C4-1 insertion mutant. C-terminal FLAG epitope-tagged versions of each allele were cloned and transformed into the C4-1 strain. Each complementing strain was used for infection of *A.c.* as in **Figure 5D**. Expression of the *L.longbeachae* allele completely restored intracellular survival of *L.p.* C4-1, whereas both the *Salmonella* and *Francisella* alleles failed to restore survival of the mutant (not shown).

Although the Lpg0730 (PerM) ortholog in *F.novicida* could not restore survival of *L.p.*, we had access to a sequence-cataloged transposon library of *F.novicida* U112 in the laboratory (Provided by F. Heffron). The library contained an insertion mutant in the *perM* locus (FTN0570). We used this strain and WT *F.novicida* U112 for infections of either *A.c.* or THP-1 macrophages. We determined that WT *F.novicida* persists in *A.c.* for over 24 h, while the FTN0570 failed to colonize amoebae (**Figure 7A**). The FTN0570 insertion mutant was also significantly attenuated for pathogenesis of THP-1 when compared to WT *F.novicida* over a 72 h time-course of infection (**Figure 7B**). These observations highlight the conservation of Lpg0730-like ABC import complexes and their requirement for optimal host cell colonization.

DISCUSSION

Details regarding physiological requirements of *L.p.* during the intracellular life-cycle are only beginning to be elucidated. Further, identification and characterization of *L.p.* factors required for intracellular survival even in the presence of a functional Dot/Icm transporter are even more limited. When we designed this screen, we anticipated that an abundance of candidates would be identified from the Dot/Icm effector protein catalog (Zhu et al., 2011). Just as *dot/icm* is required for virulence, nearly all effectors described to date are dispensable. We hypothesized that the extremely large catalog of effectors in *L.p.* (~10% of the genome) was owed to particular sets being required for colonization of a particular host. The taxonomically diverse host-range of *L.p.* could highlight specificity of particular subsets of effectors for survival in a particular host cell. We chose *A. castellanii* as host for the initial phase of the screen as a natural environmental host. *A.c.* transitions from trophozoite to cyst and reverts based upon nutrient availability (Lloyd, 2014).

Owing to our hypothesis, we indeed found a degree of host-cell specificity associated with several candidates identified in the screen, however we were surprised that only one effector, SdhA had a transposon insertion in its ORF (Laguna et al., 2006; Creasey and Isberg, 2012). Intergenic regions surrounding Lem25 and RvfA were also interrupted, but multiple complementation attempts were unsuccessful (Huang et al., 2011). Both of these mutants were more attenuated for survival in *A.c.* than in any of the metazoan host cells examined, so should be explored with greater scrutiny in the future. We also identified 12 insertions in *dot/icm* encoding genes. While none of these mutants supported translocation of Cya-RalF in the adenylate-cyclase reporter assay (not shown), it was notable that eight of the insertions were found in the DotLMN type IV coupling protein sub-complex (Sutherland et al., 2012). This provides additional evidence that separates the initial stages of substrate recognition

(DotLMN) from a superstructure that renders *L.p.* sensitive to NaCl via cytoplasm accessible conduit.

Amino acid availability to *L.p.* while in a phagosome was revealed as a key physiological constraint, similar to early studies using defined chemical media. *lpg2276* (Glu/Leu/Phe/Val dehydrogenase) was upregulated during intracellular replication in amoebae exclusively. Similarly *lpg1811* (Aspartokinase-diaminopimelate decarboxylase), important in lysine synthesis was highly upregulated in amoebae. The gene was determined dispensable in all metazoan cell types, both here and in a separate study where the *lpg1811* locus lost functionality over time when *L.p.* was continuously subjected to passage through a macrophage cell line (Ensminger et al., 2012). These results suggest that accessible lysine concentrations in the host must be distinct when comparing amoebae to macrophage. Other candidates identified in the screen can be directly correlated with particular environmental stressors encountered during colonization of protozoan cells. ProQm is a RNA-associated regulatory protein that governs activation of the ProP osmoprotectant transport system. Osmotic pressure changes are more likely to occur in the environmental host, especially in instances of desiccation upon encystment. HtpG (Hsp90) was found dispensable for survival in metazoan hosts, but was required for optimal infection of *A.c.*, perhaps a result of elevated temperature used to accelerate *A.c.* infections.

The most striking result of the screen was the identification of three independent transposon insertions in the *lpg0730* locus. Each of the insertions rendered *L.p.* completely attenuated for survival in *A.c.* and THP-1 host cells. *lpg0730* and an additional candidate *lpg0122* were both annotated to encode ABC import complex proteins (Theodoulou and Kerr, 2015). We selected these *loci* for further characterization as the identified substrates of the transporter families (YdiK and NtrD/SsrB, respectively) are variable and often ill-defined experimentally. Both *lpg0730* and *lpg0122* were demonstrated to be required for survival even in the presence of a functional Dot/Icm transporter, highlighting the importance of acquisition of particular substrates during colonization of the host cell.

Additionally, both Lpg0730 and Lpg0122 were conserved in multiple human pathogenic species that are capable of colonizing protozoa. Here we demonstrated the PerM ortholog in *F. tularensis* subsp. *novicida* was required for persistence in *A.c.* and optimal pathogenesis in THP-1 macrophages. This result supports the notion that in order to colonize amoebae, these distinct import complexes must be functional, and their activities take precedence over the contribution of other virulence mechanisms. Identification of the imported substrates will be of critical importance in order to develop methods to prevent acquisition of the particular nutrients supplied by the amoebae host.

Because the transmission of *L.p.* to humans is confined to dispersal via water droplet (aerosols), outbreaks of legionellosis are normally confined to a single source that has been compromised with infectious bacteria (Burillo et al., 2017). The barricade for controlling disease spread therefore resides predominantly upstream of the pathogen-human interface. A variety of methods are used both commercially and municipally to eliminate water-borne pathogens, including elevating temperature of contained water, lowering pH, or flushing water lines with heavy metals (Zgonc and Baideme, 2015). These treatment methods are often effective for elimination of *L.p.* and other bacteria. Contrarily, these sterilization scenarios can be insufficient to eradicate natural host amoebae; many of which can differentiate to an environmentally resistant cyst-form as part of their life-cycle (Lloyd, 2014).

Planktonic *L.p.* can be found in multiple physiological states in freshwater. Viable bacteria, which can be free-living or associated with polymicrobial biofilms, may be detected via culture on artificial media. Indeed *L.p.* has been demonstrated to remain infectious to human macrophage cell lines at low levels even after incubation at low temperature (10–15°C) in freshwater medium (Fraquil) for 6 months (Mendis et al., 2015). A second form, viable but not culturable (VBNC) could represent an adaptation of the bacterium to the nutrient-depleted environment that is freshwater. The VBNC state appears to be induced upon introduction of environmental stressors, especially temperature above 70°C (Ducret et al., 2015). VBNC *L.p.* were shown to be incapable of colonizing human airway epithelia or macrophage cell lines, but could be resuscitated when introduced to *A. polyphaga*. After completion of their intracellular life-cycle in amoebae, these protozoan-primed VBNC were now infectious to alveolar epithelial and macrophage cell lines (Epalle et al., 2014).

These results, along with the aforementioned studies comparing the infectivity of *L.p.* cultivated in laboratory medium compared to protozoan-primed bacteria, suggest that the state of the bacterium encountered at the human interface is most likely the transmissive form generated during egress from protozoa, and not the form generated through *in vitro* culture in rich medium. We felt it imperative to more thoroughly investigate the *L.p.*-protozoan interface in order to identify genetic factors required for colonization and replication of *L.p.* Because replication of *L.p.* in freshwater is restricted to the confines of protozoa, these genetic targets could be exploited to block proliferation of *L.p.* and other pathogens in amoebae, and if used in concert with current sterilization practices, could significantly reduce exposure of protozoan-primed *L.p.*, and other bacteria to humans.

MATERIALS AND METHODS

Bacterial Strains, Plasmids, Media, and Culture Conditions

All work performed in this study was approved by the OHSU Institutional Biosafety Committee (protocol #08-53). Bacterial strains, plasmids, and primers used in this study are summarized in Table S2. *L.p.* strains were cultured in ACES buffered yeast extract (AYE) or on AYE agar plates (0.2% activated charcoal) (CYEA) at 37°C. Media was supplemented with the following antibiotics where appropriate: chloramphenicol (Cm); (6.25 μg/ml), kanamycin (Kan); (20 μg/ml), streptomycin (Str); (50 μg/ml). Sucrose (5% w/v) was included in CYEA for counter-selection of recombinant plasmid pSR47S. *Escherichia coli* strains were cultured at 37°C in Luria Bertani (LB) medium supplemented with Cm (25 µg/ml), Kan (50 µg/ml), Str (50 µg/ml) where appropriate. *F. novicida* strains were cultured in Tryptic Soy Broth (TSB) supplemented with 0.1% cysteine at 37°C.

Cell Culture

Axenic *A.c.* strain Neff (ATCC 30010) and *A. polyphaga* strain (Puschkarew) Page CCAP 1501/3b (ATCC 30872) cells were propagated in supplemented PYG medium (ATCC 712) at room temperature (RT), harvested and diluted into fresh medium twice weekly. *H. vermiformis* strain CDC-19 (ATCC 50237) was propagated in SCGYEM medium (ATCC 1021) at 30°C, harvested and diluted into fresh medium weekly. Murine macrophage J774.A1 (ATCC TIB-67), human monocyte THP-1 (ATCC TIB-202) and FcγRII transgenic—Chinese Hamster ovary (CHO) cells (C. Roy laboratory) were maintained at 37°C with 5% CO_2. Undifferentiated THP-1 cells were cultured in suspension in RPMI medium supplemented with 10% fetal bovine serum (FBS). Forty-eight hours prior to infection, cells were differentiated in 24 or 96-well tissue culture plates, using phorbol 12-myristate 13-acetate (PMA). CHO FcγRII and J774.A1 were propagated in αMEM and DMEM respectively, supplemented with 10% FBS.

In Vitro Growth Studies

L.p. strains were inoculated in triplicate into 10 ml AYE or minimal media to achieve an initial $OD_{600\ nm}$ of 0.1 (Warren and Miller, 1979). Cultures were incubated on an orbital shaker (200 × rpm) at 37°C. One hundred microliters aliquots were collected from each culture at indicated times post-inoculation, suspended in 0.9 ml 1xPBS, and $OD_{600\ nm}$ were calculated via spectrophotometer.

Transposon Mutagenesis

L.p. JR32::*gfp* was generated via single-copy allelic integration of the gfp_{mut3} locus on to the chromosome immediately 3′ to the *wipA* (*lpg2718*) locus. A region encompassing sequential tandem *tac* and *icmR* promoters located 5′ to the gfp_{mut3} ORF was amplified from the plasmid pECR350 using primers 551/552. Flanking 1 kb fragments located 5′ and 3′ to the *wipA* stop codon were amplified using 549/550 and 553/554 respectively. The three PCR products were mixed (1:1:1) and used as template for short overlap extension (SOE) PCR using the 549/554 primer set. The resulting ~2.8 kb PCR product was ligated into the pSR47S vector linearized with *Bam*H1 using T4 DNA ligase (New England Biolabs). The resulting plasmid was designated pECL529. After sequence validation, pECL529 was transformed into *E. coli* DH5α λpir and introduced to JR32 via tri-parental mating scheme using *E. coli* DH5α, pRK600 as the helper strain. Transconjugants were generated through selection on CYEA Str, Kan. Colonies were isolated and plated on CYEA Str supplemented with 5% sucrose to force plasmid excision. Sucrose-resistant/kanamycin-sensitive isolates were examined using PCR to verify gfp_{mut3} insertion 3′ to the *wipA* locus. GFP positive isolates confirmed by fluorescence after growth in AYE broth supplemented with 1 mM IPTG. The resulting strain generated *JR32::*gfp_{mut3}. The transposon-containing plasmid, pNH3503 was introduced into *JR32::*$gfp_{mut3}$by electroporation. Twenty rounds of transformation and subsequent selection on CYEA Kan generated over 4,200 isolates which were randomly screened for the transposon insertion using the 573/574 primer set. Each individual insertion mutant was cultivated on CYEA in 96-well plates for 48 h at 37°C, re-suspended in stock solution (5% glycerol $_{W/V}$, 2% peptone $_{W/V}$), cataloged and stored at −80°C.

Intracellular Replication in Protozoa and Macrophages

Seventy-two hours quantitative intracellular growth assays were performed in a similar fashion to those previously described (Zamboni et al., 2003; Ninio et al., 2005). For the transposon mutagenesis screen, *A.c.* were suspended in supplemented ATCC 712 PYG media and cultured for 72 h, after which cells were harvested via centrifugation (400 × g) for 10 m. Media was aspirated and cells were suspended in infection media (3.4 mM Na citrate plus ATCC 712 salts) and seeded to tissue culture wells at 5×10^5 (0.5 ml) or 1.25×10^5 (0.125 ml) per well in 24 or 96-well format, respectively. Amoebae were allowed to adhere to plates overnight at RT. Host cells were infected with *L.p.* strains cultured to post-exponential phase in AYE broth (1 mM IPTG) at MOI = 10 (24-well) or ~10 (96-well). For the screening process, 96-well plates were floated on a 37°C water bath for 5 m and transferred to humidified incubator (37°C, 5%CO_2) for 18 h. Intracellular growth was assessed qualitatively using light and fluorescence microscopy (AMG EVOS*fl*). J774.A1 were cultured to near-confluency, harvested by centrifugation, seeded (2.5×10^5/well) in 24-well plates and incubated overnight at 37°C, 5%CO_2. One hour prior to infection, cells were washed 3x with 1x PBS, 0.5 ml fresh DMEM (10% FBS) was added to each well, and plates were returned to incubator. THP-1 monocytes were cultured for 3–5 days in RPMI (10% FBS) prior to harvest by centrifugation. Cells were resuspended in RPMI (10% FBS, 100 ng/ml PMA), seeded (2.5×10^5/well) in 24-well plates and incubated 48 h at 37°C, 5%CO_2. One hour prior to infection, cells were washed 3x with 1x PBS, after which 0.5 ml RPMI (10% FBS) was added to each well. CHO FcγRII were cultivated in αMEM (10% FBS) and processed similar to J774.A1, with the exception being the inclusion of anti-heat-killed *L.p.* polyclonal antisera (Pacific Immunology) in the αMEM replacement media 1 h prior to infection. The *A.c.* and THP-1 infections with *F. novicida* were performed in a similar manner, using bacteria cultured to post-exponential phase in TSB 0.1% cysteine.

DNA Sequencing

Insertion mutants were sequenced either by PCR using purified genomic DNA as template for primer 780, or with "arbitrary" PCR modified from O'Toole and Kolter (1998). Briefly, the 1° PCR utilized purified gDNA as template for amplification with the 571/Arb1c primer set, followed by a 2° PCR using first PCR product as a template for amplification with the 780/781 primer set.

Protein Sequence Alignment

L.p. reference amino acid sequences of polypeptides were derived from the Philadelphia 1 genome (NC_002942.5) National Center for Biotechnology Information (NCBI). Sequence alignments were performed using BLASTP with default parameters (NCBI/NLM). Phylogenetic analysis was performed by aligning *L.p.* proteins with individually selected bacterial species using BLASTP. Closest orthologous polypeptides from each species were aligned using Clustal Omega software (http://www.ebi.ac.uk/Tools/msa/clustalo/) European Molecular Biology Laboratory (EMBL).

Cya Translocation Assay

Translocation assays were performed as described previously (Nagai et al., 2005; Ninio et al., 2005). Briefly, monolayers containing 1×10^5 CHO FcγRII cells were infected with 3×10^6 opsonized *L.p.* (MOI = 30) expressing Cya hybrid proteins. After 1 h incubation at 37°C, 5% CO_2, monolayers were washed with 1x PBS and lysed. Total cAMP was extracted and quantified using cAMP Biotrak Enzymeimmumoassay System (GE Healthcare).

Transcriptional Analyses

To analyze transcription of *L.p.* genes during intracellular growth in *A. castellanii, A. polyphaga, H. vermiformis,* and J774.A1, cells were seeded in appropriate media under conditions as suited for each, and challenged with *L.p.* at an MOI = 200. The infection was synchronized by centrifuging the cells at 400 × g for 10 m and bacterial RNA was extracted immediately (t_0) and 18 h (t_{18}) post-infection. For extraction, infection buffer was removed and monolayers were disrupted with a sterile cell scraper. One milliliter of PBS was added to each well and vigorously mixed with pipette. Suspensions were transferred to a 14 ml conical tube and centrifuged for 10 m at 3,000 rpm. Supernatants were aspirated and pellets were re-suspended in 10 ml ice cold sterile H_2O (protozoan) or 10 ml ice cold sterile H_2O, 0.1% Triton X-100 (macrophage) and incubated on ice for 10 m. Lysates were centrifuged for 10 m at 4,000 × rpm. Supernatants were aspirated and pellets were resuspended in 1 ml guanidine thiocyanate buffer (GTC) (4 M guanidine thiocyanate, 0.5% Na N-lauryl sarcosine, 25 mM Na citrate, pH 7.0, 1 M β-mercaptoethanol) and transferred to a microcentrifuge tube. Suspensions were passed through a 21 gauge needle 3x and centrifuged 10 m at 15,000 × rpm. After aspiration, pellets were washed with 1xPBS, 0.1% Tween-20 and centrifuged an additional 10 m at 15,000 × rpm. After aspiration, pellets were re-suspended in 100 μl of 10 mg/ml lysozyme (Fisher Bioreagents) in 10 mM Tris, pH 8.0. Samples were incubated 15 m at RT. 750 μl of 65°C Trizol reagent (Invitrogen) was added to each sample, mixed and centrifuged for 10 m at 15,000 × rpm. The aqueous phase was transferred to a microcentrifuge tube and mixed with 500 μl of 100% ethanol. Samples were finally transferred to an RNeasy column and isolated according to manufacturer instructions, except that the in-column DNAse treatment was increased to 1 h (Qiagen). Purified RNA samples were analyzed with a Nano-drop *ND-1000* spectrophotometer and stored in aliquots at $^-$80°C.

Five Hundred nanograms of total bacterial RNA isolated from intracellular WT *L.p.* was subjected to cDNA synthesis using a Takara BluePrint RT reagent kit with random hexameric priming (Clontech) according to manufacturer instructions. After cDNA synthesis was performed in a thermocycler, samples were diluted 1:15 and stored on ice prior to qRT-PCR analysis. Primer sets were designed to target specific transcripts on the *L.p.* genome (**Figure 4**, Table S2). qRT-PCR samples were prepared using BioRad IQ SYBR Green Supermix. Reactions for each primer set probe were performed in triplicate in a Bio-Rad clear 96-well multiplate on a Bio-Rad Opticon thermocycler. Results were analyzed using Step One Software, v2.2. Relative expression, ΔΔCt analysis, was performed using *gyrB* as a reference transcript.

AUTHOR CONTRIBUTIONS

Conceptualization, methodology, funding acquisition, and project administration (EC). Investigation, data curation, validation, and formal analysis (AL, SD, RJ, GR, and EC). Visualization and original draft preparation (AL, SD, and EC). Review and Editing (EC).

ACKNOWLEDGMENTS

The authors wish to thank Drs. Hiroki Nagai and Joseph Vogel for plasmid constructs. Helen Wu, Gabrielle Ma, Taylor Toro, Ben Doron, and Dimitra Fellman contributed experimental data and technical assistance. This work was supported by grants AI123682, (National Institute of Allergy and Infectious Diseases) GM94623, (National Institute of General Medical Sciences) and by the OHSU Foundation (EC).

SUPPLEMENTARY MATERIAL

Figure S1 | Lpg0730 is conserved in multiple genera that colonize protozoa.**(A)** Clustal Omega (EMBL) phylogenetic cladogram depicting Lpg0730 (*L.p.*) and its relationship to homologous/orthologous polypeptide sequences found in the genera shown. **(B)** BLASTP amino acid sequence alignment (NCBI) of Lpg0730 (red) and homologous/orthologous polypeptide sequences found in the genera shown (multispecies). Nine representative transmembrane segments are shaded (gray). **(C)** BLASTP alignment of Lpg0730 and YdiK (*E. coli*). *indicate conserved residues. Transmembrane segments are shaded (gray). **(D)** BLASTP alignment of Lpg0730 and PerM (FTN0570) from *F. tularensis subsp. novicida*. *indicate conserved residues.

Figure S2 | Lpg0122 is conserved in multiple genera that colonize protozoa.**(A)** Clustal Omega (EMBL) phylogenetic cladogram depicting Lpg0122 (*L.p.*) and its relationship to homologous/orthologous polypeptide sequences found in the genera shown. **(B)** BLASTP amino acid sequence alignment (NCBI) of Lpg0122 (blue) and homologous/orthologous polypeptide sequences found in the genera shown (multispecies). **(C)** BLASTP alignment of Lpg0730 and TauB (*E. coli*). *indicate conserved residues.

Figure S3 | Complementation of *lpg0730*::Tn with $Lpg0730_{FLAG}$. Cultured *A.c.* were infected with indicated strain/plasmid combinations for 18 h. Representative live images were captured using light and fluorescence microscopy. Scale bar = 200 μm.

REFERENCES

Abu Kwaik, Y., Gao, L. Y., Stone, B. J., Venkataraman, C., and Harb, O. S. (1998). Invasion of protozoa by *Legionella pneumophila* and its role in bacterial ecology and pathogenesis. *Appl. Environ. Microbiol.* 64, 3127–3133.

Berger, K. H., and Isberg, R. R. (1993). Two distinct defects in intracellular growth complemented by a single genetic locus in *Legionella pneumophila*. *Mol. Microbiol.* 7, 7–19. doi: 10.1111/j.1365-2958.1993.tb01092.x

Brieland, J. K., Fantone, J. C., Remick, D. G., LeGendre, M., McClain, M., and Engleberg, N. C. (1997a). The role of *Legionella pneumophila*-infected *Hartmannella vermiformis* as an infectious particle in a murine model of Legionnaire's disease. *Infect. Immun.* 65, 5330–5333.

Brieland, J., McClain, M., Heath, L., Chrisp, C., Huffnagle, G., LeGendre, M., et al. (1996). Coinoculation with *Hartmannella vermiformis* enhances replicative *Legionella pneumophila* lung infection in a murine model of Legionnaires' disease. *Infect. Immun.* 64, 2449–2456.

Brieland, J., McClain, M., LeGendre, M., and Engleberg, C. (1997b). Intrapulmonary *Hartmannella vermiformis*: a potential niche for *Legionella pneumophila* replication in a murine model of legionellosis. *Infect. Immun.* 65, 4892–4896.

Burillo, A., Pedro-Botet, M. L., and Bouza, E. (2017). Microbiology and epidemiology of legionnaire's disease. *Infect. Dis. Clin. North Am.* 31, 7–27. doi: 10.1016/j.idc.2016.10.002

Cambronne, E. D., and Roy, C. R. (2007). The *Legionella pneumophila* IcmSW complex interacts with multiple Dot/Icm effectors to facilitate type IV translocation. *PLoS Pathog.* 3:e188. doi: 10.1371/journal.ppat.0030188

Chaulk, S. G., Smith Frieday, M. N., Arthur, D. C., Culham, D. E., Edwards, R. A., Soo, P., et al. (2011). ProQ is an RNA chaperone that controls ProP levels in *Escherichia coli*. *Biochemistry* 50, 3095–3106. doi: 10.1021/bi101683a

Cho, B. K., Federowicz, S. A., Embree, M., Park, Y. S., Kim, D., and Palsson, B. O. (2011). The PurR regulon in *Escherichia coli* K-12 MG1655. *Nucleic Acids Res.* 39, 6456–6464. doi: 10.1093/nar/gkr307

Cianciotto, N. P. (2005). Type II secretion: a protein secretion system for all seasons. *Trends Microbiol.* 13, 581–588. doi: 10.1016/j.tim.2005.09.005

Cirillo, J. D., Cirillo, S. L., Yan, L., Bermudez, L. E., Falkow, S., and Tompkins, L. S. (1999). Intracellular growth in *Acanthamoeba castellanii* affects monocyte entry mechanisms and enhances virulence of *Legionella pneumophila*. *Infect. Immun.* 67, 4427–4434.

Cirillo, J. D., Falkow, S., and Tompkins, L. S. (1994). Growth of *Legionella pneumophila* in *Acanthamoeba castellanii* enhances invasion. *Infect. Immun.* 62, 3254–3261.

Coers, J., Kagan, J. C., Matthews, M., Nagai, H., Zuckman, D. M., and Roy, C. R. (2000). Identification of Icm protein complexes that play distinct roles in the biogenesis of an organelle permissive for *Legionella pneumophila* intracellular growth. *Mol. Microbiol.* 38, 719–736. doi: 10.1046/j.1365-2958.2000.02176.x

Cormack, B. P., Valdivia, R. H., and Falkow, S. (1996). FACS-optimized mutants of the green fluorescent protein (GFP). *Gene* 173, 33–38. doi: 10.1016/0378-1119(95)00685-0

Creasey, E. A., and Isberg, R. R. (2012). The protein SdhA maintains the integrity of the Legionella-containing vacuole. *Proc. Natl. Acad. Sci. U.S.A.* 109, 3481–3486. doi: 10.1073/pnas.1121286109

Drennan, S. L., Lama, A., Doron, B., and Cambronne, E. D. (2013). Tractable mammalian cell infections with protozoan-primed bacteria. *J. Vis. Exp.* doi: 10.3791/50300

Ducret, A., Chabalier, M., and Dukan, S. (2015). Characterization and resuscitation of 'non-culturable' cells of *Legionella pneumophila*. *BMC Microbiol.* 14:3. doi: 10.1186/1471-2180-14-3

Ensminger, A. W., Yassin, Y., Miron, A., and Isberg, R. R. (2012). Experimental evolution of *Legionella pneumophila* in mouse macrophages leads to strains with altered determinants of environmental survival. *PLoS Pathog.* 8:e1002731. doi: 10.1371/journal.ppat.1002731

Epalle, T., Girardot, F., Allegra, S., Maurice-Blanc, C., Garraud, O., and Riffard, S. (2014). Viable but not culturable forms of *Legionella pneumophila* generated after heat shock treatment are infectious for macrophage-like and alveolar epithelial cells after resuscitation on *Acanthamoeba polyphaga*. *Microb. Ecol.* 69, 215–224. doi: 10.1007/s00248-014-0470-x

Fields, B. S. (1996). The molecular ecology of legionellae. *Trends Microbiol.* 4, 286–290. doi: 10.1016/0966-842X(96)10041-X

Funk, C. R., Zimniak, L., and Dowhan, W. (1992). The pgpA and pgpB genes of *Escherichia coli* are not essential: evidence for a third phosphatidylglycerophosphate phosphatase. *J. Bacteriol.* 174, 205–213. doi: 10.1128/jb.174.1.205-213.1992

Holden, E. P., Winkler, H. H., Wood, D. O., and Leinbach, E. D. (1984). Intracellular growth of *Legionella pneumophila* within *Acanthamoeba castellanii* Neff. *Infect. Immun.* 45, 18–24.

Huang, L., Boyd, D., Amyot, W. M., Hempstead, A. D., Luo, Z. Q., O'Connor, T. J., et al. (2011). The E Block motif is associated with *Legionella pneumophila* translocated substrates. *Cell. Microbiol.* 13, 227–245. doi: 10.1111/j.1462-5822.2010.01531.x

Laguna, R. K., Creasey, E. A., Li, Z., Valtz, N., and Isberg, R. R. (2006). A *Legionella pneumophila*-translocated substrate that is required for growth within macrophages and protection from host cell death. *Proc. Natl. Acad. Sci. U.S.A.* 103, 18745–18750. doi: 10.1073/pnas.0609012103

Lloyd, D. (2014). Encystment in *Acanthamoeba castellanii*: a review. *Exp. Parasitol.* 145(Suppl.), S20–S27. doi: 10.1016/j.exppara.2014.03.026

Marra, A., Blander, S. J., Horwitz, M. A., and Shuman, H. A. (1992). Identification of a *Legionella pneumophila* locus required for intracellular multiplication in human macrophages. *Proc. Natl. Acad. Sci. U.S.A.* 89, 9607–9611. doi: 10.1073/pnas.89.20.9607

Mendis, N., McBride, P., and Faucher, S. P. (2015). Short-term and long-term survival and virulence of *Legionella pneumophila* in the defined freshwater medium fraquil. *PLoS ONE* 10:e0139277. doi: 10.1371/journal.pone.0139277

Molofsky, A. B., and Swanson, M. S. (2004). Differentiate to thrive: lessons from the *Legionella pneumophila* life cycle. *Mol. Microbiol.* 53, 29–40. doi: 10.1111/j.1365-2958.2004.04129.x

Murata, T., Delprato, A., Ingmundson, A., Toomre, D. K., Lambright, D. G., and Roy, C. R. (2006). The *Legionella pneumophila* effector protein DrrA is a Rab1 guanine nucleotide-exchange factor. *Nat. Cell Biol.* 8, 971–977. doi: 10.1038/ncb1463

Nagai, H., Cambronne, E. D., Kagan, J. C., Amor, J. C., Kahn, R. A., and Roy, C. R. (2005). A C-terminal translocation signal required for Dot/Icm-dependent delivery of the Legionella RalF protein to host cells. *Proc. Natl. Acad. Sci. U.S.A.* 102, 826–831. doi: 10.1073/pnas.0406239101

Neild, A. L., and Roy, C. R. (2003). Legionella reveal dendritic cell functions that facilitate selection of antigens for MHC class II presentation. *Immunity* 18, 813–823. doi: 10.1016/S1074-7613(03)00140-7

Ninio, S., Zuckman-Cholon, D. M., Cambronne, E. D., and Roy, C. R. (2005). The Legionella IcmS-IcmW protein complex is important for Dot/Icm-mediated protein translocation. *Mol. Microbiol.* 55, 912–926. doi: 10.1111/j.1365-2958.2004.04435.x

O'Toole, G. A., and Kolter, R. (1998). Initiation of biofilm formation in *Pseudomonas fluorescens* WCS365 proceeds via multiple, convergent signalling pathways: a genetic analysis. *Mol. Microbiol.* 28, 449–461. doi: 10.1046/j.1365-2958.1998.00797.x

Pereira, C. T., Moutran, A., Fessel, M., and Balan, A. (2015). The sulfur/sulfonates transport systems in *Xanthomonas citri* pv. citri. *BMC Genomics* 16:524. doi: 10.1186/s12864-015-1736-5

Pinotsis, N., and Waksman, G. (2017). Structure of the WipA protein reveals a novel tyrosine protein phosphatase effector from *Legionella pneumophila*. *J. Biol. Chem.* 292, 9240–9251. doi: 10.1074/jbc.M117.781948

Roy, C. R. (2002). The Dot/lcm transporter of *Legionella pneumophila*: a bacterial conductor of vesicle trafficking that orchestrates the establishment of a replicative organelle in eukaryotic hosts. *Int. J. Med. Microbiol.* 291, 463–467. doi: 10.1078/1438-4221-00154

Roy, C. R., Berger, K. H., and Isberg, R. R. (1998). *Legionella pneumophila* DotA protein is required for early phagosome trafficking decisions that occur within minutes of bacterial uptake. *Mol. Microbiol.* 28, 663–674. doi: 10.1046/j.1365-2958.1998.00841.x

Segal, G., and Shuman, H. A. (1997). Characterization of a new region required for macrophage killing by *Legionella pneumophila*. *Infect. Immun.* 65, 5057–5066.

Sutherland, M. C., Nguyen, T. L., Tseng, V., and Vogel, J. P. (2012). The legionella IcmSW complex directly interacts with DotL to mediate translocation of adaptor-dependent substrates. *PLoS Pathog.* 8:e1002910. doi: 10.1371/journal.ppat.1002910

Theodoulou, F. L., and Kerr, I. D. (2015). ABC transporter research: going strong 40 years on. *Biochem. Soc. Trans.* 43, 1033–1040. doi: 10.1042/BST20150139

Tilney, L. G., Harb, O. S., Connelly, P. S., Robinson, C. G., and Roy, C. R. (2001). How the parasitic bacterium *Legionella pneumophila* modifies its phagosome and transforms it into rough ER: implications for conversion of plasma membrane to the ER membrane. *J. Cell Sci.* 114(Pt 24), 4637–4650.

Vogel, J. P., Roy, C., and Isberg, R. R. (1996). Use of salt to isolate *Legionella pneumophila* mutants unable to replicate in macrophages. *Ann. N. Y. Acad. Sci.* 797, 271–272. doi: 10.1111/j.1749-6632.1996.tb52975.x

Warren, W. J., and Miller, R. D. (1979). Growth of Legionnaires disease bacterium (*Legionella pneumophila*) in chemically defined medium. *J. Clin. Microbiol.* 10, 50–55.

Zamboni, D. S., McGrath, S., Rabinovitch, M., and Roy, C. R. (2003). *Coxiella burnetii* express type IV secretion system proteins that function similarly to components of the *Legionella pneumophila* Dot/Icm system. *Mol. Microbiol.* 49, 965–976. doi: 10.1046/j.1365-2958.2003.03626.x

Zgonc, D., and Baideme, M. (2015). Distributed treatment systems. *Water Environ. Res.* 87, 1196–1207. doi: 10.2175/106143015X14338845155624

Zhu, W., Banga, S., Tan, Y., Zheng, C., Stephenson, R., Gately, J., et al. (2011). Comprehensive identification of protein substrates of the Dot/Icm type IV transporter of *Legionella pneumophila*. *PLoS ONE* 6:e17638. doi: 10.1371/journal.pone.0017638

10

The Membrane Protein LasM Promotes the Culturability of *Legionella pneumophila* in Water

Laam Li and Sébastien P. Faucher*

Department of Natural Resource Sciences, Faculty of Agricultural and Environmental Sciences, McGill University, Montreal, QC, Canada

Edited by:
Elizabeth L. Hartland, University of Melbourne, Australia

Reviewed by:
Marina Santic', University of Rijeka, Croatia
Michele S. Swanson, University of Michigan, USA

Correspondence:
Sébastien P. Faucher
sebastien.faucher2@mcgill.ca

The water-borne pathogen *Legionella pneumophila (Lp)* strongly expresses the *lpg1659* gene in water. This gene encodes a hypothetical protein predicted to be a membrane protein using *in silico* analysis. While no conserved domains were identified in Lpg1659, similar proteins are found in many *Legionella* species and other aquatic bacteria. RT-qPCR showed that *lpg1659* is positively regulated by the alternative sigma factor RpoS, which is essential for *Lp* to survive in water. These observations suggest an important role of this novel protein in the survival of *Lp* in water. Deletion of *lpg1659* did not affect cell morphology, membrane integrity or tolerance to high temperature. Moreover, *lpg1659* was dispensable for growth of *Lp* in rich medium, and during infection of the amoeba *Acanthamoeba castellanii* and of THP-1 human macrophages. However, deletion of *lpg1659* resulted in an early loss of culturability in water, while over-expression of this gene promoted the culturability of *Lp*. Therefore, these results suggest that *lpg1659* is required for *Lp* to maintain culturability, and possibly long-term survival, in water. Since the loss of culturability observed in the absence of Lpg1659 was complemented by the addition of trace metals into water, this membrane protein is likely a transporter for acquiring essential trace metal for maintaining culturability in water and potentially in other metal-deprived conditions. Given its role in the survival of *Lp* in water, Lpg1659 was named LasM for *Legionella* aquatic survival membrane protein.

Keywords: *Legionella pneumophila*, freshwater, survival, *lpg1659*, membrane protein, metal transporter

INTRODUCTION

Legionella pneumophila (*Lp*) is a water-borne opportunistic pathogen that can infect human alveolar macrophages, resulting in a severe form of pneumonia called Legionnaires' disease (Fields et al., 2002). Recent years have seen an increase in the incidence of sporadic cases and outbreaks of Legionnaires' disease around the world, mostly associated with *Legionella*-contaminated man-made water systems (European Centre for Disease Prevention and Control, 2013; Phin et al., 2014).Since the major mode of transmission is through water, the dynamics of *Lp* in freshwater have been studied extensively. It is known that *Lp* replicates only in the presence of sufficient nutrients or permissive hosts (Fields et al., 2002). Under nutrient limitation, *Lp* can survive for a prolonged

Abbreviations: *Lp*, *Legionella pneumophila*; VBNC, viable but non-culturable; WT, wild-type.

period of time, up to 1.5 years in some cases, in varying compositions of freshwater including tap water, drinking water and creek water (Skaliy and McEachern, 1979; Schofield, 1985; Lee and West, 1991; Paszko-Kolva et al., 1992; Söderberg et al., 2008).

In an artificial freshwater model, Fraquil, *Lp* had less than a 1-log reduction in CFU counts after 20 weeks at 25°C (Li et al., 2015; Mendis et al., 2015). Moreover, *Lp* can survive for several months under a range of temperatures, pH and trace metal concentrations and remain infectious more than 6 months after exposure to Fraquil (Mendis et al., 2015). This ability for long-term survival in different kinds of water and under different conditions allows *Lp* to colonize water systems, and persist until it encounters a suitable host.

A limited number of studies have identified genes that are important for the survival of *Lp* in water. Söderberg et al. (2008) found that *lspD*, *lspE*, *lspF*, and *pilD*, which encode components of the virulence-related type II secretion system, are necessary for *Lp* to survive in water at temperatures below 17°C. In addition, the alternative sigma factor RpoS, as well as the stringent response regulators RelA and SpoT are required for the survival of *Lp* in water at 25°C (Trigui et al., 2014). In order to identify additional genes contributing to survival in water, we previously conducted a microarray analysis comparing the transcriptome of *Lp* exposed to water to that of *Lp* grown in rich medium (Li et al., 2015). Since bacteria tend to respond to environmental changes via transcriptomic reorganization (Ishihama, 2000; Hecker et al., 2009), genes that are highly up-regulated upon water exposure could be crucial for the successful adaptation and survival of *Lp* in water systems. Using this approach, we found that one highly up-regulated gene, *bdhA*, encoding 3-hydroxybutyrate dehydrogenase, is important for the long-term survival of *Lp* in water at 37°C (Li et al., 2015).

Due to the success of this approach, we selected another highly up-regulated gene, *lpg1659*, for further characterization. *lpg1659* was significantly up-regulated in *Lp* exposed to water for 2 and 6 h (Li et al., 2015). No previous studies have characterized *lpg1659*, which encodes a hypothetical protein with no putative functions. Nevertheless, we found that this protein is highly conserved in many *Legionella* species, as well as other aquatic bacteria. Therefore, we hypothesized that *lpg1659* is important for *Lp* to survive in freshwater. A deletion mutant of *lpg1659* was used to better understand its role with respect to cell structure, survival in water and growth of *Lp*. We found that Lpg1659 plays a role in the long-term culturability of *Lp* upon water exposure at 37 and 42°C. Evidence presented here suggests that it likely acts as an ion transporter, facilitating the uptake of one or more essential trace metal ions. Based on our results, the *lpg1659* gene was named *lasM* for *Legionella* aquatic survival membrane protein.

METHODS

Bacterial Strains and Culture Conditions

All *Lp* strains used in this study were constructed from JR32, a streptomycin resistant derivative of *Lp* Philadelphia-1 (Sadosky et al., 1993). The constitutively competent KS79 strain, derived from JR32, was used for the construction of the *lpg1659* mutant strain (de Felipe et al., 2008). Unless specified otherwise, *Lp* was grown on ACES-Buffered Charcoal Yeast Extract agar supplemented with 0.4 mg ml^{-1} L-cysteine, 0.25 mg ml^{-1} ferric pyrophosphate and 0.1% α-ketoglutarate (i.e., BCYEα agar) at 37°C for 3 days (Feeley et al., 1979; Edelstein, 1981). This medium was further supplemented with 25 μg ml^{-1} kanamycin or 5 μg ml^{-1} chloramphenicol, when necessary. *Escherichia coli* strains derived from DH5α were grown on Luria-Bertani agar at 37°C overnight and was supplemented with 25 μg ml^{-1} chloramphenicol, when necessary. Descriptions of the bacterial strains used in this study can be found in **Table 1**.

Construction of Mutant, Complemented, and Over-Expression Strains

To construct the *lpg1659* deletion mutant strain SPF248 (Δ*lasM*), 1 kb length sequences upstream and downstream of *lpg1659* were first amplified from the wild-type (WT) strain KS79 by PCR using Taq polymerase (Invitrogen), with the primer sets 1659_UpF/1659_UpR and 1659_DownF/1659_DownR, respectively. The kanamycin cassette was amplified from pSF6 with the primer set Kn-F/Kn-R, purified and further amplified with the primer set 1659_KnF/1659_KnR to obtain a 1 kb kanamycin fragment where the 5′ end is complementary to the 3′ end of the upstream fragment, and 3′ end is complementary to the 5′ end of downstream fragment. A 3 kb mutant allele was amplified using the three 1 kb fragments as template, Phusion DNA polymerase (NEB) and the primer set 1659_UpF/1659_DownR. The purified amplicon was then introduced into KS79 through natural transformation (de Felipe et al., 2008). The recombinants were selected for kanamycin resistance, and successful replacement of the target gene by the kanamycin cassette was validated by PCR.

To construct the pSF83 plasmid (plasM) for complementation, the target gene *lpg1659* together with 500 bp region upstream of the translation start site was amplified from KS79 using the primer set Com1659F2_SacI/Com1659R_XbaI. The amplicon and the plasmid pXDC39 were both digested with SacI and XbaI (NEB) and ligated using T4 DNA ligase (NEB). The ligation mixture was transformed into competent *E. coli* DH5α and the transformants were selected for chloramphenicol resistance. Correct insertion of the amplicon in the plasmid extracted from transformants was validated by PCR using the primer set pXDC39-F/Com1659R_XbaI. This pSF83 plasmid was then introduced into the mutant strain Δ*lasM* by electroporation as described previously (Chen et al., 2006), so as to construct the complemented strain SPF294 (Δ*lasM*+plasM). The recombinants were selected for kanamycin and chloramphenicol resistance before validation by PCR.

To construct the pSF73 plasmid (plasMi) for over-expression, the target gene *lpg1659* was first amplified from KS79 using the primer set Com1659F_SacI/Com1659R_XbaI. The amplicon and the plasmid pMMB207c were both digested with SacI and XbaI (NEB) and ligated with T4 DNA ligase (NEB). The ligation mixture was transformed into competent *E. coli* DH5α and the transformants were selected for chloramphenicol resistance. The presence of the insert (*lpg1659*) in the plasmid extracted

TABLE 1 | Bacterial strains used in this study.

Name	Relevant genotype	References
LEGIONELLA PNEUMOPHILA		
JR32	r^-m^+, Sm^R	Sadosky et al., 1993
KS79 (WT)	JR32 Δ*comR*	de Felipe et al., 2008
LELA3118 (*dotA* mutant)	JR32 *dotA*::Tn903dIllacZ	Sadosky et al., 1993
LM1376 (*rpoS* mutant)	JR32 *rpoS*::Tn903dGent, Gm^R	Hales and Shuman, 1999
SPF176 (prpoS)	LM1376 pSF49	Trigui et al., 2014
SPF248 (Δ*lasM*)	KS79 Δ*lpg1659*, Kn^R	This work
SPF294 (Δ*lasM*+plasM)	SPF248 pSF83, Kn^RCm^R	This work
SPF298 (WT+plasMi)	KS79 pSF73, Cm^R	This work
ESCHERICHIA COLI		
DH5α	*supE*44 Δ*lac*U169 (Φ80 *lacZ*ΔM15) *hsdR*17 *recA*1 *endA*1 *gyrA*96 *thi*-1 *relA*1	Invitrogen
pMMB207C	DH5α, Δ*mobA*, Cm^R	Charpentier et al., 2008
pSF6	DH5α, pGEMT-easy-*rrnb*	Faucher et al., 2011
pSF49	DH5α, pMMB207C-p_{tac}-*rpoS*, Cm^R	Trigui et al., 2014
pSF73	DH5α, pMMB207C-p_{tac}-*lpg1659*, Cm^R	This work
pSF83	DH5α, pXDC39-$p_{lpg1659}$-*lpg1659*, Cm^R	This work
pXDC39	DH5α, pMMB207c, Δ*Ptac*, Δ*lacI*, Cm^R	Xavier Charpentier

from transformants was validated by PCR using the primer set PromF/Com1659R_XbaI. This pSF73 plasmid was then introduced into KS79 by electroporation in order to construct the over-expression strain SPF298 (WT+plasMi). The recombinants were selected for kanamycin and chloramphenicol resistance before validation by PCR. The primer sequences are listed in **Table 2**.

RT-qPCR

JR32, the *rpoS* mutant and the complemented strain SPF176 were first suspended in ACES-buffered Yeast Extract (AYE) broth at an initial OD_{600} of 0.1. Fifty milliliter cultures of JR32, *rpoS* mutant, SPF176 (prpoS OFF) and SPF176 induced with 0.5 mM IPTG (prpoS ON) were grown in 250 ml Erlenmeyer flasks at 37°C shaking (250 rpm) to exponential phase (OD_{600} of 1.0). Each culture was then centrifuged and washed with Fraquil three times before suspending in Fraquil to an OD_{600} of 1.0. Fraquil is an artificial freshwater medium that does not support growth but allows long-term survival of *Lp* (Li et al., 2015; Mendis et al., 2015).The composition of Fraquil is 0.25 μM $CaCl_2$, 0.15 μM $MgSO_4$, 0.15 μM $NaHCO_3$, 10 nM K_2HPO_4, 0.1 μM $NaNO_3$, 10 nM $FeCl_3$, 1 nM $CuSO_4$, 0.22 nM $(NH_4)_6Mo_7O_{24}$, 2.5 nM $CoCl_2$, 23 nM $MnCl_2$, and 4 nM $ZnSO_4$ in ultra-pure Milli-Q water (Morel et al., 1975). Thirty milliliter of each suspension was transferred to 125 ml Erlenmeyer flask and incubated at 37°C shaking for 6 h. Samples were then collected and RNA was extracted as described previously (Li et al., 2015). One microgram of purified RNA was used for reverse transcription reactions along with a negative control without reverse transcriptase. For qPCR reactions, the 16S rRNA gene-specific primer set 16s_QF/16s_QF and the *lpg1659* gene-specific primer set 1659_QF/1659_QR were designed with the IDT primer design software (https://www.idtdna.com/Primerquest/), and their amplification efficiency were proven to be >85% (data not shown). qPCR was performed on an iQ™5 Multicolor Real-Time PCR Detection System (Bio-Rad) using iTaq universal SYBR green supermix (Bio-Rad) according to the manufacturer's protocol. The 16SrRNA gene was used as the reference gene to normalize the data. Fold change was calculated as described previously (Livak and Schmittgen, 2001) and are presented as $\log_2$ ratios.

TABLE 2 | Primer sequences used in this study.

Name	Sequence (5′–3′)*
1659_UpF	CAATCAGAACAAGGTGTGTATGG
1659_UpR	CAGTCTAGCTATCGCCATGTACGATGAGTACTGAATTCCTGC
1659_DownF	GATGCTGAAGATCAGTTGGGTCACGTCCTATCACATTCTATTACTC
1659_DownR	AGATCGATGAAGGCTTGTAGC
1659_KnF	GCAGGAATTCAGTACTCATCGTACATGGCGATAGCTAGACTG
1659_KnR	GAGTAATAGAATGTGATAGGACGTGACCCAACTGATCTTCAGCATC
1659_QF	CGGTCACTCTTTGGTATATGTC
1659_QR	CTGATTGACTGGATCGAACATC
16s_QF	AGAGATGCATTAGTGCCTTCGGGA
16s_QR	ACTAAGGATAAGGGTTGCGCTCGT
Com1659F_SacI	CCGGAGCTCGCAGGAATTCAGTACTCATCG
Com1659F2_SacI	CCGGAGCTCCACCTTTCAGATTGTTAGTCGC
Com1659R_XbaI	CGCTCTAGAGAGTAATAGAATGTGATAGGACG
Kn-F	TACATGGCGATAGCTAGACTG
Kn-R	ACCCAACTGATCTTCAGCATC
pXDC39-F	GCTTCCACAGCAATGGCATCC
PromF	CGTATAATGTGTGGAATTGTGAG

**The underlined bases indicate restriction sites.*

Bioinformatics Analysis

The hypothetical protein (accession number: YP_095686.1) encoded by *lasM* was compared to proteins encoded by other bacteria using Standard Protein BLAST (http://blast.ncbi.nlm.nih.gov/Blast.cgi?PAGE=Proteins). The NCBI CD-search was used to identify any conserved domains present in the LasM protein. Three servers from the CBS Prediction Servers (http://www.cbs.dtu.dk/services/) were then used to predict the putative function(s) of LasM. First, TMHMM Server v.2.0 was used to predict transmembrane helices (Sonnhammer et al., 1998; Krogh et al., 2001). Second, SignalP 4.1 Server was used to predict the presence and location of signal peptides (Petersen et al., 2011). Lastly, ProtFun 2.2 Server was used to predict cellular role, enzyme class as well as the gene ontology category based on the amino acid sequence (Jensen et al., 2002, 2003).

Microscopic Examination of Cell Morphology

The WT strain KS79 and the mutant strain Δ*lasM* were suspended in Fraquil at an OD_{600} of 1.0. Immediately after suspension, a wet mount was prepared and viewed at 1000× magnification using digital microscopy (Nikon Eclipse 80i). Images of 10 random microscopic fields were captured for each strain using the NIS Element software (Nikon Instruments, Inc.). The length of 10 cells in each image was estimated using the ImageJ software in order to determine the average cell length.

Extracellular Growth Assay

The KS79 and Δ*lasM* strains were suspended in AYE broth at an OD_{600} of 0.1. Twenty-five ml of each culture was transferred into three 125 ml Erlenmeyer flasks and grown at 37°C shaking. The OD_{600} of each culture was measured by a spectrophotometer once every 4 h for a period of 32 h.

Cell Lines and Infection Assays

The amoeba *Acanthamoeba castellanii* was grown to confluence in 20 ml of PYG broth (Moffat and Tompkins, 1992) in a 75 cm^2 tissue culture flask (Sarstedt) at 30°C. Before infection with *Lp*, the old medium with non-adherent amoebae was replaced with 10 ml of fresh PYG broth. The flask was then shaken sharply to release the adherent amoebae into the medium. This suspension was enumerated and diluted to 5×10^5 cells per ml. One ml was placed into each well of a 24-well plate (Sarstedt). The amoebae were allowed to adhere for 2 h before the medium was replaced with Ac buffer, which does not support the growth of *Lp* (Moffat and Tompkins, 1992). The plate was incubated at 30°C for another 2 h before infection.

The human monocyte-like cell line THP-1 was grown in 30 ml of RPMI 1640 (Life Technologies) supplemented with 10% fetal bovine serum and 2 mM glutamine at 37°C under 5% CO_2 (Kim et al., 2009). Three days prior to infection, 1 ml of THP-1 culture (5×10^5 cells per ml) was placed into each well of a 24-well plate (Sarstedt) and treated with 1×10^{-7} M phorbol 12-myristate 13-acetate (PMA) (Fisher Scientific) to induce maturation toward adherent macrophage-like cells. Subsequently, the medium was replaced by fresh RPMI without PMA 2 h before infection with *Lp*.

For the infection assays, the KS79, Δ*lasM,* and *dotA* mutant were suspended in AYE broth at an OD_{600} of 0.1 and then, diluted 10-fold to approximately 2.5×10^6 cells per ml. The *dotA* mutant is defective for intracellular growth (Roy and Isberg, 1997) and was used as a negative control. Two microliter of each bacterial suspension was added to three replicate wells of *A. castellanii* and THP-1 cells, resulting in an MOI of 0.1. The *A. castellanii* infection plate was incubated at 30°C and the intracellular growth of each strain was determined by CFU counts on BCYEα agar at 24 h intervals for 7 days; whereas the THP-1 infection plate was incubated at 37°C under 5% CO_2, and the intracellular growth was monitored at 24 h intervals for 5 days.

Survival Assays in Water

Strains grown on BCYEα agar were washed three times with Fraquil and suspended in fresh Fraquil at an OD_{600} of 0.1. One milliliter of bacterial suspension was added to 4 ml of fresh Fraquil in a 25 cm^2 plastic flask (Sarstedt). For each strain, three replicate flasks were incubated at 25, 37, and 42°C, and CFU counts were measured once per 3 weeks, once per 2 weeks and once per week, respectively. In addition, membrane integrity of the samples incubated at 42°C for 7 weeks was determined by Live/Dead staining and flow cytometry, using freshly grown KS79 as the live control and KS79 boiled in a water bath for 10 min as the dead control, as described previously (Li et al., 2015).

For the survival assay using Fraquil containing10 times trace metals, the final salt and trace metal concentration is 0.25 μM $CaCl_2$, 0.15 μM $MgSO_4$, 0.15 μM $NaHCO_3$, 10 nM K_2HPO_4, 0.1 μM $NaNO_3$, 0.1 μM $FeCl_3$, 10 nM $CuSO_4$, 2.2 nM $(NH_4)_6Mo_7O_{24}$, 25 nM $CoCl_2$, 0.23 μM $MnCl_2$, and 40 nM $ZnSO_4$.

Heat Tolerance Assays

To test tolerance to heat shock, the KS79 and Δ*lasM* strains were suspended in Fraquil at an OD_{600} of 0.1. One ml of each strain was aliquoted into 13 ml tubes (Sarstedt) in triplicate. Tubes were acclimated to 25°C for 24 h and then, transferred into a water bath set to 55°C. Samples were taken before, after 0.5 h and after 1 h of heat shock to determine the changes in CFU counts.

To test growth at an elevated temperature, the KS79 and Δ*lasM* strains were suspended in AYE broth at an OD_{600} of 0.1. Twenty-five milliliter of each culture was transferred into three 125 ml Erlenmeyer flasks and grown at 42°C shaking. The CFU of each strain was monitored daily.

RESULTS

lasM Is Positively Regulated by RpoS in Water

In a previous study, we found that *lasM* was significantly up-regulated by more than 6-fold after 2 h (Log_2 ratio = 2.66) and 24-fold after 6 h (Log_2 ratio = 4.60) of exposure to water when compared to a control grown in rich medium (**Figure 1A**; Li et al., 2015). RpoS is an alternative sigma factor and a stress response regulator important for the survival of *Lp* in water (Trigui et al., 2014). In order to test whether the expression of *lasM* is controlled by RpoS, RT-qPCR was performed using RNA

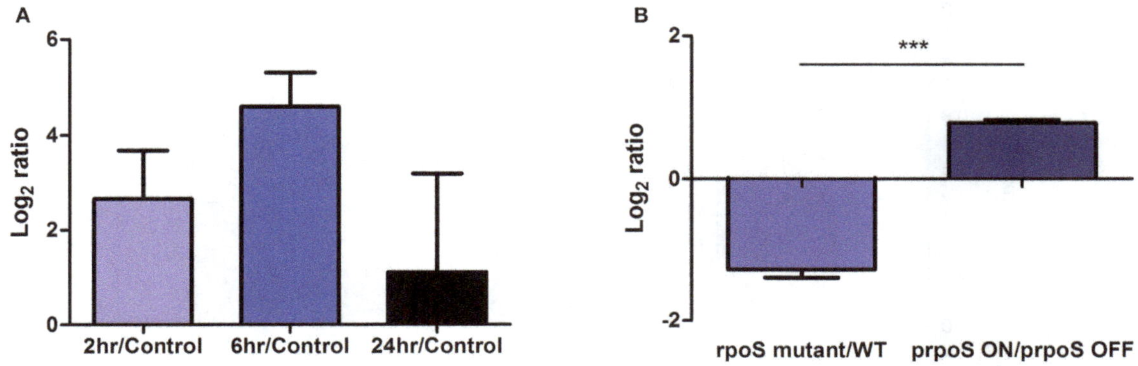

FIGURE 1 | LasM is expressed in water and is regulated by RpoS. (A) The expression of *lasM* based on a transcriptomic analysis of the WT JR32 strain exposed to water for 2, 6, and 24 h compared to the control grown to exponential phase in rich medium (adapted from Li et al., 2015 published under Creative Commons Attribution 4.0 International License; https://creativecommons.org/licenses/by/4.0/). **(B)** The RpoS-dependant regulation of *lasM* in *Lp* exposed to water for 6 h was assessed by RT-qPCR. The expression of *lasM* in the *rpoS* mutant was compared to that in the WT strain. The expression of *lasM* in the complemented strain where *rpoS* is induced with 0.5 mM IPTG (prpoS ON) was compared to that of the non-induced complement (prpoS OFF). One-tailed unpaired Student's *t*-test was used to assess significant differences between the two ratios (***$p < 0.0005$). Data shown are the mean and SD of 3 biological replicates.

from the WT, *rpoS* mutant and complemented strains that were exposed to water. The fold change of the *lasM* transcript upon water exposure was lower in the *rpoS* mutant compared to the WT, reflected in the negative Log_2 ratio of −1.3 (**Figure 1B**). In contrast, induction of RpoS expression in the mutant (prpoS ON) resulted in a positive Log_2 ratio of 0.8, indicative of an increase in *lasM* expression, when compared to its non-induced counterpart (prpoS OFF). Taken together, these results suggest that the expression of *lasM* is positively regulated by RpoS in water.

LasM Is a Conserved Protein Found in Aquatic Bacteria

According to the NCBI Protein database, the hypothetical protein encoded by *lasM* is composed of 347 amino acids. Standard Protein BLAST revealed that LasM found in the Philadelphia-1 strain of *Lp* is highly conserved in other strains, such as Paris, Corby, Alcoy and Lens (≥98% identity). Other *Legionella* species that contain a LasM homolog include *L. norrlandica* (87% identity), *L. moravica* (77% identity), *L. tucsonensis* (76% identity), and *L. longbeachae* (68% identity). LasM also shares significant homology (61–63% identity) with hypothetical or membrane proteins found in *Methylophaga nitratireducenticrescens*, *Methylophaga lonarensis*, *Moritella dasanensis*, *Endozoicomonas elysicola*, *Colwellia psychrerythraea*, *Marinobacter santoriniensis* among others. These bacteria were isolated from various aquatic environments, such as water treatment systems, soda lakes and the Arctic Ocean (Kurahashi and Yokota, 2007; Kim et al., 2008; Handley et al., 2009; Antony et al., 2012; Villeneuve et al., 2013). Therefore, the protein sequence homology suggests that LasM is a conserved protein found not only in *Legionella* species, but also in other aquatic bacteria.

Since LasM does not harbor any conserved domains, *in silico* analysis was performed using the CBS Prediction Servers to predict its putative function(s). Eight transmembrane helices were predicted using the TMHMM server (**Figure 2**), suggesting that LasM is a transmembrane protein. Moreover, its N-terminal is likely to be located on the cytoplasmic side of the membrane ($p = 0.99911$). No potential signal peptides were identified using the SignalP Server, suggesting that LasM is located in the cytoplasmic membrane. The ProtFun Server predicted LasM to be a non-enzyme ($p = 0.809$) involved in "Transport and Binding" ($p = 0.740$), most likely to be a transporter ($p = 0.409$) among 14 different Gene Ontology categories. Our *in silico* analysis suggests that LasM is a conserved membrane protein, that may be involved in the transport of an unknown substance as part of its aquatic lifestyle.

Deletion of *lasM* Does Not Affect Cellular Morphology

Since LasM is predicted to be a membrane protein, we investigated whether the deletion of *lasM* would alter the cellular morphology of *Lp*. Microscopic analysis of wet mounts of 3 days old culture suspended in Fraquil shows that both the WT and the *lasM* deletion mutant are rod shaped cells of comparable size (**Figures 3A,B**). Statistical analysis confirms that there were no significant differences ($p = 0.1659$) between the cell lengths of the two strains (**Figure 3C**). These results indicate that the absence of LasM in *Lp* does not affect cell shape or cell size.

LasM Is Dispensable for Growth *In vitro* and *In vivo*

Growth and infection assays were performed to investigate the potential role of LasM in extracellular and intracellular growth. The WT and Δ*lasM* strains produced similar growth curves in rich medium, suggesting that *lasM* is dispensable for the growth of *Lp in vitro* (**Figure 4A**). The lag phase of both strains lasted 8 h and exponential growth occurred between 8 and 20 h, followed by a late post-exponential/stationary phase.

CFU of the WT and Δ*lasM* strains increased by 4-log after 7 days of growth within the amoeba *A. castellanii*

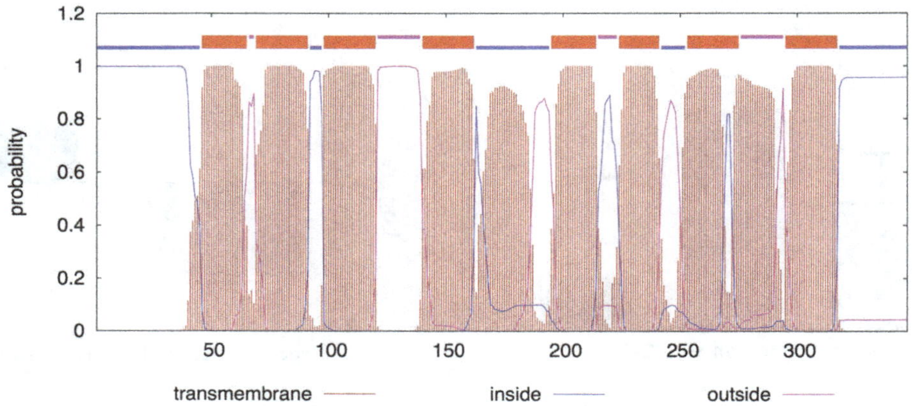

FIGURE 2 | TMHMM posterior probability for LasM. The sequence of LasM was analyzed by TMHMM Server v.2.0. The segmented line on top summarizes the most probable location of the sequence. The lower part shows the probability of belonging to different locations. Red indicates transmembrane regions, blue indicates intracellular portions of the protein and pink indicates extracellular portions.

(**Figure 4B**). In contrast, both the WT and Δ*lasM* only produced a 2.5-log increase in CFU counts after 5 days of growth within human macrophage-like THP-1 cells (**Figure 4C**). As expected, the *dotA* mutant that served as a negative control demonstrated a reduction in CFU counts in both infection models (**Figures 4B,C**). No significant differences were observed between the WT and Δ*lasM* in both infection assays, showing that deletion of *lasM* does not affect the ability of *Lp* to infect and multiply intracellularly within *A. castellanii* or within THP-1 cells.

Deletion and Over-Expression of *lasM* Affects the Culturability of *Lp* in Water

Since *lasM* is highly up-regulated in *Lp* exposed to water (Li et al., 2015) and positively regulated by RpoS, we hypothesized that it may be important for the survival of *Lp* in water. We monitored the changes in CFU counts of the WT, Δ*lasM* and the complemented strain (Δ*lasM*+plasM) during long-term exposure to water at three different temperatures. At 25°C, all three strains survived well, and no significant reduction in CFU counts was observed after 24 weeks in water (**Figure 5A**). In contrast, the CFU of all three strains dropped below detection limit after 22 weeks of water exposure at 37°C (**Figure 5B**). The Δ*lasM* strain had a faster reduction in CFU count and a significantly lower CFU than the WT starting at week 12. The complemented strain survived as well as the WT, suggesting that *lasM* is indeed important for *Lp* to maintain culturability in water. Similar trend was observed in the strains exposed to 42°C, where an early loss in culturability was only observed in Δ*lasM* but not in the WT or the complemented strain (**Figure 5C**). At this temperature, the CFU counts of all three strains decreased more rapidly than at 37°C, dropping below the detection limit after only 5 weeks of water exposure.

Since the deletion of *lasM* resulted in an early loss of culturability at both 37 and 42°C, we hypothesized that over-expression of this gene in the WT would promote the culturability of *Lp* in water. Given that the difference in culturability between the WT and Δ*lasM* strains was the greatest at 42°C, the effect of *lasM* over-expression was tested at this temperature. A plasmid containing an inducible P*tac* promoter preceding the *lasM* ORF was constructed and introduced into the WT strain (WT+plasMi). Under non-inducing conditions (WT+plasMi OFF), the decline in CFU counts over time was similar to the WT strain (**Figure 5F**). Interestingly, over-expression of LasM using IPTG (WT+plasMi ON) increased the culturability of *Lp* by 1 week. This further supports the notion that *lasM* is important for *Lp* to maintain culturability in water.

Deletion of *lasM* Does Not Affect the Culturability of *Lp* in Water Containing Excess Trace Metals

Given that LasM was predicted to be a membrane protein involved in "transport and binding" and since it was found to be important for maintaining the culturability of *Lp* in water, we hypothesized that this protein could be involved in acquiring essential nutrients that are present in low amounts in water, such as trace metals. Therefore, we tested the culturability of the WT, Δ*lasM* and the complemented strain (Δ*lasM*+plasM) in water containing 10 times of trace metals at 42°C. Surprisingly, the CFU counts declined at the same rate for all three strains throughout the 7 weeks of exposure, and the mutant no longer demonstrated an early loss of culturability as observed previously (**Figures 5C,D**). Therefore, LasM does not alter the kinetics of culturability over time when excess trace metals are present in water.

Deletion of *lasM* Does Not Affect Membrane Integrity of *Lp* Exposed to Water

After exposure to water at 42°C for 7 weeks, we analyzed the cell status of the WT, Δ*lasM,* and the complemented strain (Δ*lasM*+plasM) using Live/Dead staining and flow cytometry. Live/Dead staining differentiates between dead cells that have a damaged membrane and viable cells with an intact membrane. In this case, over 90% of each strain under investigation stained as

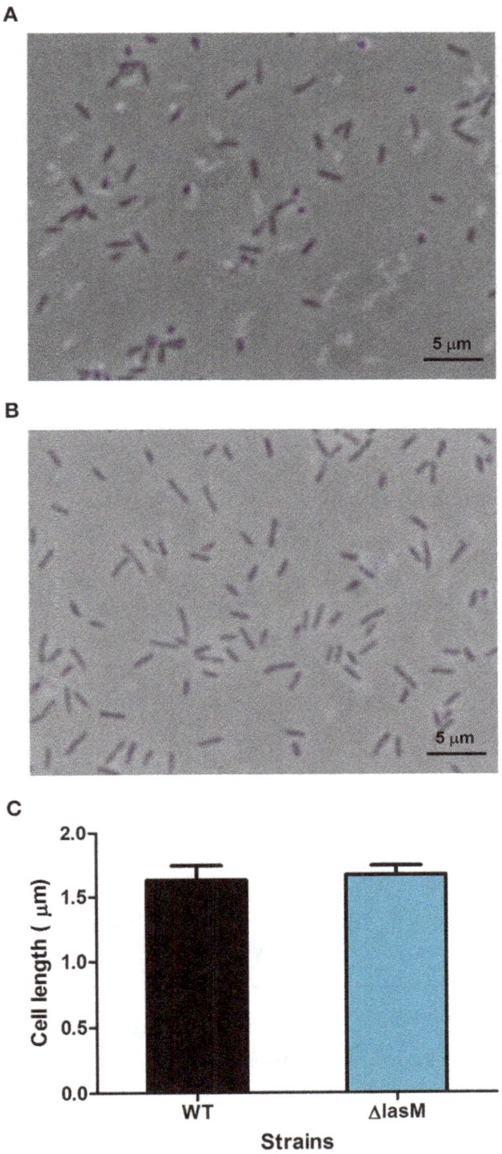

FIGURE 3 | Deletion of *lasM* does not affect cell morphology or cell size. A wet mount image of **(A)** the WT strain and **(B)** Δ*lasM* was observed under 1000× magnification. **(C)** The cell length of the WT strain and Δ*lasM* was estimated using ImageJ. Data shown are the mean and SD of the length of 10 cells per microscopic image in 10 analyzed images ($n = 100$).

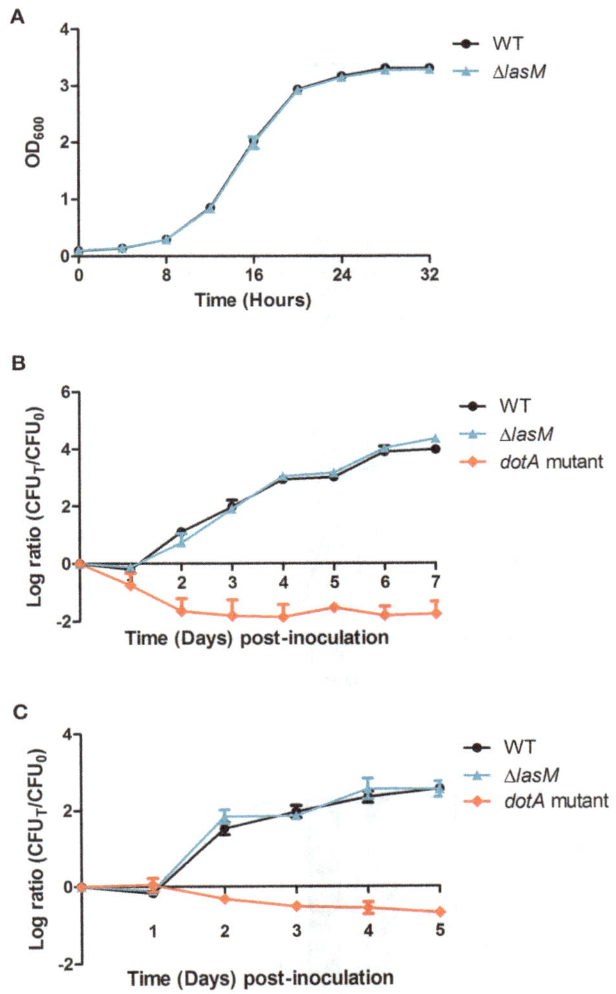

FIGURE 4 | Deletion of *lasM* does not affect the growth of *Lp in vitro* and *in vivo*. (A) Optical density at 600 nm of the WT and Δ*lasM* strains grown in rich medium for 32 h. **(B)** The amoeba *A. castellanii* or **(C)** cultured THP-1 macrophages were infected with the WT strain, Δ*lasM* or the *dotA* mutant (negative control) at an MOI of 0.1. Changes in cell titer were monitored using daily CFU counts and are presented as the log ratio of the CFU on each day (CFU_T) over the initial CFU (CFU_0). Data shown are the mean and SD of 3 biological replicates.

viable cells, and less than 2% of each population were stained as dead (**Figure 5E**). This data shows that the absence of membrane protein LasM does not significantly affect the membrane integrity of *Lp* after exposure to water.

Deletion of *lasM* Does Not Affect the Tolerance of *Lp* to High Temperature

Since early loss of culturability of the *lasM* mutant was only observed at 37 and 42°C but not at 25°C, we investigated whether the mutant was sensitive to elevated temperatures. First, we compared the tolerance of the WT and Δ*lasM* strains in water at 55°C. The CFU counts of both strains decreased to 42–47% of the initial population after 0.5 h and to 13–14% after 1 h of exposure to heat shock (**Figure 6A**). Increasing exposure time significantly reduced their CFU counts ($p < 0.0001$), but no significant differences were found between the two strains ($p = 0.5605$). Then, we compared the tolerance of both strains in rich medium at 42°C and, again, found no significant differences between their CFU counts over a period of 4 days (**Figure 6B**). Both strains grew in the first day and then their CFU counts dropped below the detection limit after 3 days. Taken together, these results suggest that deletion of *lasM* does not affect the tolerance of *Lp* to high temperature in water or in rich medium.

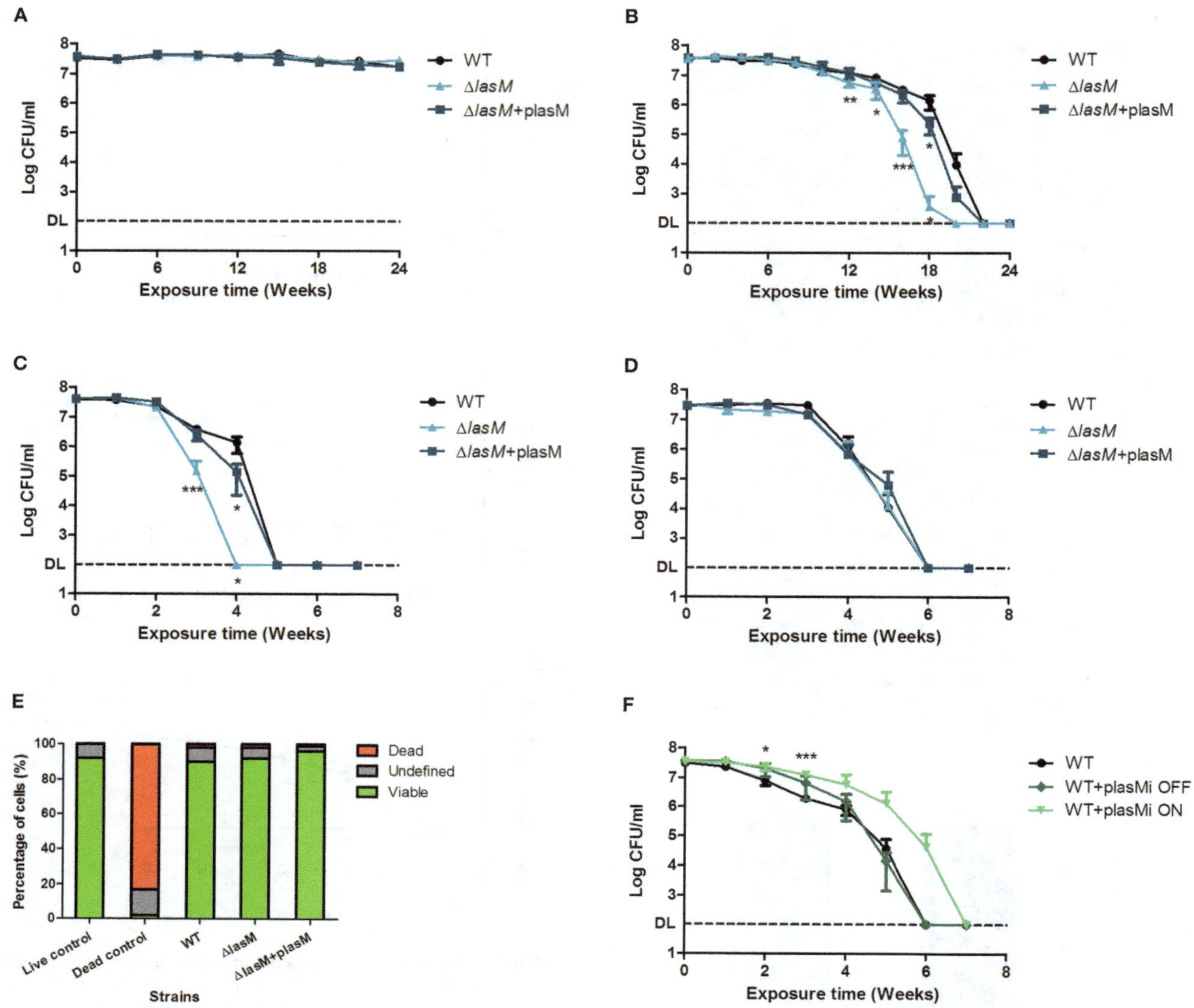

FIGURE 5 | Deletion and over-expression of *lasM* affects the culturability of *Lp* in water at 37 and 42°C. The CFU counts of the WT, Δ*lasM* and complemented strain (Δ*lasM*+plasM) in water at **(A)** 25°C, **(B)** 37°C, **(C)** 42°C, and **(D)** in water with 10 times trace metals at 42°C. **(E)** Percentage of cells in different status after exposure to water at 42°C for 7 weeks. Live/Dead staining and flow cytometry were used to analyze 5000 cells in each replicate. Freshly grown *Lp* was used as the live control and heat-killed *Lp* was used as the dead control. **(F)** CFU counts of the WT strain and the over-expression strain (WT+plasMi) in water at 42°C. WT+plasMi OFF indicates that the over-expression of *lasM* was not induced, while WT+plasMi ON indicates that the over-expression of *lasM* was induced with 1 mM IPTG. Data shown are the mean and SD of 3 biological replicates. One-tailed unpaired Student's *t*-test was used to assess significant differences against the WT (* $p < 0.05$; ** $p < 0.005$; *** $p < 0.0005$). DL indicates the detection limit.

DISCUSSION

Despite the lack of nutrients, *Lp* can survive in freshwater for a prolonged period of time and also remain infectious (James et al., 1999; Li et al., 2015; Mendis et al., 2015). In this study, we characterize *lpg1659* (*lasM*), a gene that encodes a hypothetical protein with no putative functions. *In silico* analysis shows that *lasM* encodes a membrane protein that is likely involved in transport and binding. Differential expression of membrane proteins are well documented in literature. In *E. coli*, some membrane proteins like OmpA and general porins are constitutively expressed at high levels, whereas other membrane proteins are only expressed under specific conditions (Koebnik et al., 2000). Exposure to water at varying temperatures elicited specific changes in the outer membrane protein composition of enterohemorrhagic *E. coli* (Wang and Doyle, 1998). In another waterborne pathogen *Leptospira interrogans*, the inner membrane protein LipL31 was specifically down-regulated during exposure to distilled water, while other membrane proteins were not affected (Trueba et al., 2004). In our case,

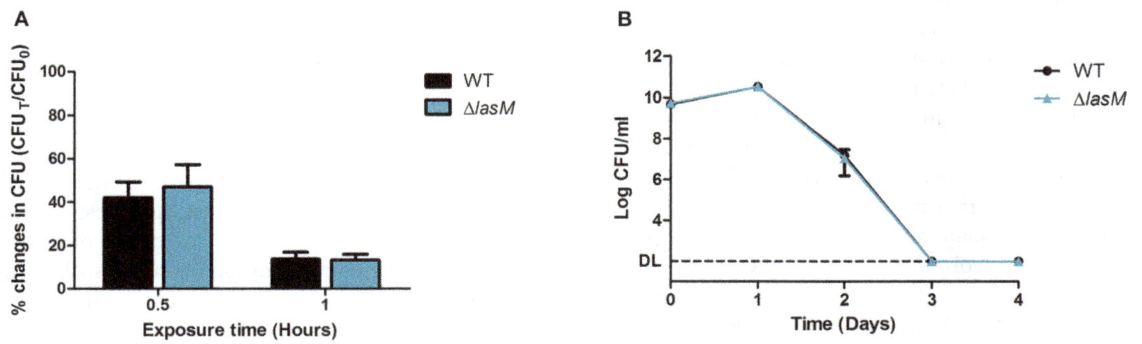

FIGURE 6 | Deletion of *lasM* does not affect the tolerance of *Lp* to high temperature. (A) The WT strain and Δ*lasM* were suspended in water and exposed to 55°C for 0.5 and 1 h. The CFU counts after heat shock (CFU_T) was compared to the CFU counts before heat shock (CFU_0). **(B)** The WT strain and Δ*lasM* were grown in rich medium at 42°C and their daily CFU counts were monitored. Data shown are the mean and SD of 3 biological replicates. DL indicates the detection limit.

the expression of the LasM membrane protein was induced significantly after 2 and 6 h of exposure to water, compared to exponential growth in AYE (Li et al., 2015). A previous study showed that *lasM* is also induced in the post-exponential phase of growth (Hovel-Miner et al., 2009). Therefore, it is likely that a signal present in Fraquil and post-exponential phase, such as starvation, induces the expression of *lasM*. Since *Lp* shuts down transcription and translation 24 h after exposure to water (Trigui et al., 2014; Li et al., 2015), genes required for survival in water are likely induced during earlier time points (i.e., the adaptation period), and are subsequently repressed. For example the *bdhA* gene, encoding 3-hydroxybutyrate dehydrogenase, is required for the long-term survival of *Lp* in water, and follows this expression pattern (Li et al., 2015).

Since the alternative sigma factor RpoS is essential for *Lp* to survive in water (Trigui et al., 2014), any genes under the control of RpoS could also be involved in the survival of *Lp* in water. For example, *bdhA* is positively regulated by RpoS in water and it was found to be important for the survival of *Lp* in water at 37°C (Li et al., 2015). According to a previous study, *lasM* is positively regulated by RpoS in rich medium (Hovel-Miner et al., 2009). Here, we report that *lasM* is also under RpoS control when *Lp* is exposed to water, further suggesting an association between LasM and water survival.

Since LasM is predicted to be a membrane protein, we hypothesized that it may play a role in maintaining cell morphology or membrane integrity. However, deletion of *lasM* affected neither the cell shape nor size of freshly grown *Lp* that was suspended in water. Moreover, the membrane integrity of *Lp* exposed to water for 7 weeks was unaffected. We further found that deletion of *lasM* does not affect the ability of *Lp* to infect host cells such as *A. castellanii* and THP-1 cells. Furthermore, it does not affect the growth of *Lp in vitro*. This is consistent with a previous study showing that the insertion of a transposon in the *lasM* gene did not result in growth advantages or disadvantages in rich medium (O'Connor et al., 2011).

Nevertheless, we observed an early loss of culturability in Δ*lasM* exposed to water at 37 and 42°C, suggesting that *lasM* is important for *Lp* to maintain culturability in warm water. These temperatures are commonly found in man-made water systems such as cooling towers that *Lp* is able to colonize (Rogers et al., 1994; Darelid et al., 2002). Live/Dead staining shows that the proportion of dead cell in the Δ*lasM* population is comparable to that of the WT, but the mutant enters a viable but non-culturable (VBNC) state earlier than the WT. VBNC cells are in a quiescent state awaiting revival or transitioning to death (Li et al., 2014). Previous studies show that *Lp* that was induced into the VBNC state under certain conditions, such as starvation and exposure to disinfectants, may be resuscitated back into culturable and infectious cells using different methods (reviewed by Li et al., 2014). For example, *Lp* that entered the VBNC status 125 days after exposure to sterilized tap water at 20°C were resuscitated by the addition of *A. castellanii* (Steinert et al., 1997). However, our attempt to resuscitate the samples using the same method failed and the VBNC cells remained non-infectious (data not shown). Differences in VBNC-inducing conditions (e.g., water, temperature, etc.) and additional factors may contribute to the failure of resuscitation (reviewed by Li et al., 2014). VBNC cells that cannot be resuscitated are considered to be in the process of dying. Therefore, we conclude that Δ*lasM* started dying at an earlier time point than the WT when exposed to water. Given that over-expression of LasM also promotes the culturability of *Lp* in water, we conclude that LasM is important for the long-term survival of *Lp* in water at temperatures above 25°C.

Early loss of culturability was not observed in Δ*lasM* exposed to water at 25°C for 24 weeks. This is mirrored in a previous study where the deletion of *bdhA* results in an early loss of culturability and causes a survival defect in *Lp* exposed to water at 37°C but not at 25°C (Li et al., 2015). In both cases, it is possible that the survival defect becomes apparent at 25°C after a longer incubation period in water. Since CFU counts drop more gradually at 25°C than at 37 and 42°C (Mendis et al., 2015), any defect resulting from the deletion of an important gene may appear at a later time point at 25°C.

Environmental stresses such as heat shock are known to increase membrane fluidity and eventually result in cell damage (Beney and Gervais, 2001; Richter et al., 2010). Therefore, it is possible that LasM is important for *Lp* to maintain culturability at a higher temperature, not necessarily in water. If the absence of the LasM membrane protein reduced the ability of *Lp* to deal with heat-induced membrane damage, the mutant would produce a faster drop in CFU counts than the WT at elevated temperatures. However, we show that deletion of *lasM* does not affect the rate of CFU reduction in *Lp* exposed to water at 55°C or the CFU changes in *Lp* grown in rich medium at 42°C, suggesting that LasM is important for maintaining culturability in water, but that it is not directly involved in the resistance of *Lp* to high temperatures.

It is noteworthy that LasM was predicted to be a transporter, albeit with low probability ($p = 0.409$). Since the metabolic rate of *Lp* increases with increasing temperature (Kusnetsov et al., 1996), more energy and resources would be needed for active metabolism at higher temperatures. Therefore, if the function of LasM is to facilitate nutrient transport, then the loss of LasM could result in a more severe defect in water at 37°C or at 42°C than at 25°C. Indeed, the absence of LasM did not affect the kinetics of culturability in water with 10 times the original amount of trace metals, suggesting that the early loss of culturability previously observed in the mutant can be complemented by excess trace metals in water, namely copper, molybdenum, cobalt, manganese, zinc, sodium and iron. This finding supports our hypothesis that LasM is a transporter of one or more of the essential trace metals present in water. In the absence of this transporter, *Lp* might not be able to acquire sufficient trace metals from water environment to maintain culturability, and possibly long-term survival. Our data show that over-expression of *lasM* in *Lp* seems to allow better acquisition of essential trace metals, helping to maintain culturability for a longer period of time. Extra trace metals in water might also increase the amount that diffused into cells, and thus, allow the mutant lacking LasM to maintain culturability as well as the WT. It is not yet clear which trace metal is transported by LasM. Based on our experiments, LasM could transport one or a combination of copper, molybdenum, cobalt, manganese, zinc and iron.

In conclusion, this study reveals that the LasM protein is important for *Lp* to acquire essential trace metals in order to maintain culturability in water, which is consistent with its most probable predicted function. Our results do not ruled out the possibility that LasM could increase the fitness of *L. pneumophila* in other settings where the concentration of metals is low or where there is fierce competition for them, such as multi-species biofilms existing in water systems. The *lasM* gene is highly up-regulated in water and positively regulated by RpoS. It encodes a novel membrane protein, which is highly conserved in many *Legionella* species and other aquatic bacteria. We postulate that LasM is an important protein for other aquatic bacteria to maintain culturability and survival in water and in conditions presenting low concentration of metals. Absence of this protein does not affect cell morphology, membrane integrity, tolerance to high temperature or the growth of *Lp*, both *in vitro* and *in vivo*. Further investigation would be required to better understanding the exact trace metal(s) being transported by LasM and the underlying mechanism.

AUTHOR CONTRIBUTIONS

LL and SF conceived and designed the experiments. LL conducted the experiments and wrote the manuscript. SF contributed in writing and review of the manuscript.

ACKNOWLEDGMENTS

We would like to thank Nilmini Mendis for reviewing the manuscript and English editing.

REFERENCES

Antony, C. P., Doronina, N. V., Boden, R., Trotsenko, Y. A., Shouche, Y. S., and Murrell, J. C. (2012). *Methylophaga lonarensis* sp. nov., a moderately haloalkaliphilic methylotroph isolated from the soda lake sediments of a meteorite impact crater. *Int. J. Syst. Evol. Microbiol.* 62, 1613–1618. doi: 10.1099/ijs.0.035089-0

Beney, L., and Gervais, P. (2001). Influence of the fluidity of the membrane on the response of microorganisms to environmental stresses. *Appl. Microbiol. Biotechnol.* 57, 34–42. doi: 10.1007/s002530100754

Charpentier, X., Faucher, S. P., Kalachikov, S., and Shuman, H. A. (2008). Loss of RNase R induces competence development in *Legionella pneumophila*. *J. Bacteriol.* 190, 8126–8136. doi: 10.1128/JB.01035-08

Chen, D.-Q., Huang, S.-S., and Lu, Y.-J. (2006). Efficient transformation of *Legionella pneumophila* by high-voltage electroporation. *Microbiol. Res.* 161, 246–251. doi: 10.1016/j.micres.2005.09.001

Darelid, J., Löfgren, S., and Malmvall, B. E. (2002). Control of nosocomial Legionnaires' disease by keeping the circulating hot water temperature above 55°C: experience from a 10-year surveillance programme in a district general hospital. *J. Hosp. Infect.* 50, 213–219. doi: 10.1053/jhin.2002.1185

de Felipe, K. S., Glover, R. T., Charpentier, X., Anderson, O. R., Reyes, M., Pericone, C. D., et al. (2008). *Legionella* eukaryotic-like type IV substrates interfere with organelle trafficking. *PLoS Pathogens* 4:e1000117. doi: 10.1371/journal.ppat.1000117

Edelstein, P. H. (1981). Improved semiselective medium for isolation of *Legionella pneumophila* from contaminated clinical and environmental specimens. *J. Clin. Microbiol.* 14, 298–303.

European Centre for Disease Prevention and Control (2013). *Annual Epidemiological Report 2013: Reporting on 2011 Surveillance Data and 2012 Epidemic Intelligence Data*. Stockholm: ECDC.

Faucher, S. P., Mueller, C. A., and Shuman, H. A. (2011). *Legionella pneumophila* transcriptome during intracellular multiplication in human macrophages. *Front. Microbiol.* 2:60. doi: 10.3389/fmicb.2011.00060

Feeley, J. C., Gibson, R. J., Gorman, G. W., Langford, N. C., Rasheed, J. K., Mackel, D. C., et al. (1979). Charcoal-yeast extract agar: primary isolation medium for *Legionella pneumophila*. *J. Clin. Microbiol.* 10, 437–441.

Fields, B. S., Benson, R. F., and Besser, R. E. (2002). *Legionella* and Legionnaires' disease: 25 years of investigation. *Clin. Microbiol. Rev.* 15, 506–526. doi: 10.1128/CMR.15.3.506-526.2002

Hales, L. M., and Shuman, H. A. (1999). The *Legionella pneumophila rpoS* gene is required for growth within *Acanthamoeba castellanii*. *J. Bacteriol.* 181, 4879–4889.

Handley, K. M., Héry, M., and Lloyd, J. R. (2009). *Marinobacter santoriniensis* sp. nov., an arsenate-respiring and arsenite-oxidizing bacterium isolated from hydrothermal sediment. *Int. J. Syst. Evol. Microbiol.* 59, 886–892. doi: 10.1099/ijs.0.003145-0

Hecker, M., Reder, A., Fuchs, S., Pagels, M., and Engelmann, S. (2009). Physiological proteomics and stress/starvation responses in *Bacillus subtilis* and *Staphylococcus aureus*. *Res. Microbiol.* 160, 245–258. doi: 10.1016/j.resmic.2009.03.008

Hovel-Miner, G., Pampou, S., Faucher, S. P., Clarke, M., Morozova, I., Morozov, P., et al. (2009). σ^S controls multiple pathways associated with intracellular multiplication of *Legionella pneumophila*. *J. Bacteriol.* 191, 2461–2473. doi: 10.1128/JB.01578-08

Ishihama, A. (2000). Functional modulation of *Escherichia coli* RNA polymerase. *Annu. Rev. Microbiol.* 54, 499–518. doi: 10.1146/annurev.micro.54.1.499

James, B. W., Mauchline, W. S., Dennis, P. J., Keevil, C. W., and Wait, R. (1999). Poly-3-hydroxybutyrate in *Legionella pneumophila*, an energy source for survival in low-nutrient environments. *Appl. Environ. Microbiol.* 65, 822–827.

Jensen, L. J., Gupta, R., Blom, N., Devos, D., Tamames, J., Kesmir, C., et al. (2002). Prediction of human protein function from post-translational modifications and localization features. *J. Mol. Biol.* 319, 1257–1265. doi: 10.1016/S0022-2836(02)00379-0

Jensen, L. J., Gupta, R., Stærfeldt, H.-H., and Brunak, S. (2003). Prediction of human protein function according to Gene Ontology categories. *Bioinformatics* 19, 635–642. doi: 10.1093/bioinformatics/btg036

Kim, E.-H., Charpentier, X., Torres-Urquidy, O., McEvoy, M. M., and Rensing, C. (2009). The metal efflux island of *Legionella pneumophila* is not required for survival in macrophages and amoebas. *FEMS Microbiol. Lett.* 301, 164–170. doi: 10.1111/j.1574-6968.2009.01813.x

Kim, H. J., Park, S., Lee, J. M., Park, S., Jung, W., Kang, J.-S., et al. (2008). *Moritella dasanensis* sp. nov., a psychrophilic bacterium isolated from the Arctic ocean. *Int. J. Syst. Evol. Microbiol.* 58, 817–820. doi: 10.1099/ijs.0.65501-0

Koebnik, R., Locher, K. P., and Van Gelder, P. (2000). Structure and function of bacterial outer membrane proteins: barrels in a nutshell. *Mol. Microbiol.* 37, 239–253. doi: 10.1046/j.1365-2958.2000.01983.x

Krogh, A., Larsson, B., Von Heijne, G., and Sonnhammer, E. L. L. (2001). Predicting transmembrane protein topology with a hidden Markov model: application to complete genomes. *J. Mol. Biol.* 305, 567–580. doi: 10.1006/jmbi.2000.4315

Kurahashi, M., and Yokota, A. (2007). *Endozoicomonas elysicola* gen. nov., sp. nov., a γ-proteobacterium isolated from the sea slug Elysia ornata. *Syst. Appl. Microbiol.* 30, 202–206. doi: 10.1016/j.syapm.2006.07.003

Kusnetsov, J. M., Ottoila, E., and Martikainen, P. J. (1996). Growth, respiration and survival of *Legionella pneumophila* at high temperatures. *J. Appl. Bacteriol.* 81, 341–347. doi: 10.1111/j.1365-2672.1996.tb03517.x

Lee, J., and West, A. (1991). Survival and growth of *Legionella* species in the environment. *J. Appl. Bacteriol.* 70, 121S–129S.

Li, L., Mendis, N., Trigui, H., and Faucher, S. P. (2015). Transcriptomic changes of *Legionella pneumophila* in water. *BMC Genomics* 16:637. doi: 10.1186/s12864-015-1869-6

Li, L., Mendis, N., Trigui, H., Oliver, J. D., and Faucher, S. P. (2014). The importance of the viable but non-culturable state in human bacterial pathogens. *Front. Microbiol.* 5:258. doi: 10.3389/fmicb.2014.00258

Livak, K. J., and Schmittgen, T. D. (2001). Analysis of relative gene expression data using real-time quantitative PCR and the $2^{\Delta\Delta CT}$ method. *Methods* 25, 402–408. doi: 10.1006/meth.2001.1262

Mendis, N., McBride, P., and Faucher, S. P. (2015). Short-term and long-term survival and virulence of *Legionella pneumophila* in the defined freshwater medium Fraquil. *PLoS ONE* 10:e0139277. doi: 10.1371/journal.pone.0139277

Moffat, J. F., and Tompkins, L. S. (1992). A quantitative model of intracellular growth of *Legionella pneumophila* in *Acanthamoeba castellanii*. *Infect. Immun.* 60, 296–301.

Morel, F. M., Westall, J. C., Reuter, J., and Chaplick, J. P. (1975). "Description of the algal growth media 'Aquil' and 'Fraquil'," in *Technical Report 16* (Cambridge: Massachusetts Institute of Technology).

O'Connor, T. J, Adepoju, Y., Boyd, D., and Isberg, R. R. (2011). Minimization of the *Legionella pneumophila* genome reveals chromosomal regions involved in host range expansion. *Proc. Natl. Acad. Sci. U.S.A.* 108, 14733–14740. doi: 10.1073/pnas.1111678108

Paszko-Kolva, C., Shahamat, M., and Colwell, R. R. (1992). Long-term survival of *Legionella pneumophila* serogroup 1 under low-nutrient conditions and associated morphological changes. *FEMS Microbiol. Lett.* 102, 45–55. doi: 10.1111/j.1574-6968.1992.tb05794.x

Petersen, T. N., Brunak, S., Von Heijne, G., and Nielsen, H. (2011). SignalP 4.0: discriminating signal peptides from transmembrane regions. *Nat. Methods* 8, 785–786. doi: 10.1038/nmeth.1701

Phin, N., Parry-Ford, F., Harrison, T., Stagg, H. R., Zhang, N., Kumar, K., et al. (2014). Epidemiology and clinical management of Legionnaires' disease. *Lancet Infect. Dis.* 14, 1011–1021. doi: 10.1016/S1473-3099(14)70713-3

Richter, K., Haslbeck, M., and Buchner, J. (2010). The heat shock response: life on the verge of death. *Mol. Cell* 40, 253–266. doi: 10.1016/j.molcel.2010.10.006

Rogers, J., Dowsett, A. B., Dennis, P. J., Lee, J. V., and Keevil, C. W. (1994). Influence of temperature and plumbing material selection on biofilm formation and growth of *Legionella pneumophila* in a model potable water system containing complex microbial flora. *Appl. Environ. Microbiol.* 60, 1585–1592.

Roy, C. R., and Isberg, R. R. (1997). Topology of *Legionella pneumophila* DotA: an inner membrane protein required for replication in macrophages. *Infect. Immun.* 65, 571–578.

Sadosky, A. B., Wiater, L. A., and Shuman, H. A. (1993). Identification of *Legionella pneumophila* genes required for growth within and killing of human macrophages. *Infect. Immun.* 61, 5361–5373.

Schofield, G. M. (1985). A note on the survival of *Legionella pneumophila* in stagnant tap water. *J. Appl. Bacteriol.* 59, 333–335. doi: 10.1111/j.1365-2672.1985.tb03327.x

Skaliy, P., and McEachern, H. V. (1979). Survival of the Legionnaires' disease bacterium in water. *Ann. Intern. Med.* 90, 662–663. doi: 10.7326/0003-4819-90-4-662

Söderberg, M. A., Dao, J., Starkenburg, S. R., and Cianciotto, N. P. (2008). Importance of type II secretion for survival of *Legionella pneumophila* in tap water and in amoebae at low temperatures. *Appl. Environ. Microbiol.* 74, 5583–5588. doi: 10.1128/AEM.00067-08

Sonnhammer, E. L. L., von Heijne, G., and Krogh, A. (1998). "A hidden Markov model for predicting transmembrane helices in protein sequences," in *ISMB 1998 Proceedings* (Menlo Park, CA: AAAI Press), 175–182.

Steinert, M., Emödy, L., Amann, R., and Hacker, J. (1997). Resuscitation of viable but nonculturable *Legionella pneumophila* Philadelphia JR32 by *Acanthamoeba castellanii*. *Appl. Environ. Microbiol.* 63, 2047–2053.

Trigui, H., Dudyk, P., Oh, J., Hong, J.-I., and Faucher, S. P. (2014). A regulatory feedback loop between RpoS and SpoT supports the survival of *Legionella pneumophila* in water. *Appl. Environ. Microbiol.* 81, 918–928. doi: 10.1128/AEM.03132-14

Trueba, G., Zapata, S., Madrid, K., Cullen, P., and Haake, D. (2004). Cell aggregation: a mechanism of pathogenic *Leptospira* to survive in fresh water. *Int. Microbiol.* 7, 35–40.

Villeneuve, C., Martineau, C., Mauffrey, F., and Villemur, R. (2013). *Methylophaga nitratireducenticrescens* sp. nov. and *Methylophaga frappieri* sp. nov., isolated from the biofilm of the methanol-fed denitrification system treating the seawater at the Montreal Biodome. *Int. J. Syst. Evol. Microbiol.* 63, 2216–2222. doi: 10.1099/ijs.0.044545-0

Wang, G., and Doyle, M. P. (1998). Survival of enterohemorrhagic *Escherichia coli* O157:H7 in water. *J. Food Prot.* 61, 662–667.

11

Formation of the *Legionella* Replicative Compartment at the Crossroads of Retrograde Trafficking

Kevin Bärlocher[†], *Amanda Welin*[†] *and Hubert Hilbi**

Institute of Medical Microbiology, University of Zürich, Zurich, Switzerland

Edited by:
Elizabeth L. Hartland,
Hudson Institute of Medical Research,
Australia

Reviewed by:
James Samuel,
Texas A&M University, United States
Isabelle Vergne,
Centre National de la Recherche
Scientifique (CNRS), France

***Correspondence:**
Hubert Hilbi
hilbi@imm.uzh.ch

[†]These authors have contributed equally to this work.

Retrograde trafficking from the endosomal system through the Golgi apparatus back to the endoplasmic reticulum is an essential pathway in eukaryotic cells, serving to maintain organelle identity and to recycle empty cargo receptors delivered by the secretory pathway. Intracellular replication of several bacterial pathogens, including *Legionella pneumophila*, is restricted by the retrograde trafficking pathway. *L. pneumophila* employs the Icm/Dot type IV secretion system (T4SS) to form the replication-permissive *Legionella*-containing vacuole (LCV), which is decorated with multiple components of the retrograde trafficking machinery as well as retrograde cargo receptors. The *L. pneumophila* effector protein RidL is secreted by the T4SS and interferes with retrograde trafficking. Here, we review recent evidence that the LCV interacts with the retrograde trafficking pathway, discuss the possible sites of action and function of RidL in the retrograde route, and put forth the hypothesis that the LCV is an acceptor compartment of retrograde transport vesicles.

Keywords: ***Dictyostelium discoideum*, effector protein, host-pathogen interaction, pathogen vacuole, retrograde transport, retromer, sorting nexin, type IV secretion**

COMPONENTS OF THE RETROGRADE TRAFFICKING PATHWAY

Eukaryotic cells employ complex endomembrane systems to process internalized material and transport newly synthesized proteins from the endoplasmic reticulum (ER) via the Golgi apparatus to target organelles and membranes (Johannes and Wunder, 2011). The sub-compartments of these endocytic and secretory (anterograde) pathways, respectively, communicate through vesicular trafficking. Endosomes represent sorting hubs, where the endocytic, secretory and lysosomal pathways meet and diverge.

Retrograde transport from endosomes back to the ER maintains organelle integrity, prevents lysosomal degradation and recycles cargo receptors of the secretory transport machinery (Johannes and Popoff, 2008; Seaman, 2012; Burd and Cullen, 2014; Lu and Hong, 2014; Progida and Bakke, 2016). Dependent on the cargo and various endogenous factors, retrograde transport can occur from different endosomal donor compartments such as early, late and recycling endosomes (EE, LE, RE) as well as the tubular endosomal network (TEN) (Bonifacino and Rojas, 2006; **Figure 1**). Each

Abbreviations: Icm/Dot, intracellular multiplication/defective organelle trafficking; LCV, *Legionella*-containing vacuole; OCRL, Oculocerebrorenal syndrome of Lowe; PI, phosphoinositide; PtdIns, phosphatidylinositol; SNARE, soluble *N*-ethylmaleimide-sensitive factor attachment receptor; SNX, sorting nexin; T4SS, type IV secretion system; TGN, *trans*-Golgi network.

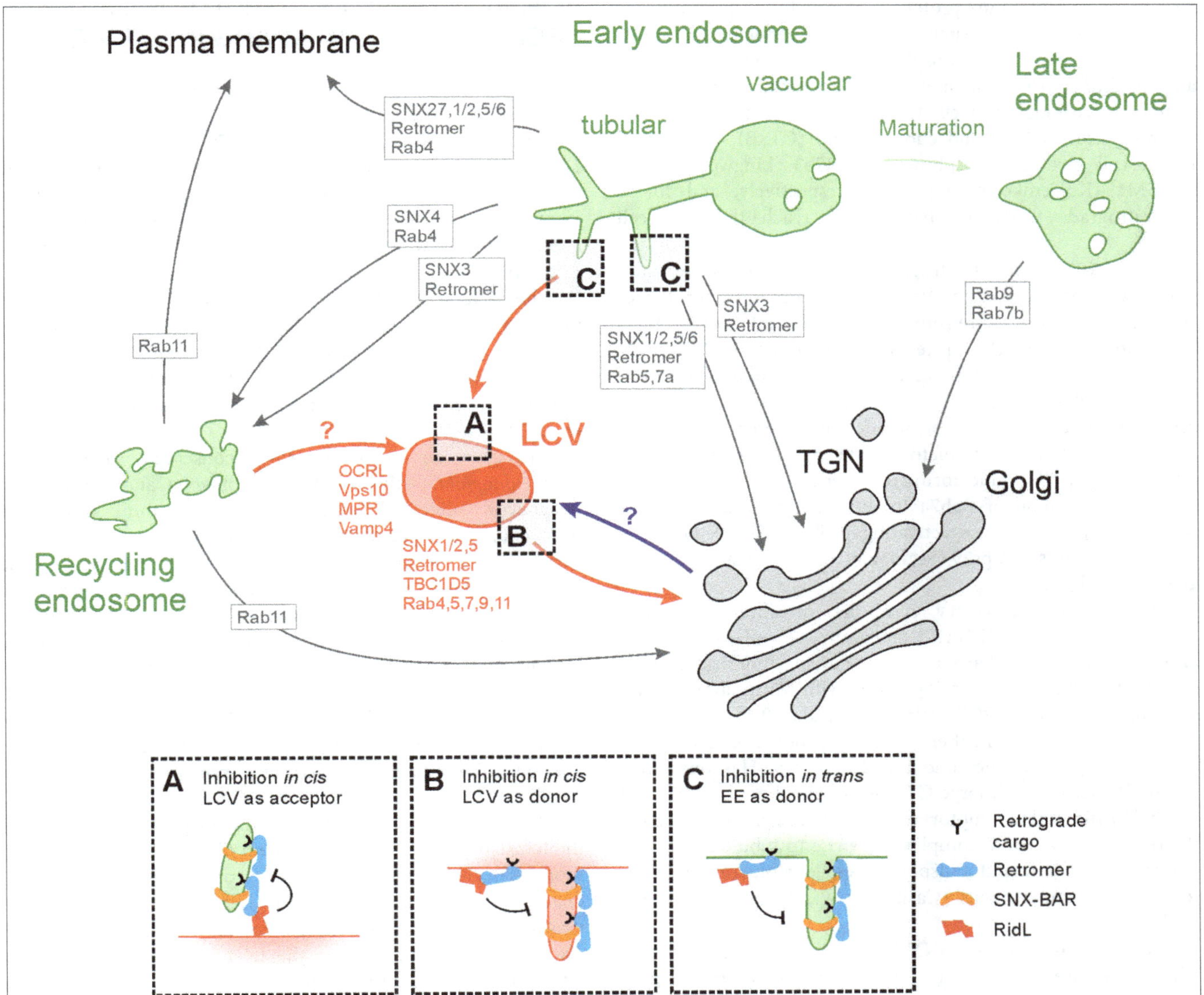

FIGURE 1 | Positioning of the *Legionella*-containing vacuole in the retrograde route. Tubular/vacuolar early endosomes (EE; green) represent a sorting hub, where endosomal, recycling and retrograde trafficking pathways emerge to deliver cargo to late and recycling endosomes, the plasma membrane or the *trans*-Golgi network (TGN). The pathways employ distinct transport machineries (indicated with gray arrows and boxes), including small Rab GTPases, the PI 5-phosphatase OCRL, the retromer and sorting nexins (SNXs). Some of these retrograde trafficking components as well as retrograde cargos (MPR, sortilin) are also found on the LCV (indicated in red), suggesting that the pathogen vacuole is participating in these pathways. Possible retrograde transport routes involving the LCV are shown with red arrows. Inhibition of retrograde trafficking by RidL targeting the retromer might occur *in cis*, considering the LCV as (A) acceptor or (B) donor compartment for retrograde transport vesicles, or (C) *in trans* with the endosome as donor compartment. A possible role in LCV formation of late secretory (anterograde) trafficking emanating from the TGN is shown in blue. For details see text.

membrane is defined by the presence of specific mono- or polyphosphorylated phosphoinositide (PI) lipids and small GTPases of the Rab family, which together determine the spatio-temporal recruitment of specific transport machineries (Di Paolo and De Camilli, 2006; Wandinger-Ness and Zerial, 2014).

The PI 5-phosphatase OCRL (Oculocerebrorenal syndrome of Lowe) is required for retrograde trafficking from EEs to the *trans*-Golgi network (TGN) and for receptor recycling between endosomes and the plasma membrane (PM) (Noakes et al., 2011; Vicinanza et al., 2011; Sharma et al., 2015). OCRL and its *Dictyostelium* homolog Dd5P4 (*D. discoideum* 5-phosphatase 4) hydrolyze PtdIns(4,5)P_2 and PtdIns(3,4,5)P_3 to yield PtdIns(4)P and PtdIns(3,4)P_2 (Zhang et al., 1995; Loovers et al., 2007). OCRL binds to clathrin-coated pits (Ungewickell et al., 2004; Mao et al., 2009) as well as to a range of Rab GTPases including Rab5 on EEs (Hyvola et al., 2006; Fukuda et al., 2008; Hou et al., 2011).

Cargo shuttled via the retrograde endosome-to-TGN pathway comprises many different membrane and soluble proteins as well as lipids. Membrane protein cargos include Vps10 domain family cargo receptors such as cation-dependent or cation-independent mannose 6-phosphate receptor (CD- or CI-MPR) and sortilin/Vps10 (Seaman et al., 1997; Seaman, 2007). Soluble protein cargos include well-studied bacterial AB toxins, such as Shiga toxin (STxB) and cholera toxin (CTxB), which bind to the cell surface glycosphingolipids Gb3 (Lingwood, 1993) or GM1 (Kuziemko et al., 1996), respectively, and hijack the retrograde trafficking pathway to reach the host cell cytosol.

A key component of retrograde trafficking is the retromer complex consisting of a Vps26-Vps29-Vps35 heterotrimer (herein referred to as "retromer") and different combinations of sorting nexin (SNX) proteins, which harbor a PI-binding PX (phox homology) domain (Cullen and Korswagen, 2011; Teasdale and Collins, 2012). Additional factors required for retromer recruitment are retrograde cargo receptors and the small GTPase Rab7a, a regulator of late endosome dynamics, in its activated, GTP-bound form (Rojas et al., 2008; Lucas et al., 2016). The binding of Rab7a to the membrane is negatively regulated by the GTPase activating protein (GAP) TBC1D5, which itself binds to Vps29 (Seaman et al., 2009; Harrison et al., 2014; Jia et al., 2016).

The retromer interacts with the membrane-associated heterodimer of SNX-BAR subfamily members SNX1/2 and SNX5/6, which harbor an additional dimerization and membrane curvature sensing BAR (Bin/Amphiphysin/Rvs) domain (Rojas et al., 2007; Wassmer et al., 2007; Frost et al., 2009). Together with further accessory proteins including the WASH complex (Gomez and Billadeau, 2009; Harbour et al., 2010), EHD1/dynamin large GTPases (Gokool et al., 2007) and p150^{glued}/microtubule motor dynein (Wassmer et al., 2009), the retromer-SNX-BAR complex gives rise to tubule formation, elongation, scission, and transport of vesicles along microtubules to receiver compartments (Carlton et al., 2004; van Weering et al., 2012).

SNXs also include the SNX-PX subfamily member SNX3, which comprises of only a PX domain, binds to Vps35-Vps26 and is necessary for retromer recruitment to the endosomal membrane and the recognition of some cargo (Chen et al., 2013; Harrison et al., 2014; Lucas et al., 2016). The SNX-FERM subfamily member SNX27 interacts with cargo, Vps26, SNX1-SNX2, and the WASH complex (Temkin et al., 2011; Steinberg et al., 2013; Gallon et al., 2014). The SNX27-retromer complex is necessary for Rab4-dependent endosome-to-PM recycling of different cargos including many transporters such as glucose transporter 1 (GLUT1) (Steinberg et al., 2013).

After un-coating of the transport carrier, different components such as tethering factors and SNARE (soluble *N*-ethylmaleimide-sensitive factor attachment receptor) proteins are necessary for selective delivery to target membranes (Progida and Bakke, 2016). Thus, eukaryotic cells employ many distinct types of cargo-specific retrograde trafficking machinery to retrieve proteins and lipids from endosomes for recycling and to maintain organelle integrity.

FORMATION AND MATURATION OF THE *LEGIONELLA*-CONTAINING VACUOLE

The opportunistic pathogen *L. pneumophila* is the clinically most relevant *Legionella* species causing the severe pneumonia Legionnaires' disease (Bangsborg, 1997; Newton et al., 2010). *Legionella* spp. are ubiquitous Gram-negative environmental bacteria that colonize both natural and technical water systems, periodically causing disease outbreaks tracing back to cooling towers, whirlpools and showers. In the environment, *L. pneumophila* survives in both extra- and intracellular niches, i.e., the bacteria colonize biofilms, but also parasitize free-living protozoa (Hilbi et al., 2011). Natural hosts for *L. pneumophila* include *Acanthamoeba*, *Hartmannella*, and *Tetrahymena* species, and the bacterium also thrives within *Dictyostelium discoideum*, a commonly employed protozoan experimental host (Hoffmann et al., 2014b).

Through an evolutionarily conserved mechanism, *L. pneumophila* avoids killing by protozoan and mammalian phagocytes alike, allowing replication within a distinct intracellular compartment termed the *Legionella*-containing vacuole (LCV) (Isberg et al., 2009; Hubber and Roy, 2010; Hilbi and Haas, 2012; Sherwood and Roy, 2013; Finsel and Hilbi, 2015). Accidental infection of humans occurs through the inhalation of contaminated aerosols, although recently, a probable person-to-person transmission was reported for the first time (Correia et al., 2016). If the alveolar macrophages fail to eliminate the invading bacteria through cell-autonomous processes, intracellular replication of the bacteria will ensue, possibly culminating in a severe pneumonia (Simon and Hilbi, 2015).

The formation of the LCV is a complex process governed by the Icm/Dot type IV secretion system (T4SS). This T4SS injects more than 300 different "effector" proteins into the host cell, dictating every step from uptake to interference with various vesicle trafficking pathways and autophagy, interaction with the ER, and finally escape from the host cell (Finsel and Hilbi, 2015; Ensminger, 2016). To this end, the Icm/Dot-translocated proteins subvert numerous host cell targets, including PI lipids, which are used by several effectors as membrane anchors, and small GTPases (Haneburger and Hilbi, 2013; Rothmeier et al., 2013; Finsel and Hilbi, 2015). LCVs avoid luminal acidification and fusion with lysosomes, but continuously and extensively communicate with multiple vesicle trafficking routes (Horwitz, 1983; Urwyler et al., 2009a; Xu et al., 2010; Zhao et al., 2017).

Interaction with the endocytic pathway is indicated by the presence of the small GTPases Rab5a, Rab7a, and Rab21 on the LCV membrane (Urwyler et al., 2009b; Hoffmann et al., 2014a). Moreover, LCVs acquire PtdIns(3)*P* within 1 min of bacterial uptake, and gradually lose this PI lipid within 2 h (Weber et al., 2014). In the endocytic pathway, PtdIns(3)*P* is a crucial regulator, necessary for recruitment of early endosomal antigen 1 (EEA1) and for downstream events leading to fusion of the phagosome with bactericidal lysosomes (Stenmark, 2009).

A hallmark of LCV maturation is the intimate interaction of the pathogen vacuole with the ER (Swanson and Isberg, 1995; Lu and Clarke, 2005; Robinson and Roy, 2006), as a consequence of Rab1-dependent recruitment of secretory vesicles at ER exit sites (Kagan and Roy, 2002; Arasaki and Roy, 2010; Arasaki et al., 2012). PtdIns(4)*P* is a major regulator of secretory vesicle trafficking through the Golgi apparatus (Jean and Kiger, 2012), required for late steps of endocytosis (Jeschke et al., 2015) and present on LCVs (Weber et al., 2006, 2014). In fact, PtdIns(4)*P* transiently localizes to LCVs independently of the Icm/Dot T4SS immediately following bacterial uptake, but is then rapidly cleared. Over the following 2 h, PtdIns(4)*P* again accumulates on LCVs in an Icm/Dot-dependent manner, preceding attachment of the ER (Weber et al., 2014). In addition to Rab1, several other small GTPases involved in the secretory pathway are present on the LCV (Urwyler et al., 2009b; Hoffmann et al., 2014a). Of these, Rab8a, Rab10, and Rab32, all implicated in Golgi to endosome trafficking, promote intracellular replication of *L. pneumophila* (Hoffmann et al., 2014a). Thus, *L. pneumophila* exploits the secretory trafficking pathway to promote formation of the replication-permissive LCV.

RESTRICTION OF INTRACELLULAR REPLICATION OF *L. PNEUMOPHILA* BY RETROGRADE TRAFFICKING

A number of intracellular bacterial pathogens subvert retrograde trafficking (Personnic et al., 2016). These include the obligate intracellular pathogen *Chlamydia trachomatis* (Aeberhard et al., 2015; Mirrashidi et al., 2015), facultative intracellular *Salmonella enterica* serovar Typhimurium (McGourty et al., 2012), and *Coxiella burnetii* (McDonough et al., 2013).

An early study using *D. discoideum* already implicated a role for retrograde trafficking during intracellular replication of *L. pneumophila* (Weber et al., 2009). This study revealed that the PI 5-phophatase OCRL/Dd5P4 restricts intracellular growth of *L. pneumophila*, while promoting the accumulation of ER on LCVs as well as the transition from "tight" to "spacious" pathogen vacuoles (Weber et al., 2009). In a more recent study, depletion of individual components of the retrograde machinery, such as the retromer subunits Vps26a/Vps26b or Vps29, or the retrograde cargo CI-MPR, led to increased intracellular replication of *L. pneumophila* (Finsel et al., 2013). Among the small Rab GTPases implicated in retrograde trafficking, Rab5a but not Rab7a, Rab9a or Rab11a restrict intracellular replication (Hoffmann et al., 2014a). Furthermore, in *D. discoideum* the TBC1D5 homolog promotes intracellular growth of *L. pneumophila* (Bärlocher et al., 2017). Since mammalian TBC1D5 functions as a Rab7 GAP and negatively regulates Rab7a, which is implicated in retrograde trafficking, these findings corroborate a function of the retrograde pathway in restricting *L. pneumophila* replication.

INTERACTION OF LCVS WITH THE RETROGRADE TRAFFICKING PATHWAY

Proteomics analysis of isolated LCVs from infected protozoan and mammalian host cells revealed components of the retrograde trafficking pathway including small GTPases (Shevchuk et al., 2009; Urwyler et al., 2009b; Hoffmann et al., 2014a; Herweg et al., 2015; Schmölders et al., 2017). Whereas the late endosomal marker Rab9 is present on *L. pneumophila* wild-type as well as Δ*icmT* mutant-containing vacuoles, Rab4, Rab5, Rab7, and Rab11 are recruited to LCVs in an Icm/Dot-dependent manner (Finsel et al., 2013; Hoffmann et al., 2014a; Bärlocher et al., 2017). Rab4 and Rab11 associate predominantly with EE/RE or with RE/TGN, and are involved in EE/RE-to-PM or RE-to-PM/TGN trafficking, respectively (Zerial and McBride, 2001; Miserey-Lenkei et al., 2007; **Figure 1**).

With respect to retromer components, Vps26, Vps29, and Vps35 as well as SNX1, SNX2, SNX5, and TBC1D5 can be detected on LCVs (Finsel et al., 2013; Bärlocher et al., 2017). Moreover, retrograde cargos including Vps10, MPR, and Vamp4 are present on the pathogen vacuole (Finsel et al., 2013; Hoffmann et al., 2014a; **Figure 1**). The CI-MPR is shuttled between endosomes and the TGN in retromer-, SNX1/2- as well as Rab7b- and Rab9-dependent pathways (Lombardi et al., 1993; Arighi et al., 2004; Rojas et al., 2007; Progida et al., 2010, 2012). The SNARE Vamp4 is a retrograde cargo predominantly found at the TGN, which cycles to the cell surface and back via EE/RE (Tran et al., 2007).

The *L. pneumophila* Icm/Dot substrate RidL (Retromer interactor decorating LCVs) was identified as a binding partner of the retromer complex, and the effector protein promotes intracellular bacterial replication (Finsel et al., 2013). Upon infection of *D. discoideum* or macrophages, RidL decorates the LCV membrane and accumulates preferentially at the bacterial poles. Pulldown and protein-lipid overlay experiments revealed that RidL interacts specifically with the retromer Vps29 subunit and the PI lipid PtdIns(3)*P*, respectively. Ectopically produced RidL inhibits retrograde trafficking of fluorescently labeled STxB in HeLa cells, as evidenced by reduced Golgi localization of the toxin. Furthermore, in macrophages infected with wild-type *L. pneumophila*, retrograde trafficking of fluorescently labeled CTxB is impaired in a RidL-dependent manner. In presence of RidL, CTxB colocalizes with an EE/RE marker (transferrin receptor), but not with markers for lysosomes (dextran) or the Golgi apparatus (GM130). In absence of RidL, however, the toxin colocalizes with both EE/RE and Golgi markers, but not with the lysosome marker, consistent with a RidL-dependent inhibition of retrograde trafficking at endosome exit sites (Finsel et al., 2013).

A recent study revealed that the 29 kDa N-terminal domain of RidL ($RidL_{2-281}$) adopts a new "foot-like" fold comprising a protruding hydrophobic β-hairpin at its "heel," which interacts with Vps29 (Bärlocher et al., 2017). In HeLa cells, the fragment $RidL_{9-258}$ co-localizes with the retromer complex and displaces the Rab7 GAP TBC1D5 from the retromer. Similarly, RidL translocated by *L. pneumophila* reduces the amount of TBC1D5 on LCVs during infection of *D. discoideum*. Thus, displacement of TBC1D5 from the Vps29 retromer subunit by the hydrophobic β-hairpin of RidL might contribute to effector function.

THE LCV AS AN ACCEPTOR COMPARTMENT IN THE RETROGRADE TRAFFICKING PATHWAY

Identification of the site of action of RidL and the position of the LCV in the retrograde trafficking route are pivotal for an understanding of effector function (**Figure 1**). Further insights into the role of RidL were obtained by studying the effects of *ridL* deletion on the levels of retrograde trafficking components on LCVs (Finsel et al., 2013). Deletion of *ridL* does not affect the localization of the retromer subunits to the LCV, but the retrograde cargos Vps10 (a *D. discoideum* homolog of mammalian sortilin) and CI-MPR decorate a larger portion of LCVs. Moreover, the late endosomal/lysosomal marker LAMP1 increasingly localizes to LCVs in the absence of RidL. A larger portion of LCVs also stain positive for SNX1 and SNX2 in the absence of RidL, possibly because RidL competes with the SNXs for PtdIns(3)*P* binding. Additionally, the level of the retrograde trafficking regulator and Rab7 GAP TBC1D5 is increased on LCVs in absence of RidL (Bärlocher et al., 2017). Collectively, these findings suggest that RidL and its inhibition of retrograde trafficking results in decreased levels of retrograde cargos and some components of the retrograde machinery on LCVs.

Regarding the position of the LCV in the retrograde trafficking route different scenarios are conceivable. The LCV might represent a donor compartment in the retrograde trafficking route. Blockage of the pathway at the LCV by RidL would prevent vesicles from budding, elongation or scission, resulting in increased levels of retrograde components on the LCV in the presence of RidL (**Figure 1**). However, this is the opposite of what was observed, and therefore, it is unlikely that the LCV is a donor compartment in the retrograde pathway.

Alternatively, the LCV might be an acceptor compartment of retrograde trafficking vesicles (**Figure 1**). Blockage of incoming vesicles by RidL would prevent tethering, un-coating, attachment or fusion, leading to decreased levels of retrograde cargo, as observed (Finsel et al., 2013). Under these circumstances, the components of the retrograde transport machinery on the LCV would likely not be directly affected, unless incomplete vesicle un-coating or "kiss-and-run" interactions take place (Trahey and Hay, 2010). Furthermore, the presence of Rab4 (EE) and Rab11 (RE) on the LCV is also in agreement with the pathogen vacuole being a compartment accepting incoming retrograde trafficking from endosomes.

Interestingly, the retrograde cargos STxB and CTxB do not accumulate on LCVs, regardless of whether RidL is present or not (Finsel et al., 2013), and thus, the pathogen vacuole seems to serve as an acceptor compartment in some retrograde routes, e.g., the CI-MPR trafficking pathway, but not in others. At this point, we cannot rule out that the LCV is an acceptor compartment in the secretory pathway. Inhibition of retrograde trafficking by RidL might also reduce anterograde delivery of cargos such as MPR or Vps10/sortilin to the LCV by affecting the whole transport cycle (Riederer et al., 1994; Arighi et al., 2004). However, given the accumulation of retrograde machinery on LCVs such an indirect scenario seems less plausible.

CONCLUSIONS AND PERSPECTIVES

The retrograde vesicle trafficking pathway appears to represent a pivotal component of cell-autonomous immunity, since several bacterial pathogens produce effector proteins subverting the pathway. *L. pneumophila* is unique in translocating an effector, RidL, which directly targets the eukaryotic retromer complex (Finsel et al., 2013). RidL binds the Vps29 retromer subunit, inhibits retrograde vesicle trafficking and promotes intracellular bacterial replication. The 29 kDa N-terminal domain of RidL adopts a novel fold including a hydrophobic β-hairpin, which binds Vps29, thereby displacing the Rab7 GAP TBC1D5 and contributing to subversion of retrograde trafficking in *L. pneumophila*-infected cells (Bärlocher et al., 2017).

The exact mode of action of RidL remains to be determined. While its 29 kDa N-terminal fragment is sufficient to displace TBC1D5, the C-terminal 102 kDa fragment of RidL likely harbors additional (perhaps catalytic) activities. Indeed, several *L. pneumophila* Icm/Dot substrates are large (>80 kDa) and represent multi-domain proteins with distinct (catalytic) activities (Itzen and Goody, 2011; Sherwood and Roy, 2013; Finsel and Hilbi, 2015). Thus, in addition to anchoring to Vps29 and displacing TBC1D5, RidL likely has other functions contributing to the inhibition of retrograde trafficking.

Moreover, the site of RidL action and the position of the LCV in the retrograde trafficking route are ill-defined at present. Cargos of the retrograde trafficking route (e.g., CI-MPR, sortilin) as well as the retromer subunits and the bacterial effector localize to LCVs. Yet, the retrograde toxin cargos STxB or CTxB cannot be detected on the LCV membrane, although their trafficking is affected by RidL. Due to its presence on the pathogen vacuole, RidL seems to act *in cis* (on the LCV) but might also function *in trans* (in a distance from the pathogen vacuole). Most available data is in agreement with the pathogen vacuole being an acceptor compartment in the retrograde trafficking pathway, such that RidL blocks the productive interaction of the LCV with incoming retrograde vesicles *in cis*, rather than inhibits the formation of retrograde transport vesicles at endosome exit sites *in trans* (**Figure 1**). Further studies will address the molecular function of RidL and the position of the LCV along the retrograde route.

AUTHOR CONTRIBUTIONS

All authors listed have made a substantial, direct and intellectual contribution to the work, and approved it for publication.

ACKNOWLEDGMENTS

Research in the laboratory of HH was supported by the Swiss National Science Foundation (SNF; 31003A_153200), the University of Zürich, the Novartis Foundation for Medical-Biological Research, the OPO foundation, and the German Ministry of Education and Research (BMBF) in the context of the EU Infect-ERA initiative (project EUGENPATH). AW was supported by a grant from the Swedish Research Council (2014-396). The funders had no role in study design, data collection and analysis, decision to publish, or preparation of the manuscript.

REFERENCES

Aeberhard, L., Banhart, S., Fischer, M., Jehmlich, N., Rose, L., Koch, S., et al. (2015). The proteome of the isolated *Chlamydia trachomatis* containing vacuole reveals a complex trafficking platform enriched for retromer components. *PLoS Pathog.* 11:e1004883. doi: 10.1371/journal.ppat.1004883

Arasaki, K., and Roy, C. R. (2010). *Legionella pneumophila* promotes functional interactions between plasma membrane syntaxins and Sec22b. *Traffic* 11, 587–600. doi: 10.1111/j.1600-0854.2010.01050.x

Arasaki, K., Toomre, D. K., and Roy, C. R. (2012). The *Legionella pneumophila* effector DrrA is sufficient to stimulate SNARE-dependent membrane fusion. *Cell Host Microbe* 11, 46–57. doi: 10.1016/j.chom.2011.11.009

Arighi, C. N., Hartnell, L. M., Aguilar, R. C., Haft, C. R., and Bonifacino, J. S. (2004). Role of the mammalian retromer in sorting of the cation-independent mannose 6-phosphate receptor. *J. Cell Biol.* 165, 123–133. doi: 10.1083/jcb.200312055

Bangsborg, J. M. (1997). Antigenic and genetic characterization of *Legionella* proteins: contributions to taxonomy, diagnosis and pathogenesis. *APMIS Suppl.* 70, 1–53.

Bärlocher, K., Hutter, C. A. J., Swart, A. L., Steiner, B., Welin, A., Hohl, M., et al. (2017). Structural insights into *Legionella* RidL-Vps29 retromer subunit interaction reveal displacement of the regulator TBC1D5. *Nat. Commun.* 8:1543. doi: 10.1038/s41467-017-01512-5

Bonifacino, J. S., and Rojas, R. (2006). Retrograde transport from endosomes to the *trans*-Golgi network. *Nat. Rev. Mol. Cell Biol.* 7, 568–579. doi: 10.1038/nrm1985

Burd, C., and Cullen, P. J. (2014). Retromer: a master conductor of endosome sorting. *Cold Spring Harb. Perspect. Biol.* 6:a016774. doi: 10.1101/cshperspect.a016774

Carlton, J., Bujny, M., Peter, B. J., Oorschot, V. M., Rutherford, A., Mellor, H., et al. (2004). Sorting nexin-1 mediates tubular endosome-to-TGN transport through coincidence sensing of high-curvature membranes and 3-phosphoinositides. *Curr. Biol.* 14, 1791–1800. doi: 10.1016/j.cub.2004.09.077

Chen, C., Garcia-Santos, D., Ishikawa, Y., Seguin, A., Li, L., Fegan, K. H., et al. (2013). SNX3 regulates recycling of the transferrin receptor and iron assimilation. *Cell Metab.* 17, 343–352. doi: 10.1016/j.cmet.2013.01.013

Correia, A. M., Ferreira, J. S., Borges, V., Nunes, A., Gomes, B., Capucho, R., et al. (2016). Probable person-to-person transmission of Legionnaires' disease. *N. Engl. J. Med.* 374, 497–498. doi: 10.1056/NEJMc1505356

Cullen, P. J., and Korswagen, H. C. (2011). Sorting nexins provide diversity for retromer-dependent trafficking events. *Nat. Cell Biol.* 14, 29–37. doi: 10.1038/ncb2374

Di Paolo, G., and De Camilli, P. (2006). Phosphoinositides in cell regulation and membrane dynamics. *Nature* 443, 651–657. doi: 10.1038/nature05185

Ensminger, A. W. (2016). *Legionella pneumophila*, armed to the hilt: justifying the largest arsenal of effectors in the bacterial world. *Curr. Opin. Microbiol.* 29, 74–80. doi: 10.1016/j.mib.2015.11.002

Finsel, I., and Hilbi, H. (2015). Formation of a pathogen vacuole according to *Legionella pneumophila*: how to kill one bird with many stones. *Cell. Microbiol.* 17, 935–950. doi: 10.1111/cmi.12450

Finsel, I., Ragaz, C., Hoffmann, C., Harrison, C. F., Weber, S., Van Rahden, V. A., et al. (2013). The *Legionella* effector RidL inhibits retrograde trafficking to promote intracellular replication. *Cell Host Microbe* 14, 38–50. doi: 10.1016/j.chom.2013.06.001

Frost, A., Unger, V. M., and De Camilli, P. (2009). The BAR domain superfamily: membrane-molding macromolecules. *Cell* 137, 191–196. doi: 10.1016/j.cell.2009.04.010

Fukuda, M., Kanno, E., Ishibashi, K., and Itoh, T. (2008). Large scale screening for novel Rab effectors reveals unexpected broad Rab binding specificity. *Mol. Cell. Proteomics* 7, 1031–1042. doi: 10.1074/mcp.M700569-MCP200

Gallon, M., Clairfeuille, T., Steinberg, F., Mas, C., Ghai, R., Sessions, R. B., et al. (2014). A unique PDZ domain and arrestin-like fold interaction reveals mechanistic details of endocytic recycling by SNX27-retromer. *Proc. Natl. Acad. Sci. U.S.A.* 111, E3604–E3613. doi: 10.1073/pnas.1410552111

Gokool, S., Tattersall, D., and Seaman, M. N. (2007). EHD1 interacts with retromer to stabilize SNX1 tubules and facilitate endosome-to-Golgi retrieval. *Traffic* 8, 1873–1886. doi: 10.1111/j.1600-0854.2007.00652.x

Gomez, T. S., and Billadeau, D. D. (2009). A FAM21-containing WASH complex regulates retromer-dependent sorting. *Dev. Cell* 17, 699–711. doi: 10.1016/j.devcel.2009.09.009

Haneburger, I., and Hilbi, H. (2013). Phosphoinositide lipids and the *Legionella* pathogen vacuole. *Curr. Top. Microbiol. Immunol.* 376, 155–173. doi: 10.1007/82_2013_341

Harbour, M. E., Breusegem, S. Y., Antrobus, R., Freeman, C., Reid, E., and Seaman, M. N. (2010). The cargo-selective retromer complex is a recruiting hub for protein complexes that regulate endosomal tubule dynamics. *J. Cell Sci.* 123, 3703–3717. doi: 10.1242/jcs.071472

Harrison, M. S., Hung, C. S., Liu, T. T., Christiano, R., Walther, T. C., and Burd, C. G. (2014). A mechanism for retromer endosomal coat complex assembly with cargo. *Proc. Natl. Acad. Sci. U.S.A.* 111, 267–272. doi: 10.1073/pnas.1316482111

Herweg, J. A., Hansmeier, N., Otto, A., Geffken, A. C., Subbarayal, P., Prusty, B. K., et al. (2015). Purification and proteomics of pathogen-modified vacuoles and membranes. *Front. Cell. Infect. Microbiol.* 5:48. doi: 10.3389/fcimb.2015.00048

Hilbi, H., and Haas, A. (2012). Secretive bacterial pathogens and the secretory pathway. *Traffic* 13, 1187–1197. doi: 10.1111/j.1600-0854.2012.01344.x

Hilbi, H., Hoffmann, C., and Harrison, C. F. (2011). *Legionella* spp. outdoors: colonization, communication and persistence. *Environ. Microbiol. Rep.* 3, 286–296. doi: 10.1111/j.1758-2229.2011.00247.x

Hoffmann, C., Finsel, I., Otto, A., Pfaffinger, G., Rothmeier, E., Hecker, M., et al. (2014a). Functional analysis of novel Rab GTPases identified in the proteome of purified *Legionella*-containing vacuoles from macrophages. *Cell. Microbiol.* 16, 1034–1052. doi: 10.1111/cmi.12256

Hoffmann, C., Harrison, C. F., and Hilbi, H. (2014b). The natural alternative: protozoa as cellular models for *Legionella* infection. *Cell. Microbiol.* 16, 15–26. doi: 10.1111/cmi.12235

Horwitz, M. A. (1983). The Legionnaires' disease bacterium (*Legionella pneumophila*) inhibits phagosome-lysosome fusion in human monocytes. *J. Exp. Med.* 158, 2108–2126. doi: 10.1084/jem.158.6.2108

Hou, X., Hagemann, N., Schoebel, S., Blankenfeldt, W., Goody, R. S., Erdmann, K. S., et al. (2011). A structural basis for Lowe syndrome caused by mutations in the Rab-binding domain of OCRL1. *EMBO J.* 30, 1659–1670. doi: 10.1038/emboj.2011.60

Hubber, A., and Roy, C. R. (2010). Modulation of host cell function by *Legionella pneumophila* type IV effectors. *Annu. Rev. Cell Dev. Biol.* 26, 261–283. doi: 10.1146/annurev-cellbio-100109-104034

Hyvola, N., Diao, A., McKenzie, E., Skippen, A., Cockcroft, S., and Lowe, M. (2006). Membrane targeting and activation of the Lowe syndrome protein OCRL1 by Rab GTPases. *EMBO J.* 25, 3750–3761. doi: 10.1038/sj.emboj.7601274

Isberg, R. R., O'connor, T. J., and Heidtman, M. (2009). The *Legionella pneumophila* replication vacuole: making a cosy niche inside host cells. *Nat. Rev. Microbiol.* 7, 13–24. doi: 10.1038/nrmicro1967

Itzen, A., and Goody, R. S. (2011). Covalent coercion by *Legionella pneumophila*. *Cell Host Microbe* 10, 89–91. doi: 10.1016/j.chom.2011.08.002

Jean, S., and Kiger, A. A. (2012). Coordination between Rab GTPase and phosphoinositide regulation and functions. *Nat. Rev. Mol. Cell Biol.* 13, 463–470. doi: 10.1038/nrm3379

Jeschke, A., Zehethofer, N., Lindner, B., Krupp, J., Schwudke, D., Haneburger, I., et al. (2015). Phosphatidylinositol 4-phosphate and phosphatidylinositol 3-phosphate regulate phagolysosome biogenesis. *Proc. Natl. Acad. Sci. U.S.A.* 112, 4636–4641. doi: 10.1073/pnas.1423456112

Jia, D., Zhang, J. S., Li, F., Wang, J., Deng, Z., White, M. A., et al. (2016). Structural and mechanistic insights into regulation of the retromer coat by TBC1d5. *Nat. Commun.* 7:13305. doi: 10.1038/ncomms13305

Johannes, L., and Popoff, V. (2008). Tracing the retrograde route in protein trafficking. *Cell* 135, 1175–1187. doi: 10.1016/j.cell.2008.12.009

Johannes, L., and Wunder, C. (2011). Retrograde transport: two (or more) roads diverged in an endosomal tree? *Traffic* 12, 956–962. doi: 10.1111/j.1600-0854.2011.01200.x

Kagan, J. C., and Roy, C. R. (2002). *Legionella* phagosomes intercept vesicular traffic from endoplasmic reticulum exit sites. *Nat. Cell Biol.* 4, 945–954. doi: 10.1038/ncb883

Kuziemko, G. M., Stroh, M., and Stevens, R. C. (1996). Cholera toxin binding affinity and specificity for gangliosides determined by surface plasmon resonance. *Biochemistry* 35, 6375–6384. doi: 10.1021/bi952314i

Lingwood, C. A. (1993). Verotoxins and their glycolipid receptors. *Adv. Lipid Res.* 25, 189–211.

Lombardi, D., Soldati, T., Riederer, M. A., Goda, Y., Zerial, M., and Pfeffer, S. R. (1993). Rab9 functions in transport between late endosomes and the *trans*-Golgi network. *EMBO J.* 12, 677–682.

Loovers, H. M., Kortholt, A., De Groote, H., Whitty, L., Nussbaum, R. L., and Van Haastert, P. J. (2007). Regulation of phagocytosis in *Dictyostelium* by the inositol 5-phosphatase OCRL homolog Dd5P4. *Traffic* 8, 618–628. doi: 10.1111/j.1600-0854.2007.00546.x

Lu, H., and Clarke, M. (2005). Dynamic properties of *Legionella*-containing phagosomes in *Dictyostelium* amoebae. *Cell. Microbiol.* 7, 995–1007. doi: 10.1111/j.1462-5822.2005.00528.x

Lu, L., and Hong, W. (2014). From endosomes to the *trans*-Golgi network. *Semin. Cell Dev. Biol.* 31, 30–39. doi: 10.1016/j.semcdb.2014.04.024

Lucas, M., Gershlick, D. C., Vidaurrazaga, A., Rojas, A. L., Bonifacino, J. S., and Hierro, A. (2016). Structural mechanism for cargo recognition by the retromer complex. *Cell* 167, 1623–1635. doi: 10.1016/j.cell.2016.10.056

Mao, Y., Balkin, D. M., Zoncu, R., Erdmann, K. S., Tomasini, L., Hu, F., et al. (2009). A PH domain within OCRL bridges clathrin-mediated membrane trafficking to phosphoinositide metabolism. *EMBO J.* 28, 1831–1842. doi: 10.1038/emboj.2009.155

McDonough, J. A., Newton, H. J., Klum, S., Swiss, R., Agaisse, H., and Roy, C. R. (2013). Host pathways important for *Coxiella burnetii* infection revealed by genome-wide RNA interference screening. *MBio* 4:e00606-12. doi: 10.1128/mBio.00606-12

McGourty, K., Thurston, T. L., Matthews, S. A., Pinaud, L., Mota, L. J., and Holden, D. W. (2012). *Salmonella* inhibits retrograde trafficking of mannose-6-phosphate receptors and lysosome function. *Science* 338, 963–967. doi: 10.1126/science.1227037

Mirrashidi, K. M., Elwell, C. A., Verschueren, E., Johnson, J. R., Frando, A., Von Dollen, J., et al. (2015). Global mapping of the Inc-human interactome reveals that retromer restricts *Chlamydia* infection. *Cell Host Microbe* 18, 109–121. doi: 10.1016/j.chom.2015.06.004

Miserey-Lenkei, S., Waharte, F., Boulet, A., Cuif, M. H., Tenza, D., El Marjou, A., et al. (2007). Rab6-interacting protein 1 links Rab6 and Rab11 function. *Traffic* 8, 1385–1403. doi: 10.1111/j.1600-0854.2007.00612.x

Newton, H. J., Ang, D. K., Van Driel, I. R., and Hartland, E. L. (2010). Molecular pathogenesis of infections caused by *Legionella pneumophila*. *Clin. Microbiol. Rev.* 23, 274–298. doi: 10.1128/CMR.00052-09

Noakes, C. J., Lee, G., and Lowe, M. (2011). The PH domain proteins IPIP27A and B link OCRL1 to receptor recycling in the endocytic pathway. *Mol. Biol. Cell* 22, 606–623. doi: 10.1091/mbc.E10-08-0730

Personnic, N., Bärlocher, K., Finsel, I., and Hilbi, H. (2016). Subversion of retrograde trafficking by translocated pathogen effectors. *Trends Microbiol.* 24, 450–462. doi: 10.1016/j.tim.2016.02.003

Progida, C., and Bakke, O. (2016). Bidirectional traffic between the Golgi and the endosomes - machineries and regulation. *J. Cell Sci.* 129, 3971–3982. doi: 10.1242/jcs.185702

Progida, C., Cogli, L., Piro, F., De Luca, A., Bakke, O., and Bucci, C. (2010). Rab7b controls trafficking from endosomes to the TGN. *J. Cell Sci.* 123, 1480–1491. doi: 10.1242/jcs.051474

Progida, C., Nielsen, M. S., Koster, G., Bucci, C., and Bakke, O. (2012). Dynamics of Rab7b-dependent transport of sorting receptors. *Traffic* 13, 1273–1285. doi: 10.1111/j.1600-0854.2012.01388.x

Riederer, M. A., Soldati, T., Shapiro, A. D., Lin, J., and Pfeffer, S. R. (1994). Lysosome biogenesis requires Rab9 function and receptor recycling from endosomes to the trans-Golgi network. *J. Cell. Biol.* 125, 573–582.

Robinson, C. G., and Roy, C. R. (2006). Attachment and fusion of endoplasmic reticulum with vacuoles containing *Legionella pneumophila*. *Cell. Microbiol.* 8, 793–805. doi: 10.1111/j.1462-5822.2005.00666.x

Rojas, R., Kametaka, S., Haft, C. R., and Bonifacino, J. S. (2007). Interchangeable but essential functions of SNX1 and SNX2 in the association of retromer with endosomes and the trafficking of mannose 6-phosphate receptors. *Mol. Cell. Biol.* 27, 1112–1124. doi: 10.1128/MCB.00156-06

Rojas, R., Van Vlijmen, T., Mardones, G. A., Prabhu, Y., Rojas, A. L., Mohammed, S., et al. (2008). Regulation of retromer recruitment to endosomes by sequential action of Rab5 and Rab7. *J. Cell Biol.* 183, 513–526. doi: 10.1083/jcb.200804048

Rothmeier, E., Pfaffinger, G., Hoffmann, C., Harrison, C. F., Grabmayr, H., Repnik, U., et al. (2013). Activation of Ran GTPase by a *Legionella* effector promotes microtubule polymerization, pathogen vacuole motility and infection. *PLoS Pathog.* 9:e1003598. doi: 10.1371/journal.ppat.1003598

Schmölders, J., Manske, C., Otto, A., Hoffmann, C., Steiner, B., Welin, A., et al. (2017). Comparative proteomics of purified pathogen vacuoles correlates intracellular replication of *Legionella pneumophila* with the small GTPase Ras-related protein 1 (Rap1). *Mol. Cell. Proteomics* 16, 622–641. doi: 10.1074/mcp.M116.063453

Seaman, M. N. (2007). Identification of a novel conserved sorting motif required for retromer-mediated endosome-to-TGN retrieval. *J. Cell Sci.* 120, 2378–2389. doi: 10.1242/jcs.009654

Seaman, M. N. (2012). The retromer complex - endosomal protein recycling and beyond. *J. Cell Sci.* 125, 4693–4702. doi: 10.1242/jcs.103440

Seaman, M. N., Harbour, M. E., Tattersall, D., Read, E., and Bright, N. (2009). Membrane recruitment of the cargo-selective retromer subcomplex is catalysed by the small GTPase Rab7 and inhibited by the Rab-GAP TBC1D5. *J. Cell Sci.* 122, 2371–2382. doi: 10.1242/jcs.048686

Seaman, M. N., Marcusson, E. G., Cereghino, J. L., and Emr, S. D. (1997). Endosome to Golgi retrieval of the vacuolar protein sorting receptor, Vps10p, requires the function of the VPS29, VPS30, and VPS35 gene products. *J. Cell Biol.* 137, 79–92. doi: 10.1083/jcb.137.1.79

Sharma, S., Skowronek, A., and Erdmann, K. S. (2015). The role of the Lowe syndrome protein OCRL in the endocytic pathway. *Biol. Chem.* 396, 1293–1300. doi: 10.1515/hsz-2015-0180

Sherwood, R. K., and Roy, C. R. (2013). A Rab-centric perspective of bacterial pathogen-occupied vacuoles. *Cell Host Microbe* 14, 256–268. doi: 10.1016/j.chom.2013.08.010

Shevchuk, O., Batzilla, C., Hägele, S., Kusch, H., Engelmann, S., Hecker, M., et al. (2009). Proteomic analysis of *Legionella*-containing phagosomes isolated from *Dictyostelium*. *Int. J. Med. Microbiol.* 299, 489–508. doi: 10.1016/j.ijmm.2009.03.006

Simon, S., and Hilbi, H. (2015). Subversion of cell-autonomous immunity and cell migration by *Legionella pneumophila* effectors. *Front. Immunol.* 6:447. doi: 10.3389/fimmu.2015.00447

Steinberg, F., Gallon, M., Winfield, M., Thomas, E. C., Bell, A. J., Heesom, K. J., et al. (2013). A global analysis of SNX27-retromer assembly and cargo specificity reveals a function in glucose and metal ion transport. *Nat. Cell Biol.* 15, 461–471. doi: 10.1038/ncb2721

Stenmark, H. (2009). Rab GTPases as coordinators of vesicle traffic. *Nat. Rev. Mol. Cell Biol.* 10, 513–525. doi: 10.1038/nrm2728

Swanson, M. S., and Isberg, R. R. (1995). Association of *Legionella pneumophila* with the macrophage endoplasmic reticulum. *Infect. Immun.* 63, 3609–3620.

Teasdale, R. D., and Collins, B. M. (2012). Insights into the PX (phox-homology) domain and SNX (sorting nexin) protein families: structures, functions and roles in disease. *Biochem. J.* 441, 39–59. doi: 10.1042/BJ20111226

Temkin, P., Lauffer, B., Jager, S., Cimermancic, P., Krogan, N. J., and Von Zastrow, M. (2011). SNX27 mediates retromer tubule entry and endosome-to-plasma membrane trafficking of signalling receptors. *Nat. Cell Biol.* 13, 715–721. doi: 10.1038/ncb2252

Trahey, M., and Hay, J. C. (2010). Transport vesicle uncoating: it's later than you think. *F1000 Biol. Rep.* 2:47. doi: 10.3410/B2-47

Tran, T. H., Zeng, Q., and Hong, W. (2007). VAMP4 cycles from the cell surface to the *trans*-Golgi network via sorting and recycling endosomes. *J. Cell Sci.* 120, 1028–1041. doi: 10.1242/jcs.03387

Ungewickell, A., Ward, M. E., Ungewickell, E., and Majerus, P. W. (2004). The inositol polyphosphate 5-phosphatase Ocrl associates with endosomes that are partially coated with clathrin. *Proc. Natl. Acad. Sci. U.S.A.* 101, 13501–13506. doi: 10.1073/pnas.0405664101

Urwyler, S., Brombacher, E., and Hilbi, H. (2009a). Endosomal and secretory markers of the *Legionella*-containing vacuole. *Commun. Integr. Biol.* 2, 107–109. doi: 10.4161/cib.7713

Urwyler, S., Nyfeler, Y., Ragaz, C., Lee, H., Mueller, L. N., Aebersold, R., et al. (2009b). Proteome analysis of *Legionella* vacuoles purified by magnetic immunoseparation reveals secretory and endosomal GTPases. *Traffic* 10, 76–87. doi: 10.1111/j.1600-0854.2008.00851.x

van Weering, J. R., Sessions, R. B., Traer, C. J., Kloer, D. P., Bhatia, V. K., Stamou, D., et al. (2012). Molecular basis for SNX-BAR-mediated assembly of distinct endosomal sorting tubules. *EMBO J.* 31, 4466–4480. doi: 10.1038/emboj.2012.283

Vicinanza, M., Di Campli, A., Polishchuk, E., Santoro, M., Di Tullio, G., Godi, A., et al. (2011). OCRL controls trafficking through early endosomes via PtdIns4,5P(2)-dependent regulation of endosomal actin. *EMBO J.* 30, 4970–4985. doi: 10.1038/emboj.2011.354

Wandinger-Ness, A., and Zerial, M. (2014). Rab proteins and the compartmentalization of the endosomal system. *Cold Spring Harb. Perspect. Biol.* 6:a022616. doi: 10.1101/cshperspect.a022616

Wassmer, T., Attar, N., Bujny, M. V., Oakley, J., Traer, C. J., and Cullen, P. J. (2007). A loss-of-function screen reveals SNX5 and SNX6 as potential components of the mammalian retromer. *J. Cell Sci.* 120, 45–54. doi: 10.1242/jcs.03302

Wassmer, T., Attar, N., Harterink, M., Van Weering, J. R., Traer, C. J., Oakley, J., et al. (2009). The retromer coat complex coordinates endosomal sorting and dynein-mediated transport, with carrier recognition by the *trans*-Golgi network. *Dev. Cell* 17, 110–122. doi: 10.1016/j.devcel.2009.04.016

Weber, S. S., Ragaz, C., and Hilbi, H. (2009). The inositol polyphosphate 5-phosphatase OCRL1 restricts intracellular growth of *Legionella*, localizes to the replicative vacuole and binds to the bacterial effector LpnE. *Cell. Microbiol.* 11, 442–460. doi: 10.1111/j.1462-5822.2008.01266.x

Weber, S. S., Ragaz, C., Reus, K., Nyfeler, Y., and Hilbi, H. (2006). *Legionella pneumophila* exploits PI(4)*P* to anchor secreted effector proteins to the replicative vacuole. *PLoS Pathog.* 2:e46. doi: 10.1371/journal.ppat.0020046

Weber, S., Wagner, M., and Hilbi, H. (2014). Live-cell imaging of phosphoinositide dynamics and membrane architecture during *Legionella* infection. *MBio* 5:e00839-13. doi: 10.1128/mBio.00839-13

Xu, L., Shen, X., Bryan, A., Banga, S., Swanson, M. S., and Luo, Z. Q. (2010). Inhibition of host vacuolar H^+-ATPase activity by a *Legionella pneumophila* effector. *PLoS Pathog.* 6:e1000822. doi: 10.1371/journal.ppat.1000822

Zerial, M., and McBride, H. (2001). Rab proteins as membrane organizers. *Nat. Rev. Mol. Cell Biol.* 2, 107–117. doi: 10.1038/35052055

Zhang, X., Jefferson, A. B., Auethavekiat, V., and Majerus, P. W. (1995). The protein deficient in Lowe syndrome is a phosphatidylinositol-4,5-bisphosphate 5-phosphatase. *Proc. Natl. Acad. Sci. U.S.A.* 92, 4853–4856. doi: 10.1073/pnas.92.11.4853

Zhao, J., Beyrakhova, K., Liu, Y., Alvarez, C. P., Bueler, S. A., Xu, L., et al. (2017). Molecular basis for the binding and modulation of V-ATPase by a bacterial effector protein. *PLoS Pathog.* 13:e1006394. doi: 10.1371/journal.ppat.1006394

12

The Toolbox for Uncovering the Functions of *Legionella* Dot/Icm Type IVb Secretion System Effectors: Current State and Future Directions

*Gunnar N. Schroeder**

Centre for Experimental Medicine, School of Medicine, Dentistry and Biomedical Sciences, Queen's University Belfast, Belfast, United Kingdom

Edited by:
Matthias P. Machner, National Institutes of Health (NIH), United States

Reviewed by:
Zhao-Qing Luo, Purdue University, United States
Hubert Hilbi, University of Zurich, Switzerland
Tamara O'Connor, School of Medicine, Johns Hopkins University, United States

***Correspondence:**
Gunnar N. Schroeder
g.schroeder@qub.ac.uk

The defective in organelle trafficking/intracellular multiplication (Dot/Icm) Type IVb secretion system (T4SS) is the essential virulence factor for the intracellular life style and pathogenicity of *Legionella* species. Screens demonstrated that an individual *L. pneumophila* strain can use the Dot/Icm T4SS to translocate an unprecedented number of more than 300 proteins into host cells, where these, so called Icm/Dot-translocated substrates (IDTS) or effectors, manipulate host cell functions to the benefit of the bacteria. Bioinformatic analysis of the pan-genus genome predicts at least 608 orthologous groups of putative effectors. Deciphering the function of these effectors is key to understanding *Legionella* pathogenesis; however, the analysis is challenging. Substantial functional redundancy renders classical, phenotypic screening of single gene deletion mutants mostly ineffective. Here, I review experimental approaches that were successfully used to identify, validate and functionally characterize T4SS effectors and highlight new methods, which promise to facilitate unlocking the secrets of *Legionella*'s extraordinary weapons arsenal.

Keywords: *Legionella*, Type IVb secretion system, Dot/Icm, effectors, toolbox, host targets, infection models, functional genomics

INTRODUCTION

Legionella pneumophila was recognized as human pathogen in 1976 after a devastating outbreak of pneumonia, termed Legionnaires' disease, at an American Legion convention (Fraser et al., 1977; McDade et al., 1977). Investigations into the epidemiological and pathological mechanisms soon established that *L. pneumophila* is a ubiquitous, facultative intracellular pathogen of protozoa (Rowbotham, 1980), which, after inhalation, can also thrive in human alveolar macrophages. Key to exploiting phagocytic hosts is its ability to evade phago-lysosomal degradation (Horwitz, 1983a). Instead the bacteria create the *Legionella* containing vacuole (LCV) (Horwitz, 1983b), which shelters them from intracellular defenses and intercepts nutrients, supporting replication.

The defective in organelle trafficking/intracellular multiplication (Dot/Icm) Type IVb secretion system (T4SS) is critical for LCV biogenesis and intracellular replication (Berger and Isberg, 1993; Segal et al., 1998). It is located at the bacterial poles and, upon membrane contact, translocates proteins into host cells (Charpentier et al., 2009; Jeong et al., 2017), which manipulate cellular processes and are therefore called effectors. Although for many Icm/Dot-translocated substrates (IDTS) an actual effect on the host awaits demonstration, they will here collectively be referred to as effectors.

Facilitated by the genome sequences of prototype *L. pneumophila* strains (Cazalet et al., 2004; Chien et al., 2004) screens for T4SS substrates established that each strain translocates more than 300 proteins (Nagai et al., 2002; Luo and Isberg, 2004; De Felipe et al., 2005, 2008; Kubori et al., 2008; Burstein et al., 2009; Huang et al., 2011; Zhu et al., 2011; Lifshitz et al., 2013). Comparative genomics of an increasing number of *L. pneumophila* isolates and more than 38 *Legionella* spp. showed that, while sharing the Dot/Icm T4SS, extensive diversity in the effector arsenals exists (Schroeder et al., 2010; Gomez-Valero et al., 2014; Burstein et al., 2016). Only 7 proteins of an estimated 608 orthologous groups of effectors across the genus are conserved in all species (Burstein et al., 2016).

Despite advances in our understanding about some effectors (Finsel and Hilbi, 2015; So et al., 2015; Qiu and Luo, 2017), we still lack knowledge about the functions of the majority. Deciphering their functions is challenging, as effectors are a heterogeneous group with limited homology to characterized proteins. This mini-review summarizes methods that were employed to characterize Dot/Icm T4SS effectors and highlights additional methods that could help uncovering the weapons which *Legionella* spp. hold in their arsenals.

CHARACTERISTICS OF DOT/ICM T4SS EFFECTORS

Work over the past 15 years revealed several characteristics of effectors. A translocation signal, which directs them to the T4SS, is commonly found in the C-terminus (Nagai et al., 2005). It consists of a pattern of 20–35 amino acids with specific biophysical properties, e.g., small polar and/or charged residues, and can include a so called E-Block motif encompassing several glutamic acid residues (Nagai et al., 2005; Kubori et al., 2008; Burstein et al., 2009; Huang et al., 2011; Lifshitz et al., 2013). Some effectors comprise an additional internal export signal (Cambronne and Roy, 2007; Jeong et al., 2015). Many effectors are large (>100 kDa), with modular architecture (**Figure 1A**), consisting of different functional domains, e.g., localization, target binding, and enzymatic activity domains. Most prominent feature, which facilitated the discovery of the first effector RalF, is the occurrence of domains with striking homology to eukaryotic proteins (Nagai et al., 2002; Cazalet et al., 2004; De Felipe et al., 2005; Gomez-Valero et al., 2011).

Integration of these characteristics and parameters, such as regulatory motifs, in machine-learning approaches enabled prediction algorithms. Several programs are available (Meyer et al., 2013; Zou et al., 2013; An et al., 2016). Dot/Icm effector-focused algorithms were applied to 38 *Legionella* spp., revealing not only 608 orthologous groups of effectors, but also 99 frequently-occurring domains, which facilitate the identification of new effectors (Burstein et al., 2009, 2016; Lifshitz et al., 2013).

PROBING TRANSLOCATION AND LOCALIZATION

Several assays for the validation of T4SS-mediated transport exist (**Figure 1B**). The visualization of endogenous effectors in host cells using antibodies and immunofluorescence (IF) or electron microscopy (EM) is the gold standard to infer physiologically accurate information (**Figure 1B.1**) and was achieved for a few effectors, e.g., SdeA, LidA, RidL, SidC, SidM, RalF, (Nagai et al., 2002; Luo and Isberg, 2004; Bardill et al., 2005; Machner and Isberg, 2006; Finsel et al., 2013). An antibody against SidC was used to visualize the reconstitution of translocation of a SidC variant lacking its translocation signal by fusion to putative effectors (Vanrheenen et al., 2006; Huang et al., 2011).

Alternatively, effectors were detected in host cell extracts by immunoblot (Vanrheenen et al., 2006; Lin et al., 2015), which in combination with fractionation steps to isolate organelles or LCVs, also informed about their subcellular localization (Ivanov et al., 2010; Hoffmann et al., 2014a; Lin et al., 2015).

As many effectors seem to be of low abundancy, several assays employ overexpression and exploit that the T4SS tolerates reporter domains fused to the N-terminus of effectors, if these do not fold rapidly into rigid structures (Amyot et al., 2013). One or multiple epitope tags [e.g., M45 (Weber et al., 2006), 4xHemagglutinin (HA) (Dolezal et al., 2012), 13xMyc (Viner et al., 2012) or 3xFlag (Isaac et al., 2015)] were employed to detect translocated effectors.

An early screen using an enzymatic reporter domain measured restoration of an antibiotic resistance gene by the Cre/*loxP* recombinase after T4SS-mediated translocation of Cre-effector fusions from a *Legionella* donor into bacterial recipients (**Figure 1B.2**) (Luo and Isberg, 2004). However, the β-lactamase TEM-1 and the calmodulin-dependent adenylate-cyclase domain of *Bordetella pertussis* toxin Cya are the most frequently used enzymatic reporters (**Figures 1B.3, 4**), providing high sensitivity by enzymatic signal amplification (TEM1: cleavage of a β-lactam fluorescence resonance energy transfer (FRET) sensor; Cya; generation of cyclic AMP) (Chen, 2004; Nagai et al., 2005; De Felipe et al., 2008).

Despite localization of several effectors by IF, SidC is the only imaged by super-resolution microscopy (Naujoks et al., 2016) and no live imaging data tracking *Legionella* effectors during infection exists. Lack of fluorescent protein tags compatible with T4SS-mediated translocation might account for this. Split fluorescent proteins, used for *Salmonella* T3SS effectors (**Figure 1B.5**) (Van Engelenburg and Palmer, 2010) and new enzyme-tags (**Figure 1B.6**) (Halo-, Snap and Clip-tags), which self-label with fluorophores that are suitable for live-, super-resolution- and electron-microscopy (Bottanelli et al., 2016; Liss et al., 2016), are promising tools to reveal the dynamics and distribution of effectors on a nanoscale.

INFECTION MODELS

Legionella effectors target fundamental processes, conserved between protozoa and mammals, resulting in a range of

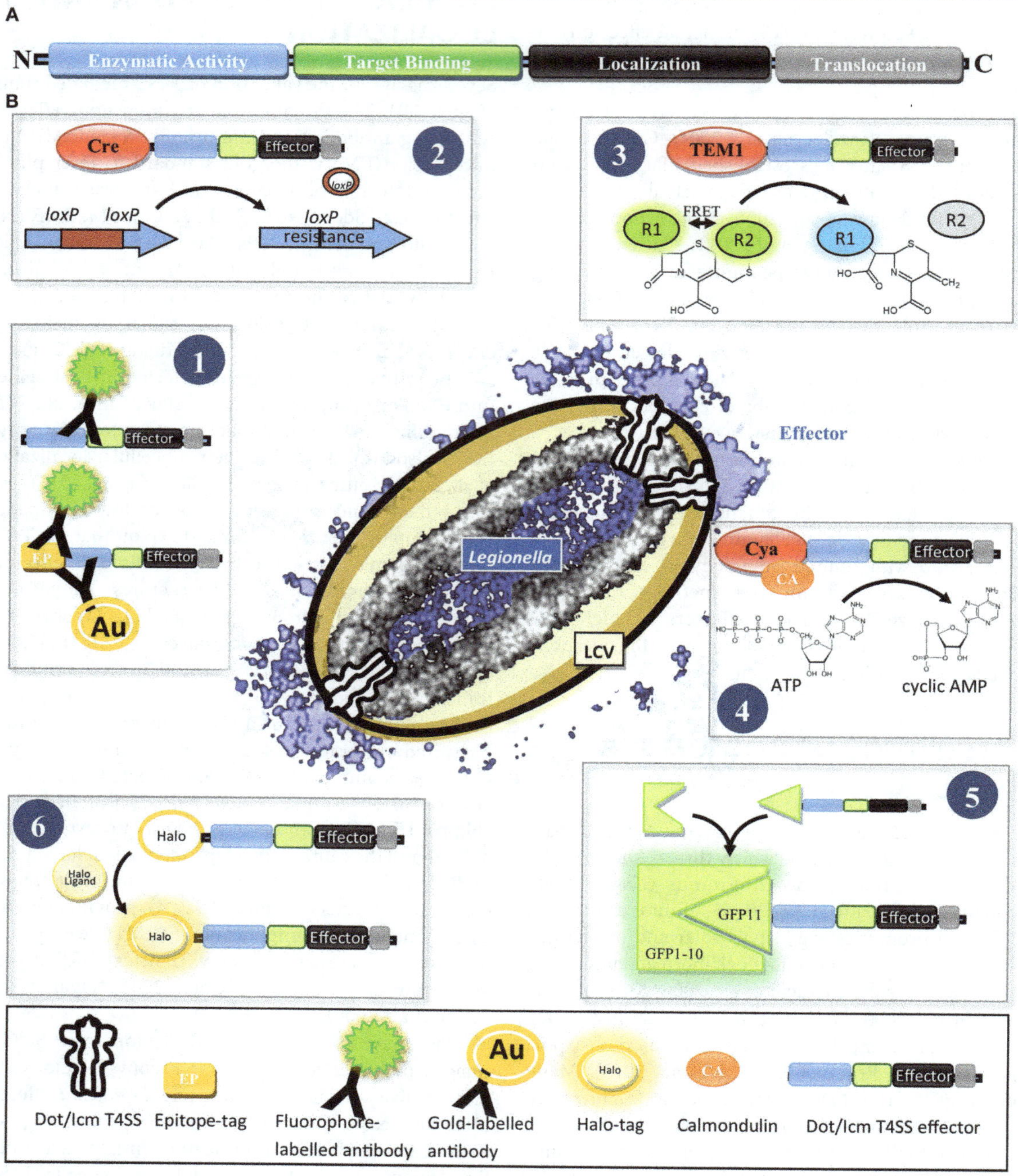

FIGURE 1 | (A) Scheme of the typical architecture of Dot/Icm T4SS effectors. Effectors often show a modular structure consisting of a translocation signal and localization, target binding and enzymatic activity domains. **(B)** Reporter systems for determining the translocation and localization of Dot/Icm T4SS effectors. (1) Fluorophore- or gold particle-conjugated antibodies specific for an effector or an epitope tag are used to detect the effector in host cell lysates by immunoblot or visualize it by immunofluorescence- or electron microscopy. (2) Cre/*loxP* recombinase system: After delivery of Cre-effector fusions into recipient cells the recombinase removes a loxP-flanked disruptor cassette from a gene reporter conferring antibiotic resistance. (3) β-lactamase (TEM-1) assay: Translocation of TEM-1-effector fusions results in hydrolysis of a green fluorescent β-lactam FRET substrate, separating FRET donor and acceptor, generating a blue fluorescent product. (4) Calmodulin-dependent adenylate-cyclase (Cya) assay: Upon arrival of a Cya-effector fusion in the host, Cya gets activated by binding calmodulin and turns over ATP to cyclic AMP, which can be quantified by ELISA. (5) Split-GFP reporter system: Effectors are fused to the small (GFP11) fragment of a split GFP. Upon delivery into cells expressing the non-fluorescent large (GFP1-10) GFP fragment, spontaneous reassembly of effector-fused GFP11 and GFP1-10 occurs, restoring fluorescence emission which can be visualized by fixed or live imaging fluorescence microscopy. (6) Halo-tag reporter: After translocation into the host the Halo-tagged effector can be ligated with versatile fluorophores for detection by conventional, super-resolution- or electron microscopy.

infection models, which, despite similarities, have each strengths, and weaknesses. *Hartmannella vermiformis, Naegleria* spp. and in particular *Acanthamoeba castellanii* and *Dictyostelium discoideum* are frequently used environmental hosts (Rowbotham, 1980; Newsome et al., 1985; Fields et al., 1990; Moffat and Tompkins, 1992; Solomon and Isberg, 2000; Hoffmann et al., 2014b). Requirements for specific effector subsets in different protozoa vary and are more stringent than in macrophages (O'Connor et al., 2011), making protozoa indispensable to study the evolutionary pressures behind the acquisition of effectors. To analyze the interaction of *Legionella* with macrophages, various mammalian [e.g., U937 (Pearlman et al., 1988), HL-60 (Marra et al., 1990), THP-1, Raw264.7 (Cirillo et al., 1994), J774 (Husmann and Johnson, 1992), M-HS (Matsunaga et al., 2001)] and insect [S2 and Kc167 (Dorer et al., 2006; Sun et al., 2013)] cell lines served as models. Moreover, non-phagocytic cells [e.g., HEp-2 (Cirillo et al., 1994), A549 (Mousnier et al., 2014), HeLa (Finsel et al., 2013), HEK293 (Losick et al., 2010), CHO (McCusker et al., 1991; Kagan and Roy, 2002)], with optional ectopic-expression of Fcγ-receptor to boost the invasion efficiency of *Legionella*, were employed. To evaluate the relevance of findings for human disease, differences in patterns of protein family expansion, e.g., Rab GTPases (Klöpper et al., 2012), and innate immune signaling, e.g., inflammasome activation (Krause and Amer, 2016), between cell lines, mice and humans need to be considered. Ultimately, results need validation in primary macrophages and *in vivo* models that approximate the complexity of the human immune system.

Insects such as *Drosophila melanogaster* (Kubori et al., 2010) and *Galleria mellonella* (Harding et al., 2012, 2013b; Aurass et al., 2013) mount innate immune responses and represent straight-forward infection models. Tests in mammals, e.g., guinea pigs, rats, rhesus monkeys, and marmosets, showed that guinea pigs develop disease similar to humans (Baskerville et al., 1983; Davis et al., 1983). Mice, with exception of A/J mice, which are defective in an NAIP5-dependent inflammasome response to flagellin, are resistant to *Legionella* (Brieland et al., 1994). Nevertheless, because of the wealth of engineered mouse strains, infections of A/J mice with wild-type or non-permissive mice with flagellin-deficient *Legionella* have become the predominant *in vivo* models and gave important insight into effector and immune biology (Brown et al., 2016). In the future, humanized mice (Walsh et al., 2017) and *ex vivo* human lung tissue models (Jäger et al., 2014) will improve our capabilities to define roles of effectors in human infection.

GENETICS APPROACHES TO DETERMINE EFFECTOR FUNCTIONS

Legionella is amenable for gene deletion by homologous recombination (Merriam et al., 1997; Bryan et al., 2011; O'Connor et al., 2011) and mutagenesis with transposons (Ott, 1994; Pope et al., 1994; Edelstein et al., 1999; O'Connor et al., 2011). Assays recording intracellular growth by colony counting or continuously, using fluorescent or bioluminescent strains, are established (Coers et al., 2007; Tiaden et al., 2013; Schroeder et al., 2015). Mixed infection competition experiments measuring performance of a wild type vs. a mutant strain achieved better resolution of differences in virulence in some cases (Ensminger et al., 2012; Finsel et al., 2013; Harding et al., 2013b). However, attenuation of strains lacking single effectors was rarely observed. Some effectors might be dispensable in a specific host; but *Legionella* also achieves resilience by deploying families of paralogue effectors, which seem functionally redundant (Cazalet et al., 2004; Chien et al., 2004).

To reduce the complexity of the effector network, O'Connor and Isberg developed two genetic approaches. Insertional mutagenesis and depletion (iMAD, **Figure 2A**) is based on the combinatorial screening of effector deletion mutants for intracellular growth in hosts, which are also host factor depleted (O'Connor et al., 2012; O'Connor and Isberg, 2014). Additive or compensatory effects of the lack of an effector and a host factor are monitored and interrogated using computational clustering and network analysis, grouping effectors with similar profiles and predicting functional redundancy.

In a second approach five genomic regions were deleted to create a minimized genome strain lacking 31% of effectors (O'Connor et al., 2011). This strain grows normally in macrophages; but is attenuated in protozoa, underlining the importance of examining several infection models. Subsequently, the minimized genome strain and intermediates lacking subsets of the genomic regions proved to be valuable tools to link Dot/Icm T4SS dependent phenotypes to a chromosomal region and, through gene-by-gene screening, to individual effectors (Choy et al., 2012; Arasaki et al., 2017; Kotewicz et al., 2017). Deletion of additional effectors could generate even more powerful strains for loss-of-function or gain-of-function experiments, in which the perturbation of host processes by individual effectors can be dissected.

HETEROLOGOUS EXPRESSION SYSTEMS FOR PHENOTYPIC ANALYSIS

Alternatives to investigating effectors during infection rely on heterologous expression and delivery. These are often technically and analytically less complex, but do not reflect physiological concentrations, microenvironment of delivery and the effects of other effectors. Microinjection of recombinant effectors, e.g., SetA (Jank et al., 2012), offers excellent control of concentration and timing of injection, enabling the characterization of toxic effectors; however requires protein purification and a microinjector. Relinquishing the tight control over the delivery, but reducing technical requirements, microbial microinjection exploits a *Yersinia enterocolitica* strain with functional type III secretion system (T3SS), but lacking effectors (Wölke et al., 2011), to deliver individual Dot/Icm effectors (Rothmeier et al., 2013). The suitability of this approach for a wide range of T4SS effectors still needs confirmation.

Ectopic-expression in mammalian cells remains the workhorse to assess effector-induced modulation of host processes and subcellular targeting by co-localization with organelle markers. Numerous studies exist. Libraries containing

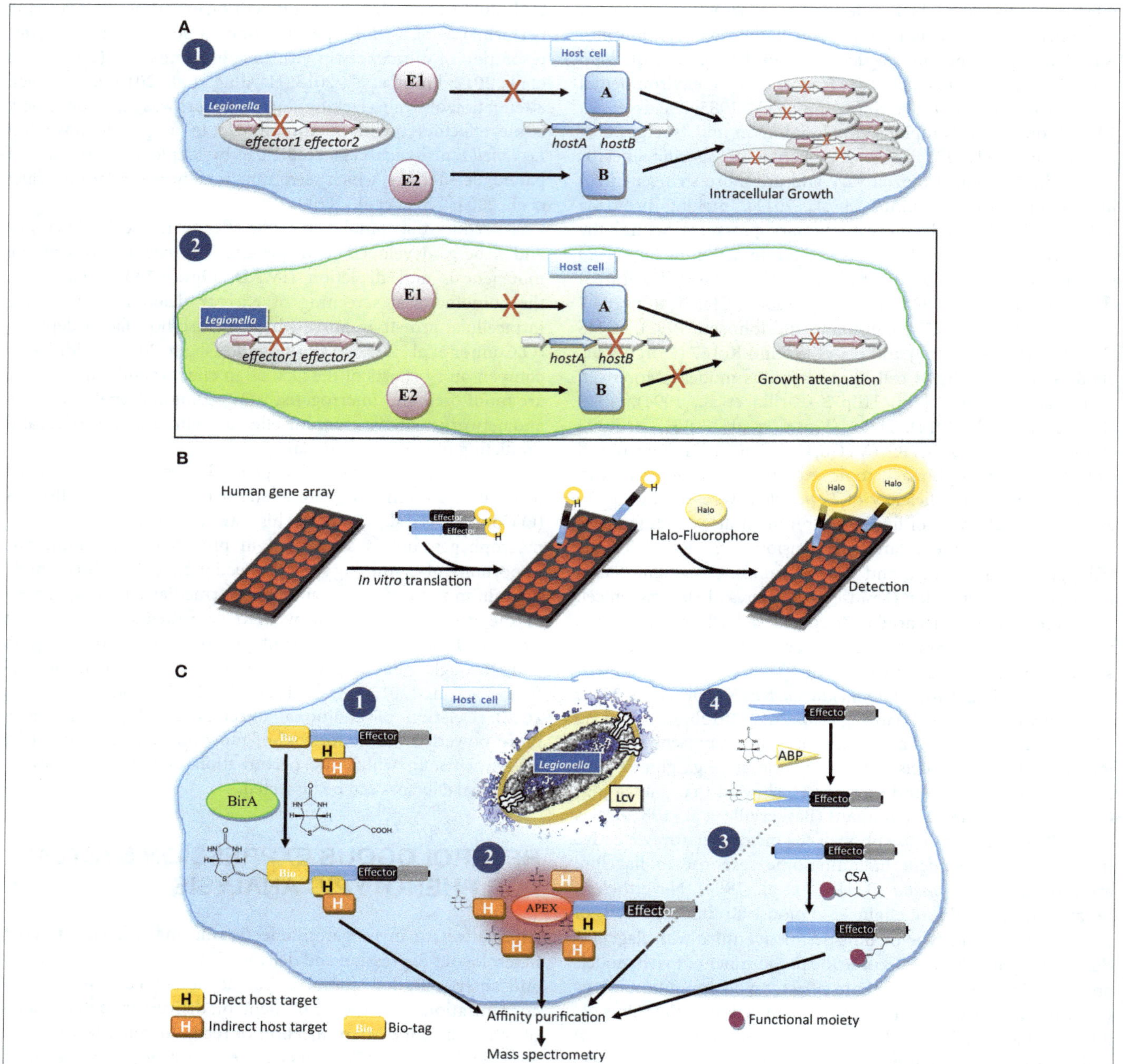

FIGURE 2 | Genetics and proteomics methods for the functional characterization of Dot/Icm T4SS effectors. **(A)** Insertional mutagenesis and depletion (iMAD) disentangles the complex network of effector-host manipulations: (1) Characterization of a *Legionella* mutant lacking a single effector (E1), which acts through host protein A, does not result in reduced intracellular growth, because a second effector (E2) induces a redundant process through host protein B. (2) Following the iMAD strategy, screening of single effector mutants in host cells, which are also depleted for host factors, eliminates redundant pathways, resulting in attenuation of the strain. **(B)** Nucleic Acid-Programmable Protein Array (NAPPA) for profiling of host cell targets of effectors: Human genes are printed as array on slides and translated *in vitro*. Recombinant Halo-tagged effector is added and after washes reacted with a fluorophore-ligand for the Halo-tag, allowing detection of the human proteins, which bound and retained the effector on the array. **(C)** Proteomics approaches for the characterization of effectors: (1) BirA/Bio-tag system: Bio-tagged effector is translocated into biotin ligase BirA expressing cells leading to biotinlylation of the Bio-tag. After optional cross-linking the bioinylated effector and bound host proteins are isolated by tandem-affinity purification for interactome analysis by MS. (2) Proximity biotinylation: Translocation of an effector fused to e.g., the peroxidase APEX or a promiscuous biotin ligase (BioID) results in biotinylation of host proteins in the proximity of the effector, which can be processed as for (1) to identify potential interactors. (3) Identifying post-translational modifications (PTMs): Infected cells are infused or metabolically-labeled with a chemical substrate analog (CSA) for a PTM-catalyzing enzyme. Host proteins and effectors, which are modified with the CSA, can be isolated after ligation of an affinity handle such as biotin to the CSA and characterized by MS. (4) Profiling the enzymatic activities of effectors with activity-based probes (ABPs): Infected cells or lysates are treated with a chemical ABP, which irreversibly binds to a specific enzyme class and contains or allows addition of an affinity handle. Effectors, which possess such enzymatic activity, react with the probe, are isolated and identified by MS.

up to 275 effectors for viral transduction or transfection were used to screen for effectors, which modulate, e.g., caspase activation (Zhu et al., 2013), the cytoskeleton (Liu et al., 2017), translation (Barry et al., 2013), or NF-kB activation (Ge et al., 2009; Losick et al., 2010).

Saccharomyces cerevisiae is an important tool to study effectors (Popa et al., 2016). Phenotypic screens identified *Legionella* effectors that subvert endosomal trafficking or are cytotoxic for yeast (Campodonico et al., 2005; Shohdy et al., 2005). The availability of yeast gene deletion and overexpression strain collections (Gelperin et al., 2005; Sopko et al., 2006; Giaever et al., 2014) bears particular potential. These have proven useful to test enzymatic activities of effectors, e.g., LegS2 or SidP, in functional complementation assays (Degtyar et al., 2009; Toulabi et al., 2013) and to profile synergistic and antagonistic genetic interactions between yeast and effector genes, allowing to infer affected pathways (Viner et al., 2012). Moreover, screening of overexpressed host proteins or effectors for suppression of effector-induced toxicity toward yeast identified host targets, effectors pairs with antagonistic activities and, so called meta-effectors, which regulate other effectors (Tan and Luo, 2011; Tan et al., 2011; Guo et al., 2014; Urbanus et al., 2016).

IDENTIFICATION OF PROTEIN TARGETS

Dissecting the molecular mechanisms underlying effector-induced phenotypes often requires the identification of host targets. Yeast two-hybrid screening is a powerful method to identify protein-protein interactions and was used for several effectors (Banga et al., 2007; Lomma et al., 2010; Harding et al., 2013a; Michard et al., 2015). Similarly, pull-down of interactors from host cell lysates using purified effector or co-immunoprecipitation (Co-IP) from cells ectopically expressing an effector bait were frequently used (Machner and Isberg, 2006; Price et al., 2009; Finsel et al., 2013; Urbanus et al., 2016). In a cell-free assay system, the Nucleic Acid-Programmable Protein Array (NAPPA, **Figure 2B**) (Yu et al., 2015), human bait gene arrays are translated *in vitro*, exposed to Halo-tagged effector and bound effector detected by ligation of a fluorophore to the Halo-tag. This system, circumventing protein isolation, promises to reveal a global view of interactors.

Despite their proven value, all above-mentioned *in vitro* and heterologous expression methods struggle with the identification of false-positive and -negative targets, because they do not reflect the unique proteomic landscape which an effector experiences when injected at the LCV membrane into a cell that responds to the infection and is manipulated by hundreds of effectors.

We established a method to determine the interactomes of effectors during infection (**Figure 2C.1**) (Mousnier et al., 2014). *Legionella* expressing an effector fused to a tandem-affinity tag including a biotinylation site (Bio-tag), are used to infect cells expressing *Escherichia coli* biotin ligase BirA. The translocated effector is biotinylated, allowing isolation of effector-host target complexes for analysis by mass spectrometry (MS). Using this approach, we identified new interactors of PieE and profiled the infection-relevant interactions of the promiscuous Rab GTPase-binding effectors SidM and LidA (Mousnier et al., 2014; So et al., 2016).

Exciting prospects for effector target discovery arise from the development of proximity-biotinylation systems (**Figure 2C.2**). These rely on promiscuous biotinylation of proteins in proximity of engineered BirA (Roux et al., 2012) or the peroxidase APEX2 (Hung et al., 2016) followed by characterization of biotinylated targets by MS. Translocation of T3SS effector-APEX2 fusions by *Chlamydia* was recently described (Rucks et al., 2017) suggesting that this could be adopted for T4SS effectors.

PROFILING POST-TRANSLATIONAL MODIFICATIONS (PTMS) AND ENZYMATIC ACTIVITIES

Effectors exploit host proteins as receptors (Gaspar and Machner, 2014) and subvert their functions, which is often achieved by post-translational modification (PTM) (Michard and Doublet, 2015). The discovery of the phosphocholination activity of AnkX and phosphoribosyl-ubiquitin ligase activity in SdeA illustrated that careful analysis of protein targets by MS is key to identify new PTMs (Mukherjee et al., 2011; Bhogaraju et al., 2016; Qiu et al., 2016). PTM specific antibodies were used e.g., to study effector-mediated phosphorylation or histone modifications (Ge et al., 2009; Rolando et al., 2013). This can be complemented by autoradiography assays using radioactive substrates, which excel in sensitivity, and are employed to study, e.g., AMPylation (Neunuebel et al., 2011; Tan and Luo, 2011) or glycosyltransferase effectors (Jank et al., 2012). Non-radioactive chemical substrate analogs (CSAs), which can be functionalized to visualize and isolate modified proteins, were developed for several PTMs (Grammel et al., 2011; Lu et al., 2012; Fischle and Schwarzer, 2016). CSAs can also reveal PTMs on effectors, as demonstrated for the post-translational lipidation of effectors (**Figure 2C.3**) (Ivanov et al., 2010; Lin et al., 2015; Schroeder et al., 2015). CSAs enable profiling of PTMs on proteome level from cell extracts, living cells or, as shown for SidM-mediated AMPylation, on NAPPA arrays (Yu et al., 2015), promising global overviews of PTMs at a coverage similar to the *Legionella*-shaped ubiquitinome (Ivanov and Roy, 2013; Bruckert and Abu Kwaik, 2015).

The discovery of new enzymatic activities is challenging as small molecules, e.g., ATP or lipids, can be substrates and/or effectors not necessarily target host proteins. Bioinformatic analysis to identify homologous enzymes and catalytic motifs is critical to find leads (Watson et al., 2005) for focused enzymatic assays, as exemplified by LpdA (lipolysis, Schroeder et al., 2015), SidF (phosphate release, Hsu et al., 2012) or LncP (nucleotide transport, Dolezal et al., 2012). Biophysical methods, e.g., differential scanning fluorimetry, allow screening of ligands (Ciulli, 2013).

For the identification of enzymes-of-interest, e.g., redundant effectors, in the *Legionella* proteome activity-based probes (ABPs) offer a solution (**Figure 2C.4**). ABPs are typically small molecules that irreversibly react with a specific enzyme class and functionalized to allow purification of modified enzymes for MS analysis. ABPs are available for many enzymes including the ubiquitin-conjugation and chromatin-modifying machineries (Willems et al., 2014; Fischle and Schwarzer, 2016; Hewings et al., 2017). ABPs will help to disentangle the redundancy problem, probe for eukaryotic-like enzymes and assign functions in new *Legionella* isolates.

SYNTHESIS

Deciphering the functions of thousands of effectors is a formidable challenge; however new genetics tools and a rapidly growing number of chemical biology and proteomics methods provide a well-suited toolbox to reveal fascinating new mechanisms of host manipulation by *Legionella*.

AUTHOR CONTRIBUTIONS

The author confirms being the sole contributor of this work and approved it for publication.

ACKNOWLEDGMENTS

I thank Dr. Aurélie Mousnier (QUB) for her critical review of the manuscript.

REFERENCES

Amyot, W. M., DeJesus, D., and Isberg, R. R. (2013). Poison domains block transit of translocated substrates via the *Legionella pneumophila* Icm/Dot system. *Infect. Immun.* 81, 3239–3252. doi: 10.1128/IAI.00552-13

An, Y., Wang, J., Li, C., Leier, A., Marquez-Lago, T., Wilksch, J., et al. (2016). Comprehensive assessment and performance improvement of effector protein predictors for bacterial secretion systems III, IV and VI. *Brief. Bioinform.* doi: 10.1093/bib/bbw100. [Epub ahead of print].

Arasaki, K., Mikami, Y., Shames, S. R., Inoue, H., Wakana, Y., and Tagaya, M. (2017). *Legionella* effector Lpg1137 shuts down ER-mitochondria communication through cleavage of syntaxin 17. *Nat. Commun.* 8:15406. doi: 10.1038/ncomms15406

Aurass, P., Schlegel, M., Metwally, O., Harding, C. R., Schroeder, G. N., Frankel, G., et al. (2013). The *Legionella pneumophila* Dot/Icm-secreted effector PlcC/CegC1 together with PlcA and PlcB promotes virulence and belongs to a novel zinc metallophospholipase C family present in bacteria and fungi. *J. Biol. Chem.* 288, 11080–11092. doi: 10.1074/jbc.M112.426049

Banga, S., Gao, P., Shen, X., Fiscus, V., Zong, W.-X., Chen, L., et al. (2007). *Legionella pneumophila* inhibits macrophage apoptosis by targeting pro-death members of the Bcl2 protein family. *Proc. Natl. Acad. Sci. U.S.A.* 104, 5121–5126. doi: 10.1073/pnas.0611030104

Bardill, J. P., Miller, J. L., and Vogel, J. P. (2005). IcmS-dependent translocation of SdeA into macrophages by the *Legionella pneumophila* type IV secretion system. *Mol. Microbiol.* 56, 90–103. doi: 10.1111/j.1365-2958.2005.04539.x

Barry, K. C., Fontana, M. F., Portman, J. L., Dugan, A. S., and Vance, R. E. (2013). IL-1α signaling initiates the inflammatory response to virulent *Legionella pneumophila in vivo*. *J. Immunol.* 190, 6329–6339. doi: 10.4049/jimmunol.1300100

Baskerville, A., Fitzgeorge, R. B., Broster, M., and Hambleton, P. (1983). Histopathology of experimental Legionnaires' disease in guinea pigs, rhesus monkeys and marmosets. *J. Pathol.* 139, 349–362. doi: 10.1002/path.1711390310

Berger, K. H., and Isberg, R. R. (1993). Two distinct defects in intracellular growth complemented by a single genetic locus in *Legionella pneumophila*. *Mol. Microbiol.* 7, 7–19. doi: 10.1111/j.1365-2958.1993.tb01092.x

Bhogaraju, S., Kalayil, S., Liu, Y., Bonn, F., Colby, T., Matic, I., et al. (2016). Phosphoribosylation of ubiquitin promotes serine ubiquitination and impairs conventional ubiquitination. *Cell* 167, 1636–1649.e13. doi: 10.1016/j.cell.2016.11.019

Bottanelli, F., Kromann, E. B., Allgeyer, E. S., Erdmann, R. S., Wood Baguley, S., Sirinakis, G., et al. (2016). Two-colour live-cell nanoscale imaging of intracellular targets. *Nat. Commun.* 7:10778. doi: 10.1038/ncomms10778

Brieland, J., Freeman, P., Kunkel, R., Chrisp, C., Hurley, M., Fantone, J., et al. (1994). Replicative *Legionella pneumophila* lung infection in intratracheally inoculated A/J mice. A murine model of human Legionnaires' disease. *Am. J. Pathol.* 145, 1537–1546.

Brown, A. S., Yang, C., Hartland, E. L., and van Driel, I. R. (2016). The regulation of acute immune responses to the bacterial lung pathogen *Legionella pneumophila* *J. Leukoc. Biol.* 101, 875–886. doi: 10.1189/jlb.4MR0816-340R

Bruckert, W. M., and Abu Kwaik, Y. (2015). Complete and ubiquitinated proteome of the *Legionella*-containing vacuole within human macrophages. *J. Proteome Res.* 14, 236–248. doi: 10.1021/pr500765x

Bryan, A., Harada, K., and Swanson, M. S. (2011). Efficient generation of unmarked deletions in *Legionella pneumophila*. *Appl. Environ. Microbiol.* 77, 2545–2548. doi: 10.1128/AEM.02904-10

Burstein, D., Amaro, F., Zusman, T., Lifshitz, Z., Cohen, O., Gilbert, J. A., et al. (2016). Genomic analysis of 38 *Legionella* species identifies large and diverse effector repertoires. *Nat. Genet.* 48, 167–175. doi: 10.1038/ng.3481

Burstein, D., Zusman, T., Degtyar, E., Viner, R., Segal, G., and Pupko, T. (2009). Genome-scale identification of *Legionella pneumophila* effectors using a machine learning approach. *PLoS Pathog.* 5:e1000508. doi: 10.1371/journal.ppat.1000508

Cambronne, E. D., and Roy, C. R. (2007). The *Legionella pneumophila* IcmSW complex interacts with multiple Dot/Icm effectors to facilitate type IV translocation. *PLoS Pathog.* 3, 1837–1848. doi: 10.1371/journal.ppat.0030188

Campodonico, E. M., Chesnel, L., and Roy, C. R. (2005). A yeast genetic system for the identification and characterization of substrate proteins transferred into host cells by the *Legionella pneumophila* Dot/Icm system. *Mol. Microbiol.* 56, 918–933. doi: 10.1111/j.1365-2958.2005.04595.x

Cazalet, C., Rusniok, C., Brüggemann, H., Zidane, N., Magnier, A., Ma, L., et al. (2004). Evidence in the *Legionella pneumophila* genome for exploitation of host cell functions and high genome plasticity. *Nat. Genet.* 36, 1165–1173. doi: 10.1038/ng1447

Charpentier, X., Gabay, J. E., Reyes, M., Zhu, J. W., Weiss, A., and Shuman, H. A. (2009). Chemical genetics reveals bacterial and host cell functions critical for type IV effector translocation by *Legionella pneumophila*. *PLoS Pathog.* 5:e1000501. doi: 10.1371/journal.ppat.1000501

Chen, J. (2004). *Legionella* effectors that promote nonlytic release from protozoa. *Science* 303, 1358–1361. doi: 10.1126/science.1094226

Chien, M., Morozova, I., Shi, S., Sheng, H., Chen, J., Gomez, S. M., et al. (2004). The genomic sequence of the accidental pathogen *Legionella pneumophila*. *Science* 305, 1966–1968. doi: 10.1126/science.1099776

Choy, A., Dancourt, J., Mugo, B., O'Connor, T. J., Isberg, R. R., Melia, T. J., et al. (2012). The *Legionella* effector RavZ inhibits host autophagy through irreversible Atg8 deconjugation. *Science* 338, 1072–1076. doi: 10.1126/science.1227026

Cirillo, J. D., Falkow, S., and Tompkins, L. S. (1994). Growth of *Legionella pneumophila in Acanthamoeba castellanii* enhances invasion. *Infect. Immun.* 62, 3254–3261.

Ciulli, A. (2013). Biophysical screening for the discovery of small-molecule ligands. *Methods Mol. Biol.* 1008, 357–388. doi: 10.1007/978-1-62703-398-5_13

Coers, J., Vance, R. E., Fontana, M. F., and Dietrich, W. F. (2007). Restriction of *Legionella pneumophila* growth in macrophages requires the concerted action of cytokine and Naip5/Ipaf signalling pathways. *Cell. Microbiol.* 9, 2344–2357. doi: 10.1111/j.1462-5822.2007.00963.x

Davis, G. S., Winn, W. C., Gump, D. W., Craighead, J. M., and Beaty, H. N. (1983). Legionnaires' pneumonia in guinea pigs and rats produced by aerosol exposure. *Chest* 83(5 Suppl.), 15S–16S.

De Felipe, K. S., Glover, R. T., Charpentier, X., Anderson, O. R., Reyes, M., Pericone, C. D., et al. (2008). *Legionella* eukaryotic-like type IV substrates interfere with organelle trafficking. *PLoS Pathog.* 4:e1000117. doi: 10.1371/journal.ppat.1000117

De Felipe, K. S., Pampou, S., Jovanovic, O. S., Pericone, C. D., Ye, S. F., Kalachikov, S., et al. (2005). Evidence for acquisition of *Legionella* type IV secretion substrates via interdomain horizontal gene transfer. *J. Bacteriol.* 187, 7716–7726. doi: 10.1128/JB.187.22.7716-7726.2005

Degtyar, E., Zusman, T., Ehrlich, M., and Segal, G. (2009). A *Legionella* effector acquired from protozoa is involved in sphingolipids metabolism and is targeted to the host cell mitochondria. *Cell. Microbiol.* 11, 1219–1235. doi: 10.1111/j.1462-5822.2009.01328.x

Dolezal, P., Aili, M., Tong, J., Jiang, J.-H., Marobbio, C. M., Lee, S. F., et al. (2012). *Legionella pneumophila* secretes a mitochondrial carrier protein during infection. *PLoS Pathog.* 8:e1002459. doi: 10.1371/journal.ppat.1002459

Dorer, M. S., Kirton, D., Bader, J. S., and Isberg, R. R. (2006). RNA interference analysis of *Legionella* in *Drosophila* cells: exploitation of early secretory apparatus dynamics. *PLoS Pathog.* 2, 315–327. doi: 10.1371/journal.ppat.0020034

Edelstein, P. H., Edelstein, M. A., Higa, F., and Falkow, S. (1999). Discovery of virulence genes of *Legionella pneumophila* by using signature tagged mutagenesis in a guinea pig pneumonia model. *Proc. Natl. Acad. Sci. U.S.A.* 96, 8190–8195. doi: 10.1073/pnas.96.14.8190

Ensminger, A. W., Yassin, Y., Miron, A., and Isberg, R. R. (2012). Experimental evolution of *Legionella pneumophila* in mouse macrophages leads to strains with altered determinants of environmental survival. *PLoS Pathog.* 8:e1002731. doi: 10.1371/journal.ppat.1002731

Fields, B. S., Nerad, T. A., Sawyer, T. K., King, C. H., Barbaree, J. M., Martin, W. T., et al. (1990). Characterization of an axenic strain of *Hartmannella vermiformis* obtained from an investigation of nosocomial Legionellosis. *J. Protozool.* 37, 581–583. doi: 10.1111/j.1550-7408.1990.tb01269.x

Finsel, I., and Hilbi, H. (2015). Formation of a pathogen vacuole according to *Legionella pneumophila*: how to kill one bird with many stones. *Cell. Microbiol.* 17, 935–950. doi: 10.1111/cmi.12450

Finsel, I., Ragaz, C., Hoffmann, C., Harrison, C. F., Weber, S., Van Rahden, V. A., et al. (2013). The *Legionella* effector RidL inhibits retrograde trafficking to promote intracellular replication. *Cell Host Microbe* 14, 38–50. doi: 10.1016/j.chom.2013.06.001

Fischle, W., and Schwarzer, D. (2016). Probing Chromatin-modifying enzymes with chemical tools. *ACS Chem. Biol.* 11, 689–705. doi: 10.1021/acschembio.5b01023

Fraser, D. W., Tsai, T. R., Orenstein, W., Parkin, W. E., Beecham, H. J., Sharrar, R. G., et al. (1977). Legionnaires' disease: description of an epidemic of pneumonia. *N. Engl. J. Med.* 297, 1189–1197. doi: 10.1056/NEJM197712012972201

Gaspar, A. H., and Machner, M. P. (2014). VipD is a Rab5-activated phospholipase A1 that protects *Legionella pneumophila* from endosomal fusion. *Proc. Natl. Acad. Sci. U.S.A.* 111, 4560–4565. doi: 10.1073/pnas.1316376111

Ge, J., Xu, H., Li, T., Zhou, Y., Zhang, Z., Li, S., et al. (2009). A *Legionella* type IV effector activates the NF-κB pathway by phosphorylating the IκB family of inhibitors. *Proc. Natl. Acad. Sci. U.S.A.* 106, 13725–13730. doi: 10.1073/pnas.0907200106

Gelperin, D. M., White, M. A., Wilkinson, M. L., Kon, Y., Kung, L. A., Wise, K. J., et al. (2005). Biochemical and genetic analysis of the yeast proteome with a movable ORF collection. *Genes Dev.* 19, 2816–2826. doi: 10.1101/gad.1362105

Giaever, G., Nislow, C., Alonso, J. M., Stepanova, A. N., Leisse, T. J., Kim, C. J., et al. (2014). The yeast deletion collection: a decade of functional genomics. *Genetics* 197, 451–465. doi: 10.1534/genetics.114.161620

Gomez-Valero, L., Rusniok, C., Cazalet, C., and Buchrieser, C. (2011). Comparative and functional genomics of *Legionella* identified eukaryotic like proteins as key players in host-pathogen interactions. *Front. Microbiol.* 2, 1–20. doi: 10.3389/fmicb.2011.00208

Gomez-Valero, L., Rusniok, C., Rolando, M., Neou, M., Dervins-Ravault, D., Demirtas, J., et al. (2014). Comparative analyses of *Legionella* species identifies genetic features of strains causing Legionnaires' disease. *Genome Biol.* 15:505. doi: 10.1186/PREACCEPT-1086350395137407

Grammel, M., Luong, P., Orth, K., and Hang, H. C. (2011). A chemical reporter for protein AMPylation. *J. Am. Chem. Soc.* 133, 17103–17105. doi: 10.1021/ja205137d

Guo, Z., Stephenson, R., Qiu, J., Zheng, S., and Luo, Z. Q. (2014). A *Legionella* effector modulates host cytoskeletal structure by inhibiting actin polymerization. *Microbes Infect.* 16, 225–236. doi: 10.1016/j.micinf.2013.11.007

Harding, C. R., Mattheis, C., Mousnier, A., Oates, C. V., Hartland, E. L., Frankel, G., et al. (2013a). LtpD is a novel *Legionella pneumophila* effector that binds phosphatidylinositol 3-phosphate and inositol monophosphatase IMPA1. *Infect. Immun.* 81, 4261–4270. doi: 10.1128/IAI.01054-13

Harding, C. R., Schroeder, G. N., Reynolds, S., Kosta, A., Collins, J. W., Mousnier, A., et al. (2012). *Legionella pneumophila* pathogenesis in the *Galleria mellonella* infection model. *Infect. Immun.* 80, 2780–2790. doi: 10.1128/IAI.00510-12

Harding, C. R., Stoneham, C. A., Schuelein, R., Newton, H., Oates, C. V., Hartland, E. L., et al. (2013b). The Dot/Icm effector SdhA is necessary for virulence of *Legionella pneumophila* in *Galleria mellonella* and A/J mice. *Infect. Immun.* 81, 2598–2605. doi: 10.1128/IAI.00296-13

Hewings, D. S., Flygare, J. A., Bogyo, M., and Wertz, I. E. (2017). Activity-based probes for the ubiquitin conjugation–deconjugation machinery: new chemistries, new tools, and new insights. *FEBS J.* 284, 1555–1576. doi: 10.1111/febs.14039

Hoffmann, C., Finsel, I., Otto, A., Pfaffinger, G., Rothmeier, E., Hecker, M., et al. (2014a). Functional analysis of novel Rab GTPases identified in the proteome of purified *Legionella*-containing vacuoles from macrophages. *Cell. Microbiol.* 16, 1034–1052. doi: 10.1111/cmi.12256

Hoffmann, C., Harrison, C. F., and Hilbi, H. (2014b). The natural alternative: protozoa as cellular models for *Legionella* infection. *Cell. Microbiol.* 16, 15–26. doi: 10.1111/cmi.12235

Horwitz, M. A. (1983a). The Legionnaires' disease bacterium (*Legionella pneumophila*) inhibits phagosome-lysosome fusion in human monocytes. *J. Exp. Med.* 158, 2108–2126. doi: 10.1084/jem.158.6.2108

Horwitz, M. A. (1983b). Formation of a novel phagosome by the Legionnaires' disease bacterium (*Legionella pneumophila*) in human monocytes. *J. Exp. Med.* 158, 1391–1331. doi: 10.1084/jem.158.4.1319

Hsu, F., Zhu, W., Brennan, L., Tao, L., Luo, Z.-Q., and Mao, Y. (2012). Structural basis for substrate recognition by a unique *Legionella* phosphoinositide phosphatase. *Proc. Natl. Acad. Sci. U.S.A.* 109, 13567–13572. doi: 10.1073/pnas.1207903109

Huang, L., Boyd, D., Amyot, W. M., Hempstead, A. D., Luo, Z. Q., O'Connor, T. J., et al. (2011). The E Block motif is associated with *Legionella pneumophila* translocated substrates. *Cell. Microbiol.* 13, 227–245. doi: 10.1111/j.1462-5822.2010.01531.x

Hung, V., Udeshi, N. D., Lam, S. S., Loh, K. H., Cox, K. J., Pedram, K., et al. (2016). Spatially resolved proteomic mapping in living cells with the engineered peroxidase APEX2. *Nat. Protoc.* 11, 456–475. doi: 10.1038/nprot.2016.018

Husmann, L. K., and Johnson, W. (1992). Adherence of *Legionella pneumophila* to guinea pig peritoneal macrophages, J774 mouse macrophages, and undifferentiated U937 human monocytes: role of Fc and complement receptors. *Infect. Immun.* 60, 5212–5218.

Isaac, D. T., Laguna, R. K., Valtz, N., and Isberg, R. R. (2015). MavN is a *Legionella pneumophila* vacuole-associated protein required for efficient iron acquisition during intracellular growth. *Proc. Natl. Acad. Sci. U.S.A.* 112, E5208–E5217. doi: 10.1073/pnas.1511389112

Ivanov, S. S., Charron, G., Hang, H. C., and Roy, C. R. (2010). Lipidation by the host prenyltransferase machinery facilitates membrane localization of *Legionella pneumophila* effector proteins. *J. Biol. Chem.* 285, 34686–34698. doi: 10.1074/jbc.M110.170746

Ivanov, S. S., and Roy, C. R. (2013). Pathogen signatures activate a ubiquitination pathway that modulates the function of the metabolic checkpoint kinase mTOR. *Nat. Immunol.* 14, 1219–1228. doi: 10.1038/ni.2740

Jäger, J., Marwitz, S., Tiefenau, J., Rasch, J., Shevchuk, O., Kugler, C., et al. (2014). Human lung tissue explants reveal novel interactions during *Legionella pneumophila* infections. *Infect. Immun.* 82, 275–285. doi: 10.1128/IAI.00703-13

Jank, T., Böhmer, K. E., Tzivelekidis, T., Schwan, C., Belyi, Y., and Aktories, K. (2012). Domain organization of *Legionella* effector SetA. *Cell. Microbiol.* 14, 852–868. doi: 10.1111/j.1462-5822.2012.01761.x

Jeong, K. C., Ghosal, D., Chang, Y.-W., Jensen, G. J., and Vogel, J. P. (2017). Polar delivery of *Legionella* type IV secretion system substrates is essential for virulence. *Proc. Natl. Acad. Sci. U.S.A.* 114, 8077–8082. doi: 10.1073/pnas.1621438114

Jeong, K. C., Sutherland, M. C., and Vogel, J. P. (2015). Novel export control of a *Legionella* Dot/Icm substrate is mediated by dual, independent signal sequences. *Mol. Microbiol.* 96, 175–188. doi: 10.1111/mmi.12928

Kagan, J. C., and Roy, C. R. (2002). *Legionella* phagosomes intercept vesicular traffic from endoplasmic reticulum exit sites. *Nat. Cell Biol.* 4, 945–954. doi: 10.1038/ncb883

Klöpper, T. H., Kienle, N., Fasshauer, D., and Munro, S. (2012). Untangling the evolution of Rab G proteins: implications of a comprehensive genomic analysis. *BMC Biol.* 10:71. doi: 10.1186/1741-7007-10-71

Kotewicz, K. M., Ramabhadran, V., Sjoblom, N., Behringer, J., Scheck, R. A., Isberg, R. R., et al. (2017). A single *Legionella* effector catalyzes a multistep ubiquitination pathway to rearrange tubular endoplasmic reticulum for replication. *Cell Host Microbe* 21, 169–181. doi: 10.1016/j.chom.2016.12.007

Krause, K., and Amer, A. O. (2016). Caspase exploitation by *Legionella pneumophila*. *Front. Microbiol.* 7, 1–10. doi: 10.3389/fmicb.2016.00515

Kubori, T., Hyakutake, A., and Nagai, H. (2008). *Legionella* translocates an E3 ubiquitin ligase that has multiple U-boxes with distinct functions. *Mol. Microbiol.* 67, 1307–1319. doi: 10.1111/j.1365-2958.2008.06124.x

Kubori, T., Shinzawa, N., Kanuka, H., and Nagai, H. (2010). *Legionella* metaeffector exploits host proteasome to temporally regulate cognate effector. *PLoS Pathog.* 6, 1–8. doi: 10.1371/journal.ppat.1001216

Lifshitz, Z., Burstein, D., Peeri, M., Zusman, T., Schwartz, K., Shuman, H. A., et al. (2013). Computational modeling and experimental validation of the *Legionella* and *Coxiella* virulence-related type-IVB secretion signal. *Proc. Natl. Acad. Sci. U.S.A.* 110, E707–E715. doi: 10.1073/pnas.1215278110

Lin, Y. H., Doms, A. G., Cheng, E., Kim, B., Evans, T. R., and Machner, M. P. (2015). Host cell-catalyzed S-palmitoylation mediates Golgi targeting of the *Legionella* ubiquitin ligase GobX. *J. Biol. Chem.* 290, 25766–25781. doi: 10.1074/jbc.M115.637397

Liss, V., Barlag, B., Nietschke, M., and Hensel, M. (2016). Self-labelling enzymes as universal tags for fluorescence microscopy, super-resolution microscopy and electron microscopy. *Sci. Rep.* 5:17740. doi: 10.1038/srep17740

Liu, Y., Zhu, W., Tan, Y., Nakayasu, E. S., Staiger, C. J., and Luo, Z.-Q. (2017). A *Legionella* effector disrupts host cytoskeletal structure by cleaving actin. *PLOS Pathog.* 13:e1006186. doi: 10.1371/journal.ppat.1006186

Lomma, M., Dervins-Ravault, D., Rolando, M., Nora, T., Newton, H. J., Sansom, F. M., et al. (2010). The *Legionella pneumophila* F-box protein Lpp2082 (AnkB) modulates ubiquitination of the host protein parvin B and promotes intracellular replication. *Cell. Microbiol.* 12, 1272–1291. doi: 10.1111/j.1462-5822.2010.01467.x

Losick, V. P., Haenssler, E., Moy, M., and Isberg, R. R. (2010). LnaB: a *Legionella pneumophila* activator of NF-kB. *Cell. Microbiol.* 12, 1083–1097. doi: 10.1111/j.1462-5822.2010.01452.x

Lu, C., Liu, K., Tan, L. P., and Yao, S. Q. (2012). Current chemical biology tools for studying protein phosphorylation and dephosphorylation. *Chemistry* 18, 28–39. doi: 10.1002/chem.201103206

Luo, Z.-Q., and Isberg, R. R. (2004). Multiple substrates of the *Legionella pneumophila* Dot/Icm system identified by interbacterial protein transfer. *Proc. Natl. Acad. Sci. U.S.A.* 101, 841–846. doi: 10.1073/pnas.0304916101

Machner, M. P., and Isberg, R. R. (2006). Targeting of host Rab GTPase function by the intravacuolar pathogen *Legionella pneumophila*. *Dev. Cell* 11, 47–56. doi: 10.1016/j.devcel.2006.05.013

Marra, A., Horwitz, M. A., and Shuman, H. A. (1990). The HL-60 model for the interaction of human macrophages with the Legionnaires' disease bacterium. *J. Immunol.* 144, 2738–2744.

Matsunaga, K., Klein, T. W., Friedman, H., and Yamamoto, Y. (2001). Alveolar macrophage cell line MH-S is valuable as an *in vitro* model for *Legionella pneumophila* infection. *Am. J. Respir. Cell Mol. Biol.* 24, 326–331. doi: 10.1165/ajrcmb.24.3.4359

McCusker, K. T., Braaten, B. A., Cho, M. W., and Low, D. A. (1991). *Legionella pneumophila* inhibits protein synthesis in Chinese hamster ovary cells. *Infect. Immun.* 59, 240–246.

McDade, J. E., Shepard, C. C., Fraser, D. W., Tsai, T. R., Redus, M. A., and Dowdle, W. R. (1977). Legionnaires' disease: isolation of a bacterium and demonstration of its role in other respiratory disease. *N. Engl. J. Med.* 297, 1197–1203. doi: 10.1056/NEJM197712012972202

Merriam, J. J., Mathur, R., Maxfield-Boumil, R., and Isberg, R. R. (1997). Analysis of the *Legionella pneumophila* fliI gene: intracellular growth of a defined mutant defective for flagellum biosynthesis. *Infect. Immun.* 65, 2497–2501.

Meyer, D. F., Noroy, C., Moumène, A., Raffaele, S., Albina, E., and Vachiéry, N. (2013). Searching algorithm for type IV secretion system effectors 1.0: a tool for predicting type IV effectors and exploring their genomic context. *Nucleic Acids Res.* 41, 9218–9229. doi: 10.1093/nar/gkt718

Michard, C., and Doublet, P. (2015). Post-translational modifications are key players of the *Legionella pneumophila* infection strategy. *Front. Microbiol.* 6:00087. doi: 10.3389/fmicb.2015.00087

Michard, C., Sperandio, D., Baïlo, N., Pizarro-Cerdá, J., LeClaire, L., Chadeau-Argaud, E., et al. (2015). The *Legionella* kinase Legk2 targets the Arp2/3 complex to inhibit actin nucleation on phagosomes and allow bacterial evasion of the late endocytic pathway. *MBio* 6, 1–14. doi: 10.1128/mBio.00354-15

Moffat, J. F., and Tompkins, L. S. (1992). A quantitative model of intracellular growth of *Legionella pneumophila* in *Acanthamoeba castellanii*. *Infect. Immun.* 60, 296–301.

Mousnier, A., Schroeder, G. N., Stoneham, C. A., So, E. C., Garnett, J. A., Yu, L., et al. (2014). A new method to determine *in vivo* interactomes reveals binding of the *Legionella pneumophila* effector PieE to multiple Rab GTPases. *MBio* 5:e01148–14. doi: 10.1128/mBio.01148-14

Mukherjee, S., Liu, X., Arasaki, K., McDonough, J., Galan, J. E., and Roy, C. R. (2011). Modulation of Rab GTPase function by a protein phosphocholine transferase. *Nature* 477, 103–106. doi: 10.1038/nature10335

Nagai, H., Cambronne, E. D., Kagan, J. C., Amor, J. C., Kahn, R. A., and Roy, C. R. (2005). A C-terminal translocation signal required for Dot/Icm-dependent delivery of the *Legionella* RalF protein to host cells. *Proc. Natl. Acad. Sci. U.S.A.* 102, 826–831. doi: 10.1073/pnas.0406239101

Nagai, H., Kagan, J. C., Zhu, X., Kahn, R. A., and Roy, C. R. (2002). A bacterial guanine nucleotide exchange factor activates ARF on *Legionella* phagosomes. *Science* 295, 679. doi: 10.1126/science.1067025

Naujoks, J., Tabeling, C., Dill, B. D., Hoffmann, C., Brown, A. S., Kunze, M., et al. (2016). IFNs modify the proteome of *Legionella*-containing vacuoles and restrict infection via IRG1-derived itaconic acid. *PLoS Pathog.* 12:e1005408. doi: 10.1371/journal.ppat.1005408

Neunuebel, M. R., Chen, Y., Gaspar, A. H., Backlund, P. S., Yergey, A., Machner, M. P., et al. (2011). De-AMPylation of the small GTPase Rab1 by the pathogen *Legionella pneumophila*. *Science* 333, 453–456. doi: 10.1126/science.1207193

Newsome, A. L., Baker, R. L., Miller, R. D., and Arnold, R. R. (1985). Interactions between *Naegleria fowleri* and *Legionella pneumophila*. *Infect. Immun.* 50, 449–452.

O'Connor, T. J., Adepoju, Y., Boyd, D., and Isberg, R. R. (2011). Minimization of the *Legionella pneumophila* genome reveals chromosomal regions involved in host range expansion. *Proc. Natl. Acad. Sci. U.S.A.* 108, 14733–14740. doi: 10.1073/pnas.1111678108

O'Connor, T. J., Boyd, D., Dorer, M. S., and Isberg, R. R. (2012). Aggravating genetic interactions allow a solution to redundancy in a bacterial pathogen. *Science* 338, 1440–1444. doi: 10.1126/science.1229556

O'Connor, T. J., and Isberg, R. R. (2014). iMAD, a genetic screening strategy for dissecting complex interactions between a pathogen and its host. *Nat. Protoc.* 9, 1916–1930. doi: 10.1038/nprot.2014.133

Ott, M. (1994). Genetic approaches to study *Legionella pneumophila* pathogenicity. *FEMS Microbiol. Rev.* 14, 161–176. doi: 10.1111/j.1574-6976.1994.tb00085.x

Pearlman, E., Jiwa, A. H., Engleberg, N. C., and Eisenstein, B. I. (1988). Growth of *Legionella pneumophila* in a human macrophage-like (U937) cell line. *Microb. Pathog.* 5, 87–95. doi: 10.1016/0882-4010(88)90011-3

Popa, C., Coll, N. S., Valls, M., and Sessa, G. (2016). Yeast as a heterologous model system to uncover Type III effector function. *PLoS Pathog.* 12:e1005360. doi: 10.1371/journal.ppat.1005360

Pope, C. D., Dhand, L., and Cianciotto, N. P. (1994). Random mutagenesis of *Legionella pneumophila* with mini-Tn10. *FEMS Microbiol. Lett.* 124, 107–111. doi: 10.1111/j.1574-6968.1994.tb07269.x

Price, C. T., Al-Khodor, S., Al-Quadan, T., Santic, M., Habyarimana, F., Kalia, A., et al. (2009). Molecular mimicry by an F-box effector of *Legionella pneumophila* hijacks a conserved polyubiquitination machinery within macrophages and protozoa. *PLoS Pathog.* 5:e1000704. doi: 10.1371/journal.ppat.1000704

Qiu, J., and Luo, Z. (2017). *Legionella* and *Coxiella* effectors: strength in diversity and activity. *Nat. Rev. Micro.* 15, 591–605. doi: 10.1038/nrmicro.2017.67

Qiu, J., Sheedlo, M. J., Yu, K., Tan, Y., Nakayasu, E. S., Das, C., et al. (2016). Ubiquitination independent of E1 and E2 enzymes by bacterial effectors. *Nature* 533, 120–124. doi: 10.1038/nature17657

Rolando, M., Sanulli, S., Rusniok, C., Gomez-Valero, L., Bertholet, C., Sahr, T., et al. (2013). *Legionella pneumophila* effector RomA uniquely modifies host chromatin to repress gene expression and promote intracellular bacterial replication. *Cell Host Microbe* 13, 395–405. doi: 10.1016/j.chom.2013.03.004

Rothmeier, E., Pfaffinger, G., Hoffmann, C., Harrison, C. F., Grabmayr, H., Repnik, U., et al. (2013). Activation of Ran GTPase by a *Legionella* effector promotes microtubule polymerization, pathogen vacuole motility and infection. *PLoS Pathog.* 9:e1003598. doi: 10.1371/journal.ppat.1003598

Roux, K. J., Kim, D. I., Raida, M., and Burke, B. (2012). A promiscuous biotin ligase fusion protein identifies proximal and interacting proteins in mammalian cells. *J. Cell Biol.* 196, 801–810. doi: 10.1083/jcb.201112098

Rowbotham, T. J. (1980). Preliminary report on the pathogenicity of *Legionella pneumophila* for freshwater and soil amoebae. *J. Clin. Pathol.* 33, 1179–1183. doi: 10.1136/jcp.33.12.1179

Rucks, E. A., Olson, M. G., Jorgenson, L. M., Srinivasan, R. R., and Ouellette, S. P. (2017). Development of a proximity labeling system to map the *Chlamydia trachomatis* inclusion membrane. *Front. Cell. Infect. Microbiol.* 7:40. doi: 10.3389/fcimb.2017.00040

Schroeder, G. N., Aurass, P., Oates, C. V., Tate, E. W., Hartland, E. L., Flieger, A., et al. (2015). *Legionella pneumophila* effector LpdA Is a palmitoylated phospholipase D virulence factor. *Infect. Immun.* 83, 3989–4002. doi: 10.1128/IAI.00785-15

Schroeder, G. N., Petty, N. K., Mousnier, A., Harding, C. R., Vogrin, A. J., Wee, B., et al. (2010). *Legionella pneumophila* strain 130b possesses a unique combination of type IV secretion systems and novel Dot/Icm secretion system effector proteins. *J. Bacteriol.* 192, 6001–6016. doi: 10.1128/JB.00778-10

Segal, G., Purcell, M., and Shuman, H. A. (1998). Host cell killing and bacterial conjugation require overlapping sets of genes within a 22-kb region of the *Legionella pneumophila* genome. *Proc. Natl. Acad. Sci. U.S.A.* 95, 1669–1674. doi: 10.1073/pnas.95.4.1669

Shohdy, N., Efe, J. A., Emr, S. D., and Shuman, H., a (2005). Pathogen effector protein screening in yeast identifies *Legionella* factors that interfere with membrane trafficking. *Proc. Natl. Acad. Sci. U.S.A.* 102, 4866–4871. doi: 10.1073/pnas.0501315102

So, E. C., Mattheis, C., Tate, E. W., Frankel, G., and Schroeder, G. N. (2015). Creating a customized intracellular niche: subversion of host cell signaling by *Legionella* type IV secretion system effectors. *Can. J. Microbiol.* 635, 617–635. doi: 10.1139/cjm-2015-0166

So, E. C., Schroeder, G. N., Carson, D., Mattheis, C., Mousnier, A., Broncel, M., et al. (2016). The Rab-binding profiles of bacterial virulence factors during infection. *J. Biol. Chem.* 291, 5832–5843. doi: 10.1074/jbc.M115.700930

Solomon, J. M., and Isberg, R. R. (2000). Growth of *Legionella pneumophila* in *Dictyostelium discoideum*: a novel system for genetic analysis of host-pathogen interactions. *Trends Microbiol.* 8, 478–480. doi: 10.1016/S0966-842X(00)01852-7

Sopko, R., Huang, D., Preston, N., Chua, G., Papp, B., Kafadar, K., et al. (2006). Mapping pathways and phenotypes by systematic gene overexpression. *Mol. Cell* 21, 319–330. doi: 10.1016/j.molcel.2005.12.011

Sun, E. W., Wagner, M. L., Maize, A., Kemler, D., Garland-Kuntz, E., Xu, L., et al. (2013). *Legionella pneumophila* infection of *Drosophila* S2 cells induces only minor changes in mitochondrial dynamics. *PLoS ONE* 8:e62972. doi: 10.1371/journal.pone.0062972

Tan, Y., Arnold, R. J., and Luo, Z.-Q. (2011). *Legionella pneumophila* regulates the small GTPase Rab1 activity by reversible phosphorylcholination. *Proc. Natl. Acad. Sci. U.S.A.* 108, 21212–21217. doi: 10.1073/pnas.11140 23109

Tan, Y., and Luo, Z.-Q. (2011). *Legionella pneumophila* SidD is a deAMPylase that modifies Rab1. *Nature* 475, 506–509. doi: 10.1038/nature10307

Tiaden, A. N., Kessler, A., and Hilbi, H. (2013). Analysis of *Legionella* infection by flow cytometry. *Methods Mol. Biol.* 954, 233–249. doi: 10.1007/978-1-62703-161-5_14

Toulabi, L., Wu, X., Cheng, Y., and Mao, Y. (2013). Identification and structural characterization of a *Legionella* phosphoinositide phosphatase. *J. Biol. Chem.* 288, 24518–24527. doi: 10.1074/jbc.M113.474239

Urbanus, M. L., Quaile, A. T., Stogios, P. J., Morar, M., Rao, C., Di, R., et al. (2016). Diverse mechanisms of metaeffector activity in an intracellular bacterial pathogen, *Legionella pneumophila*. *Mol. Syst. Biol.* 12:893. doi: 10.15252/msb.20167381

Van Engelenburg, S. B., and Palmer, A. E. (2010). Imaging type-III secretion reveals dynamics and spatial segregation of *Salmonella* effectors. *Nat. Methods* 7, 325–330. doi: 10.1038/nmeth.1437

Vanrheenen, S. M., Luo, Z. Q., O'Connor, T., and Isberg, R. R. (2006). Members of a *Legionella pneumophila* family of proteins with ExoU (Phospholipase A) active sites are translocated to target cells. *Infect. Immun.* 74, 3597–3606. doi: 10.1128/IAI.02060-05

Viner, R., Chetrit, D., Ehrlich, M., and Segal, G. (2012). Identification of two *Legionella pneumophila* effectors that manipulate host phospholipids biosynthesis. *PLoS Pathog.* 8:e1002988. doi: 10.1371/journal.ppat. 1002988

Walsh, N. C., Kenney, L. L., Jangalwe, S., Aryee, K.-E., Greiner, D. L., Brehm, M. A., et al. (2017). Humanized mouse models of clinical disease. *Annu. Rev. Pathol. Mech. Dis.* 12, 187–215. doi: 10.1146/annurev-pathol-052016-100332

Watson, J. D., Laskowski, R. A., and Thornton, J. M. (2005). Predicting protein function from sequence and structural data. *Curr. Opin. Struct. Biol.* 15, 275–284. doi: 10.1016/j.sbi.2005.04.003

Weber, S. S., Ragaz, C., Reus, K., Nyfeler, Y., and Hilbi, H. (2006). *Legionella pneumophila* exploits PI(4)P to anchor secreted effector proteins to the replicative vacuole. *PLoS Pathog.* 2:e46. doi: 10.1371/journal.ppat.0020046

Willems, L. I., Overkleeft, H. S., and Van Kasteren, S. I. (2014). Current developments in activity-based protein profiling. *Bioconjug. Chem.* 25, 1181–1191. doi: 10.1021/bc500208y

Wölke, S., Ackermann, N., and Heesemann, J. (2011). The *Yersinia enterocolitica* type 3 secretion system (T3SS) as toolbox for studying the cell biological effects of bacterial Rho GTPase modulating T3SS effector proteins. *Cell. Microbiol.* 13, 1339–1357. doi: 10.1111/j.1462-5822.2011.01623.x

Yu, X., Decker, K. B., Barker, K., Neunuebel, M. R., Saul, J., Westcott, N., et al. (2015). Host-Pathogen interaction profiling using self-assembling human protein arrays. *J. Proteome Res.* 14, 1920–1936. doi: 10.1021/pr50 13015

Zhu, W., Banga, S., Tan, Y., Zheng, C., Stephenson, R., Gately, J., et al. (2011). Comprehensive identification of protein substrates of the Dot/Icm type IV transporter of *Legionella pneumophila*. *PLoS ONE* 6:e17638. doi: 10.1371/journal.pone.0017638

Zhu, W., Hammad, L. A., Hsu, F., Mao, Y., and Luo, Z.-Q. (2013). Induction of caspase 3 activation by multiple *Legionella pneumophila* Dot/Icm substrates. *Cell. Microbiol.* 15, 1783–1795. doi: 10.1111/cmi.12157

Zou, L., Nan, C., Hu, F., and Hancock, J. (2013). Accurate prediction of bacterial type IV secreted effectors using amino acid composition and PSSM profiles. *Bioinformatics* 29, 3135–3142. doi: 10.1093/bioinformatics/btt554

13

PilY1 Promotes *Legionella pneumophila* Infection of Human Lung Tissue Explants and Contributes to Bacterial Adhesion, Host Cell Invasion, and Twitching Motility

Julia Hoppe[1], *Can M. Ünal*[1], *Stefanie Thiem*[1], *Louisa Grimpe*[1], *Torsten Goldmann*[2,3], *Nikolaus Gaßler*[4], *Matthias Richter*[5], *Olga Shevchuk*[6]* *and Michael Steinert*[1,7]*

[1] Institut für Mikrobiologie, Technische Universität Braunschweig, Braunschweig, Germany, [2] Pathology of the University Hospital of Lübeck and the Leibniz Research Center, Borstel, Germany, [3] Airway Research Center North (ARCN), Member of the German Center for Lung Research, Borstel, Germany, [4] Institut für Pathologie, Klinikum Braunschweig, Braunschweig, Germany, [5] Klinikum Salzdahlumerstraße Braunschweig, Braunschweig, Germany, [6] Center for Proteomics, University of Rijeka, Rijeka, Croatia, [7] Helmholtz Center for Infection Research, Braunschweig, Germany

Edited by:
Daniel E. Voth,
University of Arkansas for Medical Sciences, USA

Reviewed by:
Hayley J. Newton,
University of Melbourne, Australia
Vincent Joseph Starai,
University of Georgia, USA
Sunny Shin,
University of Pennsylvania, USA

***Correspondence:**
Olga Shevchuk
Olga.Shevchuk@helmholtz-hzi.de
Michael Steinert
m.steinert@tu-bs.de

Legionnaires' disease is an acute fibrinopurulent pneumonia. During infection *Legionella pneumophila* adheres to the alveolar lining and replicates intracellularly within recruited macrophages. Here we provide a sequence and domain composition analysis of the *L. pneumophila* PilY1 protein, which has a high homology to PilY1 of *Pseudomonas aeruginosa*. PilY1 proteins of both pathogens contain a von Willebrand factor A (vWFa) and a C-terminal PilY domain. Using cellular fractionation, we assigned the *L. pneumophila* PilY1 as an outer membrane protein that is only expressed during the transmissive stationary growth phase. PilY1 contributes to infection of human lung tissue explants (HLTEs). A detailed analysis using THP-1 macrophages and A549 lung epithelial cells revealed that this contribution is due to multiple effects depending on host cell type. Deletion of PilY1 resulted in a lower replication rate in THP-1 macrophages but not in A549 cells. Further on, adhesion to THP-1 macrophages and A549 epithelial cells was decreased. Additionally, the invasion into non-phagocytic A549 epithelial cells was drastically reduced when PilY1 was absent. Complementation variants of a PilY1-negative mutant revealed that the C-terminal PilY domain is essential for restoring the wild type phenotype in adhesion, while the putatively mechanosensitive vWFa domain facilitates invasion into non-phagocytic cells. Since PilY1 also promotes twitching motility of *L. pneumophila,* we discuss the putative contribution of this newly described virulence factor for bacterial dissemination within infected lung tissue.

Keywords: ***L. pneumophila*, PilY1, human lung tissue explants, adherence, invasion, twitching motility**

INTRODUCTION

Legionella pneumophila is the causative agent of the Legionnaires' disease, a severe form of pneumonia (Fraser et al., 1977; McDade et al., 1977; Fields et al., 2002). Upon transmission to the respiratory tract through aerosols containing *Legionella*, the bacteria enter and replicate within alveolar macrophages and epithelial cells (Horwitz and Silverstein, 1980; Mody et al., 1993; Jäger et al., 2014). The cellular internalization of *L. pneumophila* can be enhanced by the presence of antibodies and complement. The major outer membrane protein (MOMP) of *L. pneumophila* binds complement component C3 and C3i and mediates the uptake of the bacteria via the complement receptors CR1 and CR3 of macrophages. Phagocytosed *L. pneumophila* recruit compartments from the endoplasmatic reticulum (ER), modulate the host phosphoinositide metabolism, modify the host endocytic pathway, intercept vesicle trafficking and avoid fusion with lysosomes (Shevchuk et al., 2014). The development of the *Legionella*-containing vacuole (LCV) is pre-dominantly orchestrated by the Dot/Icm type IV secretion system, which is known to export numerous bacterial effector proteins into the host cell cytoplasm (Horwitz, 1983; Kagan and Roy, 2002; Isberg et al., 2009; Hubber and Roy, 2010). Bacterial attachment and entry into host cells are considered to be a pre-requisite for this pathogen-directed host cell modulation (Roy et al., 1998; Wiater et al., 1998; Hilbi et al., 2001; Charpentier et al., 2009). Several determinants of *L. pneumophila* are known to contribute to adherence and entry into different host cell types, including type IV pili, Hsp60, the structural toxin RtxA, the intergrin analog LaiA and the GAG binding protein Lcl and the adenylate cyclase LadC (Garduño et al., 1998; Stone and Abu Kwaik, 1998; Cirillo et al., 2001; Chang et al., 2005; Newton et al., 2008; Duncan et al., 2011).

In a previous study, we screened a *L. pneumophila* mini-Tn10 transposon library for mutants that fail to avoid fusion of their respective LCV with lysosomes (Shevchuk et al., 2014). The range of the identified mutants indicated that interference with lysosomal degradation is multifactorial. Several mutants with different insertions in the Lpc2666 gene exhibited significantly higher co-localization with lysosomal compartments and reduced replication rates in macrophages and protozoa (Shevchuk et al., 2014). The sequence analysis revealed that Lpc2666 encodes for a type IV fimbrial biogenesis PilY1-like protein that shares homology with the C-terminal domain of PilY1 of *Pseudomonas aeruginosa* and the PilC1/2 of *Neisseria gonorrhoeae, Neisseria meningitides,* and *Kingella kingae.* The PilC proteins of *K. kingae* and *Neisseria* species have already been characterized as type IV pili biogenesis factors and are known to be involved in adherence to epithelial cells (Rudel et al., 1995a,b; Scheuerpflug et al., 1999; Porsch et al., 2013). The PilY1 of *P. aeruginosa* has also been shown to be essential for type IV pilus assembly and evidently contributes to cell adhesion and virulence (Bohn et al., 2009; Heiniger et al., 2010). In addition, it has been demonstrated that the PilY- or PilC-like proteins are required for pilus stability. Accordingly, mutations in the respective genes result in the loss of the type IV pilus dependent twitching motility. Moreover, *P. aeruginosa* PilY1 participates in the regulation of a type IV pilus independent motility (Wolfgang et al., 1998; Morand et al., 2004; Bohn et al., 2009; Kuchma et al., 2010; Porsch et al., 2013).

In the present study, we analyzed the sequence and domain composition of the *L. pneumophila* PilY1. We ascertained PilY1 as an outer membrane protein that is expressed during the stationary growth phase of the bacteria. Since the *L. pneumophila* PilY1 knockout mutant exhibited defects in twitching motility as well as in host cell adherence, invasion and intracellular replication, we hypothesize that PilY1 mediates extra- and intracellular virulence mechanisms which are required for the efficient infection of human lung tissue explants (HLTEs).

MATERIALS AND METHODS

Cultivation of Bacteria and Eukaryotic Cells

L. pneumophila Corby strains and mutants were routinely cultured on buffered charcoal-yeast extract (BCYE) agar for 3–5 days. Liquid cultures were inoculated in buffered yeast extract (YEB) medium and grown at 37°C with agitation at 180 rpm to an OD600 of 3.0 with 12.5 μg/ml chloramphenicol and 500 μM IPTG or 20 μg/ml kanamycin if required. Human alveolar epithelial A549 cells (DSMZ, ACC-107) were grown in DMEM and the human monocyte cell line THP-1 (DSMZ, ACC-16) in RPMI 1640 medium, both supplemented with 10% FCS and 4.5 mM glutamine at 37°C and 5% CO_2. For the experiments, the A549 cells were seeded 18 h before infection into 24-well tissue culture plates (GreinerBioOne) at a density of 5×10^5 cells/well. The THP-1 cells were seeded in 96-well tissue culture plates (TTP) at a density of 10^5 cells/well and differentiated with phorbol myristate acetate (PMA; Sigma) to a final concentration of 100 nM for 48 h.

Site-Directed Mutagenesis of *pilY1* and Construction of Complementation Strains

The *L. pneumophila* Corby *pilY1* knockout mutant was generated by allelic exchange of the chromosomal *pilY1* gene for a kanamycin resistance cassette. For this purpose, a PCR fragment carrying the antibiotic resistance gene flanked by a 1 kb homologous region to the target locus was generated by a three-step amplification procedure (Derbise et al., 2003). In the first step, the flanking upstream and downstream regions of the *pilY1* gene and the kanamycin cassette were amplified independently. The primers PilYupR and PilYdownF contain an extension of 20 nucleotides homologous to the kanamycin resistance cassette. In the second step, the resulting three PCR products were mixed at equimolar concentrations and subjected to a second PCR to generate the *pilY1* knockout construct, which was ligated into the pGEM®-T Easy vector. The third step was required to obtain large amounts of the desired DNA. The DNA was introduced into *L. pneumophila* by natural transformation as described previously with modifications (Sexton and Vogel, 2004; Schunder et al., 2010). Briefly, 1 ml of an exponentially grown overnight culture was incubated with 1 μg of the PCR product for 3 days at 30°C without agitation. Subsequently, bacteria were grown on antibiotic selective media for 4 more days at 37°C

TABLE 1 | Plasmids and Primers used in this study.

Plasmid	Description	References
pXDC WTPilY	pXDC61-derived vector expressing PilY1	This study
pXDC ΔPilYdom	pXDC61-derived vector expressing PilY1 lacking the PilY domain (Δaa 600–1169); Cm^r	This study
pXDC ΔvWFa	pXDC61-derived vector expressing PilY1 lacking the vWFa domain (Δaa 405–526); Cm^r	This study
pXDC61	pMMB207C-derived vector; Cm^r	Charpentier et al., 2009
pGEM-T	Shuttle vector	Promega
pGEM-T PilYKm	Vector for knockout of pilY gene; Km^r	This study

Primer	5′–3′ Sequence	References
PilYup F	GTTGAATATGGCTCATAGCGTCCATGATAATCAAAACC	This study
PilYup R	CTCAAACCCAACTTTACAAAGC	This study
PilYdown F	TAAAAACCTTGCAGGAATACGG	This study
PilYdown R	TTGTAACACTGGCAGAGGCAAATTCAATAGAGGATACCC	This study
Km F	TCTGCCAGTGTTACAACCAATT	This study
Km R	ATGAGCCATATTCAACGGGA	This study
PilYconf F	GGCAGATTAATTGTAATGTCAGTGT	This study
PilYconf R	CCAGGATTTTCATTAGTCGAGTTAAT	This study
PilYcom F *XbaI*	GCGCGTCTAGATTGAATTTGCCACCAGCC	This study
PilYcom	GCGCGGAATTCGTTTTGATTATCATGGACGC	This study
R *EcoRI*	GCGCGGAATTCTATCATGGACGCTTTAAC	This study
PilYvWF F *XbaI*	TGGTCAACTAACCGAAAG	This study
PilYdom R		
PilY-vWF R	CGGTTAGTTGACCATTGCGGACAATTGCCACT	This study
SP6	TATTTAGGTGACACTATAG	Promega
T7	TAATACGACTCACTATAGGG	Promega
pXDC F	TTGACAATTAATCATCGGC	This study
pXDC R	CTGTATCAGGCTGAAAATC	This study

with 5% CO_2. Screening for mutants obtained by homologous recombination was performed by PCR using the primer pairs PilYconfF/R. For complementation of the *pilY1* mutant, the full length *pilY1* PilY and vWFa domains were amplified using the primers pilYcomF/R, PilYvWFF/comR, and PilY-vWFF/comR/domR. The resulting PCR fragments were digested with *XbaI* and *EcoRI* and cloned into the pXDC/pMMB207C vector at the position of the *blaM* orf (Charpentier et al., 2009) and electroporated into the strain *L. pneumophila* Corby Δ*pilY1*. For a complete list of plasmids and primers prefer to **Table 1**. Complementation of the strains on protein level, and the correct localization on the surface of the bacteria were confirmed by western blot using polyclonal rabbit α-PilY1 antibodies directed against a central peptide of 12 amino acids (Supplementary Information and Supplementary Figures 1 and 4).

Cellular Localization of PilY1 and Its Variants Using Triton X-100 Solubility

Cell fractionation was performed as previously described with modifications (Vincent et al., 2006; Kuchma et al., 2007, 2010). Cells were resuspended in 50 mM Tris-HCl [pH 8] with lysozyme (0.2 mg/ml^{-1}) and DNAse (1 μg/ml) and lysed in a French pressure cell (14,000 PSI). Extracts were then centrifuged 10 min at 10,000 g to remove unlysed cells and obtain whole-cell lysates. To separate the soluble cytoplasmic fraction from the total membrane (TM) fraction, whole-cell lysates were centrifuged at 100,000 g for 1 h at 4°C. The membrane pellet was resuspended in 50 mM Tris-HCl [pH 8] to yield the TM fraction. The inner membranes were solubilized by the addition of 20 mM $MgSO_4$ and 1% Triton X-100; the outer membranes were collected by additional centrifugation at 100.000 g for 1 h. To yield the outer-membrane fraction, pellets were resuspended in 50 mM Tris-HCl [pH 8] and SDS loading buffer. Protein concentrations of each fraction were determined using a Roti®-Nanoquant protein quantification assay (Roth) according to the manufacturer's instructions. For each cellular fraction, 25 μg of total protein were immunoblotted with antibodies which recognize the PilY protein, and detected by using a horseradish conjugated secondary antibody (Dianova) and the Amersham ECL Western Blotting Detection Reagent. The rabbit polyclonal PilY antibody was generated against a peptide corresponding to the PilY 359–370 aa (Eurogentec). The integrity of the cellular fractions was confirmed using the antibodies which recognize the inner membrane type I signal peptidase LepB and the outer membrane protein MOMP (Vincent and Vogel, 2006).

Alkaline phosphatase-conjugated antibodies (Thermo Scientific) were used for detection.

PilY1 Expression Analysis

To determine PilY expression, the *L. pneumophila* wild type strain was grown in YEB broth at 37°C with agitation at 180 rpm. PilY1 expression was tested with 10^9 bacteria respectively from the exponential, early stationary, stationary and late stationary growth phases. The bacteria were lysed with 7 M Urea, 20 mM Tris-HCl [pH 9], 100 mM DTT and 1% Triton X-100. Equal cell amounts were immunoblotted with antibodies that recognize the PilY protein and detected as described above. Western Blot analysis was performed using the antibody for FlaA as a marker for the stationary growth phase expression, and Mip as control for equal cell amounts (Helbig et al., 1995; Heuner et al., 1995). Alkaline phosphatase conjugated antibody (Thermo Scientific) was used for detection.

Infection of Human Lung Tissue Explants

For infection, tumor-free pulmonary tissue samples, which had been obtained from surgery patients, were infected with 10^7 bacteria/ml of early stationary phase strains of *L. pneumophila.* The tissue samples were incubated in RPMI 1640 with 10% FCS, 20 mM HEPES, and 1 mM sodium pyruvate at 37°C and 5% CO_2 for up to 48 h as described previously (Jäger et al., 2014). For the determination of the CFU per g tissue, samples from six donors were infected. At the indicated time points after infection, samples were weighed, homogenized and dilutions were plated on BCYE agar and incubated at 37°C with 5% CO_2 for 4 days.

Infection Assays with THP-1 Macrophages and A549 Epithelial Cells

The human monocytic cell line THP-1 and the human alveolar epithelial cell line A549 were pre-treated as described above and infected with *L. pneumophila* strains from the early stationary phase with a MOI 1 (multiplicity of infection) for THP-1 cells and a MOI 100 for A549 cells for 2 h in 5% CO_2 at 37°C. Cells were then washed with PBS to remove extracellular bacteria. The A549 cells were additionally treated with 100 μg/ml gentamicin for 1 h. At indicated time points following infection, cells were lysed with 0.01% Triton X-100, serial dilutions were plated on BCYE agar and incubated at 37°C with 5% CO_2 for 4 days in order to determine the CFU/ml.

Adhesion and Invasion Assay

The experiments were performed with confluent A549 cells and differentiated THP-1 macrophages as described above. Stationary phase *L. pneumophila* strains were added to THP-1 macrophages (MOI 20) and A549 cells (MOI 100). The infections were synchronized by centrifugation at 300 g for 5 min. For the adhesion assay, the cells were pre-incubated with 10 μM cytochalasin B for 1 h to prevent phagocytosis. After 30 min of co-incubation with THP-1 macrophages and after 1 h co-incubation with A549 cells, the monolayers were washed twice with PBS to remove non-adherent bacteria and lysed with 0.01% Triton X-100. Serial dilutions of the inoculum and of bacteria recovered from lysed cells were plated on BCYE agar. Results are expressed as the ratio of adherent bacteria compared to the inoculum. For the invasion assay a differential staining protocol was performed. For this, the cells were infected for 1 h (THP-1) or 2 h (A549) with bacteria that were labeled with rhodamine. Unbound bacteria were removed by two washes of PBS, and the cells were fixed with 4% PFA over night at 4°C. Following fixation, PFA was removed by three washes with PBS containing 50 mM glycin, and the samples were blocked using 10% NHS in SorC-buffer. The extracellular bacteria were additionally stained using a polyclonal rabbit α-*L. pneumophila* antibody (ABIN23774) followed by an Alexa Fluor® 488-coupled goat α-rabbit antibody (Abcam, ab15007). Mounted samples were analyzed using a Leica SP8 confocal laser scanning microscope. For each strain, at least 50 cells with associated bacteria were analyzed in three independent experiments, and the ratio of internalized bacteria were calculated. Rhodamine labeling of bacteria was performed as described previously (Shevchuk et al., 2014).

Motility Assays

Surface motility was monitored as previously described (Coil and Anné, 2009; Stewart et al., 2009). Briefly, 10 μl of early stationary phase cultures of *Legionella* strains were dropped onto fresh BCYE plates containing 1.6% (twitching motility) or 0.5% (sliding motility) agar. The inoculated plates were incubated at 30°C and growth was observed for the next 14 days. Plates were photographed, and the area in which migration had occurred was calculated with Adobe Photoshop 7.0.1.

Statistical Analysis

All experiments were performed in duplicate and repeated at least three times. Statistical analysis was performed using Student's *t*-test in GraphPad Prism 5.0. Differences were considered significant at a $p \leq 0.05$.

RESULTS

Sequence and Domain Composition Analyses of PilY1

Given the importance of PilY and PilC proteins in type IV pilus biogenesis and stability, including pilus extension and retraction in several bacterial pathogens (Orans et al., 2010; Johnson et al., 2011; Cheng et al., 2013; Porsch et al., 2013), we performed a sequence-based domain annotation of the *L. pneumophila* PilY1 protein. Based on sequence homology with *P. aeruginosa* PilY1, *N. gonorrhoeae, N. meningitides,* and *K. kingae* PilC, we predicted the PilY1 structure of *L. pneumophila* as depicted in **Figure 1.** At the N-terminus the PilY1 protein contains a transmembrane domain with a putative signal peptide cleavage site, similar to the Sec secretion signal peptide of *P. aeruginosa* PilY1 (Lewenza et al., 2005; Kuchma et al., 2010). Moreover, a 200 amino acid N-terminal region with homology to the von Willebrand factor A (vWFa) domain was identified, which is also present in the PilY1

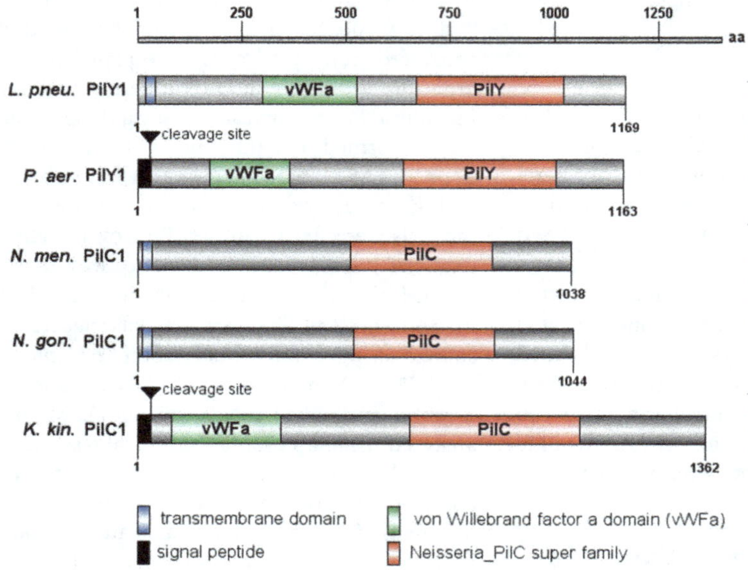

overall	L. p.	P. a.	N. m.	N. g.	K. k.
L. p.	100	30,1	20,6	20,8	19,8
P. a.		100	20	21	21,5
N.m.			100	72,2	30,1
N. g.				100	28,8
K. k.					100

vWFa	L. p.	P. a.	N. m.	N. g.	K. k.
L. p.	100	18	n.a.	n.a.	23,8
P. a.		100	n.a.	n.a.	15,3
N.m.			n.a.	n.a.	n.a.
N. g.				n.a.	n.a.
K. k.					100

Pildom	L. p.	P. a.	N. m.	N. g.	K. k.
L. p.	100	43,6	30,5	30,5	25,7
P. a.		100	31,8	32,1	29,6
N.m.			100	67,4	40,4
N. g.				100	36,6
K. k.					100

FIGURE 1 | Predicted domain organization of *L. pneumophila* Corby PilY1 protein and that of homologous proteins from *P. aeruginosa PAK, Neisseria meningitidis, Neisseria gonorrhoeae,* and *Kingella kingae.* Sequence analysis and diagrams were performed with SMART (Simple Modular Architecture Research Tool) and IBS (Illustrator of Biological Sequences; Letunic et al., 2015; Liu et al., 2015). The degree of homology between the single proteins is also shown in form of matrices for the whole protein, and the single domains for each protein (n.a., not applicable).

of *P. aeruginosa* and the PilC1 of *K. kingae*. The highest sequence similarity within this protein class was found in the C-terminal PilY or PilC domain. With 36% aa identity and 48% aa similarity, the *L. pneumophila* PilY domain is most closely related to the PilY domain of *P. aeruginosa.*

PilY1 Is Expressed during Stationary Growth Phase

L. pneumophila employs a biphasic life cycle, where it alternates between an infectious, non-replicating form, which promotes transmission to and manipulation of new host cells, and an intracellular, replicative phase, which downregulates specific virulence traits. The expression of virulence genes correlate with the post-exponential and stationary growth phases of the bacteria and can be modeled in liquid culture medium (Molofsky and Swanson, 2004). Transcriptional profiling revealed that *pilY1* expression is induced during the transmissive, post-exponential growth phase (Brüggemann et al., 2006; Faucher et al., 2011). In the present study, we compared the PilY1 expression of whole-cell lysates of *L. pneumophila* Corby during exponential (E), early stationary (ES), stationary (S), and late stationary (LS) growth phase by Western blot analysis (**Figure 2A**). The discrimination between exponential and stationary growth phases of *L. pneumophila* was confirmed using a flagellin (FlaA) specific antibody, since flagellation is a prominent feature of the stationary growth phases. As a control for equal cellular amounts in the lysates, the constitutively expressed macrophage infectivity potentiator (Mip) was used for immunoblotting. As shown in **Figure 2A**, PilY1 expression was up-regulated during stationary growth which represents the transmissive phase of infection.

PilY1 Localizes in the Outer Membrane Fraction

According to the signal peptide prediction program SignalP, the PilY1 protein harbors an N-terminal signal peptide and the subcellular localization prediction tool PSORTB 3.0.2 predicts that the PilY1 protein localizes in the outer membrane (Emanuelsson et al., 2007; Yu et al., 2010). To determine the cellular localization of the PilY1 protein experimentally, *L. pneumophila* cells from the stationary growth phase were fractionated utilizing Triton X-100 solubility and ultracentrifugation. Most Gram-negative bacterial inner membrane proteins are soluble in the non-ionic detergent Triton X-100, whereas outer-membrane proteins are typically insoluble (Nikaido, 1994; Vincent et al., 2006). With this approach whole cell lysates (WC) of *L. pneumophila* were separated into cytoplasmic (Cyt), total membrane (TM), inner membrane (IM), and outer membrane (OM) fractions. Western blotting revealed the presence of PilY1 in the whole cell lysate (WC), total membrane (TM), and outer membrane (OM) fractions (**Figure 2B**). The integrity of the membrane fractions was confirmed by detection of the inner membrane signal peptidase LepB and the major outer membrane protein MOMP (Bellinger-Kawahara and Horwitz, 1990; Chen et al., 2007). In summary, these results suggest that PilY1 is located in the outer membrane fraction of *L. pneumophila.*

L. pneumophila PilY1 Promotes Infection of Human Lung Tissue Explants

Human lung tissue explants (HLTE) with their multitude of cell types and extracellular components are well-suited

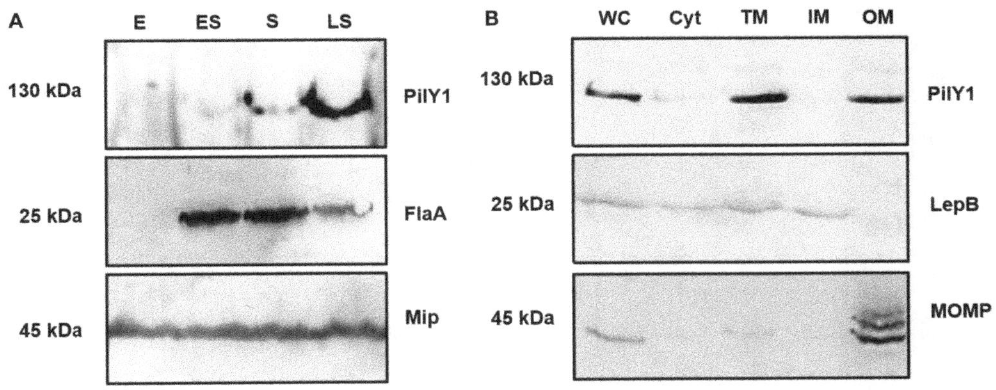

FIGURE 2 | Growth phase dependent expression of *L. pneumophila* Corby PilY1 and its subcellular localization. (A) Immunoblotting of bacterial whole-cell lysates from exponential (E), early stationary (ES), stationary (S), and late stationary (LS) growth phases with antibodies specific for PilY1, Mip (constitutive expression) and FlaA (expression during stationary growth phase). PilY1 expression was only detected in the stationary growth phase of *L. pneumophila* which represents the transmissive phase of infection. **(B)** Immunoblotting of subcellular fractions of *L. pneumophila* Corby from the stationary growth phase. Whole cell lysate (WC), cytoplasmic (Cyt), total membrane (TM), inner membrane (IM), and outer membrane (OM) fractions were immunoblotted with antibodies against PilY1, LepB (inner membrane signal peptidase), and MOMP (major outer membrane protein).

for a comprehensive investigation of extra- and intracellular pathogenicity mechanisms of *L. pneumophila* at a unique level of complexity (Jäger et al., 2014). Since PilY1 is expressed during the transmissive phase of *L. pneumophila,* we analyzed the contribition of this factor that is associated with the outer membrane during HLTE infection. Tumor-free pulmonary tissue samples had been obtained from surgery patients and inoculated with the *L. pneumophila* Corby wild type strain, the PilY1-negative *L. pneumophila ΔpilY1* mutant, the PilY1 domain-positive *L. pneumophila ΔpilY1* WTPilY1 complemented mutant, and the DotA-negative *L. pneumophila* Δ*dotA* mutant. The CFUs/g tissue were determined during 48 h postinoculation (**Figure 3A**). The *L. pneumophila ΔpilY1* mutant showed a significant growth defect in lung tissue with a four-fold reduced bacterial load compared to the *L. pneumophila* wild type strain 48 h post infection. Complementation restored the ability of the mutant to replicate within the infected tissue to a large extent. The number of DotA-negative bacteria, which cannot replicate within host cells, did not increase significantly. Interestingly, when the replication rates of the strains were compared by dividing the CFUs of respective time points (CFU_X) by the initial CFU at 2 h post infection (CFU_{2h}), no significant differences could be observed (**Figure 3D**). In summary, these results demonstrate that PilY1 promotes the infection of human lung tissue in an early state of establishing the infection.

PilY1 Affects Intracellular Replication of *L. pneumophila*

Intracellular replication in macrophages and epithelial cells is essential for *L. pneumophila* infection of lung tissue (Mody et al., 1993; Gao et al., 1998; Jäger et al., 2014). In a previous screening of mini-Tn10 transposon mutants we showed that the PilY1-negative mutants (Lpc2666) were attenuated in their ability to replicate intracellularly within U937 macrophage-like cells (Shevchuk et al., 2014). To assess the impact of *pilY1* on intracellular replication during the infection of HLTEs more specifically, THP-1 macrophage-like cells and A549 lung epithelial cells were infected with the *L. pneumophila* WT, the Δ*pilY1* mutant and the Δ*pilY1* mutant complemented with PilY1. In both host cell types, the overall intracellular replication of the Δ*pilY1* mutant was significantly attenuated after 24 and 48 h compared with the *L. pneumophila* wild type strain (**Figures 3B,C**). With a 40-fold reduction after 48 h, the strongest intracelluar growth defect was observed in the THP-1 macrophage-like cells. A closer look at the replication rates in both cell types revealed that the pronounced attenuation in replication in THP-1 cells was indeed due to a defect in intracellular multiplication (**Figure 3E**). In A549 cells the reduced infection was due to a lower intracellular load with bacteria rather than to a change in the replication rate (**Figure 3F**). Either growth defect was repaired by the complementation with the full length *pilY1* gene. These results show that PilY1 contributes to the intracellular growth of *L. pneumophila* in macrophages and epithelial cells by different means. This cell type dependent effect regarding intracellular replication could also be confirmed by immunofluorescence microscopy. In THP-1 macrophages the *ΔpilY1* mutant co-localized with the phagolysosomal marker LAMP-1 to a significantly greater extent than wild type bacteria in. On the contrary, no difference in LAMP-1 co-localization was observed in A549 cells (Supplementary Figure 3).

PilY1 Contributes to Host Cell Adherence and Invasion

Attachment to and invasion of host cells are critical steps in the cellular infection cycle of *L. pneumophila* and a wide variety of factors which are important for these processes have been identified, including type IV pili (Stone and Abu Kwaik, 1998; Cianciotto, 2001; Molmeret et al., 2004). To

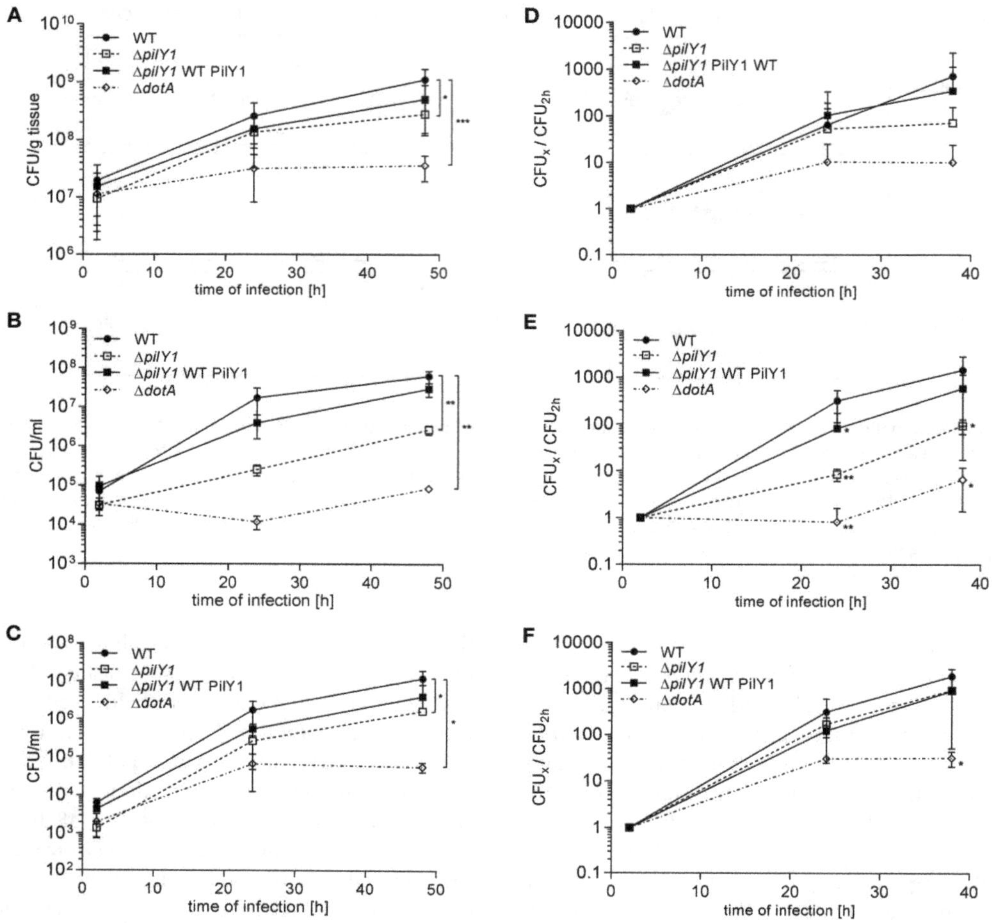

FIGURE 3 | Effect of PilY1 on *L. pneumophila* Corby infection of HLTEs, THP-1 macrophage-like cells and A549 lung epithelial cells. (A,D) HLTEs, **(B,E)** THP-1 macrophages, and **(C,F)** A549 cells were infected with *L. pneumophila* Corby wild type strain (WT), the *pilY1*-negative mutant (Δ*pilY1*), the *pilY1* complemented mutant (Δ*pilY1* WTPilY1), and the DotA deficient mutant strain (Δ*dotA*). CFUs were determined by plating serial dilutions on BCYE agar plates at the indicated time points. The replication rates during the same infection set-ups were calculated and plotted accordingly for HLTEs **(D)** THP-1 macrophages **(E)** and A549 cells **(F)** by dividing the CFU of the respective time point (CFU_X) by the CFU of the 2 h time point (CFU_{2h}). The graphs show means and standard deviations from duplicate infections with tissues from six donors **(A)** and from three independent cellular infections in duplicate **(B,C)**. Significance was assessed by applying an unpaired two-sided Student's *t*-test (*$p \leq 0.05$, **$p \leq 0.01$, ***$p \leq 0.001$).

further dissect the effect on intracellular growth, we examined the specific contribution of *pilY1* to adhesion and invasion of THP-1 macrophage-like and A549 lung epithelial cells. For the determination of adherent *L. pneumophila*, phagocytosis was prevented with the actin polymerization inhibitor Cytochalasin B. The adhesion assay revealed that the adherence of the Δ*pilY1* mutant to THP-1 macrophage-like and A549 lung epithelial cells is decreased by ~50% compared to the *L. pneumophila* wild type phenotype (**Figure 4A**). Complementation in *trans* of the Δ*pilY1* mutant with the full length *pilY1* gene restored adherence, while the truncated form of PilY1, lacking the C-terminal PilY domain (ΔPilYdom), was not able to restore the adherence defect. The complementation with a deletion variant of PilY1, lacking the vWFa domain (ΔvWFa), revealed that this domain is dispensable for adhesion of *L. pneumophila* to THP-1 macrophage-like cells.

In order to evaluate whether invasion of the bacteria was also affected by the deletion of *pilY1* differential immunofluorescence staining was performed, and the ratio of intracellular bacteria to total cell-associated bacteria was determined. Here, again cell type dependent effects could be observed, since invasion of A549 lung epithelial cells was significantly affected and reduced by 90%. In case of THP-1 macrophage-like cells no significant reduction was measured. In case of A549 cells, complementation with the vWFa domain improved the invasion efficiency (**Figure 4B**).

PilY1 Influences Surface Motility of *L. pneumophila*

In addition to adhesion, invasion, and intracellular replication, lung infection depends on dissemination of the pathogen within the tissue. With respect to *L. pneumophila* twitching motility, a type IV pilus-mediated movement and sliding

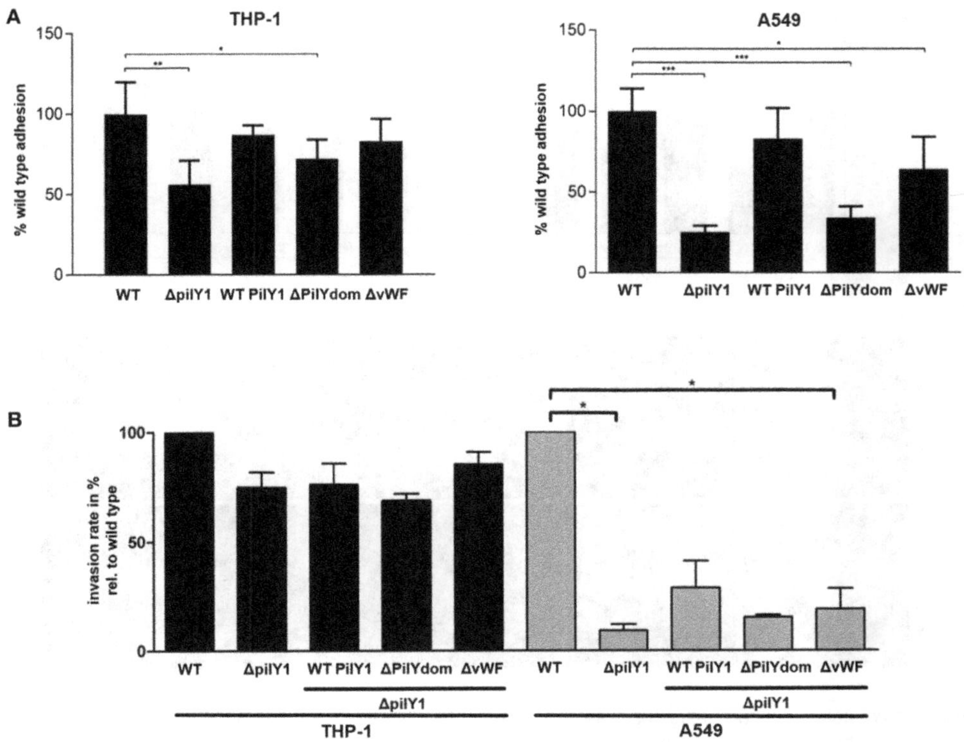

FIGURE 4 | Contribution of PilY1 domains to host cell adherence and invasion. (A) Adherence to THP-1 macrophage like cells and A549 epithelial cells was analyzed with the *L. pneumophila* Corby wild type strain (WT), the *pilY1-negative* mutant (Δ*pilY1*), the complemented *pilY1* mutants expressing the wild type PilY1 (Δ*pilY1* WTPilY1), the PilY1 lacking the C-terminal PilY1 domain (ΔPilYdom) and the PilY1 lacking the vWFa domain (ΔvWFa). Bacterial adhesion was determined 30 min (THP-1) or 1 h (A549) after inoculation and is presented as a ratio of adhering bacteria to the total number of inoculated bacteria. Bacterial uptake by phagocytosis was blocked with Cytochalasin B. **(B)** Invasion of host cells was determined 1 h (THP-1) or 2 h (A549) after inoculation with rhodamine labeled bacteria. Extracellular bacteria additionally were stained with polyclonal α-*L. penumophila* antibodies and detected with an Alexa Fluor® 488-coupled secondary antibody. Presented is the ratio of intracellular bacteria to the total number of cell-associated bacteria. Data represent the means and standard deviations from duplicates of three independent experiments (*t*-test, $^{*}p \leq 0.05$, $^{**}p \leq 0.01$, $^{***}p \leq 0.001$).

motility, a flagellum- and pilus-independent translocation, which is facilitated by secreted surfactant, were described (Coil and Anné, 2009; Stewart et al., 2009). Analogous to twitching and sliding motility assays, which were previously performed, we examined the contribution of PilY1 to both motility processes (**Figures 5A–C**). The motility of *L. pneumophila* was quantified by measuring the peripheral motility zones of bacterial colonies grown on agar plates. Interestingly, the Δ*pilY1* mutant exhibited a significantly reduced twitching motility compared to the wild type strain (**Figure 5A**). In complementation assays, the strains expressing PilY1 (WT PilY) or a truncated form with a deletion of the vWFa domain (ΔvWFA) revealed the wild type phenotype. However, the *L. pneumophila* mutant expressing a PilY1 protein without the C-terminal PilY domain (ΔPilYdom) was unable to complement the twitching motility defect of the Δ*pilY1* mutant. In contrast to the colony morphology of the wild type strain, the border in the motility zone of the Δ*pilY1* mutant was very smooth (**Figure 5C**). This different colony morphology was also observed for the complemented strain lacking the C-terminal PilY domain (ΔPilYdom).

In the sliding motility assay, the spreading area of the Δ*pilY1* mutant was greater than that of the wild type strain (**Figures 5B,C**). In addition to an enhanced number and size of the lobes in the sliding motility zone, we also observed increased surfactant secretion on the agar surface. The genetic complementation demonstrated that only the mutant strains expressing PilY1 exhibited the smaller wild type motility. Complementation with the ΔPilYdom protein resulted in an intermediate phenotype, while complementation without the vWFa domain (ΔvWFa) resulted in sliding motility similar to that of the Δ*pilY1* mutant phenotype. In conclusion, these data indicate that PilY1 positively influences twitching motility, but confines sliding motility.

DISCUSSION

PilY1 proteins are conserved within many genera of gamma- and beta-proteobacteria, including pathogens with a broad host spectrum (Rahman et al., 1997; Morand et al., 2001, 2004; Kehl-Fie et al., 2008; Siryaporn et al., 2014). Sequence and domain composition analyses revealed that the PilY1-like protein of

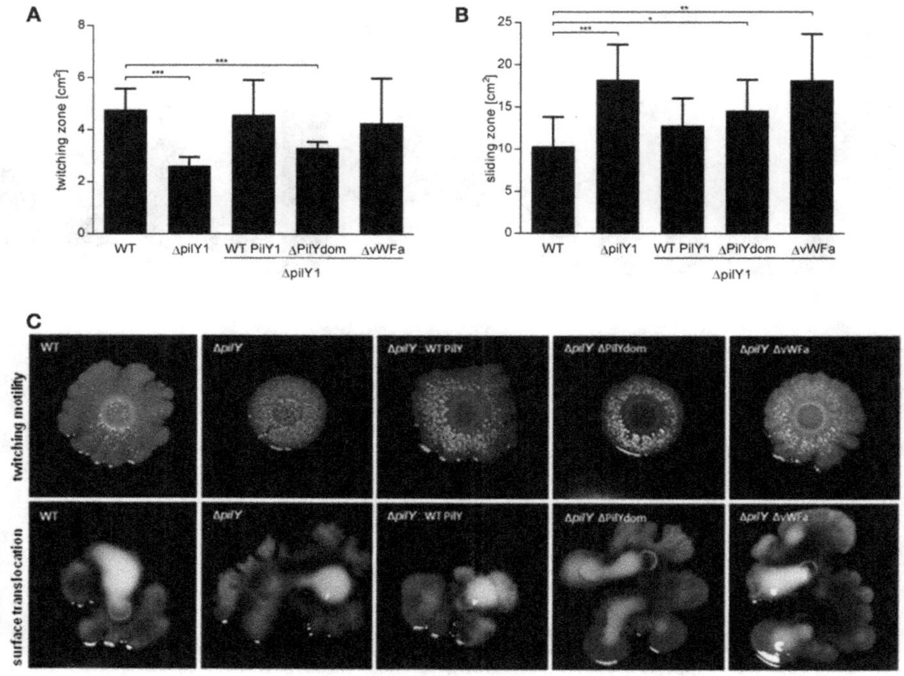

FIGURE 5 | Contribution of PilY1 domains to twitching and sliding motility. For **(A)** twitching and **(B)** sliding motility assays stationary phase liquid cultures of *L. pneumophila* Corby wild type (WT), the *pilY1-negative* mutant (Δ*pilY1*), the complemented *pilY1* mutant strains expressing the wild type PilY1 (Δ*pilY1* WTPilY1), the PilY1 lacking the C-terminal PilY1 domain (ΔPilYdom) and the PilY1 lacking vWFa domain (ΔvWFa) were spotted on BCYE agar plates and grown at 30°C for 14 days. **(C)** Images of twitching motility and sliding translocation are representatives of three independent experiments. The central raised ring represents the initial inoculum and the outer rings show the motility zone. Migration areas were calculated with Adobe Photoshop 7.0.1. (*t*-test *$p \leq 0.05$, **$p \leq 0.01$, ***$p \leq 0.001$).

L. pneumophila has the highest homology to the *P. aeruginosa* PilY1. This protein, which is exposed on the cell surface, activates virulence features such as pilus biogenesis, twitching motility, secretion of secondary metabolites, and biofilm formation (Bohn et al., 2009; Kuchma et al., 2010; Siryaporn et al., 2014). In *P. aeruginosa* the minor pilins (FimU, PilVWXE) prime type IVa pilus assembly and promote the surface display of the PilY1 adhesin (Giltner et al., 2010; Heiniger et al., 2010; Luo et al., 2015; Nguyen et al., 2015). *L. pneumophila* possesses a comparable *pilY1* pilin gene cluster; an orthologous set of pilin-like proteins controls type IV pilus dynamics and the localization of PilC in *N. gonorrhoeae* (Winther-Larsen et al., 2005).

Interestingly, PilY1 proteins of *L. pneumophila*, *P. aeruginosa*, and *K. kingae* have a domain that shares homology with the eukaryotic mechano-sensitive vWFa domain and therefore could serve as a sensor for the mechanical cues associated with surface contact. PilY1 of *P. aeruginosa* is up-regulated upon surface contact, which is necessary for surface-activated virulence. Furthermore, deletion of the putatively mechanosensitive vWFa domain places PilY1 in a constitutively active state, inducing virulence in cells, which are not attached to the surface (Siryaporn et al., 2014). Apparently, shear forces exerted on surface-attached *P. aeruginosa* cells shift the surface adhesion PilY1 into an active stretched state and thus inducing virulence toward a broad range of hosts without relying upon chemical recognition of any specific host factor. Accordingly, we hypothesize that *L. pneumophila* PilY1 may exert related functions in pathogenicity.

L. pneumophila pathogenesis in lung tissue typically depends on intracellular growth of the pathogen and bacterial dissemination within the tissue. In the present study, we showed that *L. pneumophila* PilY1 is a cell-surface-associated protein which is expressed during the highly virulent stationary growth phase. The presence of PilY1 on the bacterial surface during the transmissive phase of the pathogen seems to be a pre-requisite for a putative mechanosensitive virulence factor or regulator. Since *L. pneumophila* encounters many different surface structures during human infection, we analyzed the effect of PilY1 on the infection of HLTEs. This complex infection model, which comprises a multitude of cell types and extracellular components, including lung epithelial cells, macrophages, and extracellular matrix (ECM), revealed that the *L. pneumophila ΔpilY1* mutant has a significant growth defect. In part, this defect seems to be due to the attenuated intracellular growth of the PilY1-negative mutant in phagocytic host cells. This is suggested by previous infections of U937-macrophage-like cells with the PilY1-negative Mini-Tn10 transposon mutant Lpc2666 (Shevchuk et al., 2014). In the present study, we further substantiated this conclusion using infection experiments with site-directed mutants in THP-1 macrophage-like cells and A549 lung epithelial cells. The intracellular growth defects that were observed further confirmed a multifaceted effect of PilY1 on infection processes ranging from adherence and uptake in non-phagocytic to additionally intracellular replication in phagocytic cells. Thus, it is noteworthy that in HLTEs adherence and invasion related processes strongly influence the outcome of the infection of intracellularly replicating *L. pneumophila*.

Studies performed with *P. aeruginosa*, *Neisseria*, and *K. kingae* revealed a strong influence of PilY1 and PilC on adherence to epithelial cells (Kehl-Fie et al., 2008; Cheng et al., 2013; Porsch et al., 2013). Since intracellular growth of *L. pneumophila* is a result of adherence, invasion, and intracellular replication, we further dissected the functional contributions of PilY1 to these processes. We found that PilY1 of *L. pneumophila* is required for host cell adherence to THP-1 macrophage-like and A549 lung epithelial cells. As adhesion is critical for host cell invasion and both processes are critical for the intracellular life cycle, these effects, including the resulting decrease in growth rate within phagocytic host cells may be linked to each other. This is certainly true for non-phagocytic A549 epithelial cells and in a broader sense for the HLTE infection model, where absence of PilY1 resulted in a drastic decrease in invasive capacity. Complementation of PilY1-negative mutants with single domain deletion variants revealed that the C-terminal PilY1 domain was essential for adherence to THP-1 and A549 cells, whereas the vWFa domain contributes to invasion of non-phagocytic cells. These results resemble observations made in *P. aeruginosa* where the deletion of the calcium binding C-terminal PilC domain abrogated surface-induced virulence, mimicking the loss of the full-length PilY1 protein, whereas PilY1 with a vWFa domain deletion induced virulence in a constitutive manner (Orans et al., 2010; Siryaporn et al., 2014). In analogy, we assume that the vWFa domain of *L. pneumophila* may also have primarily regulatory functions.

For interpretation of the cellular infection data of the PilY1-negative strains, it should be taken into consideration that the intracellular growth defect may be multifactorial and not only related to type IV pili biogenesis. Originally the corresponding Mini-Tn10 transposon mutant Lpc2666 was screened for its failure to inhibit the fusion of LCVs with lysosomes; fluorescence microscopy demonstrated a significantly higher co-localization ratio with lysosomes, an observation that could be confirmed with the targeted *pilY1* knock-out mutant of the current study (Shevchuk et al., 2014, and Supplementary Figure 2). Furthermore, when repeated in THP-1 macrophage-like and A549 lung epithelial cells, these studies confirmed the differential involvement of PilY1 during intracellular replication. In macrophages but not in epithelial cells PilY1 contributed to avoidance of phagolysosomal maturation as measured by LAMP-1 co-localization (**Figure 3** and Supplementary Figure 3). The inhibition of the fusion of LCVs with lysosomes is an intracellular process which is temporally and spatially distinct from the initial type-IV pili-mediated attachment to the host cell surface. Thus, while the effects of adherence and invasion can be explained solely by the intimate attachment of the type IV-pili to the host cell surface, the avoidance of lysosomal degradation could also include some kind of interference with host cell type specific signaling or even a type IV-pili-independent regulation of other bacterial virulence factors.

To further dissect the PilY1-mediated virulence mechanisms and to support our hypothesis that PilY1 plays similar roles in *L. pneumophila* and *P. aeruginosa*, we analyzed the surface motility of *L. pneumophila*. Surface motility allows pathogens to migrate over mucosal epithelia and to disperse to other anatomical sites within the host (Kearns, 2010; Taylor and Buckling, 2011; Burrows, 2012). *L. pneumophila* and *P. aeruginosa* can generally exhibit three distinct surface-associated motile behaviors, namely, flagellum-propelled swarming motility, type IV pilus-mediated twitching motility, and surfactant-mediated passive sliding motility (Bohn et al., 2009; Coil and Anné, 2009; Stewart et al., 2009). In *P. aeruginosa*, the minor pilins PilW and PilX, as well as PilY1 participate in c-di-GMP-mediated repression of swarming motility. Moreover, it has been shown that biofilm formation and swarming motility are inversely regulated (Kuchma et al., 2012). In twitching motility of *P. aeruginosa*, PilY1 functions as a calcium-dependent regulator which is essential for type IV pilli biogenesis (Carbonnelle et al., 2006; Heiniger et al., 2010; Orans et al., 2010; Johnson et al., 2011; Cheng et al., 2013; Porsch et al., 2013). Sliding motility of *P. aeruginosa* is observed in the absence of type IV pili and responds to many of the same regulatory proteins and environmental cues as swarming motility (Murray and Kazmierczak, 2008). We demonstrated that *Legionella* PilY1 positively influences twitching motility, but down-regulates sliding motility. The positive effect of PilY1 on twitching motility can be explained by a facilitated type IV pilus assembly. The observed suppression of sliding motility, however, cannot simply be explained by pilus biogenesis. Possible explanations are that PilY1-negative mutants are more leaky for surfactants, or that type IV pili-mediated attachment hampers sliding motility, or that PilY1 down- or up-regulates other effector molecules of *L. pneumophila*.

CONCLUSION

In the present study, we characterized PilY1 as a new virulence factor of *L. pneumophila*, a factor which influences intracellular growth and surface motility of the pathogen. We propose that PilY1 functions as an adhesion factor, which not only influences bacterial uptake by host cells, but also the successful establishment of the early LCV by providing close cell-cell contact for effective effector translocation. Whether or not PilY is a master mediator in the hierarchy of virulence regulation, as was proposed for *P. aeruginosa*, remains speculative. The observed effects of PilY1 on adherence, invasion, and twitching motility of *L. pneumophila* can theoretically be explained by their contribution to type IV pili biogenesis. The interference with lysosomal degradation and the suppression of sliding motility have not yet been studied extensively enough to classify them as type IV pili-independent. Similarly, it is an open question if the vWFa domain of PilY1 exerts regulatory functions on the intracellular growth cycle and on tissue dissemination of the pathogen via assembly of type IV pili. Since bacterial cells which are attached to the surface are subjected to larger shear forces than planktonic cells when liquid flows over them, it is appealing to speculate on a mechanosensing function of the vWFa domain. For *P. aeruginosa,* the detection of mechanical cues associated with surface attachment is independent of both the respective surface and the host (Siryaporn et al., 2014; Ellison and Brun, 2015). This enables the pathogen to induce virulence toward a broad range of hosts. Future studies will show if *L. pneumophila* utilizes a similar mechanism to infect a plethora of protozoa species and human cell types.

AUTHOR CONTRIBUTIONS

JH was involved in all experimental procedures and statistical analysis. CÜ participated in the cloning experiments and contributed to the design of the study. ST and LG performed co-localization and immunofluorescence studies. TG, NG, and MR were responsible for the generation of lung tissue explants (this study is in accordance with the Helsinki declaration and approved by the ethic committee of the Hannover Medical School; No: 2235-2014) and contributed their expertise with the HLTE infection model. OS and MS were involved in the experimental design, data analysis and preparation of the manuscript.

ACKNOWLEDGMENTS

We thank Barbara Schulz for critically reading the manuscript.

REFERENCES

Bellinger-Kawahara, C., and Horwitz, M. A. (1990). Complement component C3 fixes selectively to the major outer membrane protein (MOMP) of *Legionella pneumophila* and mediates phagocytosis of liposome-MOMP complexes by human monocytes. *J. Exp. Med.* 172, 1201–1210. doi: 10.1084/jem.172.4.1201

Bohn, Y. S. T., Brandes, G., Rakhimova, E., Horatzek, S., Salunkhe, P., Munder, A., et al. (2009). Multiple roles of *Pseudomonas aeruginosa* TBCF10839 PilY1 in motility, transport and infection. *Mol. Microbiol.* 71, 730–747. doi: 10.1111/j.1365-2958.2008.06559.x

Brüggemann, H., Hagman, A., Jules, M., Sismeiro, O., Dillies, M.-A., Gouyette, C., et al. (2006). Virulence strategies for infecting phagocytes deduced from the *in vivo* transcriptional program of *Legionella pneumophila. Cell. Microbiol.* 8, 1228–1240. doi: 10.1111/j.1462-5822.2006.00703.x

Burrows, L. L. (2012). *Pseudomonas aeruginosa* twitching motility: type IV pili in action. *Annu. Rev. Microbiol.* 66, 493–520. doi: 10.1146/annurev-micro-092611-150055

Carbonnelle, E., Helaine, S., Nassif, X., and Pelicic, V. (2006). A systematic genetic analysis in *Neisseria meningitidis* defines the Pil proteins required for assembly, functionality, stabilization and export of type IV pili. *Mol. Microbiol.* 61, 1510–1522. doi: 10.1111/j.1365-2958.2006.05341.x

Chang, B., Kura, F., Amemura-Maekawa, J., Koizumi, N., and Watanabe, H. (2005). Identification of a novel adhesion molecule involved in the virulence of *Legionella pneumophila. Infect. Immun.* 73, 4272–4280. doi: 10.1128/IAI.73.7.4272-4280.2005

Charpentier, X., Gabay, J. E., Reyes, M., Zhu, J. W., Weiss, A., and Shuman, H. A. (2009). Chemical genetics reveals bacterial and host cell functions critical for type IV effector translocation by *Legionella pneumophila. PLoS Pathog.* 5:e1000501. doi: 10.1371/journal.ppat.1000501

Chen, J., Reyes, M., Clarke, M., and Shuman, H. A. (2007). Host cell-dependent secretion and translocation of the LepA and LepB effectors of *Legionella pneumophila. Cell. Microbiol.* 9, 1660–1671. doi: 10.1111/j.1462-5822.2007.00899.x

Cheng, Y., Johnson, M. D. L., Burillo-Kirch, C., Mocny, J. C., Anderson, J. E., Garrett, C. K., et al. (2013). Mutation of the conserved calcium-binding motif in *Neisseria gonorrhoeae* PilC1 impacts adhesion but not piliation. *Infect. Immun.* 81, 4280–4289. doi: 10.1128/IAI.00493-13

Cianciotto, N. P. (2001). Pathogenicity of *Legionella pneumophila. Int. J. Med. Microbiol.* 291, 331–343. doi: 10.1078/1438-4221-00139

Cirillo, S. L., Bermudez, L. E., El-Etr, S. H., Duhamel, G. E., and Cirillo, J. D. (2001). *Legionella pneumophila* entry gene rtxA is involved in virulence. *Infect. Immun.* 69, 508–517. doi: 10.1128/IAI.69.1.508-517.2001

Coil, D. A., and Anné, J. (2009). Twitching motility in *Legionella pneumophila. FEMS Microbiol. Lett.* 293, 271–277. doi: 10.1111/j.1574-6968.2009.01532.x

Derbise, A., Lesic, B., Dacheux, D., Ghigo, J. M., and Carniel, E. (2003). A rapid and simple method for inactivating chromosomal genes in Yersinia. *FEMS Immunol. Med. Microbiol.* 38, 113–116. doi: 10.1016/S0928-8244(03)00181-0

Duncan, C., Prashar, A., So, J., Tang, P., Low, D. E., Terebiznik, M., et al. (2011). Lcl of *Legionella pneumophila* is an immunogenic GAG binding adhesin that promotes interactions with lung epithelial cells and plays a crucial role in biofilm formation. *Infect. Immun.* 79, 2168–2181. doi: 10.1128/IAI.01304-10

Ellison, C., and Brun, Y. V. (2015). Mechanosensing: a regulation sensation. *Curr. Biol.* 25, R113–R115. doi: 10.1016/j.cub.2014.12.026

Emanuelsson, O., Brunak, S., von Heijne, G., and Nielsen, H. (2007). Locating proteins in the cell using TargetP, SignalP and related tools. *Nat. Protoc.* 2, 953–971. doi: 10.1038/nprot.2007.131

Faucher, S. P., Mueller, C. A., and Shuman, H. A. (2011). *Legionella pneumophila* transcriptome during intracellular multiplication in human macrophages. *Front. Microbiol.* 2:60. doi: 10.3389/fmicb.2011.00060

Fields, B. S., Benson, R. F., and Besser, R. E. (2002). Legionella and Legionnaires' disease: 25 years of investigation. *Clin. Microbiol. Rev.* 15, 506–526. doi: 10.1128/CMR.15.3.506-526.2002

Fraser, D. W., Tsai, T. R., Orenstein, W., Parkin, W. E., Beecham, H. J., Sharrar, R. G., et al. (1977). Legionnaires' disease: description of an epidemic of pneumonia. *N. Engl. J. Med.* 297, 1189–1197. doi: 10.1056/NEJM197712012972201

Gao, L. Y., Stone, B. J., Brieland, J. K., and Abu Kwaik, Y. (1998). Different fates of *Legionella pneumophila* pmi and mil mutants within macrophages and alveolar epithelial cells. *Microb. Pathog.* 25, 291–306. doi: 10.1006/mpat.1998.0237

Garduño, R. A., Garduño, E., and Hoffman, P. S. (1998). Surface-associated hsp60 chaperonin of *Legionella pneumophila* mediates invasion in a HeLa cell model. *Infect. Immun.* 66, 4602–4610.

Giltner, C. L., Habash, M., and Burrows, L. L. (2010). *Pseudomonas aeruginosa* minor pilins are incorporated into type IV pili. *J. Mol. Biol.* 398, 444–461. doi: 10.1016/j.jmb.2010.03.028

Heiniger, R. W., Winther-Larsen, H. C., Pickles, R. J., Koomey, M., and Wolfgang, M. C. (2010). Infection of human mucosal tissue by *Pseudomonas aeruginosa* requires sequential and mutually dependent virulence factors and a novel pilus-associated adhesin. *Cell. Microbiol.* 12, 1158–1173. doi: 10.1111/j.1462-5822.2010.01461.x

Helbig, J. H., Ludwig, B., Lück, P. C., Groh, A., Witzleb, W., and Hacker, J. (1995). Monoclonal antibodies to Legionella Mip proteins recognize genus- and species-specific epitopes. *Clin. Diagn. Lab. Immunol.* 2, 160–165.

Heuner, K., Bender-Beck, L., Brand, B. C., Lück, P. C., Mann, K. H., Marre, R., et al. (1995). Cloning and genetic characterization of the flagellum subunit gene (flaA) of *Legionella pneumophila* serogroup 1. *Infect. Immun.* 63,

2499–2507.

Hilbi, H., Segal, G., and Shuman, H. A. (2001). Icm/dot-dependent upregulation of phagocytosis by *Legionella pneumophila*. *Mol. Microbiol.* 42, 603–617. doi: 10.1046/j.1365-2958.2001.02645.x

Horwitz, M. A. (1983). The Legionnaires' disease bacterium (*Legionella pneumophila*) inhibits phagosome-lysosome fusion in human monocytes. *J. Exp. Med.* 158, 2108–2126. doi: 10.1084/jem.158.6.2108

Horwitz, M. A., and Silverstein, S. C. (1980). Legionnaires' disease bacterium (*Legionella pneumophila*) multiples intracellularly in human monocytes. *J. Clin. Invest.* 66, 441–450. doi: 10.1172/JCI109874

Hubber, A., and Roy, C. R. (2010). Modulation of host cell function by *Legionella pneumophila* type IV effectors. *Annu. Rev. Cell Dev. Biol.* 26, 261–283. doi: 10.1146/annurev-cellbio-100109-104034

Isberg, R. R., O'Connor, T. J., and Heidtman, M. (2009). The *Legionella pneumophila* replication vacuole: making a cosy niche inside host cells. *Nat. Rev. Microbiol.* 7, 13–24. doi: 10.1038/nrmicro1967

Jäger, J., Marwitz, S., Tiefenau, J., Rasch, J., Shevchuk, O., Kugler, C., et al. (2014). Human lung tissue explants reveal novel interactions during *Legionella pneumophila* infections. *Infect. Immun.* 82, 275–285. doi: 10.1128/IAI.00703-13

Johnson, M. D. L., Garrett, C. K., Bond, J. E., Coggan, K. A., Wolfgang, M. C., and Redinbo, M. R. (2011). *Pseudomonas aeruginosa* PilY1 binds integrin in an RGD- and calcium-dependent manner. *PLoS ONE* 6:e29629. doi: 10.1371/journal.pone.0029629

Kagan, J. C., and Roy, C. R. (2002). Legionella phagosomes intercept vesicular traffic from endoplasmic reticulum exit sites. *Nat. Cell Biol.* 4, 945–954. doi: 10.1038/ncb883

Kearns, D. B. (2010). A field guide to bacterial swarming motility. *Nat. Rev. Microbiol.* 8, 634–644. doi: 10.1038/nrmicro2405

Kehl-Fie, T. E., Miller, S. E., and St. Geme, J. W. (2008). *Kingella kingae* expresses type IV pili that mediate adherence to respiratory epithelial and synovial cells. *J. Bacteriol.* 190, 7157–7163. doi: 10.1128/JB.00884-08

Kuchma, S. L., Ballok, A. E., Merritt, J. H., Hammond, J. H., Lu, W., Rabinowitz, J. D., et al. (2010). Cyclic-di-GMP-mediated repression of swarming motility by *Pseudomonas aeruginosa*: the *pilY1* gene and its impact on surface-associated behaviors. *J. Bacteriol.* 192, 2950–2964. doi: 10.1128/JB.01642-09

Kuchma, S. L., Brothers, K. M., Merritt, J. H., Liberati, N. T., Ausubel, F. M., and O'Toole, G. A. (2007). BifA, a cyclic-Di-GMP phosphodiesterase, inversely regulates biofilm formation and swarming motility by *Pseudomonas aeruginosa* PA14. *J. Bacteriol.* 189, 8165–8178. doi: 10.1128/JB.00586-07

Kuchma, S. L., Griffin, E. F., and O'Toole, G. A. (2012). Minor pilins of the type IV pilus system participate in the negative regulation of swarming motility. *J. Bacteriol.* 194, 5388–5403. doi: 10.1128/JB.00899-12

Letunic, I., Doerks, T., and Bork, P. (2015). SMART: recent updates, new developments and status in 2015. *Nucleic Acids Res.* 43, D257–D260. doi: 10.1093/nar/gku949

Lewenza, S., Gardy, J. L., Brinkman, F. S. L., and Hancock, R. E. W. (2005). Genome-wide identification of *Pseudomonas aeruginosa* exported proteins using a consensus computational strategy combined with a laboratory-based PhoA fusion screen. *Genome Res.* 15, 321–329. doi: 10.1101/gr.3257305

Liu, W., Xie, Y., Ma, J., Luo, X., Nie, P., Zuo, Z., et al. (2015). IBS: an illustrator for the presentation and visualization of biological sequences. *Bioinformatics* 31, 3359–3361. doi: 10.1093/bioinformatics/btv362

Luo, Y., Zhao, K., Baker, A. E., Kuchma, S. L., Coggan, K. A., Wolfgang, M. C., et al. (2015). A hierarchical cascade of second messengers regulates *Pseudomonas aeruginosa* surface behaviors. *mBio* 6:e02456-14. doi: 10.1128/mBio.02456-14

McDade, J. E., Shepard, C. C., Fraser, D. W., Tsai, T. R., Redus, M. A., and Dowdle, W. R. (1977). Legionnaires' disease: isolation of a bacterium and demonstration of its role in other respiratory disease. *N. Engl. J. Med.* 297, 1197–1203. doi: 10.1056/NEJM197712012972202

Mody, C. H., Paine, R., Shahrabadi, M. S., Simon, R. H., Pearlman, E., Eisenstein, B. I., et al. (1993). *Legionella pneumophila* replicates within rat alveolar epithelial cells. *J. Infect. Dis.* 167, 1138–1145. doi: 10.1093/infdis/167.5.1138

Molmeret, M., Bitar, D. M., Han, L., and Abu Kwaik, Y. (2004). Cell biology of the intracellular infection by *Legionella pneumophila*. *Microbes Infect.* 6, 129–139. doi: 10.1016/j.micinf.2003.11.004

Molofsky, A. B., and Swanson, M. S. (2004). Differentiate to thrive: lessons from the *Legionella pneumophila* life cycle. *Mol. Microbiol.* 53, 29–40. doi: 10.1111/j.1365-2958.2004.04129.x

Morand, P. C., Bille, E., Morelle, S., Eugène, E., Beretti, J.-L., Wolfgang, M., et al. (2004). Type IV pilus retraction in pathogenic Neisseria is regulated by the PilC proteins. *EMBO J.* 23, 2009–2017. doi: 10.1038/sj.emboj.7600200

Morand, P. C., Tattevin, P., Eugene, E., Beretti, J. L., and Nassif, X. (2001). The adhesive property of the type IV pilus-associated component PilC1 of pathogenic Neisseria is supported by the conformational structure of the N-terminal part of the molecule. *Mol. Microbiol.* 40, 846–856. doi: 10.1046/j.1365-2958.2001.02452.x

Murray, T. S., and Kazmierczak, B. I. (2008). *Pseudomonas aeruginosa* exhibits sliding motility in the absence of type IV pili and flagella. *J. Bacteriol.* 190, 2700–2708. doi: 10.1128/JB.01620-07

Newton, H. J., Sansom, F. M., Dao, J., Cazalet, C., Bruggemann, H., Albert-Weissenberger, C., et al. (2008). Significant role for ladC in initiation of *Legionella pneumophila* infection. *Infect. Immun.* 76, 3075–3085. doi: 10.1128/IAI.00209-08

Nguyen, Y., Sugiman-Marangos, S., Harvey, H., Bell, S. D., Charlton, C. L., Junop, M. S., et al. (2015). *Pseudomonas aeruginosa* minor pilins prime type IVa pilus assembly and promote surface display of the PilY1 adhesin. *J. Biol. Chem.* 290, 601–611. doi: 10.1074/jbc.M114.616904

Nikaido, H. (1994). Isolation of outer membranes. *Methods Enzymol.* 235, 225–234. doi: 10.1016/0076-6879(94)35143-0

Orans, J., Johnson, M. D. L., Coggan, K. A., Sperlazza, J. R., Heiniger, R. W., Wolfgang, M. C., et al. (2010). Crystal structure analysis reveals Pseudomonas PilY1 as an essential calcium-dependent regulator of bacterial surface motility. *Proc. Natl. Acad. Sci. U.S.A.* 107, 1065–1070. doi: 10.1073/pnas.0911616107

Porsch, E. A., Johnson, M. D. L., Broadnax, A. D., Garrett, C. K., Redinbo, M. R., and St. Geme, J. W. (2013). Calcium binding properties of the *Kingella kingae* PilC1 and PilC2 proteins have differential effects on type IV pilus-mediated adherence and twitching motility. *J. Bacteriol.* 195, 886–895. doi: 10.1128/JB.02186-12

Rahman, M., Källström, H., Normark, S., and Jonsson, A. B. (1997). PilC of pathogenic Neisseria is associated with the bacterial cell surface. *Mol. Microbiol.* 25, 11–25. doi: 10.1046/j.1365-2958.1997.4601823.x

Roy, C. R., Berger, K. H., and Isberg, R. R. (1998). *Legionella pneumophila* DotA protein is required for early phagosome trafficking decisions that occur within minutes of bacterial uptake. *Mol. Microbiol.* 28, 663–674. doi: 10.1046/j.1365-2958.1998.00841.x

Rudel, T., Boxberger, H. J., and Meyer, T. F. (1995b). Pilus biogenesis and epithelial cell adherence of *Neisseria gonorrhoeae* pilC double knock-out mutants. *Mol. Microbiol.* 17, 1057–1071. doi: 10.1111/j.1365-2958.1995.mmi_1706 1057.x

Rudel, T., Scheurerpflug, I., and Meyer, T. F. (1995a). Neisseria PilC protein identified as type-4 pilus tip-located adhesin. *Nature* 373, 357–359. doi: 10.1038/373357a0

Scheuerpflug, I., Rudel, T., Ryll, R., Pandit, J., and Meyer, T. F. (1999). Roles of PilC and PilE proteins in pilus-mediated adherence of *Neisseria gonorrhoeae* and *Neisseria meningitidis* to human erythrocytes and endothelial and epithelial cells. *Infect. Immun.* 67, 834–843.

Schunder, E., Adam, P., Higa, F., Remer, K. A., Lorenz, U., Bender, J., et al. (2010). Phospholipase PlaB is a new virulence factor of *Legionella pneumophila*. *Int. J. Med. Microbiol.* 300, 313–323. doi: 10.1016/j.ijmm.2010.01.002

Sexton, J. A., and Vogel, J. P. (2004). Regulation of hypercompetence in *Legionella pneumophila*. *J. Bacteriol.* 186, 3814–3825. doi: 10.1128/JB.186.12.3814-3825.2004

Shevchuk, O., Pägelow, D., Rasch, J., Döhrmann, S., Günther, G., Hoppe, J., et al. (2014). Polyketide synthase (PKS) reduces fusion of *Legionella pneumophila*-containing vacuoles with lysosomes and contributes to bacterial competitiveness during infection. *Int. J. Med. Microbiol.* 304, 1169–1181. doi: 10.1016/j.ijmm.2014.08.010

Siryaporn, A., Kuchma, S. L., O'Toole, G. A., and Gitai, Z. (2014). Surface attachment induces *Pseudomonas aeruginosa* virulence. *Proc. Natl. Acad. Sci. U.S.A.* 111, 16860–16865. doi: 10.1073/pnas.1415712111

Stewart, C. R., Rossier, O., and Cianciotto, N. P. (2009). Surface translocation by *Legionella pneumophila*: a form of sliding motility that is dependent upon type II protein secretion. *J. Bacteriol.* 191, 1537–1546. doi: 10.1128/JB.01531-08

Stone, B. J., and Abu Kwaik, Y. (1998). Expression of multiple pili by *Legionella pneumophila*: identification and characterization of a type IV pilin gene and its role in adherence to mammalian and protozoan cells. *Infect. Immun.* 66, 1768–1775.

Taylor, T. B., and Buckling, A. (2011). Selection experiments reveal trade-offs between swimming and twitching motilities in *Pseudomonas aeruginosa*. *Evol. Int. J. Organ. Evol.* 65, 3060–3069. doi: 10.1111/j.1558-5646.2011.01376.x

Vincent, C. D., Friedman, J. R., Jeong, K. C., Buford, E. C., Miller, J. L., and Vogel, J. P. (2006). Identification of the core transmembrane complex of the Legionella Dot/Icm type IV secretion system. *Mol. Microbiol.* 62, 1278–1291. doi: 10.1111/j.1365-2958.2006.05446.x

Vincent, C. D., and Vogel, J. P. (2006). The *Legionella pneumophila* IcmS-LvgA protein complex is important for Dot/Icm-dependent intracellular growth. *Mol. Microbiol.* 61, 596–613. doi: 10.1111/j.1365-2958.2006.05243.x

Wiater, L. A., Dunn, K., Maxfield, F. R., and Shuman, H. A. (1998). Early events in phagosome establishment are required for intracellular survival of *Legionella pneumophila*. Infect. *Immun.* 66, 4450–4460.

Winther-Larsen, H. C., Wolfgang, M., Dunham, S., van Putten, J. P. M., Dorward, D., Løvold, C., et al. (2005). A conserved set of pilin-like molecules controls type IV pilus dynamics and organelle-associated functions in *Neisseria gonorrhoeae*. *Mol. Microbiol.* 56, 903–917. doi: 10.1111/j.1365-2958.2005.04591.x

Wolfgang, M., Park, H. S., Hayes, S. F., van Putten, J. P., and Koomey, M. (1998). Suppression of an absolute defect in type IV pilus biogenesis by loss-of-function mutations in pilT, a twitching motility gene in *Neisseria gonorrhoeae*. *Proc. Natl. Acad. Sci. U.S.A.* 95, 14973–14978. doi: 10.1073/pnas.95.25.14973

Yu, N. Y., Wagner, J. R., Laird, M. R., Melli, G., Rey, S., Lo, R., et al. (2010). PSORTb 3.0: improved protein subcellular localization prediction with refined localization subcategories and predictive capabilities for all prokaryotes. *Bioinformatics* 26, 1608–1615. doi: 10.1093/bioinformatics/btq249

14

Polymorphism in the *Yersinia* LcrV Antigen Enables Immune Escape from the Protection Conferred by an LcrV-Secreting *Lactococcus Lactis* in a Pseudotuberculosis Mouse Model

Catherine Daniel, Amélie Dewitte, Sabine Poiret, Michaël Marceau, Michel Simonet, Laure Marceau, Guillaume Descombes, Denise Boutillier, Nadia Bennaceur, Sébastien Bontemps-Gallo, Nadine Lemaître and Florent Sebbane*

Université de Lille, CNRS, Inserm, CHU Lille, Institut Pasteur de Lille, U1019 - UMR 8204 - Center for Infection and Immunity of Lille, Lille, France

Edited by:
Fabio Bagnoli,
GlaxoSmithKline (Italy), Italy

Reviewed by:
Wayne Robert Thomas,
Telethon Kids Institute, Australia
Anne Derbise,
Institut Pasteur, France
Deborah Anderson,
University of Missouri, United States
Yinon Levy,
Israel Institute for Biological Research (IIBR), Israel
Vladimir L. Motin,
University of Texas Medical Branch at Galveston, United States

***Correspondence:**
Florent Sebbane
florent.sebbane@inserm.fr

Yersinioses caused by *Yersinia pestis, Yersinia pseudotuberculosis,* and *Yersinia enterocolitica* are significant concerns in human and veterinary health. The link between virulence and the potent LcrV antigen has prompted the latter's selection as a major component of anti-*Yersinia* vaccines. Here, we report that (i) the group of *Yersinia* species encompassing *Y. pestis* and *Y. pseudotuberculosis* produces at least five different clades of LcrV and (ii) vaccination of mice with an LcrV-secreting *Lactococcus lactis* only protected against *Yersinia* strains producing the same LcrV clade as that of used for vaccination. By vaccinating with engineered LcrVs and challenging mice with strains producing either type of LcrV or a LcrV mutated for regions of interest, we highlight key polymorphic residues responsible for the absence of cross-protection. Our results show that an anti-LcrV-based vaccine should contain multiple LcrV clades if protection against the widest possible array of *Yersinia* strains is sought.

Keywords: plague, *Yersinia pestis*, vaccine, LcrV, immune escape, polymorphism, probiotic, *Lactoccus lactis*

INTRODUCTION

Yersinia pestis, *Yersinia pseudotuberculosis,* and *Yersinia enterocolitica* are bacterial pathogens in humans and animals (1). *Y. pestis* causes plague while *Y. enterocolitica* causes bowel disease and *Y. pseudotuberculosis* causes tuberculosis-like symptoms in animals or Far East scarlet-like fever if it infects humans. However, some patients with enteric yersiniosis also develop immunopathological complications, such as reactive polyarthritis and erythema nodosum (2, 3). Furthermore, fatal hematogenous spreading of *Y. enterocolitica* and *Y. pseudotuberculosis* is often reported in immunocompromised individuals or patients with iron overload (3–5). Although plague outbreaks are reported in both hemispheres, outbreaks of human enteric yersiniosis occur mostly in the northern hemisphere (i.e., northern Europe, the United States, Canada, Russia, and Japan) (6, 7). It is noteworthy that human enteric yersiniosis is among the five foodborne illnesses with the greatest economic burden (8). In addition to the diseases' impact on human health, all three yersinioses are associated with a major socio-economic burden because they cause recurrent, fatal epidemics among livestock and in zoological and wildlife parks (9–17). Yersinioses are thus considered to be significant human and veterinary health problems. Unsurprisingly, research geared toward the production of anti-yersiniosis vaccines is now underway.

Today's anti-*Yersinia* vaccine development programs are focused on the major virulence factor LcrV, also referred to as V antigen (18–20). LcrV is a polymorphic protein (21–24). Most amino-acid variations occur in the second half-part of the protein, within which a hypervariable region has been identified (25). The residues at positions 51 and 63 are known to be highly polymorphic (26). Indeed, LcrV antigens have been classified into different types, subtypes and variants on the basis of (i) the length of the hypervariable region, (ii) the polymorphisms at residues 51 and 63, and (iii) other subtle sequence differences (21, 23, 25, 26). Furthermore, there is debate as to whether immunization using a single V type, subtype or even variant confers cross-protection against challenge with a *Yersinia* strain producing another V type, subtype or variant (22, 24, 25, 27–30). For instance, passive immunization of mice with rabbit anti-sera raised against purified V antigen from *Y. pestis* protected against plague and pseudotuberculosis (27). Moreover, Motin et al. also showed that anti-V antigen (from *Y. pseudotuberculosis*) is effective in providing passive immunity against plague and pseudotuberculosis but clearly ineffective against an infection by *Y. enterocolitica* indicating that a potential problem with cross-protection may exist (27). Furthermore, passive immunization with rabbit anti-sera raised against purified V antigen from *Y. pestis* protected against *Y. pestis* strains in which *lcrV* had been exchanged with that of *Y. enterocolitica* (24). In contrast, we have reported that active immunization with *Lactococcus lactis* secreting the V antigen from *Y. pseudotuberculosis* confers protection against pseudotuberculosis but not against plague (28). Hence, the present study was designed to establish why there was an absence of cross-protection after active immunization with an LcrV-producing *L. lactis*.

MATERIALS AND METHODS

Bacterial Strains, Plasmids, and Growth Conditions

We used *Y. pseudotuberculosis* strains (from our own collection and isogenic mutants generated in the present study), *L. lactis* strain MG1363, *Escherichia coli* SM10 λpir, the lactococcal pNZYR expression vector conferring resistance to chloramphenicol, and the pCVD442 suicide vector conferring resistance to ampicillin and sensitivity to sucrose. *Y. pseudotuberculosis, E. coli,* and *L. lactis* were, respectively, cultured in lysogeny broth at 28°C, lysogeny broth at 37°C and M17 medium supplemented with 0.5% glucose at 30°C. Growth media were supplemented with chloramphenicol (10 μg/ml) or ampicillin (100 μg/ml) when required. Custom pNZYR vectors (Eurogentec) containing the *lcrVs* of interest (translationally fused with the *L. lactis usp45* secretion signal sequence and under the control of the *Pusp45* promoter) were generated as previously described (28). Mutants of *Y. pseudotuberculosis* in which *lcrV* is mutated in the region of interest were generated by allelic exchange after mating *Y. pseudotuberculosis* with *E. coli* strain SM10 λpir harboring the pCVD442 plasmid containing an insert of interest as previously described (31, 32). Inserts of interest were obtained by overlapping PCR using the primer sets 5′-ACATGCATGC GTCATGGTTCTTCAGTTTTA-3′/5′-TATAAGAGTATGAGT TTTTCAGATTACCCAACGCCCCGGT-3′ and 5′-TTGG GTAATCTGAAAAACTCATACTCTTATAATAAAGATA-3′/5′-AAAAAGAGCTCCGTTGAGCATGGCGATAGTT-3′ and *Y. pestis* 195/P DNA template. Amplicons were digested with *SphI/SacI* endonucleases thanks to the restriction sites found in the primers (see underlined bases), purified then ligated to the *SphI/SacI*-restricted pCVD442 via standard T4 DNA ligase procedures. Mutations were verified by nucleotide sequencing. The *Y. pseudotuberculosis* strains producing the mutated LcrV were found to be fully virulent in mice using the same protocol as described below.

Evaluation of the Efficacy of Immunization

Groups ($n = 9$–12) of 8-week-old, female BALB/c mice (Charles River) were immunized intranasally with sterile, phosphate-buffered saline (PBS) containing 10^9 CFU of *L. lactis* producing (or not) the V antigen of interest, as previously described (28). Immunization was repeated on days 2, 22, 23, 43, and 44. One week after the final boost, serum was collected from animals and the anti-V titers were measured as previously described (28). Notably, microtiter well plates (Nunc-Immuno Plate) were coated with purified recombinant *Y. pestis* LcrV in PBS (pH 7.2) at 50 ng per well. Following overnight incubation, wells were blocked in PBS with 0.05% Tween 20, 3% bovine serum albumin (BSA). Blood samples were tested using 2 fold serial dilutions (from 1:50 and 1:2) in PBS with 0.05% Tween 20, 1% BSA. Specific antibody binding to LcrV was detected using horseradish peroxidase-conjugated rat anti-mouse IgG (BD Biosciences). The IgG antibody titer is given in terms of dilutions as described previously (28). Briefly, the OD450 was measured with an Elx800GUV automated microplate reader (Bio-Tek Instruments Inc., Vinooski, VT). End-point titers were calculated as the reciprocal of the dilution producing the same OD450 as three times the background using the KC4 program (Kineticalc for Windows, Bio-Tek Instruments). Lastly, 2 weeks after the final boost, mice were inoculated intravenously with 300 μl of PBS containing 10^3 *Y. pseudotuberculosis* (a dose that we assessed to be above the LD50 for all the strains used prior challenge of immunized mice). The mice's survival was monitored daily for 3 weeks.

Sequencing, Translation, and Phylogenetic Analysis

DNA from *Y. pseudotuberculosis* strains was purified using the Nucleospin kit (Macherey-Nagel) and then used to amplify *lcrV* by PCR using the primer sets 5′-TCACCGCGCA AAATTATTGC-3′; 5′-TTGTCTGCGATAAGCTCTT G-3′. Purified amplicons were sequenced with the primers used for amplification and the following primers: 5′-CCTAGCTTAT TTTCTACCCG-3′; 5′-GAACCGGGGCGTTGGGTAATC-3′; 5′-GTTGGTTGTCATAATGACCGCC-3′; 5′-CTAACCA

AGTCGTTGAGCGG-3′. Nucleotide sequences were translated using the ExPaSy Translate tool (http://web.expasy.org/translate/). Lastly, a phylogenic tree based on LcrV sequences was generated using the (http://phylogeny.lirmm.fr/phylo_cgi/index.cgi) web server (33).

Evaluation of Plasmid Stability

The stability of the various expression vectors containing the *lcrV* gene of interest and conferring resistance to chloramphenicol in *L. lactis* MG1363 was checked by growing bacteria in M17 broth. After 10 subcultures (~100 generations) in the absence of any selective pressure, bacteria were plated on M17 agar plates with or without chloramphenicol, in order to count chloramphenicol-resistant and -sensitive CFUs. Furthermore, the secretion of LcrV from three randomly clones grown on M17 agar without chloramphenicol was determined by Western-blot analysis.

Immunobloting and SDS-PAGE

For *Y. pseudotuberculosis*, overnight cultures of bacteria grown in LB at 28°C were diluted 1:20 into fresh LB containing 2.5 mM $CaCl_2$. After 2 h incubation at 37°C, the optical density (OD) at 600 nm of the cultures was measured, then bacterial cells were collected after centrifugation at 4°C for 20 min and suspended in PBS such as each suspension of interest reaches an equivalent bacterial load. Bacterial suspensions were admixed with Laemmli buffer with β mercaptoethanol. Whole-cell lysates were separated by sodium dodecyl sulfate-polyacrylamide gel electrophoresis (SDS-PAGE). Notably, two gels were run at the same time. One gel was stained with Coomassie blue dye to control the loading and the other gel was used to electrophoretically transfer proteins to a nitrocellulose membrane. Immunoblot analyses were performed using polyclonal rabbit antibodies to *Y. pestis* LcrV (as primary antibody) (34) and an anti-rabbit IgG conjugated to horseradish peroxidase (as secondary antibody). Immunoreactive proteins were visualized and quantified by chemiluminescence using the LAS-3000 apparatus (Fujifilm) and the software Multi-Gauge (Fujifilm), respectively. The signal intensities reported here have a linear relationship with antigen loading (data not shown). For recombinant *L. lactis* secreting LcrV, cultures from an overnight growth at 37°C and with equivalent ODs were centrifuged and bacterial culture supernatants were filtered through 0.22 μm Millex HA membranes. Quantity of LcrV secreted by *L. lactis* was determined by chemiluminescence as described above for *Y. pseudotuberculosis*. It was further confirmed by SDS-PAGE and Coomassie blue staining using a culture supernatant prepared another a day than that of used for quantification by chemiluminescence.

Statistical Analysis

The Mann-Whitney *U* test and the log-rank test were used to compare antibody titers and survival curves, respectively.

RESULTS

L. lactis Secreting LcrV From *Y. pestis* Does Not Confer Protection Against *Y. pseudotuberculosis* Strain 2777

Although the LcrV proteins from *Y. pestis* 195/P and *Y. pseudotuberculosis* strain 2777 share more than 97% sequence identity, we previously reported that mice immunized with *L. lactis* secreting the LcrV from *Y. pseudotuberculosis* strain 2777 were protected against pseudotuberculosis but not against plague (28). This surprising finding suggested that active immunization with LcrV from *Y. pestis* should be ineffective against *Y. pseudotuberculosis*. To test this hypothesis, we generated a strain of *L. lactis* that secreted LcrV from *Y. pestis* (V_{Yp}) and then assessed whether this new strain afforded protection against *Y. pseudotuberculosis*. To this end, groups of mice were immunized intranasally with *L. lactis* secreting V_{Yp} or that does not synthetize any LcrV (control strain), according to the previously reported protocol (28). As an additional control, a group of mice was immunized with *L. lactis* secreting LcrV from *Y. pseudotuberculosis* (V_{Ypst}). Two weeks after the end of the immunization protocol, mice were challenged intravenously with a lethal dose of *Y. pseudotuberculosis* strain 2777. The proportion of animals developing fatal pseudotuberculosis was similar in mice vaccinated with V_{Yp} and in unvaccinated mice (**Figure 1A**). This contrasted with the higher survival rate observed in animals vaccinated with V_{Ypst} (**Figure 1A**). The absence of cross protection was confirmed in an independent experiment.

Vaccination With the LcrV From *Y. pseudotuberculosis* Strain 2777 With a Single Substitution at Position 255 Does Not Confer Protection Against *Y. pseudotuberculosis* Strain 2777

The sequences of the LcrV proteins from *Y. pseudotuberculosis* strain 2777 and *Y. pestis* strain 195/P used in the above immunization experiments differ in the region considered to contain a plague-protective epitope (27, 35–37). In particular, the *Y. pestis* LcrV region lacks a threonine and has three residue substitutions: E205A, V226M, and N255D (**Figure 1B**). It is noteworthy that the missing threonine and the substituted residue 226 are located within the 16-amino acid LcrV segment (residues 218–234), which is determinant for protein conformation and is referred to here as the "conformational segment" (CS) (27, 38). These residues are part of the immunodominant epitopes recognized by H-2d CD4 T cells and known be the major protective epitope (35, 39, 40). Furthermore, the Asn-255 in the *Y. pestis* LcrV is known to be a key residue for the binding of a plague-protective monoclonal antibody, since the substitution N255D greatly reduces the antibody's affinity for LcrV (41). Given that substitution of E205A may not impact protein conformation, we hypothesized that the residue substitutions in the conformational segment and/or at position 255 were responsible for the absence of cross-protective immunity observed in our experiments. To test this hypothesis, we generated *L. lactis* strains secreting a

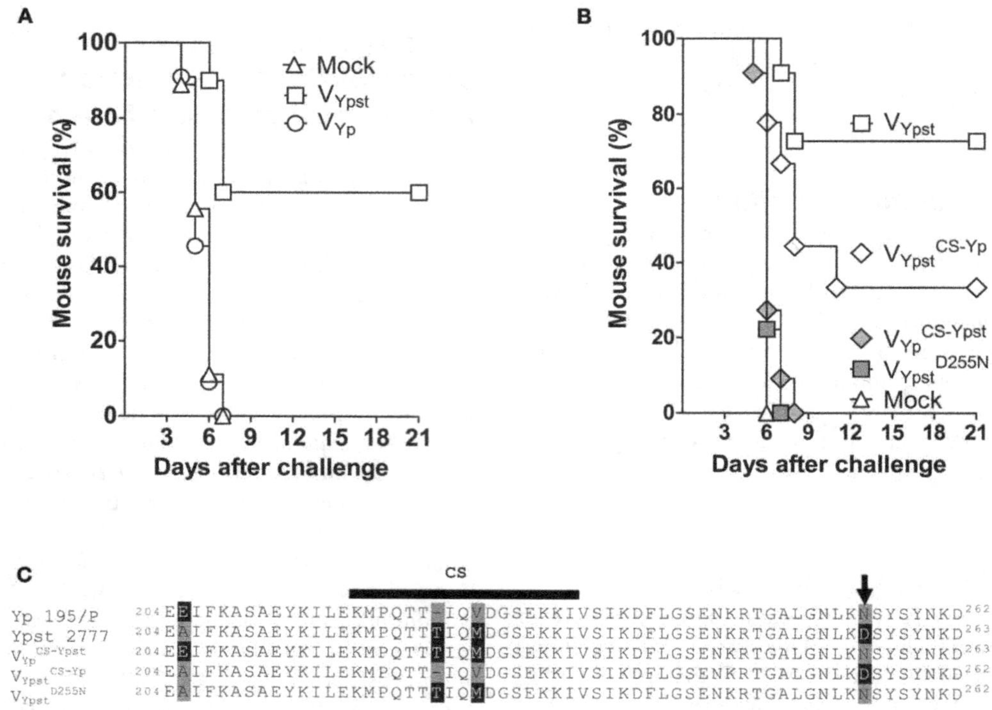

FIGURE 1 | Subtle substitutions in the V antigen used for vaccination enable *Yersinia pseudotuberculosis* to escape an immune response against LcrV. **(A,C)** The survival rate of mice ($n = 9$–11) vaccinated intranasally with *L. lactis* alone (Mock), or *L. lactis* secreting the V antigen from *Y. pseudotuberculosis* strain 2777 (V_{Ypst}), the V antigen from *Y. pestis* strain 195/P (V_{Yp}), V_{Ypst} with a substitution Asp to Asn at the position 255 (${V_{Ypst}}^{D255N}$), or V_{Ypst} or V_{Yp} in which the "conformational segment" (CS) had been exchanged (${V_{Ypst}}^{CS-Yp}$ and ${V_{Yp}}^{CS-Ypst}$). Mice were significantly protected ($p < 0.05$ in a log-rank test) against *Y. pseudotuberculosis* strain 2777 only when they were vaccinated with V_{Ypst} or ${V_{Ypst}}^{CS-Yp}$ **(A,C)**. The difference in protection observed in mice vaccinated with V_{Ypst} and ${V_{Ypst}}^{CS-Yp}$ was not significant ($p > 0.07$ in a log-rank test). **(B)** Alignments of the protective regions of interest in LcrV proteins. The arrowhead indicates the residue at position 255.

hybrid *Y. pestis*/*Y. pseudotuberculosis* LcrV protein in which the conformational segments had been exchanged (yielding ${V_{Ypst}}^{CS-Yp}$ and ${V_{Yp}}^{CS-Ypst}$) or with an Asp to Asn substitution at position 255 in V_{Ypst} (yielding ${V_{Ypst}}^{D255N}$) (**Figure 1B**). After immunization as described above, we assessed the degree of protection conferred against a lethal intravenous challenge of *Y. pseudotuberculosis* strain 2777 in the various groups of mice. All mice immunized with *L. lactis* alone or *L. lactis* secreting ${V_{Yp}}^{CS-Ypst}$ and ${V_{Ypst}}^{D255N}$ died, furthermore, the time courses of mortality were similar ($p > 0.05$) in the three groups (**Figure 1C**). Hence, ${V_{Yp}}^{CS-Ypst}$ and ${V_{Ypst}}^{D255N}$ did not confer any protection against pseudotuberculosis. In contrast, immunization with *L. lactis* secreting V_{Ypst} and ${V_{Ypst}}^{CS-Yp}$ afforded protection (**Figure 1C**). However, the survival rate was higher in mice immunized with V_{Ypst} than with ${V_{Ypst}}^{CS-Yp}$ (~75 and ~35%, respectively). Similar results were obtained in an independent experiment: absence of cross-protection with ${V_{Yp}}^{CS-Ypst}$ and ${V_{Ypst}}^{D255N}$ and, the survival rates of mice vaccinated with V_{Ypst} and ${V_{Ypst}}^{CS-Yp}$ were 50 vs. 20%, respectively. However, the difference in protection between mice vaccinated with V_{Ypst} and ${V_{Ypst}}^{CS-Yp}$ against pseudotuberculosis in both experiments was not statistically significant ($p > 0.07$ in a Gehan-Wilcoxon test). Taken as a whole, these data indicate that a single substitution at position 255 of LcrV seems to abolish the vaccine efficacy.

Protection Against Pseudotuberculosis by Vaccination With LcrV Is Dependent on the *Y. pseudotuberculosis* LcrV Clade

Although the above-mentioned difference in protection resulting from the exchange of the conformational segment (i.e., vaccination with ${V_{Ypst}}^{CS-Yp}$) was not statistically significant, we suspected that it was biologically significant. We thought that the difference in protection was due to the fact that the amino-acid sequences of the conformational segments in the *Y. pseudotuberculosis* 2777 and *Y. pestis* 195/P strains are almost identical (**Figure 1B**). Hence, we hypothesized that *L. lactis* secreting the LcrV from *Y. pseudotuberculosis* strain 2777 (hereafter referred to as V_B) would not confer protection against strains producing a LcrV composed of (i) a conformational segment strongly differing from that of V_B and (ii) the same residue as V_B at position 255 (an Asp) since this latter seems important for immune escape. To test this hypothesis, we first sought to identify strains producing a V antigen that meet our criteria. To this end, we sequenced *lcrV* from our collection of 43 *Y. pseudotuberculosis* strains. Amino-acid sequence and phylogenetic analyses listed several strains of interest which cluster into five major clades (V_A to V_{D2}) (**Figure 2A**). LcrV from each of the five clades had its own distinct conformational segment. The V_A, V_B, V_{D1} and V_{D2} antigens have an Asp at position 255, whereas the V_C antigen has an Asn (**Figure 2B**). Hence, we immunized mice with *L. lactis* secreting V_B and measured the animals' survival after inoculation with *Y. pseudotuberculosis* strain 2889, AH, and 2790 producing a V_D antigen and the strain 2777 producing the V_B antigen used for vaccination. As an additional and independent experiment, we monitored survival in mice immunized with *L. lactis* secreting V_B after challenging with the 2781 strain whose V antigen sequence is identical to that used for vaccination. Even though the vaccinated mice mounted a strong anti-V response (data not shown), the vaccination neither significantly decreased the mortality rate nor affected the kinetics of mortality in mice inoculated with strains 2889, AH, and 2790 (**Figure 2C** and **Supplementary Figure 1A**). In contrast, more than 90% of mice infected with the strain 2777 and 2781 survived (**Figure 2C** and **Supplementary Figure 1B**). This total absence of cross-protection between V_{Ypst} variants supports the hypothesis whereby the residues in the conformational domain influence protective immunity.

LcrV From *Y. pestis* Confers Protection Against *Y. pseudotuberculosis* Strains Producing the Same LcrV Clade as *Y. pestis*

A phylogenetic analysis of *Y. pestis* strains CO92 and 195/P and our *Y. pseudotuberculosis* strains revealed that *Y. pseudotuberculosis* LcrV clade C was closely related to *Y. pestis* LcrV (**Figure 2A**). Further sequence analysis showed that LcrV proteins from this particular clade have the same conformational domain sequence and an Asn residue at position 255 as LcrV from *Y. pestis* (i.e., $V_{Yp} = V_C$). Bearing in mind that the residues of the conformational segment and at position 255 might influence protective immunity (**Figures 1A,C, 2C** and **Supplementary Figure 1**), we predicted that LcrV from *Y. pestis* (i.e., V_C) should protect against *Y. pseudotuberculosis* strains producing a V_C antigen but not another V clade. In line with our prediction, mice vaccinated with the V_C antigen were protected against pseudotuberculosis caused by the strain 2843 producing a V_C antigen but not against the strain 2889 which synthesizes a V_D antigen (see the red arrowheads in **Figures 2A,D**). Lastly, it is noteworthy to highlight that vaccination with V_C (V_{Yp}) did not confer protection against a lethal challenge of *Y. pseudotuberculosis* strain 2777 (**Figure 1A**), which produces V_B (**Figure 2A**). In conclusion, our above and present data show that vaccination of mice with the LcrV antigen protects against *Yersinia* strains producing the same LcrV clade as that used for vaccination.

Polymorphism in the LcrV Antigen Enables Immune Escape

The absence of cross-protection reported in the different above sections does not appear to be dependent on the serotype of the *Y. pseudotuberculosis* strain used [**Figure 2A** (see the red arrowheads)]. It did not correlate either with the virulence of the strain used (**Figure 2C**). Lastly, it was not associated with the LcrV antigen production level between the different *Y. pseudotuberculosis* strains (**Figure 3A**). Nor does it appear to result from any instability of the lactococcal plasmid harboring the genetic construct of interest since fewer than 1% of *L. lactis* subcultured 10 times in the absence of selective pressure lost the plasmid. Lastly, it does not appear to result from an apparent difference in the amounts of V antigen secreted by the *L. lactis* strains (**Figure 3B** and **Supplementary Figure 2**). In agreement with these *in vitro* data, vaccinated mice had similar anti-V IgG titers regardless of the vaccine strain used (**Figures 3C,D**). In other words, our data support the idea that LcrV polymorphism enables escape from anti-V immunity. To confirm this conclusion, we generated isogenic mutants of *Y. pseutoduberculosis* strain 2777 in which V_B has been mutated such as the mutant strains produce either a V_c antigen or a V_B antigen with an Asp to Asn substitution at position 255 (i.e., V_B^{D255N}). Next, we compared the ability of the mutant and parental strains to escape immunity in mice vaccinated with the V_B antigen. Results showed that mice were protected against the parental strain but not against the isogenic mutants (**Figure 4**). In conclusion, LcrV antigen polymorphism enables immune escape and a single substitution at position 255 is sufficient for this escape.

DISCUSSION

Comparative sequence analysis of LcrV from the three pathogenic *Yersinia* species reveals two major types, three subtypes and several variants of the V antigen (21, 23, 25, 26). Here, we report there are at least five V variants produced by the group of *Yersinia* species encompassing *Y. pestis* and *Y. pseudotuberculosis* (**Figure 2A**). The results of former vaccination studies using purified recombinant V antigen produced in *E. coli* suggested that protection against all types of yersiniosis does not depend on the type, subtype or variant of V antigen used for vaccination (22, 41). Our previous results (28) and present data do not support the latter conclusion. Indeed, vaccination using LcrV-secreting *L. lactis* did not provide cross-protective immunity between different V antigen variants or even between different V subtypes and types (**Figures 1, 2, 4** and **Supplementary Figure 1**). Furthermore, isogenic mutants producing LcrV mutated for residues of interest escape from the protection conferred by the original LcrV (**Figure 4**). However, our data are reminiscent (to some extent) of those recorded in passive immunization experiments. Cross-protection between different V types and subtypes was not observed unless a large amount of antibody was provided (22, 24, 25, 42, 43). Hence, it is conceivable that vaccination using recombinant *L. lactis* does not generate a sufficiently cross-protective antibody titer. Unfortunately, the fact that studies showing cross-protection after vaccination with recombinant V-antigen produced in *E. coli* did not report the antibody titers (22, 41) makes it impossible to evaluate the above-mentioned hypothesis at present. The discrepancy might also be due to differences in the cellular

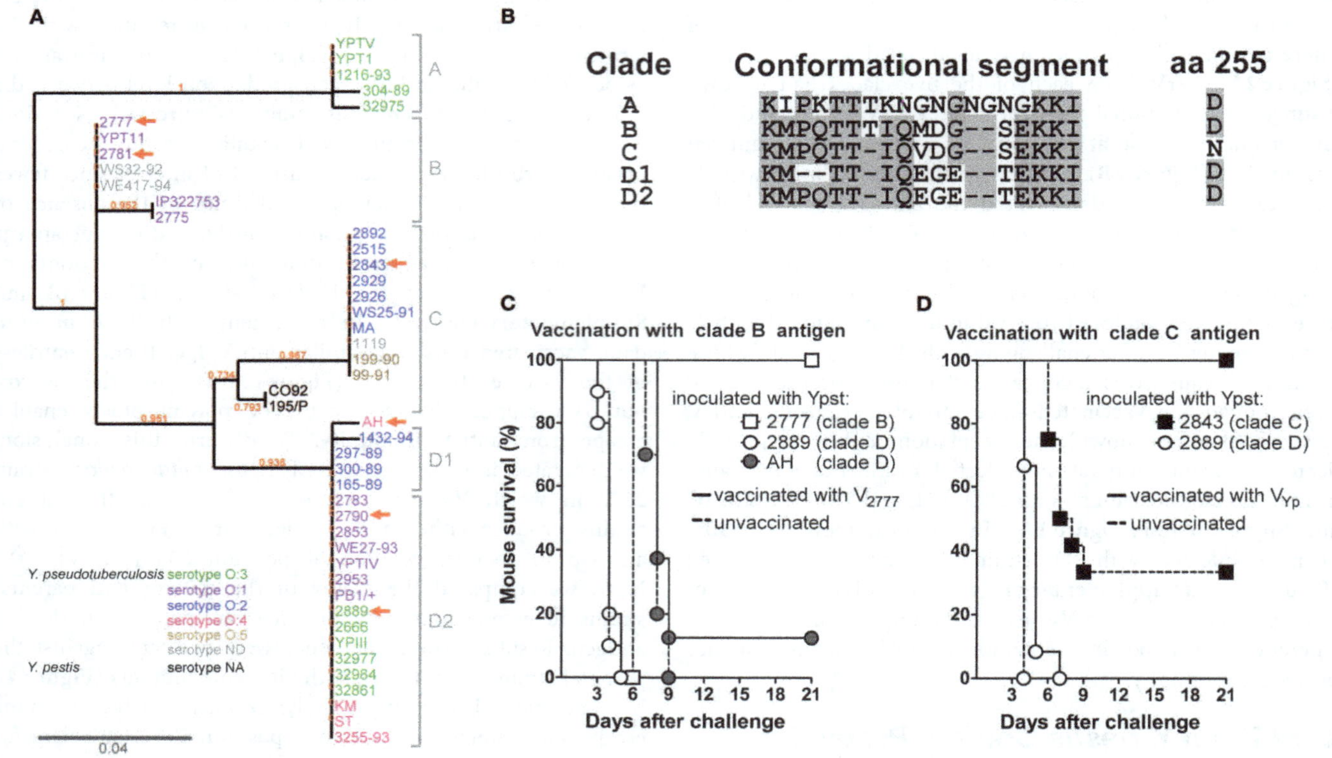

FIGURE 2 | The different clades of V antigen and their respective degrees of protective immunity. **(A)** The phylogeny of the V antigen. The tree was generated from predicted amino-acid sequences of 43 strains of *Y. pseudotuberculosis* and 2 strains of *Y. pestis* (C092 and 195/P), using "One-Click" phylogenetic analysis software (http://phylogeny.lirmm.fr/phylo_cgi/index.cgi) (33). The bootstrap values are indicated for each branch. The scale bar indicates the amount of change. The length of a branch is proportional to the number of substitutions per site. Strains that belong to the same clade are indicated by the letter **(A–D)** shown on the right side of the tree. However, the clade D has been subdivided into two groups on the basis of the sequence of the CS, as shown in **(B)**. Red arrowheads indicate the V antigen and the strains used in the experiments are shown in panels **C,D** and in the **Supplementary Figure 1**. Lastly, the serotype of each strain is given; ND, not determined; NA, not applicable. The amino-acid sequences of all the strains used to generate the tree are given in the **Supplementary Material**. **(B)** Alignments of the CS and the residue at position 255 of LcrV from different clades. **(C,D)** The survival rate of mice ($n = 9$–10) vaccinated intranasally with *L. lactis* alone (Mock) or *L. lactis* secreting the V antigen from *Y. pseudotuberculosis* strain 2777 (V_{2777}; clade B) or the V antigen from *Y. pestis* strain 195/P (V_{Yp}; clade C), after challenge with strains 2777 (clade B), 2889, AH (both clade D) (C), 2843 (clade C), or 2889 (clade D) **(D)**. Mice were significantly protected ($p < 0.05$ in a log-rank test) against *Y. pseudotuberculosis* strains 2777 and 2843.

response generated by the vaccination using purified antigen vs. recombinant *L. lactis*.

The absence of cross-protection reported here indicates that polymorphism in the *Yersinia* LcrV antigen enables immune escape. LcrV is a polymorphic antigen within which the most effectively protective region encompasses residues 135 to 262 (37). The major protective epitope is thought to correspond to the highly polymorphic segment (residues 218–234) (22, 25, 27). However, this latter assumption had not been confirmed because the major epitope's residues had not been identified *per se* (22, 24). It is worth noting that immunization with a V antigen lacking the residues 240–271 provides almost no protection (44). In our present experiments, we used (i) *L. lactis* strains secreting a V antigen in which the hypervariable segment had been exchanged and (ii) *Y. pseudotuberculosis* strains producing a V antigen whose hypervariable segment differed significantly from that used for vaccination. Our results suggest that polymorphism in the hypervariable segment enables escape from an active immune response against LcrV (**Figures 1, 2**). One could consider that the absence of cross-protection reported here reflects factors that differ from one strain to another. Our previous (28) and present work show that vaccination with V_B antigen protects against strains producing V_B but not V_C or V_D antigen [**Figure 2C, Supplementary Figure 1,** and (28)] and vaccination with V_C antigen protects against strains producing Vc but not V_B or V_D antigen (**Figures 1A, 2D**). Hence, if the absence of cross-protection reflects factors that differ from one strain to another, our data would indicate that strains producing a V_B, V_C and V_D antigen each produce a distinct, very efficient factor that enables escape from immunity directed against a V-antigen from a clade other than their own (**Figures 1A,C, 2C,D** and **Supplementary Figure 1**). However, the production of several (at least three) distinct, efficient factors for immune evasion is unlikely. Nonetheless, we acknowledge that vaccination assays using isogenic mutant strains will definitively end such a debate. However, regardless of the result of such a study, our data indicate

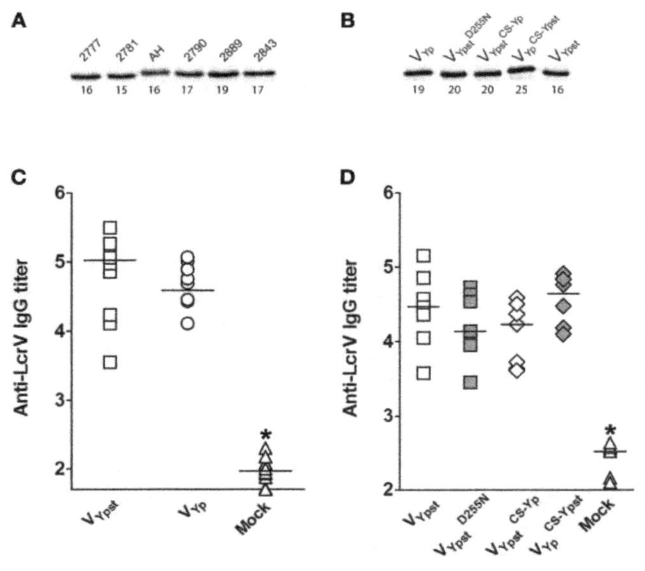

FIGURE 3 | *Y. pseudotuberculosis* strains produce similar amounts of V antigen, and recombinant *L. lactis* strains secrete similar amounts of V antigen and induce similar anti-V IgG antibody titers in vaccinated mice. An immunoblot of **(A)** whole-cell lysate of *Y. pseudotuberculosis* strains and **(B)** culture supernatants from *L. lactis* secreting the V antigen from *Y. pestis* strain 195/P (V_{Yp}), the V antigen from *Y. pseudotuberculosis* strain 2777 (V_{Ypst}), V_{Ypst} with an Asp to Asn substitution at position 255 ($V_{Ypst}{}^{D255N}$), or V_{Ypst} or V_{Yp} in which the "conformational segment" (CS) had been swapped ($V_{Ypst}{}^{CS-Yp}$ and $V_{Yp}{}^{CS-Ypst}$). The number beneath each band indicates a relative density value (expressed as a percentage), which was calculated by dividing the density of the band of interest by the sum of the density of all the bands shown on the blot (See **Supplementary Figure 3** for raw data). **(C,D)** Antibody titers shown in the panels were measured using sera from the mice used in the experiments shown in **Figures 1A,C**, respectively. Horizontal lines indicate the mean titers. Regardless of the V antigen used for vaccination, the antibody titer against the V antigen of the mock group is significantly different than the antibody titers from mice vaccinated with a V antigen ($^*p < 0.003$ a Mann-Whitney *U* test). The antibody titers measured in the various groups of mice did not differ significantly as a function of the V antigen used ($p > 0.13$ in a Mann-Whitney *U* test).

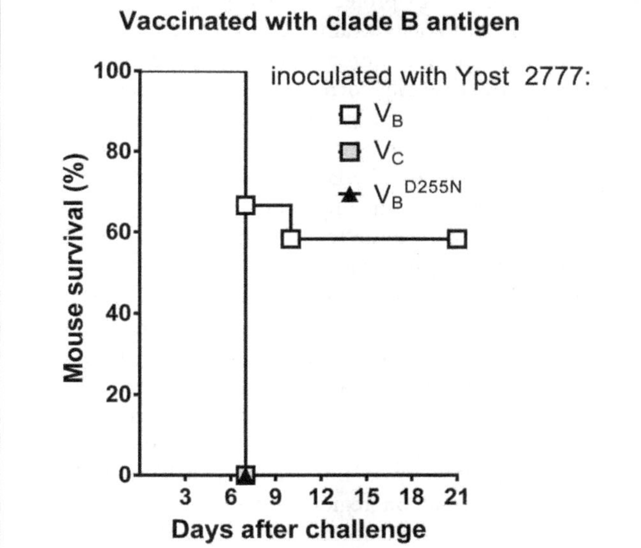

FIGURE 4 | Subtle polymorphism in the *Yersinia* LcrV antigen enables immune escape from the protection conferred by LcrV. The survival rate of mice (n = 9–11) vaccinated intranasally with *L. lactis* secreting the V_B antigen from strain 2777 after challenge with strain 2777 or its isogenic mutants in which V_B had been mutated such as they synthetize a V_C antigen or a V_B antigen with an Asp to Asn substitution at position 255 (V_B^{D255N}). Mice were significantly protected against strain 2777 but not against its isogenic mutants ($p < 0.05$ in a log-rank test).

that a V-based vaccine should be composed of several V antigens if it is to protect the host against a wide array of *Yersinia* strains or at least *Y. pseudotuberculosis* strains.

Although the role of polymorphism in the conformational segment in immune escape remains to be definitively proven, there is no doubt that residue at position 255 plays a critical role in the immune escape. Indeed, a single mutation at position 255 in the V antigen used for vaccination abolishes the efficacy of the vaccine and the same mutation in *Y. pseudotuberculosis* unables the mutant to escape from protection conferred by the original antigen (**Figures 1C, 4**). Interestingly, the three-dimensional structure of LcrV reveals that residues 218–234 and 255 are close to each other in space (38) and so might form a conformational epitope. However, the residues might also be part of two distinct epitopes that lead to the generation of cooperative antibodies. Consistent with the latter idea, a substitution at residue 255 annihilates the binding of the protective plague monoclonal antibody (mAb) 7.3 [which recognizes an epitope that does not contain the hypervariable segment (41)].

We previously reported that mucosal vaccination of mice with *L. lactis* secreting LcrV induced cell- and antibody-mediated protective immunity against *Y. pseudotuberculosis* infection in the mouse and the protection is long-lasting. We further reported that activated $CD4^+$ T lymphocytes are necessary for vaccine-induced protection, whereas $CD8^+$ T cells may have a moderate role. Several LcrV epitopes recognized by CD4 T-cell have been described in the literature for *Y. pestis* (39, 40, 45–47). Interestingly, residue at position 255 is the last residue of a strong T-cell epitope (GSENKRTGALGNLKN) for BALB/c mice (46, 47). It is unknown whether the last N residue of this epitope is important for vaccination using peptides. However, if it is important, our data suggest that vaccination should be performed using a mixture of peptides containing different residues at the end of the epitope. Moreover, further work should be performed to support this hypothesis because previous authors used linear peptides in their studies whereas we used the whole LcrV protein.

Although an anti-V humoral response is thought to be involved in protection, it is considered that the antiserum's ability to neutralize *Yersinia*-mediated toxicity against macrophages is more strongly correlated with vaccine efficacy than the anti-V antibody titer (48). In this context, the titer of native antibodies that bind the epitopes recognized experimentally by the monoclonal antibodies mAb7.3 and mAb-BA5 might be correlated with the degree of protection, since both mAb7.3 and mAb-BA5 protect against plague by neutralizing *Y. pestis*-mediated toxicity against host cells (22, 24). However, anti-V serum's ability to compete with mAb7.3 in binding to LcrV does not necessarily correlate with protection (49, 50)—suggesting that one should quantify the titer of antibodies that bind to mAb-BA5's epitope (rather than mAB7.3's epitope). Our data do not support the latter hypothesis, since mAb-BA5 recognizes an epitope (encompassing residues 196–224) that did not appear to contribute to protection in our experiments. In fact, our data suggest that only the titer of antibodies that recognize polymorphic epitopes comprising the hypervariable region and/or residue 255 is correlated with vaccine efficacy (**Figure 1C**). In other words, an antiserum's ability to neutralize *Yersinia*-mediated toxicity against macrophages may only be correlated with protection against a specific V variant. Unfortunately, setting up a vaccine efficacy assay based on macrophage toxicity is not trivial, considering that this methodology is based on regions whose residue composition (i) varies from one strain to another within the same species and (ii) may have changed during the evolution of the targeted strain.

The above conclusion contrasts with the observed correlation between anti-V titers on one hand and protection on the other (and even cross-protection between V antigen types) (22, 43, 51, 52). This discrepancy could be explained by an antibody titer threshold model in which a vaccine's protective/cross-protective efficacy is only correlated with the antibody titer when the latter is over a threshold value. Below this value, only the serum's ability to inhibit *Yersinia*-mediated toxicity against macrophages would be a marker of protection. However, data from this assay would probably be highly relevant for a targeted V variant (as mentioned above).

Lastly, our present data (in combination with previous results) strongly suggest that V-based vaccines using purified antigens or recombinant strains (such as *L. lactis*) as the antigen vehicle should be composed of multiple V antigens. The supporting evidence is as follows: (i) vaccination in humans using one LcrV variant may not necessarily induce the antibody titer threshold conferring cross-protection, since there is a difference in anti-V titers of several orders of magnitude between volunteers enrolled in clinical trial (53) and patients having experienced yersiniosis (54), (ii) only the most protective antibodies (which, according to our data, might be those recognizing polymorphic epitopes) persist over time (53), (iii) during the course of bacterial evolution, amino-acid substitutions may have occurred within the polymorphic, protective epitope, and (iv) the current assays thought to predict protection may be too restrictive (for the reasons described above).

In conclusion, we were able to identify at least one key residue composing the major protective epitopes in V antigen. Our present results indicate that a V-based vaccine should be composed of several V antigens if it is to protect the host against a wide array of *Yersinia* strains. However, this type of vaccine may fail to protect the host against strains producing a newly evolved V antigen. Consistently, the data further suggest that a vaccine efficacy assay other than that based on anti-V titer measurement and bacterial-mediated toxicity against macrophages must now be developed. Our general conclusion is based (i) on our previous study in which we reported that V antigen from *Y. pseudotuberculosis* does not confer protection against bubonic plague (28), and the present study using a pseudotuberculosis infection model and mice vaccinated with *L. lactis* secreting V antigens from *Y. pseudotuberculosis* and *Y. pestis*. Further work using *Y. pestis* and *Y. enterocolitica* with their respective LcrVs vaccine but also different animal models shall be done to fulfill our conclusions.

AUTHOR CONTRIBUTIONS

CD and FS: conceived and designed the experiments, analyzed the data, and wrote the paper. CD, AD, SP, MM, MS, LM, GD, DB, NB, SB-G, and NL: performed the experiments.

ACKNOWLEDGMENTS

We thank Jim Bliska for kindly providing the purified recombinant *Y. pestis* LcrV protein and for critical reading of the manuscript. We also thank Susan Straley for kindly providing the rabbit polyclonal anti-LcrV antibody. We thank Maud Merchez for technical assistance.

REFERENCES

1. Butler T. *Plague and Other Yersinia Infections*New York, NY; London: Plenum Medical Book Company (1983). doi: 10.1007/978-1-4684-8422-9
2. Drummond N, Murphy BP, Ringwood T, Prentice MB, Buckley JF, Fannin S. *Yersinia enterocolitica*: a brief review of the issues relating to the zoonotic pathogen, public health challenges, and the pork production chain. *Foodborne Pathog Dis.* (2012) 9:179–89. doi: 10.1089/fpd.2011.0938
3. Kaasch AJ, Dinter J, Goeser T, Plum G, Seifert H. *Yersinia pseudotuberculosis* bloodstream infection and septic arthritis: case report and review of the literature. *Infection.* (2012) 40:185–90. doi: 10.1007/s15010-011-0160-2
4. Piroth L, Meyer P, Bielefeld P, Besancenot JF. [*Yersinia bacteremia* and iron overload]. *Rev Med Interne.* (1997) 18:932–8. doi: 10.1016/S0248-8663(97)80112-9
5. Bergmann TK, Vinding K, Hey H. Multiple hepatic abscesses due to *Yersinia enterocolitica* infection secondary to primary haemochromatosis. *Scand J Gastroenterol.* (2001) 36:891–5. doi: 10.1080/003655201750313450
6. Kangas S, Takkinen J, Hakkinen M, Nakari UM, Johansson T, Henttonen H, et al. *Yersinia pseudotuberculosis* O:1 traced to raw carrots, Finland. *Emerg Infect Dis.* (2008) 14:1959–61. doi: 10.3201/eid1412.080284
7. Vincent P, Leclercq A, Martin L, Duez JM, Simonet M, Carniel E. Sudden onset of pseudotuberculosis in humans, France, 2004-05. *Emerg Infect Dis.* (2008) 14:1119–22. doi: 10.3201/eid1407.071339
8. Scharff RL. Economic burden from health losses due to foodborne illness in the United States. *J Food Prot.* (2012) 75:123–31. doi: 10.4315/0362-028X.JFP-11-058
9. Baskin GB, Montali RJ, Bush M, Quant TJ, Smith E. Yersiniosis in capture exotic mammals. *J Am Vet Med Assoc.* (1977) 171:908–12.
10. Chapman DI, Chapman NJ, Atherton JG, Platt H. Yersiniosis in a freeliving fallow deer. *Vet Rec.* (1979) 105:573–4.
11. Jerret IV, Sleek KJ, Robertson BI. Yersiniosis in farmed deer. *Aust Vet J.* (1990) 67:212–4. doi: 10.1111/j.1751-0813.1990.tb07763.x
12. Welsh RD, Ely RW, Holland RJ. Epizootic of *Yersinia pseudotuberculosis* in a wildlife park. *J Am Vet Med Assoc.* (1992) 201:142–4.
13. Bielli M, Lauzi S, Pratelli A, Martini M, Dall'ara P, Bonizzi L. Pseudotuberculosis in marmosets, tamarins, and Goeldi's monkeys (Callithrichidae/Callimiconidae) housed at a European zoo. *J Zoo Wildl Med.* (1999) 30:532–6.

14. Iwata T, Une Y, Okatani AT, Kato Y, Nakadai A, Lee K, et al. Virulence characteristics of *Yersinia pseudotuberculosis* isolated from breeding monkeys in Japan. *Vet Microbiol.* (2008) 129:404–9. doi: 10.1016/j.vetmic.2007.11.029
15. Gaydos JK, Zabek E, Raverty S. *Yersinia pseudotuberculosis* septicemia in a beaver from Washington State. *J Wildl Dis.* (2009) 45:1182–6. doi: 10.7589/0090-3558-45.4.1182
16. Wobeser G, Campbell GD, Dallaire A, Mcburney S. Tularemia, plague, yersiniosis, and Tyzzer's disease in wild rodents and lagomorphs in Canada: a review. *Can Vet J.* (2009) 50:1251–6.
17. Warth JF, Biesdorf SM, De Souza C. *Yersinia pseudotuberculosis* O III causes diarrhea in Brazilian cattle. *Adv Exp Med Biol.* (2012) 954:107–10. doi: 10.1007/978-1-4614-3561-7_13
18. Smiley ST. Current challenges in the development of vaccines for pneumonic plague. *Expert Rev Vaccines.* (2008) 7:209–21. doi: 10.1586/14760584.7.2.209
19. Demeure CE. Live vaccines against plague and pseudotuberculosis. In: Carniel E, Hinnebusch BJ, editors. *Yersinia: Systems Biology and Controls.* Northfolk: Caister Academic Press (2010). p.143–68.
20. Williamson ED, Oyston PC. Protecting against plague: towards a next-generation vaccine. *Clin Exp Immunol.* (2013) 172:1–8. doi: 10.1111/cei.12044
21. Motin VL, Pokrovskaya MS, Telepnev MV, Kutyrev VV, Vidyaeva NA, Filippov AA, et al. The difference in the lcrV sequences between *Y. pestis* and *Y. pseudotuberculosis* and its application for characterization of *Y. pseudotuberculosis* strains. *Microb Pathog.* (1992) 12:165–75. doi: 10.1016/0882-4010(92)90050-X
22. Schmidt A, Schaffelhofer S, Muller K, Rollinghoff M, Beuscher HU. Analysis of the *Yersinia enterocolitica* 0:8 V antigen for cross protectivity. *Microb Pathog.* (1999) 26:221–33. doi: 10.1006/mpat.1998.0268
23. Anisimov AP, Dentovskaya SV, Panfertsev EA, Svetoch TE, Kopylov P, Segelke BW, et al. Amino acid and structural variability of *Yersinia pestis* LcrV protein. *Infect Genet Evol.* (2010) 10:137–45. doi: 10.1016/j.meegid.2009.10.003
24. Miller NC, Quenee LE, Elli D, Ciletti NA, Schneewind O. Polymorphisms in the lcrV gene of *Yersinia enterocolitica* and their effect on plague protective immunity. *Infect Immun.* (2012) 80:1572–82. doi: 10.1128/IAI.05637-11
25. Roggenkamp A, Geiger AM, Leitritz L, Kessler A, Heesemann J. Passive immunity to infection with *Yersinia* spp. mediated by anti-recombinant V antigen is dependent on polymorphism of V antigen. *Infect Immun.* (1997) 65:446–51.
26. Sing A, Reithmeier-Rost D, Granfors K, Hill J, Roggenkamp A, Heesemann J. A hypervariable N-terminal region of *Yersinia* LcrV determines Toll-like receptor 2-mediated IL-10 induction and mouse virulence. *Proc Natl Acad Sci USA.* (2005) 102:16049–54. doi: 10.1073/pnas.0504728102
27. Motin VL, Nakajima R, Smirnov GB, Brubaker RR. Passive immunity to yersiniae mediated by anti-recombinant V antigen and protein A-V antigen fusion peptide. *Infect Immun.* (1994) 62:4192–201.
28. Daniel C, Sebbane F, Poiret S, Goudercourt D, Dewulf J, Mullet C, et al. Protection against *Yersinia pseudotuberculosis* infection conferred by a *Lactococcus lactis* mucosal delivery vector secreting LcrV. *Vaccine.* (2009) 27:1141–4. doi: 10.1016/j.vaccine.2008.12.022
29. Singh AK, Kingston JJ, Murali HS, Batra HV. A recombinant bivalent fusion protein rVE confers active and passive protection against *Yersinia enterocolitica* infection in mice. *Vaccine.* (2014) 32:1233–9. doi: 10.1016/j.vaccine.2014.01.044
30. Singh AK, Kingston JJ, Gupta SK, Batra HV. Recombinant bivalent fusion protein rVE induces CD4+ and CD8+ T-cell mediated memory immune response for protection against *Yersinia enterocolitica* infection. *Front Microbiol.* (2015) 6:1407. doi: 10.3389/fmicb.2015.01407
31. Miller VL, Mekalanos JJ. A novel suicide vector and its use in construction of insertion mutations: osmoregulation of outer membrane proteins and virulence determinants in *Vibrio cholerae* requires *toxR*. *J Bacteriol.* (1988) 170:2575–83. doi: 10.1128/jb.170.6.2575-2583.1988
32. Sebbane F, Bury-Mone S, Cailliau K, Browaeys-Poly E, De Reuse H, Simonet M. The *Yersinia pseudotuberculosis* Yut protein, a new type of urea transporter homologous to eukaryotic channels and functionally interchangeable *in vitro* with the Helicobacter pylori UreI protein. *Mol Microbiol.* (2002) 45:1165–74. doi: 10.1046/j.1365-2958.2002.03096.x
33. Dereeper A, Guignon V, Blanc G, Audic S, Buffet S, Chevenet F, et al. Phylogeny.fr: robust phylogenetic analysis for the non-specialist. *Nucleic Acids Res.* (2008) 36:W465–9. doi: 10.1093/nar/gkn180
34. Nilles ML, Williams AW, Skrzypek E, Straley SC. *Yersinia pestis* LcrV forms a stable complex with LcrG and may have a secretion-related regulatory role in the low-Ca2+ response. *J Bacteriol.* (1997) 179:1307–16. doi: 10.1128/jb.179.4.1307-1316.1997
35. Hill J, Leary SE, Griffin KF, Williamson ED, Titball RW. Regions of *Yersinia pestis* V antigen that contribute to protection against plague identified by passive and active immunization. *Infect Immun.* (1997) 65:4476–82.
36. Quenee LE, Berube BJ, Segal J, Elli D, Ciletti NA, Anderson D, et al. Amino acid residues 196-225 of LcrV represent a plague protective epitope. *Vaccine.* (2009) 28:1870–6. doi: 10.1016/j.vaccine.2009.11.076
37. Vernazza C, Lingard B, Flick-Smith HC, Baillie LW, Hill J, Atkins HS. Small protective fragments of the *Yersinia pestis* V antigen. *Vaccine.* (2009) 27:2775–80. doi: 10.1016/j.vaccine.2009.03.011
38. Derewenda U, Mateja A, Devedjiev Y, Routzahn KM, Evdokimov AG, Derewenda ZS, et al. The structure of *Yersinia pestis* V-antigen, an essential virulence factor and mediator of immunity against plague. *Structure.* (2004) 12:301–6. doi: 10.1016/j.str.2004.01.010
39. Parent MA, Berggren KN, Mullarky IK, Szaba FM, Kummer LW, Adamovicz JJ, et al. *Yersinia pestis* V protein epitopes recognized by CD4 T cells. *Infect Immun.* (2005) 73:2197–204. doi: 10.1128/IAI.73.4.2197-2204.2005
40. Shim HK, Musson JA, Harper HM, Mcneill HV, Walker N, Flick-Smith H, et al. Mechanisms of major histocompatibility complex class II-restricted processing and presentation of the V antigen of *Yersinia pestis*. *Immunology.* (2006) 119:385–92. doi: 10.1111/j.1365-2567.2006.02447.x
41. Hill J, Leary S, Smither S, Best A, Pettersson J, Forsberg A, et al. N255 is a key residue for recognition by a monoclonal antibody which protects against *Yersinia pestis* infection. *Vaccine.* (2009) 27:7073–9. doi: 10.1016/j.vaccine.2009.09.061
42. Une T, Brubaker RR. Roles of V antigen in promoting virulence and immunity in yersiniae. *J Immunol.* (1984) 133:2226–30.
43. Anderson GW Jr., Leary SE, Williamson ED, Titball RW, Welkos SL, Worsham PL, et al. Recombinant V antigen protects mice against pneumonic and bubonic plague caused by F1-capsule-positive and -negative strains of *Yersinia pestis*. *Infect Immun.* (1996) 64:4580–5.
44. Overheim KA, Depaolo RW, Debord KL, Morrin EM, Anderson DM, Green NM, et al. LcrV plague vaccine with altered immunomodulatory properties. *Infect Immun.* (2005) 73:5152–9. doi: 10.1128/IAI.73.8.5152-5159.2005
45. Do Y, Park CG, Kang YS, Park SH, Lynch RM, Lee H, et al. Broad T cell immunity to the LcrV virulence protein is induced by targeted delivery to DEC-205/CD205-positive mouse dendritic cells. *Eur J Immunol.* (2008) 38:20–9. doi: 10.1002/eji.200737799
46. Khan AA, Babu JP, Gupta G, Rao DN. Identifying B and T cell epitopes and studying humoral, mucosal and cellular immune responses of peptides derived from V antigen of *Yersinia pestis*. *Vaccine.* (2008) 26:316–32. doi: 10.1016/j.vaccine.2007.11.028
47. Uppada JB, Khan AA, Bhat AA, Deshmukh R, Rao DN. Humoral immune responses and protective efficacy of sequential B- and T-cell epitopes of V antigen of *Yersinia pestis* by intranasal immunization in microparticles. *Med Microbiol Immunol.* (2009) 198:247–56. doi: 10.1007/s00430-009-0124-7
48. Zauberman A, Cohen S, Levy Y, Halperin G, Lazar S, Velan B, et al. Neutralization of *Yersinia pestis*-mediated macrophage cytotoxicity by anti-LcrV antibodies and its correlation with protective immunity in a mouse model of bubonic plague. *Vaccine.* (2008) 26:1616–25. doi: 10.1016/j.vaccine.2008.01.033
49. Garmory HS, Griffin KF, Brown KA, Titball RW. Oral immunisation with live aroA attenuated Salmonella enterica serovar Typhimurium expressing the *Yersinia pestis* V antigen protects mice against plague. *Vaccine.* (2003) 21:3051–7. doi: 10.1016/S0264-410X(03)00112-9
50. Garmory HS, Freeman D, Brown KA, Titball RW. Protection against plague afforded by immunisation with DNA vaccines optimised for expression of the *Yersinia pestis* V antigen. *Vaccine.* (2004) 22:947–57. doi: 10.1016/j.vaccine.2003.09.020

51. Reed DS, Martinez MJ. Respiratory immunity is an important component of protection elicited by subunit vaccination against pneumonic plague. *Vaccine.* (2006) 24:2283–9. doi: 10.1016/j.vaccine.2005.11.047
52. Stacy S, Pasquali A, Sexton VL, Cantwell AM, Kraig E, Dube PH. An age-old paradigm challenged: old baboons generate vigorous humoral immune responses to LcrV, a plague antigen. *J Immunol.* (2008) 181:109–15. doi: 10.4049/jimmunol.181.1.109
53. Williamson ED, Flick-Smith HC, Lebutt C, Rowland CA, Jones SM, Waters EL, et al. Human immune response to a plague vaccine comprising recombinant F1 and V antigens. *Infect Immun.* (2005) 73:3598–608. doi: 10.1128/IAI.73.6.3598-3608.2005
54. Li B, Du C, Zhou L, Bi Y, Wang X, Wen L, et al. Humoral and cellular immune responses to *Yersinia pestis* infection in long-term recovered plague patients. *Clin Vaccine Immunol.* (2012) 19:228–34. doi: 10.1128/CVI.05559-11

15

Legionella Effector AnkX Disrupts Host Cell Endocytic Recycling in a Phosphocholination-Dependent Manner

Samual C. Allgood[†], Barbara P. Romero Dueñas[†], Rebecca R. Noll, Colleen Pike, Sean Lein and M. Ramona Neunuebel[*]

Department of Biological Sciences, University of Delaware, Newark, DE, United States

Edited by:
Hayley J. Newton, University of Melbourne, Australia

Reviewed by:
Gunnar Neels Schroeder, Queen's University Belfast, Ireland
Stacey Gilk, Indiana University School of Medicine, United States

***Correspondence:**
M. Ramona Neunuebel
neunr@udel.edu

[†]These authors have contributed equally to this work.

The facultative intracellular bacterium *Legionella pneumophila* proliferates within amoebae and human alveolar macrophages, and it is the causative agent of Legionnaires' disease, a life-threatening pneumonia. Within host cells, *L. pneumophila* establishes a replicative haven by delivering numerous effector proteins into the host cytosol, many of which target membrane trafficking by manipulating the function of Rab GTPases. The *Legionella* effector AnkX is a phosphocholine transferase that covalently modifies host Rab1 and Rab35. However, a detailed understanding of the biological consequence of Rab GTPase phosphocholination remains elusive. Here, we broaden the understanding of AnkX function by presenting three lines of evidence that it interferes with host endocytic recycling. First, using immunogold transmission electron microscopy, we determined that GFP-tagged AnkX ectopically produced in mammalian cells localizes at the plasma membrane and tubular membrane compartments, sites consistent with targeting the endocytic recycling pathway. Furthermore, the C-terminal region of AnkX was responsible for association with the plasma membrane, and we determined that this region was also able to bind the phosphoinositide lipids PI(3)P and PI(4)P *in vitro*. Second, we observed that mCherry-AnkX co-localized with Rab35, a regulator of recycling endocytosis and with major histocompatibility class I protein (MHC-I), a key immunoregulatory protein whose recycling from and back to the plasma membrane is Rab35-dependent. Third, we report that during infection of macrophages, AnkX is responsible for the disruption of endocytic recycling of transferrin, and AnkX's phosphocholination activity is critical for this function. These results support the hypothesis that AnkX targets endocytic recycling during host cell infection. Finally, we have demonstrated that the phosphocholination activity of AnkX is also critical for inhibiting fusion of the *Legionella*-containing vacuole (LCV) with lysosomes.

Keywords: ***L. pneumophila*, AnkX, endocytic recycling, phagosome maturation, phosphocholination**

INTRODUCTION

Legionella pneumophila (strain Philadelphia-1, hereafter *Legionella*) is a Gram-negative, facultative intracellular bacterium that is regarded as an important cause of hospital- and community-acquired pneumonia (CDC, 2011; Viasus et al., 2013). In nature, *Legionella* proliferates within protozoa; however, human infection can occur following inhalation of contaminated aerosolized water droplets (Horwitz and Silverstein, 1980; Segal and Shuman, 1999). In the lung, alveolar macrophages engulf *Legionella* by phagocytosis, but are unable to degrade it through the usual process of phagosome maturation, which entails the sequential fusion of the phagosome with endocytic compartments and ultimately with the lysosome (Clemens et al., 2000). Instead, *Legionella* remains enclosed in a plasma membrane-derived compartment known as the *Legionella*-containing vacuole (LCV) that undergoes drastic remodeling into a compartment resembling the endoplasmic reticulum (Tilney et al., 2001). This process is dependent on the Dot/Icm specialized type IV secretion system (T4SS) that translocates over 300 bacterial effector proteins into the host cytosol (Berger and Isberg, 1993; Segal and Shuman, 1999; Luo and Isberg, 2004; Segal, 2013). Because of functional redundancy among *Legionella* effectors (O'Connor et al., 2011), the molecular strategies important for *Legionella*'s escape from phagolysosomal maturation are not yet well-understood. AnkX was first identified as one of several *Legionella* effector proteins that harbor ankyrin repeats (Pan et al., 2008). Mainly found in eukaryotes and some intracellular pathogens, ankyrin repeats are commonly involved in protein-protein interactions (Li et al., 2006), although for AnkX, whether these interactions involve any host components is unknown. The *Legionella* Δ*ankX* mutant is impaired in its ability to escape phagosome maturation, and more than 50% of the LCVs in Δ*ankX*-infected macrophage fuse with the lysosomal marker LAMP-1 (Pan et al., 2008). This is a dramatic effect considering that functional redundancy among *Legionella* effectors often masks phenotypes caused by the deletion of single genes. Although, existing data indicate that AnkX prevents microtubule-dependent vesicular transport (Pan et al., 2008), the precise mechanism through which AnkX inhibits LCV fusion with lysosomes has not yet been defined.

A subsequent study revealed that AnkX contains an N-terminal FIC domain (filamentation induced by cAMP; Roy and Mukherjee, 2009). The FIC domain is found in proteins from bacteria to humans and it typically catalyzes adenylylation or phosphocholination (Worby et al., 2009; Yarbrough et al., 2009; Mukherjee et al., 2011). AnkX's FIC domain catalyzes phosphocholination, which is the covalent addition of a phosphocholine moiety to a serine or a threonine residue of Rab GTPases (Mukherjee et al., 2011). Rab GTPases are small GTPases that function as molecular switches alternating between an active GTP-bound form and an inactive GDP-bound form (Stenmark, 2009). These proteins play central roles in defining the identity of membrane compartments and controlling membrane trafficking (Zerial and McBride, 2001). Growing evidence suggests that *Legionella* effector proteins tap into the host's membrane transport network by manipulating the activity of Rab GTPases either through mimicry of host regulatory proteins that control the nucleotide-bound state of Rab GTPases or by post-translational modification (Stein et al., 2012; Sherwood and Roy, 2013).

Rab1, a key regulator of ER-to-Golgi traffic, is recruited to the LCV early during infection, and multiple effectors are at work during infection to tightly control Rab1 activity (Machner and Isberg, 2006, 2007; Brombacher et al., 2009; Neunuebel et al., 2011, 2012; Tan and Luo, 2011; Mihai Gazdag et al., 2013; Mousnier et al., 2014). AnkX phosphocholinates inactive (GDP-bound) Rab1 which prevents its activation by guanine exchange factors (GEF) as well as its membrane extraction by the guanine dissociation inhibitor (GDI; Goody et al., 2012). In addition to Rab1, AnkX also phosphocholinates Rab35 and renders it unable to interact with Connecdenn, a protein that activates Rab35 by functioning as a GEF (Mukherjee et al., 2011). Despite these important biochemical observations, the biological functions of Rab GTPase phosphocholination have not yet been determined. In RAW 264.7 murine macrophages phosphocholination of Rab1 did not contribute to recruitment or retention of Rab1 on the LCV in this system (Hardiman and Roy, 2014). However, it is not yet known why AnkX targets Rab35 during infection.

Rab35 shares extensive sequence homology with Rab1; however, these Rab GTPases regulate distinct membrane transport pathways (Chua et al., 2010). Rab35 controls endocytic recycling of cargo between the plasma membrane and early endosomes (Klinkert and Echard, 2016) and has been implicated in phagosome maturation in both mammalian (Egami et al., 2011) and protozoan cells (Verma and Datta, 2017). An increasing number of bacterial pathogens have been reported to target endocytic recycling to promote their intracellular survival, including enterohemorrhagic *E. coli* (Furniss et al., 2016), uropathogenic *E. coli* (Dikshit et al., 2015), *Coxiella* (Larson and Heinzen, 2017), and *Chlamydia* (Ouellette and Carabeo, 2010; Larson and Heinzen, 2017). Given the known cellular functions of Rab35, we hypothesized that by targeting Rab35, AnkX would interfere with both endocytic recycling and phagosome maturation. Here, we demonstrate that *Legionella* disrupts endocytic recycling during macrophage infection. Moreover, we show that AnkX plays a significant role in disrupting endocytic recycling, and we reveal that its phosphocholination activity is critical for this function. We also found that in the absence of its phosphocholination activity, AnkX was unable to efficiently prevent acquisition of lysosomal markers on the LCV. We conjecture that AnkX interferes with phagosome maturation by disrupting endocytic recycling.

RESULTS

AnkX Localizes to the Plasma Membrane and Peripheral Endosomes

A previous study reported that AnkX was distributed in a punctate pattern when produced ectopically in mammalian cells (Pan et al., 2008). Immunostaining with an early endosome marker showed that GFP tagged AnkX did not co-localize with early endosomes (Mukherjee et al., 2011). We hypothesized that AnkX's cellular destination may be, at least in part, determined by the localization of its enzymatic substrates, Rab1 and Rab35. However, AnkX did not localize at the Golgi where Rab1 is predominantly found (Mukherjee et al., 2011), and we reasoned that perhaps AnkX is co-opted at cellular locations where Rab35 is present, namely the plasma membrane and endosomes. We fused AnkX to an N-terminal GFP tag and analyzed its subcellular localization within transiently transfected COS-1 cells using immunogold transmission electron microscopy. We determined that GFP-AnkX was associated with the plasma membrane, nascent endosomes, and tubular compartments at the periphery of the cell, reminiscent of the subcellular localization of recycling endosomes (**Figure 1**, Supplementary Figures 1, 2). These membrane compartments shuttle cellular components to the plasma membrane and commonly assume a tubular morphology (van Ijzendoorn, 2006).

The C-Terminal Half of AnkX Associates with Plasma Membrane

We then investigated which region of AnkX was responsible for localization to the plasma membrane. To address this question we generated truncated variants of AnkX fused to an N-terminal mCherry tag as follows: AnkX1-490 (containing the FIC domain and 4 ankyrin repeats), AnkX688-949 (containing 2 ankyrin repeats), AnkX721-949 (containing 1 ankyrin repeat), AnkX491-809 and AnkX491-949 (containing the same 8 ankyrin repeats), and AnkX810-949 (containing no ankyrin repeats; **Figure 2A**). These fragments were selected based on secondary structure predictions and available structural information (Campanacci et al., 2013). To establish which of these fragments localize at the plasma membrane, we transiently transfected HeLa cells with constructs carrying the *mCherry*-fused *ankX* truncations or full length *ankX*. We then stained the cells with Wheat Germ Agglutinin, a carbohydrate-specific dye that marks the plasma membrane, and analyzed the localization of mCherry-tagged AnkX and AnkX fragments in relation to the plasma membrane (**Figure 2B**). As expected based on the TEM results, mCherry-AnkX co-localized with WGA at the plasma membrane. In singly transfected cells, we did observe that mCherry-AnkX was not uniformly distributed along the plasma membrane, and instead it assumed a rather patchy distribution, favoring one side of the cell (Supplementary Figure 4). In 5% of the cells, mCherry-AnkX did not appear to be present at the plasma membrane, at least not in levels detectable by confocal microscopy (Supplementary Figure 4). Among all other constructs visualized by confocal microscopy only cells producing mCherry-AnkX491-949 or mCherry-AnkX688-949 displayed fluorescence at the plasma membrane. HeLa cells producing either mCherry-AnkX721-949 or mCherry-AnkX810-949 displayed fluorescence predominantly in the cytosol, although the latter also displayed lower signal in the nucleus. The FIC domain-containing mCherry-AnkX1-490 variant was dispersed throughout the cytosol and was also present on vesicular compartments, whereas mCherry-AnkX491-809 was present in the cytosol, and, surprisingly also in the nucleus. Therefore, the C-terminal half of AnkX possesses properties that provide AnkX with the ability to bind the plasma membrane, but also to bind nuclear components. To further analyze the cellular distribution of mCherry-tagged AnkX variants, we then performed cellular

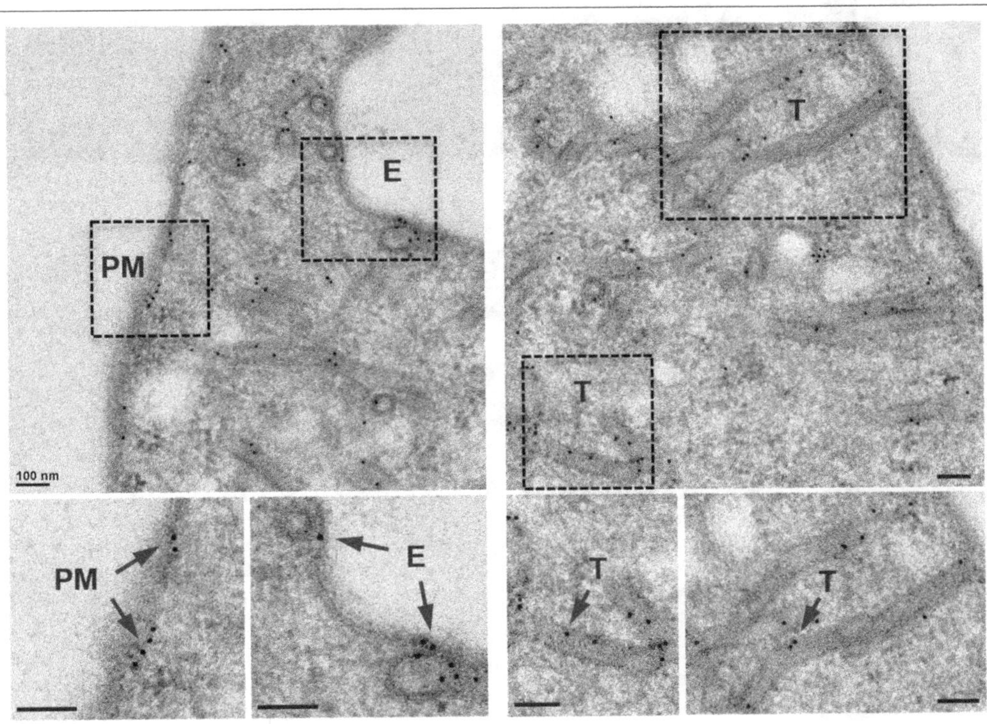

FIGURE 1 | AnkX localizes to the plasma membrane, nascent endosomes, and tubular compartments. Representative TEM images showing immunogold localization of GFP-AnkX in COS-1 cells. Fixed cells were stained with a polyclonal rabbit anti-GFP antibody. Areas highlighted by rectangles (dashed line) on the top panels are magnified in the bottom panels. Arrows indicate immunogold labeled GFP-AnkX localized to endosomes (E), plasma membrane (PM), or tubules (T). Scale bar is 100 nm.

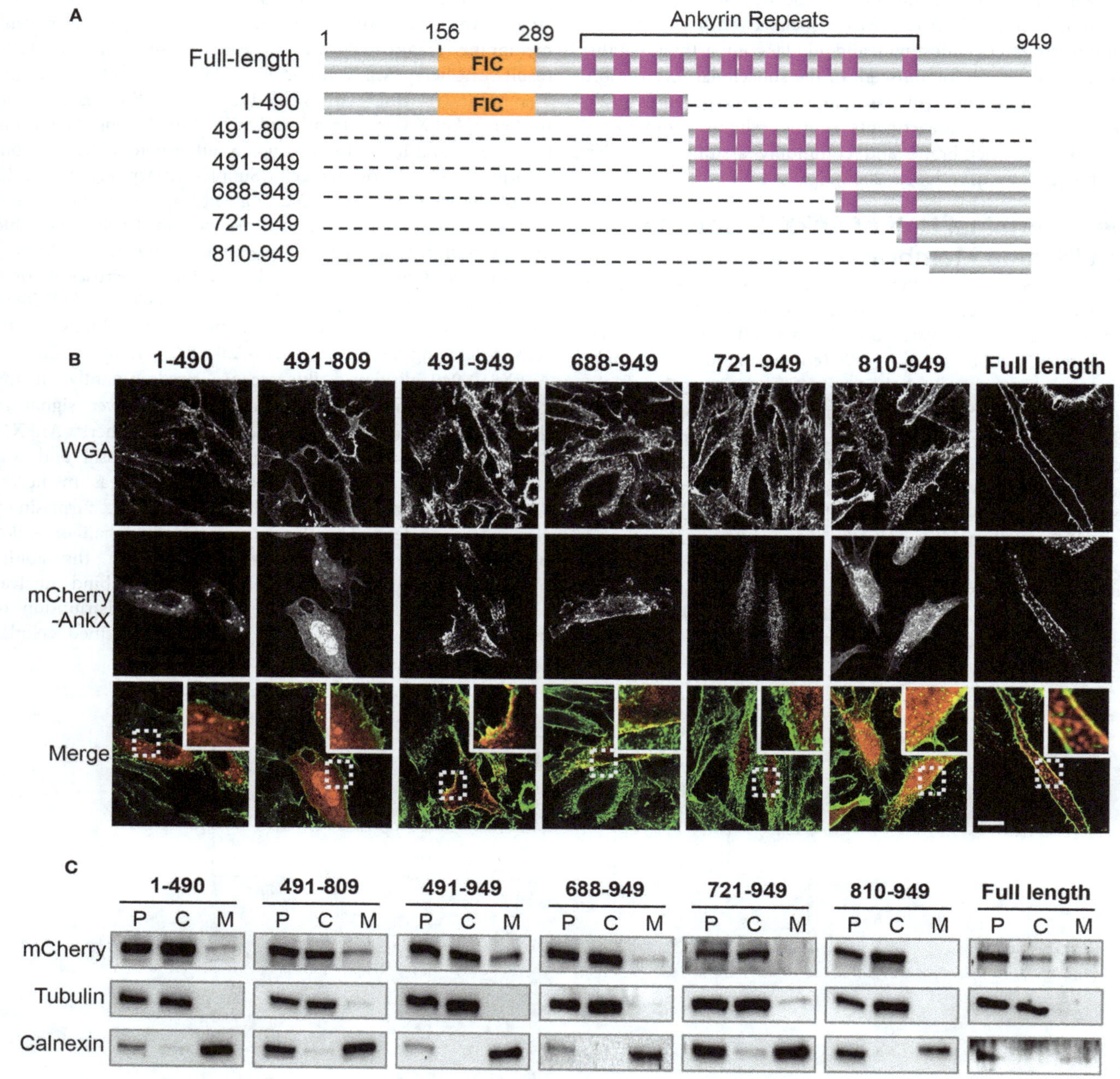

FIGURE 2 | AnkX domain mapping. **(A)** Schematic of AnkX domain variants (The FIC domain, orange; 12 ankyrin repeats, purple; dashed lines, deleted region). **(B)** Subcellular localization of AnkX fragments. HeLa cells transiently producing mCherry-AnkX or mCherry-AnkX variants were incubated with the plasma membrane dye WGA-488 post-fixation and intracellular localization was determined by confocal microscopy. Images show one z-slice of the confocal stack. **(C)** Cellular fraction of HEK293T cells producing mCherry-AnkX or mCherry-AnkX variants. The post-nuclear supernatant (P) was separated by ultracentrifugation into cytosolic (C) and membrane (M) fractions. The presence of mCherry-AnkX variants in each fraction was assessed by immunoblotting using a mCherry antibody. Tubulin and calnexin were used as markers for the cytosolic and membrane fractions, respectively.

fractionation studies using HEK293T cells producing these protein variants. We chose to employ HEK293T cells for this approach due to their increased transfection efficiency with AnkX-derived constructs. Cellular fractionation studies confirmed that mCherry-AnkX, mCherry-AnkX491-949, and mCherry-AnkX688-949 were membrane-associated (**Figure 2C**). A lighter band was detected for mCherry-AnkX688-949 in the membrane fraction compared to cells producing mCherry-AnkX491-949. It is possible that this shortened fragment had a weaker interaction with the plasma membrane that was disrupted during ultracentrifugation. As expected based on our confocal microscopy results, mCherry-AnkX721-949 and mCherry-AnkX810-949 were found in the cytosolic fraction. Both mCherry-AnkX1-490 and mCherry-AnkX491-809 were present in the cytosolic and membrane fractions indicating that these fragments are potentially also present on endosomes.

To identify the molecular mechanism underlying AnkX localization at the plasma membrane, we analyzed AnkX's ability to bind phosphoinositide lipids. Multiple *Legionella* effectors were reported to specifically recognize and bind phosphoinositide lipids (Hilbi et al., 2011). Most of the phosphoinositide binding effectors identified thus far bind PI(4)P and/or PI(3)P (Hilbi et al., 2011). Given that the LCV is enriched in PI(4)P, phosphoinositide binding is thought to serve as an anchoring mechanism for *Legionella* effectors that attach to the cytosolic surface of the LCV after translocation into the host cell. Additionally, this mechanism can facilitate targeting of particular host membrane compartments (Pizarro-Cerda et al., 2014). To evaluate AnkX's ability to bind phosphoinositide lipids, we performed a protein-lipid overlay assay using purified GST-AnkX and a commercially available nitrocellulose membrane spotted with phosphatidylinositol and all seven phosphoinositide species (**Figure 3A**). Through this approach we determined that AnkX bound mainly PI(3)P and PI(4)P (**Figure 3B**). To identify which region of AnkX was responsible for phosphoinositide binding, we generated two GST-tagged AnkX variants: GST-AnkX1-490 and GST-AnkX491-949 (Supplementary Figure 3). Our protein-lipid overlay assay showed that the C-terminal region of AnkX was capable of phosphoinositide binding while the N-terminal region was not (**Figure 3B**). We concluded that the C-terminal half of AnkX harbors the elements necessary to mediate association with the plasma membrane.

AnkX Targets Rab35-Positive Compartments

AnkX's localization at the plasma membrane and its ability to phosphocholinate Rab35 suggested that AnkX could target Rab35-positive recycling endosomes. To determine whether AnkX associates with these endosomes, we transiently co-transfected HeLa cells with a plasmid encoding GFP-Rab35 and a plasmid encoding either mCherry-AnkX or mCherry as a control. By confocal microscopy, we observed that the GFP-Rab35 signal was present in the cytosol, membrane compartments, the plasma membrane, and in some cells the nucleus (**Figure 4A**); nuclear localization of GFP-Rab35 has previously been observed, but this is believed to be caused by nuclear translocation of GFP alone (Kouranti et al., 2006). In agreement with our electron microscopy data, mCherry-AnkX was found to be present at the plasma membrane and on intracellular membrane compartments. When the two proteins were co-produced they co-localized on vacuolar compartments that resembled enlarged endosomes. Similar structures were observed in cells producing GFP-Rab35 in the absence of AnkX (**Figure 4A**).

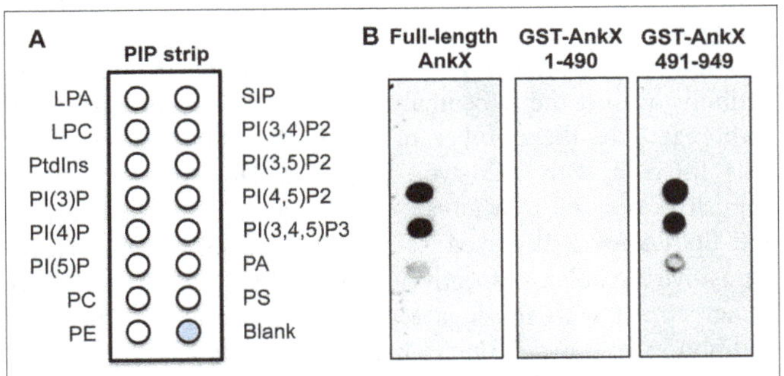

FIGURE 3 | AnkX interacts with phosphoinositide lipids via its C-terminal region. **(A)** Schematic of lipids spotted on the phosphoinositide strip (LPA, lysophosphatidic acid; LPC, lysophosphocholine; PtdIns, phosphatidylinositol; PE, phosphatidylethanolamine; PC, phosphatidylcholine; SIP, sphingosine-1-phosphate; P, phosphate; P2, biphosphate; P3, triphosphate; PA, phosphatidic acid; PS, phosphatidylserine). **(B)** Protein-lipid overlay assay reveals that a C-terminal fragment of AnkX preferentially binds PI(3)P and PI(4)P. Untagged full-length AnkX was detected with an AnkX antibody and a GST antibody was used to detect the GST-tagged truncated versions of AnkX.

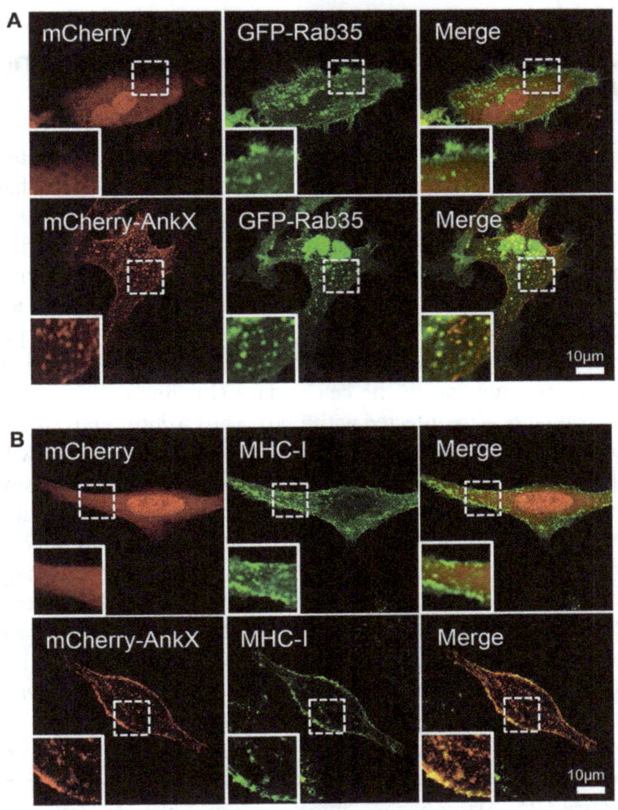

FIGURE 4 | AnkX co-localizes with Rab35 and MHC-I, a Rab35-dependent cargo. **(A)** Confocal images of HeLa cells co-transfected with plasmids encoding mCherry and GFP-Rab35 or plasmids encoding mCherry-AnkX and GFP-Rab35. Insets show magnified view of an area of interest. Bottom panel insets highlight an area where fluorescent signal from the two proteins co-occurred on intracellular vacuolar compartments. **(B)** HeLa cells ectopically producing mCherry or mCherry-AnkX were allowed to internalize monoclonal fluorescently labeled MHC-I antibodies for 30 min, followed by an acid wash. Bottom panel insets highlight the compartments where mCherry-AnkX co-occurred with internalized MHC-I.

Rab35 regulates recycling of protein cargo between the plasma membrane and endosomes, and therefore, we next sought to determine whether AnkX targets Rab35-positive recycling endosomes. The major histocompatibility Class I (MHC-I) enters cells through clathrin-independent endocytosis and then either returns back to the plasma membrane via recycling tubules or is degraded by lysosomes (Caplan et al., 2002). Loss of Rab35 activity has been shown to inhibit MHC-I recycling, indicating that its recycling back to the plasma membrane is, at least in part, dependent on Rab35 (Allaire et al., 2010). To determine whether AnkX targets recycling tubules that are carrying MHC-I, we performed an antibody uptake assay, whereby HeLa cells producing mCherry or mCherry-AnkX were incubated with MHC-I antibody conjugated with Alexa 488 fluorophore to mark endosomes that internalize MHC-I. Using confocal microscopy, we observed that mCherry-AnkX co-localized extensively with MHC-I at the plasma membrane and on tubular compartments (**Figure 4B**). These results support the hypothesis that AnkX targets Rab35-positive recycling endosomes in human cells.

AnkX Disrupts Endocytic Recycling during Macrophage Infection

A series of biochemical approaches have shown that phosphocholination prevents activation of Rab35. Therefore, we hypothesized that during infection AnkX would have the same effect, and thus disrupt the normal dynamics of recycling endosomes. To determine whether AnkX affects endocytic recycling during macrophage infection, we performed a transferrin-recycling assay. Transferrin is routinely used as a marker for endocytic recycling; it remains bound to its receptor throughout the recycling pathway and is released outside the cell upon returning to the cell surface (Mellman, 1996). U937 macrophages were infected with Lp01, Lp01Δ*dotA*, Lp01Δ*ankX*, Lp01Δ*ankX* complemented with a plasmid carrying wild-type *ankX*, or Lp01Δ*ankX* complemented with a plasmid carrying $ankX_{H229A}$, a point mutant in the catalytic motif of the FIC domain that renders the protein catalytically inactive (Mukherjee et al., 2011). At 1 h post-infection, macrophages were subjected to a pulse-chase sequence with fluorescently labeled transferrin followed by incubation with unlabeled transferrin (**Figure 5A**). After the pulse step, U937 cells retained comparable levels of transferrin regardless of which *Legionella* strain they were infected with (Supplementary Figure 6). We then quantified the amount of fluorescently labeled transferrin remaining after a 30-min chase with unlabeled transferrin (**Figure 5B**). Macrophages infected with Lp01 displayed the highest level of fluorescence, indicating increased retention of transferrin within these cells. This result supports the idea that endocytic recycling of transferrin was blocked at an early stage of infection. In contrast, macrophages infected with the Lp01Δ*dotA* translocation-deficient mutant showed the lowest fluorescence signal among the five conditions, indicating that endocytic recycling of transferrin was robust in the absence of translocated *Legionella* effectors. Macrophages infected with the Lp01Δ*ankX* strain displayed levels of fluorescence that were significantly lower than those of cells infected with Lp01, and, thus, in the absence of AnkX, labeled transferrin along with its receptor were recycled back to the plasma membrane. We also noted that, although significantly reduced in Lp01Δ*ankX*-infected cells, the amount of labeled transferrin was 1.6 times higher than that detected in macrophages infected with the Lp01Δ*dotA* mutant (**Figure 5B**). When macrophages were infected with the *ankX* complemented Lp01Δ*ankX* strain, retention of labeled transferrin increased to a level similar to that of the Lp01-infected cells. Interestingly, macrophage infection with the Lp01Δ*ankX* ectopically producing the catalytically inactive $AnkX^{H229A}$ resulted in diminished retention of labeled transferrin compared to macrophages infected with the Lp01Δ*ankX* strain producing active AnkX (**Figure 5B**). This change was not attributable to a difference in the production of AnkX or $AnkX_{H229A}$ within the Lp01Δ*ankX* background since we verified that the two proteins were produced at comparable levels prior to macrophage infection (Supplementary Figure 5). Based on these results we concluded that the phosphocholination activity of AnkX was responsible for disruption of endocytic recycling. Notably, transferrin was not present at the LCV, consistent with previous findings that this compartment lacks transferrin (Joshi et al., 2001).

Phosphocholination Activity of AnkX Is Required for Avoidance of Phagosome Maturation

The Lp01Δ*ankX* mutant has previously been shown to be impaired in its ability to escape phagosome maturation early following macrophage infection (Pan et al., 2008). Our results above indicated that the phosphocholination activity of AnkX is important for exerting its biological function. Therefore, we hypothesized that phosphocholination could play a key role in the avoidance of phagolysosome maturation. To test this hypothesis, we infected U937 macrophages with Lp01, Lp01Δ*dotA*, Lp01Δ*ankX*, Lp01Δ*ankX* complemented with a plasmid carrying wild-type *ankX*, and Lp01Δ*ankX* complemented with a plasmid carrying $ankX_{H229A}$. At 2 h post-infection, macrophages were fixed, immunostained with an antibody against the lysosomal marker LAMP-2, and quantified with regard to the number of LAMP-2 positive LCVs. U937 cells infected with wild-type displayed 18(±5)% of LAMP-2 positive vacuoles, whereas cells infected with Lp01Δ*dotA* and Lp01Δ*ankX* displayed 77.3(±9)% and 63(±4.3)% LAMP-2 positive vacuoles, respectively (**Figure 5C**). These results are in agreement with findings reported by Pan et al. (2008). In addition, we showed that when macrophages were infected with the Lp01Δ*ankX* strain complemented producing AnkX the percentage of LAMP-2 positive LCVs decreased to 31.6(±6)%. In contrast, infection with the Lp01Δ*ankX* strain complemented with catalytically inactive $AnkX^{H229A}$ did not rescue the phenotype and showed that 70.6(±7.3)% LCVs were LAMP-2 positive, similar to the percentage observed in cells infected with the Lp01Δ*ankX* or the Lp01Δ*dotA* strains. Therefore, we concluded that AnkX's phosphocholination activity plays a critical role in avoidance of phagosome maturation.

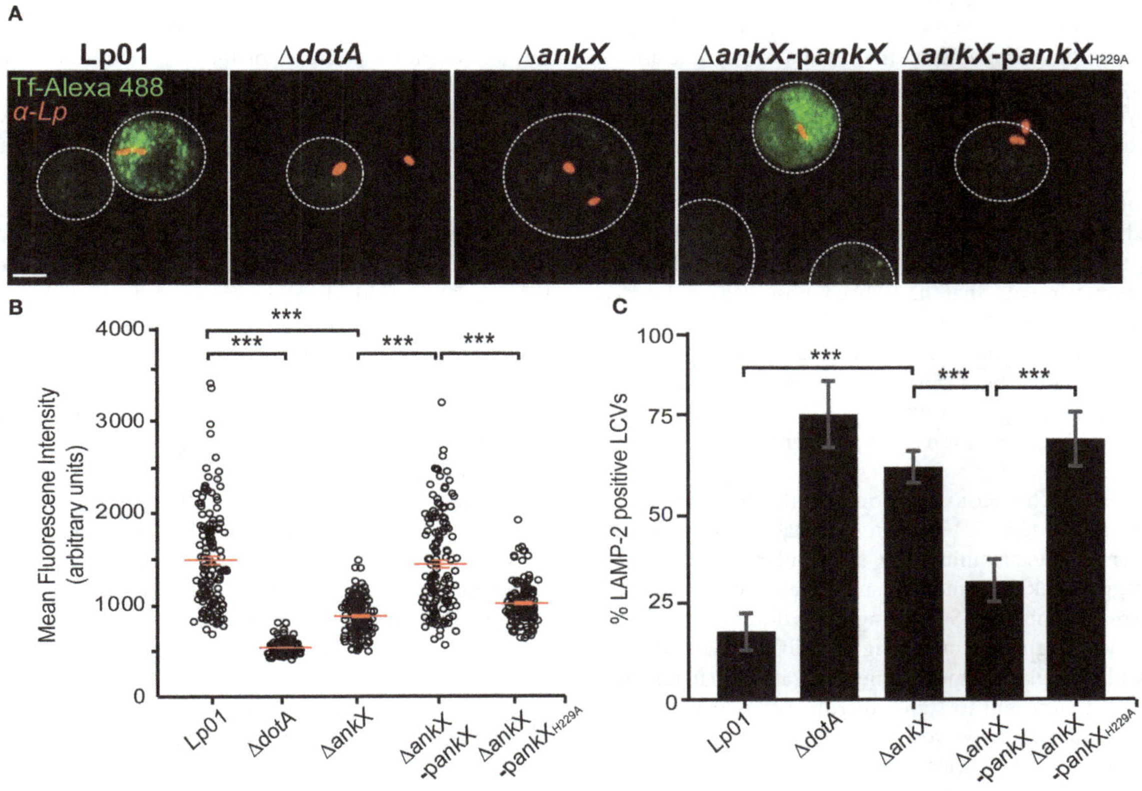

FIGURE 5 | *Legionella* blocks host endocytic recycling and prevents accumulation of lysosomal markers on the LCV in a phosphocholination-dependent manner. **(A)** Representative images of human transferrin-Alexa 488 fluorescent signal in U937 cells infected with *Legionella* strains as labeled. At 1 h post-infection, U937 cells were incubated with transferrin-Alexa 488 for 60 min, then washed and incubated with unlabeled transferrin for 30 min. Cells were fixed and immunostained with anti-*Legionella* antibodies and visualized by confocal microscopy; scale bar, 5μm. **(B)** The graph displays the amount of fluorescence retained by infected U937 cells. The mean and standard error of the mean from three independent experiments are indicated for each condition. Statistical significance was determined by one-way ANOVA followed by Tukey Kramer *post-hoc* test. The asterisks denote a $p < 0.001$. **(C)** Quantification of LAMP-2 positive LCVs in U937 cells infected with various strains of *Legionella*. Bar graph displays the percentage of LVCs that were LAMP-2 positive at 2 h post-infection. Statistical significance was determined using the Student's *t*-test with a *p*-value cut off of < 0.05. The asterisks denote a $p < 0.001$.

DISCUSSION

In eukaryotic cells, membrane transport plays essential roles in regulating signaling, metabolism, immunity, and interactions with the extracellular environment, and therefore, it is a major target for intracellular pathogens. *Legionella* manipulates the host's membrane transport system by delivering effector proteins that hijack Rab GTPases, a large class of small GTPases (over 60 in humans) performing essential roles in membrane trafficking (Stenmark, 2009). Delineating membrane transport pathways crucial for *Legionella* survival is complicated by the functional redundancy among *Legionella* effectors, and usually deletion of individual *Legionella* genes encoding effectors that target Rab GTPases does not markedly affect bacterial survival (O'Connor et al., 2011). Despite this hurdle, a previous study showed that deletion of *ankX* alone significantly increased fusion of LCVs with lysosomes revealing that AnkX plays an important role in safeguarding the LCV from phagosome maturation (Pan et al., 2008). However, the mechanistic details of how AnkX is involved in this process are not precisely understood. Here we narrow in on the biological consequences of AnkX activity during macrophage infection and demonstrate that AnkX disrupts endocytic recycling and phagolysosome biogenesis in a phosphocholination-dependent manner.

The *Legionella* effector AnkX functions as a phosphocholine transferase, and of the six Rab GTPases previously tested, Rab1 and Rab35 were efficiently phosphocholinated by AnkX (Mukherjee et al., 2011). Although the biochemical details of Rab1 and Rab35 phosphocholination are well-understood, the biological consequences of Rab1 and Rab35 phosphocholination remain elusive. Rab1 is heavily recruited to the LCV early during infection (Kagan et al., 2004), however AnkX is not required for this process (Hardiman and Roy, 2014). The importance of Rab35 manipulation by *Legionella* effectors has not yet been explored. Rab35 regulates a number of cellular processes including endocytic recycling and cytokinesis, exosome release, phagocytosis, cell migration, immunological synapse formation and neurite outgrowth (Klinkert and Echard, 2016). Because it is involved in shuttling protein cargo between the plasma membrane and early endosomes through recycling endosomes, Rab35 localizes predominantly at the plasma membrane and on tubular recycling endosomes (Chua et al., 2010). Using electron microscopy, we found that GFP-tagged AnkX had a localization pattern similar to that of Rab35 (**Figure 1**). Moreover, in HeLa cells mCherry-AnkX co-localized with GFP-Rab35 on vacuolar compartments (**Figure 4A**). These compartments could be enlarged early, sorting, or recycling endosomes, where Rab35 is known to localize. Alternatively, they may be similar to vacuolar compartments that have been observed to form as a consequence of Rab35 inactivation (Kouranti et al., 2006). If this were the case the formation of these Rab35-positive vacuoles would likely be a consequence of AnkX's inhibitory effect on Rab35 activity. However, other indirect effects cannot be excluded until confirmation by experiments using mCherry-$AnkX^{H229A}$.

We next asked whether association of AnkX with the plasma membrane was due to its interaction with Rab35. We addressed this question by determining the subcellular localization of mCherry-tagged AnkX fragments in mammalian cells in relation to the plasma membrane. Surprisingly, we found that the N-terminal AnkX fragment containing the FIC domain did not associate with the plasma membrane, indicating that targeting of Rab35 is not sufficient to stably recruit AnkX to the plasma membrane (**Figure 2**). However, mCherry-AnkX1-490 can still localize to intracellular vesicles. A potential explanation for this observation is that AnkX's substrate specificity is more relaxed allowing other Rab GTPases to be modified, and perhaps the C-terminal region required for binding the plasma membrane limits AnkX's activity to specific compartments. Further insight into AnkX's substrate preference could clarify this issue.

The AnkX491-949 fragment contained the elements necessary to mediate plasma membrane binding. We showed that this fragment was also able to bind PI(3)P and PI(4)P (**Figure 3**), although it is not yet clear if phosphoinositide binding alone is sufficient to mediate association with the plasma membrane. Furthermore, binding of multiple phosphoinositide species may suggest that AnkX has functional roles at several stages of infection since *Legionella* actively changes the phosphoinositide composition of the LCV (Weber et al., 2014). Within seconds following uptake of *Legionella* by amoebae, the early phagosome membrane displays PI(3,4,5)P, immediately followed by a rise in PI(3)P levels. Subsequently, the vacuolar membrane gradually loses these lipids and by 2 h post-infection becomes enriched in PI(4)P. PI(3)P- and PI(4)P-binding regions of *Legionella* effectors are thought to mediate anchoring to the LCV (Weber et al., 2006; Hilbi et al., 2011); they may also function in conjunction with Rab GTPase binding domains as a two-pronged mechanism to specifically target membrane compartments. The precise region of AnkX that recognizes PI(3)P and PI(4)P remains to be identified. Intriguingly, the AnkX688-949 fragment was present at both the plasma membrane and on endosomes, whereas the AnkX721-949 fragment missing just one ankyrin repeat lost its membrane localization (**Figure 2B**). It is possible, therefore, that this ankyrin repeat or other elements in this region mediates the binding. We hypothesize that a multifactorial mechanism regulates AnkX interaction with the plasma membrane, wherein both Rab35 and phosphoinositide binding are both important, but an additional component may be necessary to stabilize the interaction.

In COS-7 cells, a Rab35 siRNA knockdown causes enlargement of early endosomes (Allaire et al., 2010), and, intriguingly, heterologous production of AnkX in mammalian cells results in a similar phenotype (Mukherjee et al., 2011). Given AnkX's co-localization with Rab35 (**Figure 4A**) and its known inhibitory effect on Rab35 activity *in vitro*, we reasoned that AnkX would disrupt pathways regulated by this Rab GTPase. Rab35 has been implicated in the recycling of various cargo proteins back to the plasma membrane including the MHC-I, MHC-II, and the T-cell receptor (Patino-Lopez et al., 2008; Walseng et al., 2008; Allaire et al., 2010). Our MHC-I internalization and recycling assay showed that AnkX co-occurred with MHC-I, supporting the notion that AnkX targets Rab35-dependent endocytic recycling (**Figure 4B**). MHC-I is involved in presenting intracellular antigens to cytotoxic T cells such that cells infected with intracellular pathogens can be eliminated (Mantegazza et al., 2013). It is plausible that *Legionella* disrupts recycling of these receptors in order to modulate host immune responses during infection to prevent the death of its host cell.

In light of these findings, we hypothesized that AnkX manipulates endocytic recycling during infection of macrophages by targeting Rab35. Our data showed that, indeed, endocytic recycling of transferrin, a widely used marker for this transport pathway, is blocked early during macrophage infection (**Figures 5A,B**). The inhibitory effect was dependent on *Legionella* effectors since infection with a translocation-deficient mutant did not disrupt transferrin recycling. Infection with the Lp01 Δ*ankX* mutant showed a sharp decrease in accumulation of transferrin-positive vesicles, indicating that AnkX is largely responsible for disruption of endocytic recycling. This phenotype was reversed when the mutant was complemented with ectopically expressed *ankX*, but not when it was complemented with $ankX_{H229A}$. The necessity of an enzymatically active AnkX highlighted the key role of phosphocholination in disruption of transferrin recycling. Importantly, this result implies that phosphocholination of Rab35, and perhaps other host proteins, leads to disruption of endocytic recycling. Notably, the absence of AnkX did not abrogate *Legionella*'s ability to block endocytic recycling. This observation supports the idea that other *Legionella* effectors may also interfere with endocytic recycling.

Targeting of endocytic recycling appears to be a common strategy for intracellular pathogens. For example, uropathogenic *E. coli* (UPEC) has been proposed to manipulate endosomal recycling and degradative lysosomal fusion by modulating Rab35 to survive within bladder epithelial cells (Dikshit et al., 2015), and enterohemorrhagic *E. coli* (EHEC) modulates Rab35 recycling by secretion of the effector protein EspG in order to prevent maturation of recycling endosomes during infection (Furniss et al., 2016). More recently, a high-content imaging study showed that transferrin uptake gradually increased throughout infection of HeLa cells with *Coxiella burnetii* or *Chlamydia trachomatis* suggesting that bacterial infection expanded the endosomal system to increase capacity for endocytic material (Larson and Heinzen, 2017). The *Shigella* effector protein, IpgD, is responsible for recruiting Rab11-positive vesicles to the invasion site to assist in the rupture of the *Shigella*-containing vacuole, a step required for bacterial escape into the cytosol (Mellouk et al., 2014).

An important question is why *Legionella* would manipulate transferrin recycling during infection. It is unlikely that the pathogen is attempting to transport iron to the LCV as *Legionella* genetically encodes and translocates its own iron-sequestering pathway, which has been shown to be important for infection (Joshi et al., 2001; Chatfield et al., 2012; Isaac et al., 2015). Instead, we propose that *Legionella* effectors target Rab35 to counteract phagosome maturation. Upon phagocytosis into macrophages, *Legionella* immediately prevents phagosome maturation, avoiding fusion of early endosomes with the LCV. Rab35 is involved in formation of the phagocytic cup in macrophages (Egami et al., 2011). Silencing of Rab35 by RNA interference drastically reduced the rate of FcγR-mediated phagocytosis in macrophages, revealing that Rab35 is involved in the early stage of phagocytosis. Interestingly, in *Entamoeba histolytica*, an intestinal protozoan parasite, Rab35 is not only required for phagocytic cup formation, but also for biogenesis of phagolysosomal compartments (Verma and Datta, 2017). In light of these results, we hypothesized that similar to UPEC, *Legionella* manipulates Rab35 to subvert both endosomal recycling and degradative lysosomal fusion. Therefore, we anticipated that AnkX's phosphocholination activity would be crucial for its previously established ability to prevent fusion of lysosomes with the LCV. Our results confirmed that in the absence of AnkX, *Legionella* cannot efficiently escape phagosome maturation (**Figure 5C**). Moreover, we showed that phosphocholination played a critical role in *Legionella*'s ability to escape phagosome maturation. Given that amoebae are the natural host of *Legionella*, targeting of Rab35 to modulate phagosome maturation would be a desirable benefit for the bacterium. In human cells, where Rab35 regulates recycling of proteins involved in innate immunity, targeting of Rab35 by *Legionella* could offer an added benefit for the bacterium, and perhaps facilitate evasion of immune surveillance.

In conclusion, our data indicate that *Legionella* effectors inhibit endocytic recycling at the early stage of infection, and that the phosphocholination activity of AnkX plays a significant role in targeting this membrane transport pathway. Overall, our study reveals the biological relevance of AnkX-catalyzed phosphocholination during infection of human macrophages and provides new insight into subversion of membrane transport pathways by *Legionella* effectors. A complete perspective of how *Legionella* effectors target endocytic recycling merits further investigation and promises to reveal important details about survival of *Legionella* and other intracellular pathogens.

MATERIALS AND METHODS

Strains, Tissue Culture Cells, and Media

L. pneumophila strains Lp01 (*hsdR rpsL*) and Lp01Δ*dotA* (T4SS$^-$) are derivatives of *L. pneumophila* strain Philadelphia-1 (Berger and Isberg, 1993). The wild-type Lp01, Lp01Δ*ankX* mutant, Lp01Δ*dotA*, and the pJB1806-*ankX* plasmid were a generous gift from Dr. Craig Roy. All bacterial strains used in this study are listed in Supplementary Table 1. The pJB1806-$ankX_{H229A}$ shuttle vector was generated by quick-change mutagenesis using oligonucleotides listed in Supplementary Table 2 using a standard protocol. *Legionella* strains were grown in AYE liquid media or on CYE solid media and maintained as described (Feeley et al., 1979). The pJB1806-*ankX* and pJB1806-$ankX_{H229A}$ shuttle vectors were transformed into the Lp01Δ*ankX* strain to complement the deletion strain with catalytically active or inactive AnkX, respectively. Bacterial growth media was supplemented with 25 μg/ml chloramphenicol to maintain pJB1806 plasmids in the Lp01Δ*ankX* strain and 1 mM IPTG was added to the medium to induce gene expression. HeLa tissue culture cells were maintained in RPMI 1640 medium supplemented with 2 mM L-glutamine and 10% FBS, and incubated in 5% CO_2 at 37°C. COS-1 and HEK293T cells were grown in DMEM medium supplemented with 2 mM L-glutamine supplemented with 10% FBS. U937 cells were grown in RPMI 1640 medium supplemented with 10% FBS and were differentiated by supplementing the medium with 10 ng/ml 12-O -tetradecanoylphorbol-13-acetate (TPA) for 16 h.

Antibodies and Plasmids

Antibodies were purchased from BioLegend (MHC-I, clone w6/32 IgG2a), ThermoFisher Scientific (Goat anti-Rat Alexa 350, A-21093; Goat anti-Rat Texas Red, T-6392; Goat anti-Mouse Alexa 488, A-11001; Mouse monoclonal anti-GST, MA4-004; Rabbit polyclonal anti-mCherry, Pa5-34974, HRP-conjugated anti-mouse antibody, NA931), Abcam (Rabbit polyclonal anti-LAMP-2; ab37024; Rabbit polyclonal anti-GFP antibody, ab6556), Enzo (Rabbit polyclonal Calnexin, ADI-SPA-860), ProteinTech (Mouse Monoclonal Alpha Tubulin, 66031-1-Ig), and Jackson ImmunoResearch (Rabbit IgG ChromePure, 011-000-003). The rat polyclonal anti-Legionella antibody and the rabbit polyclonal anti-AnkX used for detection of purified AnkX in **Figure 3** were a kind gift from Dr. Matthias Machner (NIH). We also thank Dr. Zhao-Qing Luo (Purdue University) for the rabbit polyclonal anti-AnkX antibody used for detection of AnkX in *Legionella* lysate (Supplementary Figure 5).

Full-length *ankX* and truncations were amplified using oligonucleotides in Supplementary Table 2 and were inserted by recombination into pDONR221™. The inserted fragments were then recombined into the mammalian expression vector 362-pCS-Cherry-DEST (Addgene plasmid #13075) or the pcDNA6.2/N-EmGFP-DEST vector (ThermoFisher Scientific) using the Gateway™ cloning technology (ThermoFisher Scientific). Plasmids encoding GST-tagged AnkX variants were generated by using the pDONR221-$ankX_{1-490}$ and pDONR221-$ankX_{491-949}$ donor vectors to recombine *ankX* into the destination plasmid pDEST15™ (ThermoFisher Scientific). The pEGFP-C1-Rab35 plasmid was a kind gift from Dr. Julie Donaldson (NIH). Plasmids used in this study are listed in Supplementary Table 1.

Confocal Microscopy

HeLa cells grown on 12 mm cover glass (Fisherbrand™) were transiently transfected with plasmids encoding fluorescently-tagged constructs using the Lipofectamine 3000 reagent (ThermoFisher Scientific). At 14 h post-transfection cells were washed with PBS and fixed with 4% paraformaldehyde in PBS for 20 min at room temperature. To visualize the plasma membrane, cells were treated with a Wheat Germ Agglutinin Alexa Fluor 488 conjugate (ThermoFisher Scientific) at 5 μg/ml for 10 min at 37°C. Coverslips were mounted using ProLong Diamond anti-fade reagent (ThermoFisher Scientific) and image acquisition was performed using a ZEISS 710 confocal microscope, a 63× Plan-Apochromatic oil immersion objective (numerical aperture of 1.4), and the ZEN 2012 software (Carl ZEISS MicroImaging).

Transmission Electron Microscopy

COS-1 cells transiently producing GFP-AnkX were washed briefly with PBS and fixed overnight with 4% paraformaldehyde and 0.5% glutaraldehyde in 0.1 M sodium cacodylate buffer pH 7.4 at 4°C. Cells were then washed with 0.1 M sodium cacodylate buffer pH 7.4, scraped, and pelleted. The cell pellet was enrobed with 4% low melting point agarose and cut into small 1–2 mm^3 cubes. The cells were washed with 0.1 M sodium cacodylate buffer pH 7.4, dehydrated in an ascending ethanol series and infiltrated with LR White resin. Samples were embedded in gelatin capsules and polymerized by UV at 4°C for 2 days. The samples were sectioned on a Reichert-Jung Ultracut E ultramicrotome, and 60–70 nm thick sections were collected onto formvar/carbon coated 200 mesh nickel grids. *Immunogold Labeling*—Samples were blocked with 0.05 M glycine for 15 min and Aurion goat blocking solution for 30 min before being incubated on drops of anti-GFP antibody diluted to 2.8 μg/ml in Aurion 0.1% BSA-c 7.4 for 1 h. Control grids were incubated on drops of ChromePure rabbit IgG (Jackson ImmunoResearch, Cat No. 011-000-003) diluted to 2.8 μg/ml in Aurion 0.1% BSA-c 7.4. Grids were washed on six drops of Aurion 0.1% BSA-c and incubated on drops of Aurion goat anti-rabbit IgG conjugated to 10 nm gold diluted 1:20 in Aurion 0.1% BSA-c for 2 h. Grids were washed on six drops of Aurion 0.1% BSA-c, 3 drops of PBS, fixed on drops of 2% glutaraldehyde in PBS, and then washed on five drops of Nanopure water. The grids were then post-stained with 2% uranyl acetate in 50% methanol and Reynolds' lead citrate. The samples were examined with a ZEISS Libra 120 transmission electron microscope operating at 120 kV, and images were acquired with a Gatan Ultrascan 1000 CCD camera.

Cellular Fractionation

HEK293T cells producing mCherry-AnkX or -AnkX variants were scraped, washed, resuspended in PBS supplemented with protease inhibitors, and lysed by 30 passes through a 27″ gauge needle. The post-nuclear supernatant (PNS) was obtained via centrifugation at 15,000×g for 10 min at 4°C. To obtain the cytosolic fraction, the PNS was spun at 51,000 rpm for 45 min at 4°C using a TLA-100 rotor in a Beckman ultracentrifuge. PBS was added to the pellet and spun again to wash away cytosolic contaminants. The pellet was resuspended with an equal volume of 2% NP-40 in PBS and collected as the membrane fraction. The PNS, cytosolic fraction, and membrane fraction were analyzed via SDS-PAGE and immunoblot.

Detection and Quantification of LAMP-2-Positive LCVs

U937 macrophages were challenged with *Legionella* strains at a multiplicity of infection (MOI) of 50 and spun at 200 × g for 5 min and then incubated at 37°C. At 2 h post-infection cells were washed three times with PBS and fixed in 4% paraformaldehyde for 20 min. To distinguish extra- and intracellular bacteria a two-step immunostaining procedure was performed. Without macrophage permeablization, *Legionella* was detected with an anti-*Legionella* rat primary antibody (1:3,000) followed by Cascade Blue-conjugated Goat anti-Rat IgG antibody (1:1,000). Cells were then permeabilized with ice-cold methanol for 20 s and internalized bacteria were detected anti-*Legionella* rat primary (1:3,000) and Texas Red-conjugated Goat α-Rat IgG (1:1,000).

For detection of LAMP-2-positive LCVs, infected U937 cells were immunostained with anti-LAMP-2-rabbit primary antibody (1:150) washed three times with PBS and fluorescently labeled with Alexa 488-rabbit secondary antibody (1:1,000). Coverslips were mounted using ProLong Diamond anti-fade reagent (ThermoFisher Scientific). The percentage of LAMP-2-positive LCVs was determined by scoring 100 cells per coverslip with three replicates for each specific condition. The unpaired two-sided Student's *t*-test was performed with a *p*-value cut off of < 0.05 to determine where there was a statistically significant difference in localization to the LCV.

Transferrin Recycling Assay

To determine efficiency of recycling, U937 cells were grown on glass coverslips in a 24-well plate and following differentiation into macrophages cells were infected with *Legionella* strains Lp01, Lp01*ΔdotA*, Lp01Δ*ankX*, Lp01Δ*ankX* harboring pBJ1806-*ankX*, and Lp01Δ*ankX* harboring pJB18060-$ankX_{H229A}$ at an MOI of 50 for 1 h. Infected macrophages were then allowed to uptake fluorescently labeled human transferrin (Tf-Alexa-488; ThermoFisher Scientific) at 5 μg/ml in RPMI for 1 h. Cells were either fixed to measure the fluorescence intensity after the pulse step or they were washed with PBS containing unlabeled human transferrin (100 μg/ml) three times and then incubated further with unlabeled transferrin at 100 μg/ml in RPMI for 30 min. Cells were fixed and immunostained to identify infected cells and visualized by confocal microscopy as described above. All parameters for imaging remained consistent throughout the experiment. Using ImageJ-FIJI software (Schindelin et al., 2012, 2015) the mean fluorescence intensity of Tf-Alexa-488 in 50 infected cells from three replicates for each condition was measured. To account for variation in cell size, the mean fluorescence intensity was determined by normalizing each cell's fluorescence intensity to the total area of the cell. A one-way ANOVA was performed with a $p < 0.01$ to determine if any strains showed a significant difference in recycling. *Post-Hoc* Tukey-Kramer test was performed with a *p*-value cut off of < 0.01 in order to identify which pairs of infected cells were significantly different from each other.

Production and Purification of Recombinant Proteins

Recombinant HaloTag-AnkX was produced in the Single Step (KRX) competent *E. coli* strain (Promega), and AnkX was purified using the HaloTag protein purification system as previously described (Neunuebel et al., 2012). Briefly, *E. coli* cells producing HaloTag-AnkX were lysed using an LV10 microfluidizer (Microfluidics). The cell lysate was spun at 25,000 × g for 20 min, and pre-equilibrated HaloLink resin were incubated with the supernatant for 2 h at 4°C. The resin was then washed and AnkX was cleaved off the resin using His-tagged tobacco etch virus (TEV) protease (Promega) for 2 h at 4°C. HisLink resin (Promega) was used to remove the TEV protease from the supernatant.

AnkX fragments were produced as GST fusion proteins in *E. coli* BL21 (DE3) at 25°C overnight after induction with 0.5 mM isopropyl-β-dithiogalactopyranoside (IPTG). *E. coli* cells producing GST-tagged variants were harvested and resuspended in PBS supplemented with 1 mM $MgCl_2$ and 1 mM β-mercaptoethanol (PBS-MM) followed by lysis using the LV10 microfluidizer (Microfluidics). The cell lysate was centrifuged at 24,000×g for 35 min, and the supernatant was incubated with pre-equilibrated Glutathione Sepharose 4B (GE Healthcare) for 2 h at 4°C. The resin was washed three times with PBS-MM, and proteins were eluted with 50 mM Tris-HCl (pH 8) containing 10 mM reduced glutathione (Sigma). Glutathione was removed by passing the eluate through a Zeba column (ThermoFisher Scientific).

Protein-Lipid Overlay Assay

Protein-lipid overlay assays were performed using commercially available phosphoinositide strips (Echelon Biosciences Inc.). Nitrocellulose membranes pre-spotted with different phospholipids were blocked with 3% non-fat milk in PBST [PBS and 0.1% Tween-20 (v/v) pH 7.5] for 1 h at room temperature. The blocked membranes were incubated with AnkX, GST-AnkX1-490, or GST-AnkX491-949 (0.5 μg/ml in blocking buffer) overnight at 4°C. Protein binding to lipids was visualized with an anti-GST antibody (1:2,000) or an anti-AnkX antibody, and an HRP-conjugated anti-mouse antibody (1:5,000) using a ChemiDoc Touch Imaging system (Bio-Rad).

MHC-I Internalization Assay

HeLa cells were grown on cover slips and transiently transfected with mCherry-AnkX using Lipofectamine 2000 (ThermoFisher Scientific) reagent for 16 h. Following transfection cells were incubated with an anti-MHC-I antibody conjugated to Alexa 488 (1:100) for 30 min at 37°C and 5%CO_2. After internalization, cells were rinsed twice with PBS and washed with acid (0.5% acetic acid, 0.5 M NaCl, pH 3.0) for 20 s to remove remaining surface-bound antibody. The acid stripping was followed by a PBS wash and final rinse with RPMI. Cells were fixed with 4% paraformaldehyde in PBS for 20 min at room temperature and visualized by confocal microscopy.

AUTHOR CONTRIBUTIONS

MRN conceived the project, designed the experiments, performed the Rab35 and MHC-I co-localization experiments, and wrote the manuscript. SL and CP purified proteins and performed the protein-lipid over assays. RRN contributed to the cellular fractionation assays and analysis of the data. SCA and BPR performed the rest of the experiments, analyzed the data, and wrote the manuscript.

ACKNOWLEDGMENTS

We thank Drs. Shannon Moodla and Jeff Caplan from the Delaware Biotechnology Institute for assistance with immunogold TEM. We also thank Dr. Robin Morgan for critical review of this manuscript.

REFERENCES

Allaire, P. D., Marat, A. L., Dall'Armi, C., Di Paolo, G., McPherson, P. S., and Ritter, B. (2010). The Connecdenn DENN domain: a GEF for Rab35 mediating cargo-specific exit from early endosomes. *Mol. Cell* 37, 370–382. doi: 10.1016/j.molcel.2009.12.037

Berger, K. H., and Isberg, R. R. (1993). Two distinct defects in intracellular growth complemented by a single genetic locus in *Legionella pneumophila*. *Mol. Microbiol.* 7, 7–19. doi: 10.1111/j.1365-2958.1993.tb01092.x

Brombacher, E., Urwyler, S., Ragaz, C., Weber, S. S., Kami, K., Overduin, M., et al. (2009). Rab1 guanine nucleotide exchange factor SidM is a major phosphatidylinositol 4-phosphate-binding effector protein of *Legionella pneumophila*. *J. Biol. Chem.* 284, 4846–4856. doi: 10.1074/jbc.M807505200

Campanacci, V., Mukherjee, S., Roy, C. R., and Cherfils, J. (2013). Structure of the Legionella effector AnkX reveals the mechanism of phosphocholine transfer by the FIC domain. *EMBO J.* 32, 1469–1477. doi: 10.1038/emboj.2013.82

Caplan, S., Naslavsky, N., Hartnell, L. M., Lodge, R., Polishchuk, R. S., Donaldson, J. G., et al. (2002). A tubular EHD1-containing compartment involved in the recycling of major histocompatibility complex class I molecules to the plasma membrane. *EMBO J.* 21, 2557–2567. doi: 10.1093/emboj/21.11.2557

CDC (2011). Legionellosis — United States, 2000–2009. *MMWR Morb. Mortal. Wkly. Rep.* 60, 1083–1086.

Chatfield, C. H., Mulhern, B. J., Viswanathan, V. K., and Cianciotto, N. P. (2012). The major facilitator superfamily-type protein LbtC promotes the utilization of the legiobactin siderophore by *Legionella pneumophila*. *Microbiology* 158(Pt 3), 721–735. doi: 10.1099/mic.0.055533-0

Chua, C. E., Lim, Y. S., and Tang, B. L. (2010). Rab35–a vesicular traffic-regulating small GTPase with actin modulating roles. *FEBS Lett.* 584, 1–6. doi: 10.1016/j.febslet.2009.11.051

Clemens, D. L., Lee, B. Y., and Horwitz, M. A. (2000). Deviant expression of Rab5 on phagosomes containing the intracellular pathogens *Mycobacterium tuberculosis* and *Legionella pneumophila* is associated with altered phagosomal fate. *Infect. Immun.* 68, 2671–2684. doi: 10.1128/IAI.68.5.2671-2684.2000

Dikshit, N., Bist, P., Fenlon, S. N., Pulloor, N. K., Chua, C. E., Scidmore, M. A., et al. (2015). Intracellular uropathogenic, *E. coli* exploits host Rab35 for iron acquisition and survival within urinary bladder cells. *PLoS Pathog.* 11:e1005083. doi: 10.1371/journal.ppat.1005083

Egami, Y., Fukuda, M., and Araki, N. (2011). Rab35 regulates phagosome formation through recruitment of ACAP2 in macrophages during FcgammaR-mediated phagocytosis. *J. Cell Sci.* 124(Pt 21), 3557–3567. doi: 10.1242/jcs.083881

Feeley, J. C., Gibson, R. J., Gorman, G. W., Langford, N. C., Rasheed, J. K., MacKel, D. C., et al. (1979). Charcoal-yeast extract agar: primary isolation medium for *Legionella pneumophila*. *J. Clin. Microbiol.* 10, 437–441.

Furniss, R. C., Slater, S., Frankel, G., and Clements, A. (2016). Enterohaemorrhagic, *E. coli* modulates an ARF6:Rab35 signaling axis to prevent recycling endosome maturation during infection. *J. Mol. Biol.* 428, 3399–3407. doi: 10.1016/j.jmb.2016.05.023

Goody, P. R., Heller, K., Oesterlin, L. K., Muller, M. P., Itzen, A., and Goody, R. S. (2012). Reversible phosphocholination of Rab proteins by *Legionella pneumophila* effector proteins. *EMBO J.* 31, 1774–1784. doi: 10.1038/emboj.2012.16

Hardiman, C. A., and Roy, C. R. (2014). AMPylation is critical for Rab1 localization to vacuoles containing *Legionella pneumophila*. *MBio* 5:e01035-13. doi: 10.1128/mBio.01035-13

Hilbi, H., Weber, S., and Finsel, I. (2011). Anchors for effectors: subversion of phosphoinositide lipids by legionella. *Front. Microbiol.* 2:91. doi: 10.3389/fmicb.2011.00091

Horwitz, M. A., and Silverstein, S. C. (1980). Legionnaires' disease bacterium (*Legionella pneumophila*) multiples intracellularly in human monocytes. *J. Clin. Invest.* 66, 441–450. doi: 10.1172/JCI109874

Isaac, D. T., Laguna, R. K., Valtz, N., and Isberg, R. R. (2015). MavN is a *Legionella pneumophila* vacuole-associated protein required for efficient iron acquisition during intracellular growth. *Proc. Natl. Acad. Sci. U.S.A.* 112, E5208–E5217. doi: 10.1073/pnas.1511389112

Joshi, A. D., Sturgill-Koszycki, S., and Swanson, M. S. (2001). Evidence that Dot-dependent and -independent factors isolate the *Legionella pneumophila* phagosome from the endocytic network in mouse macrophages. *Cell. Microbiol.* 3, 99–114. doi: 10.1046/j.1462-5822.2001.00093.x

Kagan, J. C., Stein, M. P., Pypaert, M., and Roy, C. R. (2004). Legionella subvert the functions of Rab1 and Sec22b to create a replicative organelle. *J. Exp. Med.* 199, 1201–1211. doi: 10.1084/jem.20031706

Klinkert, K., and Echard, A. (2016). Rab35 GTPase: a central regulator of phosphoinositides and F-actin in endocytic recycling and beyond. *Traffic* 17, 1063–1077. doi: 10.1111/tra.12422

Kouranti, I., Sachse, M., Arouche, N., Goud, B., and Echard, A. (2006). Rab35 regulates an endocytic recycling pathway essential for the terminal steps of cytokinesis. *Curr. Biol.* 16, 1719–1725. doi: 10.1016/j.cub.2006.07.020

Larson, C. L., and Heinzen, R. A. (2017). High-content imaging reveals expansion of the endosomal compartment during *Coxiella burnetii* parasitophorous vacuole maturation. *Front. Cell. Infect. Microbiol.* 7:48. doi: 10.3389/fcimb.2017.00048

Li, J., Mahajan, A., and Tsai, M. D. (2006). Ankyrin repeat: a unique motif mediating protein-protein interactions. *Biochemistry* 45, 15168–15178. doi: 10.1021/bi062188q

Luo, Z. Q., and Isberg, R. R. (2004). Multiple substrates of the *Legionella pneumophila* Dot/Icm system identified by interbacterial protein transfer. *Proc. Natl. Acad. Sci. U.S.A.* 101, 841–846. doi: 10.1073/pnas.0304916101

Machner, M. P., and Isberg, R. R. (2006). Targeting of host Rab GTPase function by the intravacuolar pathogen *Legionella pneumophila*. *Dev. Cell* 11, 47–56. doi: 10.1016/j.devcel.2006.05.013

Machner, M. P., and Isberg, R. R. (2007). A bifunctional bacterial protein links GDI displacement to Rab1 activation. *Science* 318, 974–977. doi: 10.1126/science.1149121

Mantegazza, A. R., Magalhaes, J. G., Amigorena, S., and Marks, M. S. (2013). Presentation of phagocytosed antigens by MHC class I and II. *Traffic* 14, 135–152. doi: 10.1111/tra.12026

Mellman, I. (1996). Endocytosis and molecular sorting. *Annu. Rev. Cell Dev. Biol.* 12, 575–625. doi: 10.1146/annurev.cellbio.12.1.575

Mellouk, N., Weiner, A., Aulner, N., Schmitt, C., Elbaum, M., Shorte, S., et al. (2014). Shigella subverts the host recycling compartment to rupture its vacuole. *Cell Host Microbe* 16, 517–530. doi: 10.1016/j.chom.2014.09.005

Mihai Gazdag, E., Streller, A., Haneburger, I., Hilbi, H., Vetter, I. R., Goody, R. S., et al. (2013). Mechanism of Rab1b deactivation by the *Legionella pneumophila* GAP LepB. *EMBO Rep.* 14, 199–205. doi: 10.1038/embor.2012.211

Mousnier, A., Schroeder, G. N., Stoneham, C. A., So, E. C., Garnett, J. A., Yu, L., et al. (2014). A new method to determine *in vivo* interactomes reveals binding of the *Legionella pneumophila* effector PieE to multiple rab GTPases. *MBio* 5:e01148-14. doi: 10.1128/mBio.01148-14

Mukherjee, S., Liu, X., Arasaki, K., McDonough, J., Galan, J. E., and Roy, C. R. (2011). Modulation of Rab GTPase function by a protein phosphocholine transferase. *Nature* 477, 103–106. doi: 10.1038/nature10335

Neunuebel, M. R., Chen, Y., Gaspar, A. H., Backlund, P. S. Jr., Yergey, A., and Machner, M. P. (2011). De-AMPylation of the small GTPase Rab1 by the pathogen *Legionella pneumophila*. *Science* 333, 453–456. doi: 10.1126/science.1207193

Neunuebel, M. R., Mohammadi, S., Jarnik, M., and Machner, M. P. (2012). *Legionella pneumophila* LidA affects nucleotide binding and activity of the host GTPase Rab1. *J. Bacteriol.* 194, 1389–1400. doi: 10.1128/JB.06306-11

O'Connor, T. J., Adepoju, Y., Boyd, D., and Isberg, R. R. (2011). Minimization of the *Legionella pneumophila* genome reveals chromosomal regions involved in host range expansion. *Proc. Natl. Acad. Sci. U.S.A.* 108, 14733–14740. doi: 10.1073/pnas.1111678108

Ouellette, S. P., and Carabeo, R. A. (2010). A functional slow recycling pathway of transferrin is required for growth of chlamydia. *Front. Microbiol.* 1:112. doi: 10.3389/fmicb.2010.00112

Pan, X., Luhrmann, A., Satoh, A., Laskowski-Arce, M. A., and Roy, C. R. (2008). Ankyrin repeat proteins comprise a diverse family of bacterial type IV effectors. *Science* 320, 1651–1654. doi: 10.1126/science.1158160

Patino-Lopez, G., Dong, X., Ben-Aissa, K., Bernot, K. M., Itoh, T., Fukuda, M., et al. (2008). Rab35 and its GAP EPI64C in T cells regulate receptor recycling and immunological synapse formation. *J. Biol. Chem.* 283, 18323–18330. doi: 10.1074/jbc.M800056200

Pizarro-Cerda, J., Kuhbacher, A., and Cossart, P. (2014). Phosphoinositides and host-pathogen interactions. *Biochim. Biophys. Acta* 1851, 911–918. doi: 10.1016/j.bbalip.2014.09.011

Roy, C. R., and Mukherjee, S. (2009). Bacterial FIC Proteins AMP Up Infection. *Sci. Signal.* 2:pe14. doi: 10.1126/scisignal.262pe14

Schindelin, J., Arganda-Carreras, I., Frise, E., Kaynig, V., Longair, M., Pietzsch, T., et al. (2012). Fiji: an open-source platform for biological-image analysis. *Nat. Methods* 9, 676–682. doi: 10.1038/nmeth.2019

Schindelin, J., Rueden, C. T., Hiner, M. C., and Eliceiri, K. W. (2015). The ImageJ ecosystem: an open platform for biomedical image analysis. *Mol. Reprod. Dev.* 82, 518–529. doi: 10.1002/mrd.22489

Segal, G. (2013). Identification of legionella effectors using bioinformatic approaches. *Methods Mol. Biol.* 954, 595–602. doi: 10.1007/978-1-62703-161-5_37

Segal, G., and Shuman, H. A. (1999). *Legionella pneumophila* utilizes the same genes to multiply within Acanthamoeba castellanii and human macrophages. *Infect. Immun.* 67, 2117–2124.

Sherwood, R. K., and Roy, C. R. (2013). A Rab-centric perspective of bacterial pathogen-occupied vacuoles. *Cell Host Microbe* 14, 256–268. doi: 10.1016/j.chom.2013.08.010

Stein, M. P., Muller, M. P., and Wandinger-Ness, A. (2012). Bacterial pathogens commandeer Rab GTPases to establish intracellular niches. *Traffic* 13, 1565–1588. doi: 10.1111/tra.12000

Stenmark, H. (2009). Rab GTPases as coordinators of vesicle traffic. *Nat. Rev. Mol. Cell Biol.* 10, 513–525. doi: 10.1038/nrm2728

Tan, Y., and Luo, Z. Q. (2011). *Legionella pneumophila* SidD is a deAMPylase that modifies Rab1. *Nature* 475, 506–509. doi: 10.1038/nature10307

Tilney, L. G., Harb, O. S., Connelly, P. S., Robinson, C. G., and Roy, C. R. (2001). How the parasitic bacterium *Legionella pneumophila* modifies its phagosome and transforms it into rough ER: implications for conversion of plasma membrane to the ER membrane. *J. Cell Sci.* 114(Pt 24), 4637–4650.

van Ijzendoorn, S. C. (2006). Recycling endosomes. *J. Cell Sci.* 119(Pt 9), 1679–1681. doi: 10.1242/jcs.02948

Verma, K., and Datta, S. (2017). The monomeric GTPase Rab35 regulates phagocytic cup formation and phagosomal maturation in *Entamoeba histolytica*. *J. Biol. Chem.* 292, 4960–4975. doi: 10.1074/jbc.M117.775007

Viasus, D., Di Yacovo, S., Garcia-Vidal, C., Verdaguer, R., Manresa, F., Dorca, J., et al. (2013). Community-acquired *Legionella pneumophila* pneumonia: a single-center experience with 214 hospitalized sporadic cases over 15 years. *Medicine* 92, 51–60. doi: 10.1097/MD.0b013e31827f6104

Walseng, E., Bakke, O., and Roche, P. A. (2008). Major histocompatibility complex class II-peptide complexes internalize using a clathrin- and dynamin-independent endocytosis pathway. *J. Biol. Chem.* 283, 14717–14727. doi: 10.1074/jbc.M801070200

Weber, S. S., Ragaz, C., Reus, K., Nyfeler, Y., and Hilbi, H. (2006). *Legionella pneumophila* exploits PI(4)P to anchor secreted effector proteins to the replicative vacuole. *PLoS Pathog.* 2:e46. doi: 10.1371/journal.ppat.0020046

Weber, S., Wagner, M., and Hilbi, H. (2014). Live-cell imaging of phosphoinositide dynamics and membrane architecture during Legionella infection. *MBio* 5:e00839-13. doi: 10.1128/mBio.00839-13

Worby, C. A., Mattoo, S., Kruger, R. P., Corbeil, L. B., Koller, A., Mendez, J. C., et al. (2009). The fic domain: regulation of cell signaling by adenylylation. *Mol. Cell* 34, 93–103. doi: 10.1016/j.molcel.2009.03.008

Yarbrough, M. L., Li, Y., Kinch, L. N., Grishin, N. V., Ball, H. L., and Orth, K. (2009). AMPylation of Rho GTPases by Vibrio VopS disrupts effector binding and downstream signaling. *Science* 323, 269–272. doi: 10.1126/science.1166382

Zerial, M., and McBride, H. (2001). Rab proteins as membrane organizers. *Nat. Rev. Mol. Cell Biol.* 2, 107–117. doi: 10.1038/35052055

16

Regulatory Principles Governing *Salmonella* and *Yersinia* Virulence

Marc Erhardt*[1] *and Petra Dersch*[2]

[1] Young Investigator Group Infection Biology of Salmonella, Helmholtz Centre for Infection Research, Braunschweig, Germany, [2] Department of Molecular Infection Biology, Helmholtz Centre for Infection Research, Braunschweig, Germany

Edited by:
Dongsheng Zhou,
Beijing Institute of Microbiology and Epidemiology, China

Reviewed by:
Amit Kumar Tyagi,
The University of Texas MD Anderson Cancer Center, USA
James M. Slauch,
University of Illinois, USA

****Correspondence:***
Petra Dersch,
Department of Molecular Infection Biology, Helmholtz Centre for Infection Research, Inhoffenstrasse 7, 38124 Braunschweig, Germany
petra.dersch@helmholtz-hzi.de

Enteric pathogens such as *Salmonella* and *Yersinia* evolved numerous strategies to survive and proliferate in different environmental reservoirs and mammalian hosts. Deciphering common and pathogen-specific principles for how these bacteria adjust and coordinate spatiotemporal expression of virulence determinants, stress adaptation, and metabolic functions is fundamental to understand microbial pathogenesis. In order to manage sudden environmental changes, attacks by the host immune systems and microbial competition, the pathogens employ a plethora of transcriptional and post-transcriptional control elements, including transcription factors, sensory and regulatory RNAs, RNAses, and proteases, to fine-tune and control complex gene regulatory networks. Many of the contributing global regulators and the molecular mechanisms of regulation are frequently conserved between *Yersinia* and *Salmonella*. However, the interplay, arrangement, and composition of the control elements vary between these closely related enteric pathogens, which generate phenotypic differences leading to distinct pathogenic properties. In this overview we present common and different regulatory networks used by *Salmonella* and *Yersinia* to coordinate the expression of crucial motility, cell adhesion and invasion determinants, immune defense strategies, and metabolic adaptation processes. We highlight evolutionary changes of the gene regulatory circuits that result in different properties of the regulatory elements and how this influences the overall outcome of the infection process.

Keywords: virulence regulation, pathogenicity factors, motility, transcription factors, post-transcriptional modifications, environmental control systems regulatory RNAs, riboswitches, metabolic adaptation

Introduction

Gastrointestinal infections by pathogenic *Enterobacteriaceae* represent a serious economic and health problem worldwide. They cause severe diarrheal diseases, which are still a leading cause of death among children under five. In developed countries, the incidence of foodborne outbreaks is less problematic, but remains substantial and constitutes a significant socioeconomic burden. Globalization of food supply, introduction and persistence of the pathogens in unknown environmental niches, frequent environmental changes leading to rapid evolution of newly emerging variants, and the development of antibiotic-resistant *Enterobacteriaceae* which use the intestinal tract as main reservoir are reasons why foodborne intestinal diseases by these pathogens remain a global public health problem. The worldwide rise of antibiotic-resistant strains represents a serious threat for the treatment of gastrointestinal pathogens. Furthermore, antibiotic therapies of most gastrointestinal infections are (i) ineffective as they are unable to improve clinical symptoms or shorten duration of shedding, are (ii) skewing of the intestinal commensal community (dysbiosis) which was found to support inflammatory diseases, e.g., by enteric pathogens or other usually harmless commensals, and are (iii) associated with an increased risk of serious complications in cases such as enterohemorrhagic *Escherichia coli* (EHEC).

Many cases of gastrointestinal infections in Europa and North America are caused by members of the genus *Salmonella* and *Yersinia*. The enteric representatives of both genera are gram-negative, facultative anaerobic and motile bacteria, which are usually transmitted by infected animals, or contaminated food or water via the fecal–oral route. Enteropathogenic *Salmonella* and *Yersinia* species can occupy many different environmental habitats, persist in certain domestic and wild animal reservoir hosts (in particular cattle, swine, poultry, wild birds, pet reptiles), and are routinely isolated from ground water, soil, plants, and insects (Fredriksson-Ahomaa et al., 2006; Hoelzer et al., 2011; Fredriksson-Ahomaa, 2012; Wiedemann et al., 2014). Undercooked chicken and eggs are considered the major infection source of *salmonellae*, whereas undercooked pork, vegetables, and lettuce are responsible for most *Yersinia* infections (Hoelzer et al., 2011; Fredriksson-Ahomaa, 2012). Enteric *Salmonella* and *Yersinia* species cause various gut-associated symptoms (e.g., enteritis, ileitis, colitis, vomitting, intestinal cramping, and inflammatory diarrhea), and in rare cases they can lead to systemic infections and induce extra-intestinal sequelae such as fatal respiratory, hepatic, spleen, and/or neurological damage, erythema nodosum and reactive arthritis (Bottone, 1999; Koornhof et al., 1999; Ohl and Miller, 2001). Inflammatory diarrhea, which is typically associated with the highly invasive *Yersinia* species *Y. enterocolitica* and *Y. pseudotuberculosis*, and enteroinvasive *Salmonella* serotypes is characterized by an acutely increased vascular permeability and the recruitment of neutrophils which result in the formation of tissue exudates and necrotic lesions (Autenrieth and Firsching, 1996; Tsolis et al., 2008). Invasion of the bacteria from the intestinal tract into underlying lymphatic follicles (Peyer's patches) can lead to the ulceration of the tissue and abdominal pain that mimics appendicitis. The detection of the invasive bacteria by the innate immune system also results in the production of pyrogenic cytokines (IL-1β, IL-6, TNF-α and interferons) that increase the host's body temperature (typhoid/enteric fever) (Tsolis et al., 2008; Gu et al., 2013).

Enteric *Yersinia* and *Salmonella* serotypes are well adapted for survival in a variety of external environments and persistence in various host animals. The bacteria adjust rapidly to several extremely different and stressful environments during the invasion process, such as gastric acidity, increased osmolarity, changing nutrient/ion availability, and competition with the host's microbiota in the intestine (Fabrega and Vila, 2013; Heroven and Dersch, 2014). In order to succeed in invading the host, they employ a wide variety of virulence factors, such as flagella, host colonization factors (fimbriae, adhesins/invasins), and many secreted toxins and effector proteins. These molecules are needed to propel the bacteria to target cells, attach to, infect and survive inside host cells and protect the pathogen against host immune responses (**Figure 1**). To establish a successful infection, process, these virulence factors and adequate metabolic pathways must be expressed at the correct spatiotemporal conditions. Consequently, a substantial cross-talk between different pathogenicity elements and physiological adaptation processes has to exist. The correct spatiotemporal expression of the various virulence constituents is achieved by regulation at the transcriptional, post-transcriptional and post-translational levels by a myriad of transcription factors, nucleoid-associated proteins, regulatory small RNAs (sRNAs) and RNases. Moreover, a growing set of small signaling molecules and different proteases controls the activity of enzymes and virulence-relevant factors during the infection process. The majority of the contributing regulators and control mechanisms are highly conserved between *salmonellae* and *yersiniae*, but the regulatory network composition and architecture are often changed. In the following chapters we will compare regulatory factors and global networks controlling early and late stages of the infection process of *Salmonella* and *Yersinia* to highlight conserved regulatory elements, but also changes that provided an important source for the divergence of the different genera with respect to their virulence-relevant physiological and pathogenic properties.

Regulatory Circuits Controlling Early Stages of Host Colonization

Survival in the Intestinal Lumen

The gastrointestinal tract of mammals is a rich source of food for enteric pathogens as it contains a large variety of simple and polymeric sugars, amino acids, peptides, and proteins, as well as lipids and other carbon- and nitrogen-containing metabolites. The availability of the nutrients varies significantly between hosts, different sections of the intestinal tract and different stages of the infection process, as it depends strongly on (i) the diet of the host, (ii) the composition and activity of the intestinal microbiota, which includes 10^{14} well-adapted bacteria from more than 400 species, (iii) absorption by the intestinal cells, and (iv) inflammation and hypoxic conditions induced by the host immune response during the progress of an infection (Nizet and Johnson, 2009; Rohmer et al., 2011). Moreover, the host limits the pathogen's access to essential ions, including magnesium, iron, and zinc (Brown et al., 2008; Abu Kwaik and Bumann, 2013; Zhang and Rubin, 2013). Consequently, *Salmonella* and *Yersinia* must either grow on non-utilized energy sources or metabolic end products, or metabolize nutrients and import ions much more efficiently to ensure maximal fitness and competiveness against the intestinal microbiota.

An important common characteristic of enteropathogenic *Salmonella* and *Yersinia* strains is that they possess a highly complex metabolism with many redundant or alternative catabolic and biosynthetic pathways. This allows them to utilize a large variety of organic energy sources and renders them very flexible and robust against sudden changes of nutrient availability. Another basis for their success is their ability to select the best energy sources among provided nutrients to gain the most benefit for lowest costs to optimize their biological fitness (Dandekar et al., 2014; Heroven and Dersch, 2014). Furthermore, they possess a plethora of sophisticated sensing and signal transduction strategies to adapt to variation in

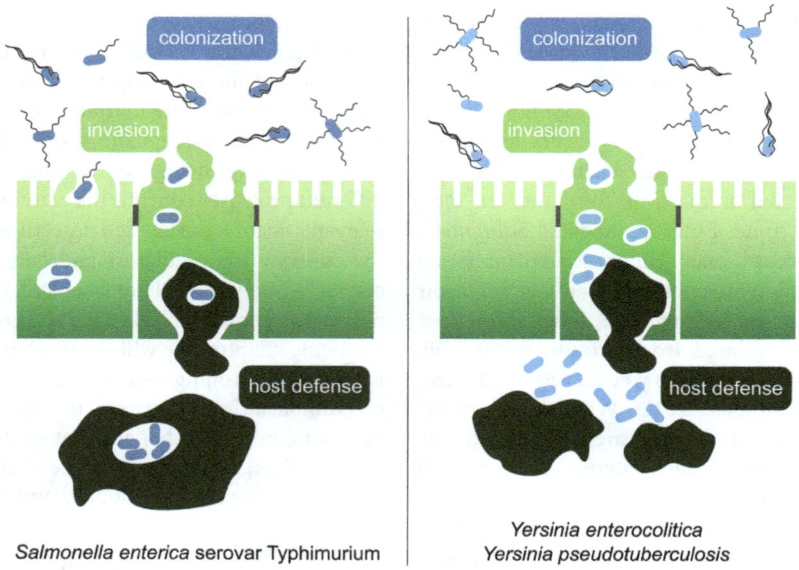

FIGURE 1 | The course of infection by enteric *Yersinia* species and *Salmonella enterica* serovar Typhimurium. Enteropathogenic *Salmonella* Typhimurium utilizes flagellar motility for host colonization in the gut lumen and directed movement toward the epithelial layer. *Salmonella* attaches to the intestinal epithelium using a variety of adhesins and secretion of effector proteins via the Spi-1 encoded injectisome device for invasion of M cells or other enterocytes. The bacteria survive and replicate within *Salmonella*-containing vacuoles inside epithelial cells and phagocytes. The enteropathogenic *Yersinia* species *Y. enterocolitica* and *Y. pseudotuberculosis* are ingested via contaminated food and enter the lymphatic system through the M cells in the small intestine. In the lymphatic tissues they reside predominantly outside of the cells.

the nutrient composition during the course of an infection. Moreover, a large number of conserved regulatory proteins has been identified over the past years, which are implicated in metabolic control (**Table 1**), and a steadily increasing number of post-transcriptional control systems, including non-coding RNAs, and small signaling molecules such as ppGpp, cAMP, and c-di-GMP has been identified as additional elements controlling virulence and metabolism.

Sensing Intestinal Stimuli by Two-Component Regulatory Systems

External stimuli, which are relevant for metabolic adaptation and virulence are frequently sensed by bacterial two component systems (TCSs) and converted into an adaptive cellular response. Among the most virulence-relevant TCS are the pleiotropic PhoP/PhoQ, EnvZ/OmpR, and BarA/SirA(UvrY) systems (Groisman, 2001; Groisman and Mouslim, 2006) (**Figures 2** and **3**). They are composed of the membrane-bound sensor kinases PhoQ, EnvZ, and BarA that sense the environmental signals and phosphorylate the cytoplasmic response regulator PhoP, OmpR, and UvrY(SirA). The TCS PhoP/PhoQ responds to low magnesium, low pH environments, and host-secreted cationic antimicrobial peptides (Groisman, 2001). The global regulator PhoP controls a very complex network of genes, whereby the individual *Yersinia* and *Salmonella* PhoP regulons have considerable differences: (i) in the molecular architecture of the regulatory sequences and promoters, and (ii) in amino acid alterations in the conserved PhoP regulator itself. This enables both PhoP response regulators to retain the ability to transcribe the core members of the regulon in both pathogens and also allows inclusion of newly acquired genes into the ancestral regulatory circuit (Perez and Groisman, 2009). In *Salmonella*, the PhoP/PhoQ system is essential for virulence and survival within macrophages (Miller et al., 1989). The response regulator PhoP represses *hilA* and the *prg* (PhoP-repressed genes) genes (Pegues et al., 1995; Bajaj et al., 1996), whereas transcription of PhoP-activated genes (*pag*) required for survival within macrophages is activated (Miller et al., 1989; Belden and Miller, 1994). PhoP also controls expression of *Salmonella* pathogenicity island-2 (Spi-2) by binding to the *ssrB* promoter region and the 5′-UTR of the *spiR* transcript (Belden and Miller, 1994). In contrast, in *yersiniae* the PhoP/PhoQ system has been found to promote proliferation of human pathogenic *yersiniae* within professional phagocytes *in vitro*, but its role during pathogenesis is less defined (Grabenstein et al., 2004, 2006; Flamez et al., 2007). Recent studies indicate that strain-specific differences that remodel expression of PhoP-dependent control functions appear to influence the overall outcome of the virulence phenotype (Grabenstein et al., 2004; Bozue et al., 2011; Nuss et al., 2014; Pisano et al., 2014). The PhoP/PhoQ system of *Yersinia* was shown to control modification of lipid A linked to antimicrobial peptide resistance and promotes survival and proliferation in macrophages and neutrophils (Grabenstein et al., 2004; Reines et al., 2012). Another important virulence control system, which is under PhoP/PhoQ control, is the carbon storage regulator (Csr) system coordinating the expression of important *Yersinia* adhesion factors (e.g., invasin), motility and multiple virulence-relevant metabolic pathways (see also Conclusion and

TABLE 1 | Comparison of the protein homology of major regulatory factors of *Salmonella enterica* serovar Typhimurium (ST) and *Yersinia pseudotuberculosis* (Ypt) using Basic Local Alignment Search Tool (BLAST) analysis.

Gene name ST strain LT2	Locus tag ST strain LT2	Gene name Ypt strain YPIII	Locus tag Ypt strain YPIII	Sequence coverage (%)	Amino acids identity (%)	Mode of regulation	Virulence pathway
BarA	STM2958	BarA	YPK_3451	99	59	TCS	Colonization, invasion, host defense
ClpP	STM0448	ClpP	YPK_3234	100	89	Protease	Colonization, invasion
CpxA/CpxS	STM4058	CpxA	YPK_4133	99	81	TCS	Host defense
CpxR	STM4059	CpxR	YPK_4132	100	89	TCS	Host defense
Crp	STM3466	Crp	YPK_0248	100	99	Transcription	Colonization, invasion, host defense
CsrA	STM2826	CsrA	YPK_3372	100	95	Translation	Colonization, invasion, host defense
Dam	STM3484	dam	YPK_0228	96	70	Translation	Invasion
DnaK	STM0012	DnaK	YPK_3594	100	92	Protein stability	Colonization
EnvZ	STM3501	EnvZ	YPK_0173	97	88	TCS	Colonization, invasion, host defense
FimZ	STM0549		*YPK_2269*	98	48	Transcription	Colonization
FimZ	STM0549		*YPK_2499*	98	31		
Fis	STM3385	Fis	YPK_0452	100	98	Transcription	Colonization, invasion, host defense
FlhC	STM1924.S	FlhC	YPK_1746	100	82	Transcription	Colonization
FlhD	STM1925	FlhD	YPK_1745	97	76	Transcription	Colonization
FliA	STM1956	FliA	YPK_2380	99	83	Transcription	Colonization, invasion
FliT	STM1962	FliT	YPK_2384	85	36	Protein stability	Colonization
FliZ	STM1955	FliZ	YPK_2378	91	55	Transcription, translation	Colonization, invasion
H-NS	STM1751	H-NS	YPK_2074	98	87	Transcription	Colonization, invasion, host defense
HdfR/ YifA	STM3897		*YPK_4064*	98	62	Transcription	Colonization
Hha	STM0473	YmoA	YPK_3214	93	82	Transcription	Colonization, invasion, host defense
HilA	STM2876		NA	NA	NA	Transcription	Invasion
HilC/ SirC	STM2867		NA	NA	NA	Transcription	Invasion
HilD	STM2875		NA	NA	NA	Transcription	Colonization, invasion, host defense
HilE	STM4509.S		*YPK_0803*	83	30	Protein activity	Invasion
IhfA	STM1339		YPK_1826	98	94	Transcription	Host defense
InvF	STM2899		NA	NA	NA	Transcription	Invasion
IscR/ YfhP	STM2544	IscR	YPK_1275	88	79	Transcription	Host defense
Lon	STM0450	Lon	YPK_3232	100	91	Protease	Colonization, invasion
LrhA	STM2330	RovM	YPK_1559	92	72	Transcription	Colonization, invasion
NA	NA	LcrQ/YscM	pYV0089	NA	NA	Translation	Host defense
NA	NA	YopD	pYV0054	NA	NA	Translation	Host defense
OmpR	STM3502	OmpR	YPK_0172	100	99	TCS	Colonization, invasion, host defense
PhoB	STM0397	PhoB	YPK_3276	100	90	TCS	Colonization, invasion
PhoP	STM1231	PhoP	YPK_1715	99	79	TCS	Invasion, host defense
PhoQ	STM1230	PhoQ	YPK_1714	96	62	TCS	Invasion, host defense

(Continued)

TABLE 1 | Continued

Gene name ST strain LT2	Locus tag ST strain LT2	Gene name Ypt strain YPIII	Locus tag Ypt strain YPIII	Sequence coverage (%)	Amino acids identity (%)	Mode of regulation	Virulence pathway
PhoR	STM0398	PhoR	YPK_3275	99	72	TCS	Colonization, invasion
QseB/ YgiX	STM3177		*YPK_3741*	98	48	TCS	Colonization
QseC/ YgiY	STM3178		*YPK_3742*	61	34	TCS	Colonization
RcsB	STM2270	RcsB	YPK_2843	100	92	TCS	Colonization, invasion, host defense
RcsD/ YojN	STM2269	RcsD	YPK_2842	98	47	TCS	Colonization, invasion, host defense
RflM/EcnR	STM4337	RcsA	YPK_1671	93	29	Transcription	Colonization
RhaS	STM4048	LcrF	pYV0076	96	22	Transcription	Host defense
Rnase E/rne	STM1185	Rnase E	YPK_1677	100	64	Translation	Invasion, host defense
RscC	STM2271	RscC	YPK_2844	99	58	TCS	Colonization, invasion, host defense
RtsA	STM4315		NA	NA	NA	Transcription	Invasion
RtsB	STM4314		*YPK_3033*	62	41	Transcription	Colonization, invasion
RtsB	STM4314		*YPK_1465*	56	35		
SdiA	STM1950	YpsR	YPK_1655	91	29	Transcription	Colonization
SdiA	STM1950	YtbR	YPK_0791	81	28	Transcription	Colonization
SirA	STM1947	SirA	YPK_2356	99	84	TCS	Colonization, invasion, host defense
SlyA	STM1444	RovA	YPK_1876	98	75	Transcription	Colonization, invasion, host defense
SpiR	STM1392		*YPK_3918*	98	44	TCS	Host defense
SscA	STM1399	LcrH	pYV0056	83	27	Translation	Host defense
SsrB	STM1391	SsrB	YPK_3919	96	55	TCS	Host defense
YdiV	STM1344		*YPK_1543*	91	22	Protein stability	Colonization, invasion
YhjH	STM3611		*YPK_0103*	91	49	Protein activity	Colonization

Sequence coverage reports the length of the identified homologous protein sequence with the indicated amino acid identity. In the majority, the true orthologs of the query protein sequences are listed, whereas close homologs with unknown regulatory function and low sequence identity are given in italics. NA = not available.

Outlook; Heroven et al., 2008; Bücker et al., 2014; Nuss et al., 2014).

The TCS OmpR/EnvZ was originally identified as regulator system controlling the outer membrane porins in response to osmolarity and temperature, but later it was found that it also plays an essential role in controlling the expression of *Yersinia* and *Salmonella* virulence functions (Dorman et al., 1989; Dorrell et al., 1998; Lee et al., 2000; Brzostek et al., 2012). In *Yersinia*, it is required for serum resistance, survival within macrophages, biofilm formation, and it influences the adhesion/invasion abilities (e.g., invasin, Ail). It was further shown to control expression of the AcrAB-TolC multidrug efflux pump, urease, and a type-VI secretion system (T6SS-4) (Raczkowska et al., 2011a, 2015; Brzostek et al., 2012; Reines et al., 2012; Gueguen et al., 2013; Skorek et al., 2013; Zhang et al., 2013). Recently, it was also demonstrated that OmpR of *Yersinia* is involved in the control of motility and flagellation by activation of the flagellar operon *flhDC* (Hu et al., 2009; Raczkowska et al., 2011b). In *Salmonella*, the EnvZ-OmpR TCS also regulates flagellar and virulence genes, but in this pathogen OmpR represses *flhDC* in response to extracellular osmolarity. It has been further implicated in post-transcriptional regulation of the Spi-1-encoded regulator HilD (Ellermeier et al., 2005; Golubeva et al., 2012) and functions as an activator of Spi-2 virulence genes (Lee et al., 2000; see also Regulation of Motility, Attachment and Invasion of the Intestinal Epithelium).

Another virulence-relevant TCS, the BarA/UvrY (SirA) system, is activated by metabolic end products (e.g., short-chain fatty acids such as format, acetate) and an imbalance of the TCA cycle (Takeuchi et al., 2009; Chavez et al., 2010). This signal transduction system is conserved in many γ-proteobacteria and controls the expression of small regulatory RNAs with unpaired GGA motifs, which are part of the Csr system (Heeb and Haas, 2001; Lapouge et al., 2007). In *Yersinia* and *Salmonella* the BarA/UvrY(SirA) system controls the expression of cell invasion genes. It is required for the expression of the *Salmonella* Spi-1 invasion genes, and transcription of these genes was shown to be restored in a *barA* mutant by the addition of acetate, but not with butyrate. This indicated that the concentration and composition of short-chain fatty acids in the distal ileum provides a signal for

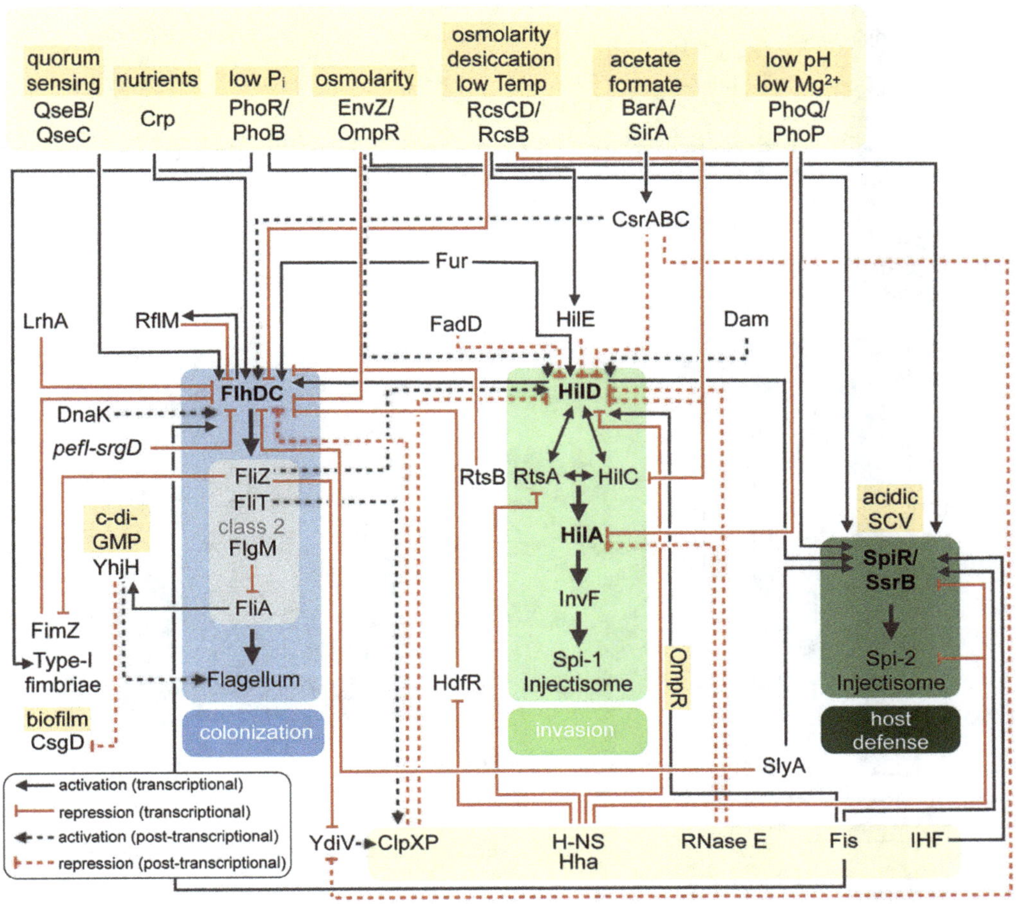

FIGURE 2 | Regulatory networks controlling *Salmonella* virulence factors. Overview of regulatory and environmental factors that control expression of the major pathogenicity traits of *Salmonella* needed for host colonization, invasion and host defense. Transcriptional or post-transcriptional regulatory effects are listed.

productive infection by *Salmonella* (Altier et al., 2000; Lawhon et al., 2002). SirA(UvrY) indirectly activates Spi-1 gene expression by controlling expression and function of Csr-type regulatory RNAs (CsrB, CsrC). CsrB and CsrC inhibit the function of the RNA-binding protein CsrA, which in turn prevents translation of *hilD*, a major activator of Spi-1 (see also Attachment and Invasion of the Intestinal Epithelium, **Figure 2**; Altier et al., 2000; Teplitski et al., 2006; Martinez et al., 2011). Similar to *Salmonella*, BarA/UvrY(SirA) controls expression of the CsrB RNA of *Yersinia*, which also regulates the expression of host cell adhesion and invasion genes (e.g., invasin), however, through a very different signal transduction cascade (see Attachment and Invasion of the Intestinal Epithelium, **Figure 3**).

Metabolic Adaptation by Global Regulatory Systems of the Carbon Metabolism

Several global regulators, which are highly conserved between *Salmonella* and *Yersinia*, are known that govern multiple cascades of control elements and complex networks in order to manage metabolic adaptation and coordinate this process with pathogenicity mechanisms. Among them is the cAMP receptor protein Crp. This regulator helps the pathogens to rank available C-sources in order to optimize their metabolism (Gorke and Stulke, 2008; Poncet et al., 2009). Primarily, the bacteria check for the availability of readily digestible sugars such as glucose and fructose. In the absence of these efficiently utilizable sugars, the signal metabolite cAMP is produced by the adenylate cyclase, which then binds to Crp to form an active cAMP-Crp complex (Hanamura and Aiba, 1991; Ishizuka et al., 1994). This metabolic sensor complex was recently shown to control more than 6% of the genes in *Y. pseudotuberculosis* (Heroven et al., 2012b; Nuss et al., 2015). *crp* mutants of *Y. pseudotuberculosis* and *Y. enterocolitica* are strongly attenuated, and comparative transcriptomics, metabolomics, and phenotypic microarrays showed that Crp is required for growth on different carbon sources and promotes survival under carbon, nitrogen, and phosphate limitation (Petersen and Young, 2002; Heroven et al., 2012b; Nuss et al., 2015). In a recent study using RNA-Seq we could also demonstrate a massive remodeling of the Crp-controlled network in response to temperature and discovered Crp as a transcriptional master regulator of numerous conserved regulatory RNAs, which adjust the *Yersinia*

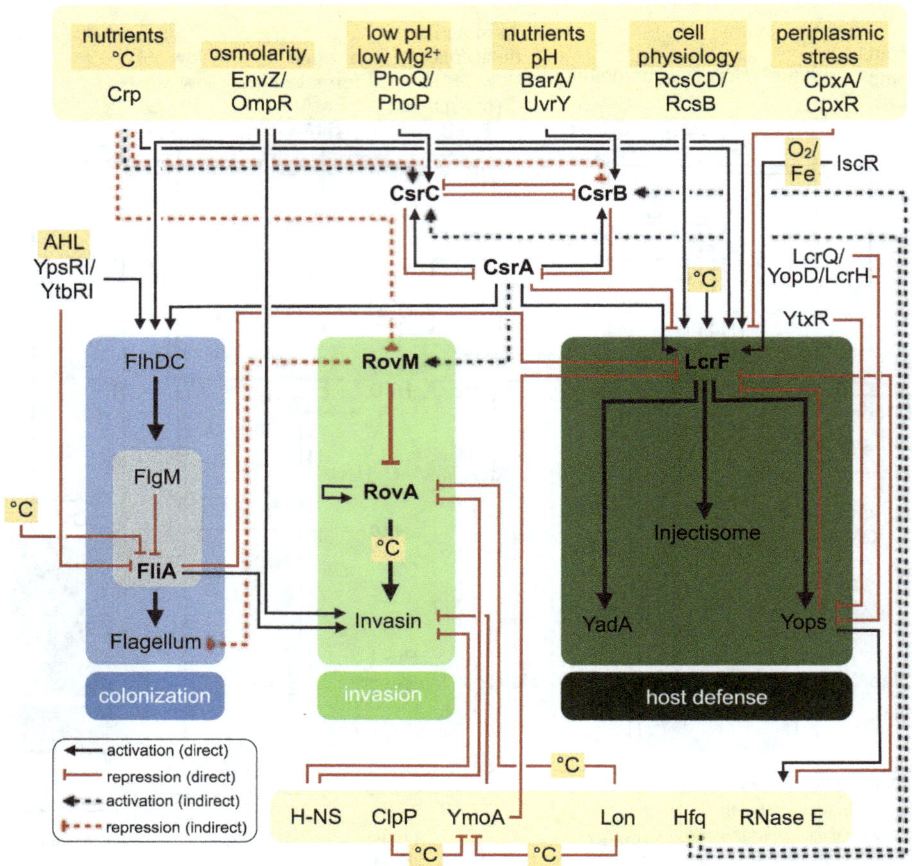

FIGURE 3 | Regulatory networks controlling *Yersinia* virulence factors. Regulatory networks controlling expression of motility, adhesion and injectisome virulence traits in *Yersinia*. Direct or indirect regulatory effects of various factors are indicated.

metabolism and fitness to the requirements of their life-style in the intestine (Nuss et al., 2015). Moreover, Crp was found to link nutrient status/carbon metabolism and the regulation of virulence factors, e.g., via the control of the switching of the two Csr-RNAs CsrB and CsrC of *Y. pseudotuberculosis* (Heroven et al., 2012b). Similarly, Crp of *Salmonella enterica* serovar Typhimurium was shown to coregulate carbon metabolism and virulence directly and indirectly through the Csr system (see below and Section Attachment and Invasion of the Intestinal Epithelium; Teplitski et al., 2006), indicating that this regulator fulfills a similar role in this enteric pathogen. In fact, a *crp* mutant of *Salmonella* is completely attenuated in mice (Zhang et al., 1997).

As mentioned above, the other crucial global metabolic control system of *Yersinia* and *Salmonella*, the Csr system, is constituted of the global regulatory RNA-binding protein CsrA and the two antagonizing non-coding sRNAs CsrB and CsrC that bind and sequester multiple CsrA dimers. This prevents CsrA binding to lower affinity mRNA targets, which results in altered mRNA translation and/or stability (Heroven et al., 2012a; Romeo et al., 2012, 2013; Vakulskas et al., 2015). The CsrA protein interacts generally with hairpin structures located in the 5′-untranslated regions of its target genes that possess a GGA motif within the single-stranded loop with conserved flanking regions (RUACAR**GGA**) (Vakulskas et al., 2015). For instance, the binding of a CsrA dimer to one GGA-containing site was shown to allow bridging and binding to another site overlapping the ribosome-binding site of a target gene which resulted in translation repression (Mercante et al., 2009). CsrA has a global influence on the *Yersinia* transcriptome, and in total approximately 20% of the CsrA-dependent genes of *Y. pseudotuberculosis* are involved in a myriad of metabolic processes and are controlled in response to ions and availability of carbon sources (Heroven et al., 2012a; Bücker et al., 2014). Similarly, the absence of the *csrA* gene in *Salmonella* also greatly reduces mRNA levels of a large number of metabolic genes, e.g., genes for the utilization of maltose, propanediol, and ethanolamine (common carbon sources/nutrients of the intestinal tract), tetrathionate metabolism, and hydrogen sulfide production. Most importantly, these metabolic functions enable *Salmonella* to use tetrathionate as a terminal electron acceptor in the intestinal tract and allow the microbe to grow on degradation products of the microbiota (Lawhon et al., 2003; Winter et al., 2010).

Regulation of Motility

Flagella-mediated motility is used by *Salmonella* and *Yersinia* for movement in the intestinal lumen. The chemotactic motility is required for efficient colonization in the intestinal tract, host cell invasion and induction of pathogenesis. It allows the pathogen to benefit from the increased nutrient availability in the inflamed intestine, and initiates host cell contact and invasion (Young et al., 2000; Stecher et al., 2004, 2008). Flagella are rotary motility organelles and rotation of the rigid flagellar filaments provides the pathogens with propulsion forces for chemotactic movement through liquids (swimming) and highly viscous environments or surfaces (swarming) (Chevance and Hughes, 2008). The bacterial flagellum is assembled by a flagellar-specific type-III secretion system (f-T3SS) that is highly homologous to the virulence-associated type-III secretion systems (v-T3SS) of the injectisome devices (Abby and Rocha, 2012). Enteropathogenic *yersiniae* and *S. enterica* serovar Typhimurium produce flagella during colonization of the intestinal mucosa, but repress flagella production after invasion of the intestinal epithelial layer. The structural subunits, forming the flagellar filament, are critical because they are potent inducers of the innate immune system of the host (Young et al., 2000; Hayashi et al., 2001). Regulation of flagella production is tightly linked with the expression of the virulence-associated injectisome devices (v-T3SS) employed by many Gram-negative pathogens to manipulate host cells (Iyoda et al., 2001; Horne and Pruss, 2006; Mouslim and Hughes, 2014) (see also Attachment and Invasion of the Intestinal Epithelium).

The correct spatiotemporal assembly of the flagellum is a complex process that involves temporal regulation of more than 60 genes organized into a transcriptional hierarchy of three promoter classes (Kapatral et al., 2004; Chevance and Hughes, 2008; Anderson et al., 2010). The first levels of this hierarchy are similar between *Yersinia* and *Salmonella*. On top of the transcriptional regulatory cascade, many environmental stimuli are integrated at the level of the flagellar master regulatory operon, *flhDC*, which is expressed from a σ^{70}-dependent flagellar Class 1 promoter (Yanagihara et al., 1999; Young et al., 1999; Kapatral et al., 2004). The active $FlhD_4C_2$ heteromultimeric complex (FlhDC) directs σ^{70}-RNA polymerase (RNAP) to transcribe from flagellar Class 2 promoters (Liu and Matsumura, 1994; Wang et al., 2006), which regulate (i) the expression of the flagellar hook basal body (HBB) complex including the f-T3SS components needed for the export of all subsequent extra-cytoplasmic subunits and (ii) two regulators, the flagella-specific alternative sigma factor, σ^{28} (encoded by *fliA*) and its cognate anti-sigma factor, FlgM (Iriarte et al., 1995; Liu and Matsumura, 1995; Ding et al., 2009). Transcription from flagellar Class 3 promoters is specific for σ^{28}-RNAP and occurs only after a functional HBB structure has been assembled (Karlinsey et al., 2000). The binding of FlgM to σ^{28} prevents premature σ^{28}-RNAP-dependent Class 3 transcription and thereby mediates a temporal sensing of the assembly state of the flagellum (Ohnishi et al., 1992; Chadsey et al., 1998). Prior to completion of the HBB structure, the f-T3SS exports only early secretion substrates and does not recognize the late secretion substrate FlgM. Upon HBB completion, the f-T3SS undergoes a switch in secretion specificity and FlgM is recognized as a late secretion substrate and secreted from the cell, thereby allowing for σ^{28}-dependent transcription from Class 3 promoters (Hughes et al., 1993; Kutsukake, 1994). Gene products transcribed from Class 3 promoters include structural components of the chemotaxis machinery, the filament, and motor-force generators. Overall the hierarchy of flagellar expression is similar in *Yersinia*. However, several differences include (i) the presence of three tandem flagellin genes (*fleA, fleB, fleC*) in *Y. enterocolitica*, (ii) a different set of flagellar genes that are controlled directly by FlhDC and FliA (Class IIIa), and (iii) different expression levels/operon structure of certain flagellar genes, e.g., *fliA* and *fliZ* (Horne and Pruss, 2006). Moreover, although a direct interaction of the *Y. pseudotuberculosis* FlgM protein to σ^{28} was confirmed and shown to repress expression of the single flagellin gene (*fliC*) in *Y. pseudotuberculosis*, a *flgM* mutant of *Y. enterocolitica* and *Y. pseudotuberculosis* is non-motile. This suggested that the flagellin genes might be post-transcriptionally regulated (Kapatral et al., 1996; Ding et al., 2009). In addition, other differences exist in the regulatory hierarchies driven by FliA and FlgM when compared to *Salmonella*, and additional sigma factors apart from σ^{28} (e.g., σ^{54}) were found to be involved in the control of flagellar biosynthesis (Ding et al., 2009).

Expression of the flagella is tightly regulated and several layers of autoregulation exist that fine-tune expression of the flagellar system. The regulatory circuits are very well characterized in *Salmonella*. One regulatory circuit involves autoregulation of *flhDC* gene transcription. After activation of flagellar synthesis, the FlhDC complex initiates a regulatory feedback loop by activating transcription of its own repressor, RflM (Singer et al., 2013), which provides a mechanism to fine-tune *flhDC* expression levels. In *Salmonella*, the Class 2 gene product FliT also has been shown to have a dual function as an f-T3SS secretion chaperone of the filament cap protein FliD and as a regulator of FlhDC activity (Fraser et al., 1999; Yamamoto and Kutsukake, 2006). FliT negatively regulates FlhDC activity by inhibiting promoter binding of the FlhDC complex and increases ClpXP-dependent proteolysis of the FlhC subunit (Yamamoto and Kutsukake, 2006; Sato et al., 2014). Secretion of its cognate secretion substrate FliD after HBB completion releases FliT and thereby provides a mechanism to fine-tune flagellar numbers in response to the growth rate (Aldridge et al., 2010). Although the organization of the flagellar genes in operons and the arrangement along the chromosome differs between *Salmonella* and enteropathogenic *yersiniae*, the order and sequence of the flagellar genes, including *fliT* and *fliD* is mostly conserved (Horne and Pruss, 2006; Ding et al., 2009), suggesting that the overall functions might be maintained.

Another Class 2 gene product, FliZ, is a positive regulator of flagellar Class 2 and Class 3 gene transcription, which increases FlhDC activity by two distinct mechanisms (Kutsukake et al., 1999) (**Figure 2**). First, FliZ is a transcriptional repressor of the EAL-domain protein YdiV (Wada et al., 2011b). The YdiV protein binds to the FlhD subunit and prevents binding of FlhDC to Class 2 promoters by targeting FlhDC complex to ClpXP-mediated proteolytic degradation (Wozniak et al., 2009; Takaya et al., 2012). Expression of YdiV is high under nutrient starvation in *Salmonella* and low under high nutrient conditions

(Wada et al., 2011a). Accordingly, *Salmonella* is only motile when nutrients are plentiful in contrast to many other bacteria (Wada et al., 2012). The second mechanism by which FliZ activates Class 2 and 3 gene transcription in *Salmonella* is via positive feed-forward regulation of invasion gene expression encoded on Spi-1 (Lucas et al., 2000; Iyoda et al., 2001). FliZ is needed for efficient induction of the Spi-1 system via post-translational activation of the Spi-1 regulatory protein HilD, which in turn activates *flhDC* gene transcription (Iyoda et al., 2001; Chubiz et al., 2010; Mouslim and Hughes, 2014; Singer et al., 2014).

Although much less is known about the regulatory components implicated in the control of *Yersinia* motility and flagella formation, a similar close regulatory cross-circuit between the flagellar (f-T3SS) and *Yersinia* virulence plasmid (pYV)-encoded v-T3SS/injectisome regulons was detected (**Figure 3**). However, the components implicated in the cross-talk appear to differ significantly. An upregulation of the *Yersinia* v-T3SS/*yop* regulon was observed in an *flhDC* mutant of *Y. enterocolitica* (Bleves et al., 2002). Several groups further reported that certain injectisome genes contain a consensus FliA promoter sequence, and Horne et al. observed a negative effect of the flagellar-specific alternative sigma factor σ^{28} on the expression of four injectisome genes (Iriarte et al., 1995; Kapatral et al., 1996; Horne and Pruss, 2006). This regulation seems to be indirect and mediated by the transcriptional master activator LcrF/VirF, which activates transcription of the majority of all v-T3SS and associated *yop* effector genes (Horne and Pruss, 2006). In addition, FliA was found to induce expression of the primary invasin gene *invA*, which is encoded between two flagellar operons (Badger and Miller, 1998; Horne and Pruss, 2006). Vice versa, this coregulation ensures coexpression of flagellar-mediated motility and invasin-promoted host cell invasion by enteric *yersiniae.*

As flagellar synthesis consumes a significant proportion of the cell's biosynthetic capacities a plethora of environmental signals and a myriad of negative and positive control mechanisms are integrated on the level of *flhDC* gene expression, translation and FlhDC complex stability that ultimately decide the commencement of flagellar biosynthesis. The *flhDC* promoter of *Salmonella* displays a complex structure and multiple transcriptional start sites have been identified in recent transcriptome studies (Kröger et al., 2012, 2013). Transcriptional activity of the different promoters in *Salmonella* depends on the presence of different transcriptional activators such as cAMP-Crp, Fur, Fis (P1 promoter), HilD (P5 promoter), and the quorum sensing two-component regulatory system QseBC (Komeda et al., 1976; Yanagihara et al., 1999; Campoy et al., 2002; Kelly et al., 2004; Bearson et al., 2010; Mouslim and Hughes, 2014; Singer et al., 2014). Moreover, *flhDC* gene expression is also under negative control of the phosphorelay system RcsCDB, the LysR-family protein RflM/EcnR, the Spi-1-encoded regulator RtsB, the LysR-type regulator LrhA (RovM in *Yersinia*), and SlyA (RovA in *Yersinia*) with the P1 and P5 promoters being the main regulatory targets (Yanagihara et al., 1999; Wang et al., 2007; Singer et al., 2013; Mouslim and Hughes, 2014). The majority of these regulators is subjected to growth phase control and exerts its control at different stages during the bacterial growth phase. This dynamic separates the transcriptional activation of *flhDC* into a role for flagellar production during early growth phase from a role in virulence gene expression at later growth phase (Mouslim and Hughes, 2014). Finally, fimbriae-associated genes were also implicated in negative regulation of the *flhDC* operon. The *pefI* and *srgD* genes are encoded on the *Salmonella* virulence plasmid as part of the *p*lasmid-*e*ncoded *f*imbriae (*pef*) locus and function together in repression of *flhDC* transcription (Wozniak et al., 2009). FimZ is an activator of type 1 fimbriae and was shown to represses *flhDC*, as well as Spi-1 genes, which indicates that cross-talk between the flagellar, fimbriae and Spi-1 regulatory systems is of importance during the transition from a motile, planktonic lifestyle to intestinal colonization and persistence (Saini et al., 2010).

Much less is known about the regulation of *flhDC* expression in *Yersinia*, but varying expression levels under different environmental conditions (Nuss et al., 2014) suggest a similar complex regulation pattern. In fact, several transcriptional regulators are already known to influence flagella formation and motility. Analysis of *flhDC* expression in *Y. enterocolitica* revealed a role of OmpR/EnvZ in the positive control of the flagellar operon, which is in contrast to the negative role observed in *E. coli* (Hu et al., 2009). Furthermore, the two hierarchical quorum sensing LuxRI orthologs (YpsRI and YtbRI) of *Yersinia* control swimming motility via regulation of *flhDC* and *fliA*, whereby *N*-acylhomoserine lactones (AHLs) synthesized via YtbI activate *flhDC*, in conjunction with YpsR, but repress *fliA* in conjunction with YtbR and YpsR. A mutant analysis further indicated that quorum sensing regulates motility both positively (via YtbRI) and negatively (via YpsRI) (Atkinson et al., 1999, 2008). A recent study addressing gene regulation by the quorum sensing proteins YenR and YenI of *Y. enterocolitica* further demonstrated that YenR represses expression of *yenS*, encoding two non-translated RNAs 169 and 105 nucleotides long that share the same 5′ end, which play a stimulatory role in swarming motility (Tsai and Winans, 2011). Similar to *Salmonella*, the LrhA homolog RovM of *Yersinia* was found to repress motility by reducing the number of flagella in *Y. pseudotuberculosis*, whereas the RovM-dependent SlyA homolog RovA had no influence (Heroven and Dersch, 2006).

On the post-transcriptional level, it has been shown that also the carbon storage regulator CsrA is required for motility in both enteropathogenic *Yersinia* species (Heroven et al., 2008). For *Y. pseudotuberculosis* it was found that CsrA interacts directly with the 5′-untranslated region of the *flhDC* mRNA and stimulated FlhDC synthesis most likely by protecting the mRNA from RNase E cleavage similar to what has been observed in *E. coli* (Heroven et al., 2008). Recently, it has also been shown that CsrA is involved in the activation of the FlhDC/FliA cascade in *Salmonella*. Regulation of *flhDC* expression via CsrA is consistent with reports that the BarA/SirA TCS represses the flagellar regulon indirectly, most likely by regulating *csrB*, which in turn is an antagonist of CsrA activity (Teplitski et al., 2003, 2006). In *Salmonella*, CsrA is known to control the expression of the specific phosphodiesterase YhjH (STM3611), governing the synthesis of (3′-5′)-cyclic-diguanosine monophosphate (c-di-GMP), which reciprocally regulates flagella function and production of biofilm matrix components (Jonas et al., 2010). Furthermore, it has been reported that the BarA/SirA TCS represses the flagellar regulon indirectly, most likely by regulating *csrB* (Teplitski et al., 2003, 2006).

Temperature is a key environmental cue for *yersiniae*, and unlike *S. enterica* serovar Typhimurium, both enteropathogenic *Yersinia* species are not motile at 37°C. Loss of motility at body temperature is due to a rapid repression of the flagellar operons (Kapatral et al., 1996). Temperature-dependent regulation of the flagellar genes appears to occur through σ^{28}/FliA, as a rapid reduction of *fliA* mRNA was observed at a temperature-upshift, whereas the *flhDC* master operon is transcribed in a temperature-independent manner (Kapatral et al., 2004; Nuss et al., 2015). Notably, in a *fliA*-deficient mutant, also the temperature-dependent expression levels of all plasmid-encoded v-T3SS/*yop* genes were considerably reduced (Horne and Pruss, 2006). This indicated that FliA further contributes to temperature regulation of crucial virulence genes, most likely through is influence on *lcrF/virF*.

Attachment and Invasion of the Intestinal Epithelium

Enteric pathogens such as *salmonellae* and *yersiniae* are armed with a set of classical virulence factors (e.g., adhesins, invasins) that promote tight attachment of the bacteria to the mucosal surface of the intestinal tract and induce their passage through the intestinal epithelial layer into underlying lymphatic tissues. During the early stages of the infection both enteropathogenic *Yersinia* species as well as *S. enterica* serovar Typhimurium bind to and invade into M cells of the epithelium, overlaying the Peyer's patches in the most distal part of the ileum (Grutzkau et al., 1990; Jones et al., 1994). The surface-exposed outer membrane protein invasin is the most efficient adhesion and internalization factor of enteropathogenic *yersiniae* (Marra and Isberg, 1997; Grassl et al., 2003; Handley et al., 2005). The invasin protein, which is required for the colonization of the intestinal tract of mice and pigs (Schaake et al., 2013), is in most enteric *Yersinia* strains predominantly produced at moderate temperature during stationary phase mimicking the free-living or food-associated lifestyle (Isberg et al., 1988; Pepe et al., 1994). This expression pattern seems to guarantee rapid internalization into M cells shortly after ingestion.

The MarR-type regulator RovA, a dimeric winged-helix DNA-binding protein of the SlyA/Hor/Rap family activates *invA* transcription in response to temperature (Revell and Miller, 2000; Nagel et al., 2001; Heroven et al., 2004). A detailed structure-functional analysis of RovA revealed that RovA is an intrinsic temperature-sensing protein – a protein thermometer – in which thermally induced conformational changes of a small loop in the dimerization domain (i) interfere with the DNA-binding capacity of the regulator, and (ii) render the regulatory protein more susceptible to proteolytic degradation by the Lon protease (Herbst et al., 2009; Quade et al., 2012; Uliczka and Dersch, 2012). An analysis comparing the invasion properties of *Y. enterocolitica* O:3 and O:8 strains revealed that RovA stability is enhanced in *Y. enterocolitica* O:3 strains due to a P98S substitution in RovA. *Y. enterocolitica* O:3 is an emerging pathogen using pigs with a higher body temperature (39–41°C) as preferred reservoir hosts, and this stabilized RovA variant was found to improve persistence of the pathogen in the porcine intestinal tract (Schaake et al., 2014; Valentin-Weigand et al., 2014). Most interestingly, SlyA, the close homolog of RovA from *Salmonella* with a very similar structure, is not a thermosensor and remains fully active and stable at 37°C. Introduction of only three amino acid substitutions, reflecting evolutionary replacements in SlyA, were sufficient to eliminate the thermosensing properties of RovA and prevent degradation (Quade et al., 2012). This indicated that only minor changes can transform a thermotolerant regulator into a thermosensor that allows adjustment of virulence and fitness determinants in response to the temperature of their environment. Strikingly, in contrast to RovA, SlyA is not involved in invasion or the colonization of the small intestine, but it is required for the survival and replication in host phagocytes and destruction of M cells during later stages of the infection (Libby et al., 1994; Daniels et al., 1996; Buchmeier et al., 1997; Watson et al., 1999; see Regulatory Circuits Controlling Later Stages of Infection and Defense Systems against the Host's Immune Response).

Transcription of *rovA* and the RovA-regulated *invA* gene is also subjected to silencing by the ancestral nucleoid-structuring protein H-NS (Heroven et al., 2004; Ellison and Miller, 2006) (**Figure 3**), a global regulator implicated in the xenogeneic repression of many virulence and physiological genes acquired by horizontal gene transfer (Navarre et al., 2006; Dorman, 2007). H-NS and RovA bind preferentially to AT-rich regions and their binding sites are superimposed in both the *invA* and *rovA* regulatory region (Heroven et al., 2004). Active RovA alleviates transcriptional repression by H-NS and stimulates the activity of the RNAP (Tran et al., 2005). This anti-silencing mechanism seems common in *Yersinia*, as several other genes have been identified, that are controlled by both regulators (Cathelyn et al., 2007). It is also observed in many other *Enterobacteriaceae*, including *Salmonella*, where H-NS promotes silencing of invasion gene regulators encoded on Spi-1, apparently to buffer fitness costs that were associated with acquisition and expression of host colonization factors (Olekhnovich and Kadner, 2007).

H-NS is also capable to form heterodimers with members of the Hha/YmoA family. Hha and YmoA are small basic proteins that have been shown to participate in the modulation of virulence gene regulation in different Gram-negative bacteria (Madrid et al., 2006; Stoebel et al., 2008). Formation of YmoA-H-NS complexes has been shown to modulate expression of invasion and several other H-NS-dependent virulence genes in *Yersinia* (Ellison et al., 2003; Heroven et al., 2007) and in *Salmonella*. However, the molecular mechanism how YmoA/Hha modulates H-NS is not well understood since its influence varies considerably among the H-NS-dependent genes in the different microorganisms. Nonetheless, it is evident that the global influence of H-NS enables the pathogens to link this set of virulence-relevant genes with other environmental control systems. For instance, the H-NS modulator YmoA is preferentially degraded at 37°C by the ClpP and Lon proteases in *Yersinia* (Jackson et al., 2004), and in *Salmonella* it has been shown that transcription of the *hns* gene – which influences expression of the Spi-1-encoded invasion genes through the regulator HilD – is repressed by the iron homeostasis regulator Fur (Troxell et al., 2011). Metal-bound Fur has also been shown to bind to an AT-rich region of the *hilD* promoter to stimulate *hilD* transcription (Ellermeier and Slauch, 2008; Teixido et al., 2011).

The LysR-type regulator RovM (LrhA in *Salmonella*) was identified as another repressor of *rovA* expression, and full repression of *rovA* expression can only occur through the cooperation of RovM with H-NS (Heroven and Dersch, 2006). RovM synthesis is only activated under nutrient limiting conditions, and this strongly suggested that dependency of *rovA* and *invA* expression on the availability of nutrients is mainly mediated through RovM (Heroven et al., 2007). In fact, later attempts to unravel the molecular regulatory mechanism of *rovM* revealed that the nutrient and ion-controlled Csr system is responsible for *rovM* expression in response to the supply of carbon sources of the growth media (Heroven et al., 2008) (**Figure 3**). Overexpression of UvrY, CsrB, and CsrC resulted in a strong decrease of RovM levels and an increase of *rovA* transcription, and high amounts of the RovA protein can be detected in a *csrA* mutant in minimal media in which *rovA* expression is normally fully repressed (Heroven et al., 2008; Bücker et al., 2014). A detailed expression analysis further revealed that in contrast to the *Salmonella* Csr system, both CsrB and CsrC are counter-regulated and respond to different TCS in response to metabolites (UvrY/BarA-CsrB) or ions (PhoP/PhoQ-CsrC) and are oppositely controlled by the Crp protein (Heroven et al., 2008; Nuss et al., 2014). Apart of its influence on the RovM-RovA-InvA regulatory cascade, the *Yersinia* Csr system and Crp are part of a large adaptive response network, adjusting metabolic, and physiological processes, stress adaptation and virulence gene expression in response to changing environmental conditions (Heroven et al., 2012a; LeGrand et al., 2015; Nuss et al., 2015; Vakulskas et al., 2015). A recent study elucidated a tight connection between pathogenicity and core metabolism by integrated transcriptome and ^{13}C-fluxome analysis of *Y. pseudotuberculosis*, and identified the pyruvate-TCA cycle node as a focal point for controlling the host colonization and virulence of *Yersinia* which is tightly controlled by the interplay of Crp and the Csr system (Bücker et al., 2014).

In contrast to *Yersinia*, initial invasion and transcytosis of *Salmonella* through the intestinal layer requires the expression of the Spi-1-encoded injectisome system, which is controlled by a very distinct regulatory cascade (**Figure 2**). The appropriate environmental conditions to allow invasion are sensed by a myriad of regulators encoded both inside and outside of the pathogenicity island (PAI). A key regulator of Spi-1 is the OmpR/ToxR family protein HilA, and all environmental signals that sense optimal conditions for Spi-1 expression are integrated on the level of HilA expression (Bajaj et al., 1996; Ellermeier and Slauch, 2007). HilA is a positive regulator that primarily activates transcription of genes encoding structural components of the Spi-1 injectisome system, including the *prg/org* and *inv/spa* operons (Lee et al., 1992; Bajaj et al., 1995) and other virulence genes outside Spi-1, including the Spi-4 encoded *sii* operon (Gerlach et al., 2007). *hilA* gene transcription is controlled by a complex feed-forward loop comprised of the three AraC-like regulatory proteins HilC, HilD, and RtsA encoded on Spi-1 or on a separate 15 kb island (Schechter and Lee, 2001; Ellermeier and Slauch, 2003; Ellermeier et al., 2005) (**Figure 2**). Each of the three regulators can independently bind and activate the *hilA* promoter, their own gene, and the genes of the respective other regulators by counteracting silencing mediated by H-NS and Hha (YmoA in *Yersinia*) (Olekhnovich and Kadner, 2002; Ellermeier et al., 2005; Olekhnovich and Kadner, 2006). It has been proposed that the feed-forward loop system of HilC, HilD, and RtsA functions as a switch that controls *hilA* expression by modulating the threshold of HilD protein required for the ultimate HilA activation. In this context, HilD is the key activator of *hilA* transcription and various environmental signals are known to influence HilD production and activity, whereby RtsA and HilC function as amplifiers of the activating signal (Ellermeier and Slauch, 2007). Moreover, HilD coordinates *hilA* transcription with the expression of other crucial infection-relevant systems, e.g., it induces expression of the flagellar operons through activation of *flhDC* transcription and controls expression of Spi-2 genes (see also the Sections Regulation of Motility and Regulatory Circuits Controlling Later Stages of Infection and Defense Systems against the Host's Immune Response).

The temporal coordination of flagellar and Spi-1 gene expression plays an important role during the initial phase of the *Salmonella* infection cycle when the bacteria are initially motile in the lumen and subsequently turn on Spi-1 virulence genes needed for invasion of the epithelial layer. This is achieved by the flagellar-encoded regulator FliZ, which simultaneously represses type 1 fimbriae and activates Spi-1 via HilD (Iyoda et al., 2001; Chubiz et al., 2010). In the following stages flagellar gene expression is downregulated on the level of the *flhDC* promoter by the combined action of multiple regulators, including the Spi-1-dependent RtsB protein and the type 1 fimbriae regulator FimZ (Ellermeier and Slauch, 2003; Saini et al., 2010; Mouslim and Hughes, 2014; Singer et al., 2014). At later stages of the infection, synthesis of Spi-1 stops and expression type 1 fimbriae is upregulated (Saini et al., 2010). It is presumed that the expression of adhesive structures is needed for colonization and persistence of bacteria that were unable to breach the intestinal epithelium during the initial infection (Saini et al., 2010).

Based on the importance of HilA and HilD for Spi-1 gene expression, it is not surprising that various regulatory factors control their synthesis and activity in response to environmental conditions and ion/nutrient availability (**Figure 2**). Among them are the *Salmonella*-specific negative regulator HilE, which interacts with HilD and affects its activity, and multiple conserved regulatory factors: (i) the acyl-CoA synthetase FadD, (ii) the EnvZ/OmpR, PhoP/PhoQ and the Rcs phosphorelay systems controlling *hilD* or *hilA* transcription (Bajaj et al., 1995; Pegues et al., 1995; Garcia-Calderon et al., 2007), (iii) the DNA adenine methylase Dam and the degradosome (RNaseE), which regulates *hilA* and/or *hilD* mRNA stability and translation (Fahlen et al., 2000; Boddicker et al., 2003; Lopez-Garrido et al., 2014), (iv) the Lon and ClpXP protease (Kage et al., 2008), and (v) the BarA/SirA-CsrABC signal cascade in which CsrA negatively affects *hilD* translation by binding the ribosome-binding site of the *hilD* mRNA (Baxter et al., 2003; Ellermeier et al., 2005; Chubiz et al., 2010; Martinez et al., 2011). Apart from H-NS and Hha the *f*actor for *i*nversion *s*timulation (Fis), also controls Spi-1 gene expression. Fis-mediated activation of the Spi-1 genes occurs through different regulatory modes: Fis binds to promoters of Spi-1 encoded genes, or controls conserved upstream regulators such as OmpR to influence the Spi-1 regulator HilD. Fis binds preferentially to supercoiled DNA, and Fis-promoted activation coincides with a high level of nucleoid supercoiling under anaerobic conditions in the gut during the initial infection of *Salmonella* (Wilson et al., 2001; Wang et al., 2013).

Regulatory Circuits Controlling Later Stages of Infection and Defense Systems against the Host's Immune Response

Efficient dissemination and long-term persistence of *S. enterica* serovar Typhimurium and both enteric *Yersinia* species in deeper tissues require the expression of an additional set of virulence genes, which allow them to resist host immune responses. The most important defense system of *yersiniae* is encoded on the virulence plasmid pYV. It includes the *yadA* adhesin gene, the *ysc* genes encoding a v-T3SS/injectisome and its dedicated *yop* effector genes (Cornelis et al., 1998), which are important to survive and multiply in the lymphoid tissues of their host. This integrated defense system prevents uptake and elimination of the extracellular pathogens by professional phagocytes. It disarms their phagocytic function, disrupts their communications, and induces their apoptosis by the injection of the Yop effector proteins which interfere with the cytoskeletal structure and certain signal transduction pathways (Bliska et al., 2013). All these virulence genes belong to the same regulon and their transcription is activated by the AraC-type regulator LcrF(VirF) in response to temperature and host cell contact (Bolin et al., 1988; Pettersson et al., 1996). A subset of the genes is also regulated by the presence of Ca^{2+}, a phenomenon, which is referred to as *l*ow *c*alcium *r*esponse (LCR) (Goguen et al., 1984; Straley et al., 1993). *ysc*/*yop* genes were also found to be regulated by proteins that antagonize LcrF(VirF). One example is YtxR which competes with LcrF(VirF) for binding to the *yopE* and *yopH* promoters (Axler-Diperte et al., 2006). Another regulator is LcrQ (YscM1 and YscM2 in *Y. enterocolitica*), a factor, which is implicated in a feedback circuit that represses *ysc-yop* gene expression when Yop secretion is inhibited, e.g., in the absence of host cell contact. LcrQ(YscM1/YscM2) in cooperation with the YopD-LcrH complex interacts with 5′-UTRs of multiple *ysc*/*yop* mRNAs to block translation. Upon host cell contact LcrQ(YscM1/YscM2) is secreted by the Ysc/Yop v-T3SS machinery, repression is relieved due to lower LcrQ concentration in the cytoplasm and results in the upregulation of the *yop* and *ysc* genes (Stainier et al., 1997; Anderson et al., 2002; Chen and Anderson, 2011; Kopaskie et al., 2013).

Thermal control of the Ysc/Yop regulon is mediated through regulation of *lcrF*(*virF*) and this involves repression of LcrF(VirF) synthesis by the thermosensitive modulator YmoA and an RNA thermometer. The YmoA-H-NS complex was shown to repress *lcrF* transcription at moderate temperatures, but this repression is eliminated at 37°C due to rapid degradation of the YmoA protein by the Lon and ClpP proteases (Jackson et al., 2004; Heroven et al., 2007; Böhme et al., 2012). Most likely, this derepression is supported by thermally induced changes of the pYV DNA topology (Cornelis et al., 1989; Michiels et al., 1991; Rohde et al., 1994, 1999). Furthermore, it has been demonstrated that the 5′-UTR of the *lcrF*(*virF*) mRNA forms a temperature-sensitive two-stem-loop structure (RNA thermometer) at moderate temperatures in which the Shine-Dalgarno sequence is sequestered in the stem of the second hairpin and prevents *lcrF* mRNA translation (Hoe and Goguen, 1993; Böhme et al., 2012). However, an upshift to 37°C within the host leads to partial denaturation, and opening of the structure allows efficient ribosome binding and *lcrF* mRNA translation (Böhme et al., 2012). Besides temperature, LcrF(VirF) is also affected by other environmental signals. A genetic screen led to the identification of the transcriptional regulator IscR, which modulates gene transcription depending on the coordination of its 2Fe–2S clusters which can be influenced by oxidative stress, oxygen limitation, or iron availability (Miller et al., 2014). Similar to the *Salmonella* v-T3SS regulator HilD, LcrF expression seems also under control of several conserved TCSs: (i) the Rcs phosphorelay which is used to adapt their cell physiology in response to perturbations of the cell envelope (Li et al., 2015), (ii) the BarA/UvrY(SirA) controlling the Csr system in response to carbon sources (LeGrand et al., 2015), and (iii) the CpxAR system that responds to extra-cytoplasmatic stress (Liu et al., 2012). As mentioned above (see Regulation of Motility), LcrF synthesis is also under negative control of the flagellar sigma factor FliA/σ^{28}, and this inverse regulation of flagellar and v-T3SS genes ensures that expression of the immune defense apparatus is repressed when *Yersinia* uses flagellar motility to colonize external habits at temperatures below 30°C (Horne and Pruss, 2006).

In contrast to *yersiniae*, which persist and replicate predominantly extracellularly in lymphatic tissues, *salmonellae* can actively invade, survive and proliferate efficiently within so-called *Salmonella*-containing *v*acuoles (SCVs) inside the eukaryotic cytoplasm (**Figure 1**). The establishment and integrity of the SCV membrane is ensured by the action of effectors translocated via another horizontally acquired injectisome device encoded on Spi-2. Expression of Spi-2 and downregulation of Spi-1 gene expression marks an important transition from the invasion mode to the intracellular survival stage that is crucially dependent on the correct spatiotemporal action of a variety of activators and repressors of gene expression. The main regulatory proteins involved in the expression of Spi-2 include the two-component systems SpiR(SsrA)/SsrB, PhoP/PhoQ, OmpR/EnvZ, as well as the DNA-binding proteins HilD, SlyA (RovA in *Yersinia*), RcsB, the iron regulator Fur, and the nucleoid-associated proteins H-NS, Hha (YmoA in *Yersinia*), YdgT, IHF, and Fis (Fass and Groisman, 2009; Choi et al., 2014). The membrane-embedded *Salmonella*-specific sensor kinase SpiR(SsrA) and the response regulator SsrB are the primary activators essential for Spi-2 expression (Walthers et al., 2007), which is induced within the acidic environment of the SCV (Miao et al., 2002; Mulder et al., 2015). SsrB binds to promoter regions of all Spi-2-encoded gene clusters and counteracts H-NS-mediated silencing (Walthers et al., 2007). SsrB was also recently shown to downregulate the production of flagellar components in SCVs in macrophages (Brown et al., 2014), which would diminish flagellin-dependent stimulation of the NLRC4 inflammasome (Franchi et al., 2006; Miao et al., 2006). In addition to SsrB, also the SlyA protein (an ortholog of RovA of *Yersinia*) is implicated in the activation of several Spi-2 genes, mostly to overcome H-NS-mediated silencing (Okada et al., 2007). In contrast to RovA, SlyA does not function as a thermosensor and acts as a negative regulator of flagellar genes at 37°C via repression of *flhDC* (Quade et al., 2012; Mouslim and Hughes, 2014). In addition, two other homologous nucleoid-associated proteins, YdgT and Hha (YmoA in *Yersinia*) which form heterocomplexes with H-NS prevent Spi-2 gene expression and appear to be important in particular when the bacteria are extracellular (Coombes et al., 2005; Silphaduang et al., 2007).

Similar to the injectisome machinery of *Yersinia*, expression of the Spi-2 encoded injectisome genes of *Salmonella* seems to depend on the DNA topology and nucleoid-associated factors that modulate the transcription of the genes by the introduction of conformational changes. One of the modulators, the global DNA-bending protein *i*ntegration *h*ost *f*actor (IHF) is required for the expression of motility, Spi-1 and Spi-2 genes during the transition from exponential to stationary growth (Mangan et al., 2006). Also Fis, which also promotes Spi-1 gene expression (see Attachment and Invasion of the Intestinal Epithelium), is needed for full expression of Spi-2. Fis binds directly to the promoter regions of *spiR* and *ssaG* (Kelly et al., 2004; Lim et al., 2006) and its expression correlates with Spi-2 gene expression inside macrophages (O Croinin et al., 2006). Oxidative stress sensed during later intracellular infection stages results in relaxation of the nucleoid, which activates expression from Spi-2 promoters. Fis seem to accelerate relaxation and stabilizes the promoter topology of *ssrAB* to allow timely expression of Spi-2 genes (O Croinin et al., 2006; Dillon and Dorman, 2010).

Spi-2 is also controlled by several TCSs that are also implicated in the control of the colonization and invasion factors. Spi-2 genes are activated by the PhoP/PhoQ system that is essential for virulence and survival within macrophages (Miller et al., 1989). The response regulator PhoP controls Spi-2 by binding to the *ssrB* promoter region and the 5′-UTR of the *spiR* transcript (Bijlsma and Groisman, 2005). The OmpR/EnvZ two-component system also functions as an activator of Spi-2 by direct binding to both promoter regions of the *spiR*/*ssrB* system (Lee et al., 2000; Feng et al., 2003). The phosphorelay system RcsCDB has been shown to have a dual regulatory role as repressor of flagellar (*flhDC*) gene expression and activator of Spi-2 expression (Wang et al., 2007).

Expression of the Spi-2 genes is further tightly coordinated with expression of the Spi-1 genes. Cross-talk between the different PAIs is mainly mediated by the Spi-1-encoded regulator protein HilD. HilD antagonizes silencing via H-NS of both, the *hilA* and *spiR* promoters, but significantly higher levels of HilD protein are required for *spiR* promoter binding. This may explain why HilD regulates Spi-1 and Spi-2 differentially (Bustamante et al., 2008). While the physiological role of the HilD-mediated cross-talk between Spi-1 and Spi-2 remains to be further analyzed *in vivo* inside host cells, it has been proposed that coordinated regulation of Spi-1 and Spi-2 gene expression involves also IHF (as described above) which influences HilD expression (Fass and Groisman, 2009).

Conclusion and Outlook

Crucial to the *Yersinia* and *Salmonella*'s capability to cause a successful infection is their ability to coordinate the expression of a plethora of virulence genes with numerous metabolic and stress adaptation functions required for survival in the different tissues during the course of the infection. To optimize their biological fitness, which is essential to compete with the intestinal microbiota and defend detrimental host responses, they only synthesize pathogenicity factors when they are needed. The importance of the physiological control of virulence-relevant determinants is supported by the fact that their constitutive or deregulated expression can strongly attenuate virulence. In order to govern environmental changes, host stresses and competition, the pathogens generally employ signal transduction systems that sense and respond to particular environmental parameters and host signals by modifying the level and/or activity of transcriptional networks and post-transcriptional control systems including regulatory RNAs, modulator proteins and signal molecules to fine-tune the expression of pathogenicity factors.

Pathogenicity genes are often organized in clusters or large genomic islands on the chromosome or on plasmids which were acquired via horizontal gene transfer. The genetic linkage of the determinants on these so-called PAIs allows the bacteria to coordinately control a set of functionally related virulence factors within a complex regulatory network. Most frequently the virulence operons of the PAIs are under direct control of pathogen-specific transcriptional master regulators (mostly of the AraC family), which are encoded on the PAI (e.g., HilA, HilD in *Salmonella*, LcrF in *Yersinia*). In addition to expression of the PAI-encoded genes, also loci outside the PAI can be controlled by the PAI master regulators (e.g., HilD controls flagellar and Spi-2 gene expression). This superimposed control and cross-talk between the different virulence systems enables the pathogens to coordinate the dynamic and order of virulence gene expression in response to the conditions imposed on the pathogen by the colonized host niches. The environmental conditions include availability of carbon/energy sources, oxygen, growth phase, pH, and osmolarity and are transmitted to the regulatory network via ancestral global TCS (e.g., EnvZ/OmpR, PhoP/PhoQ, and BarA/UvrY(SirA). In addition, global regulatory proteins/systems such as Crp, CsrABC, SlyA(RovA), and LrhA(RovM), which adjust virulence gene expression with stress responses, physiological features (biofilm formation/motility) and crucial metabolic functions are included into this network. Another important conserved regulatory feature is the silencing of pathogenicity gene clusters by the nucleoid-associated protein H-NS due to its binding preference to DNA with a higher %AT content. H-NS promoted silencing prevents unwanted virulence gene expression under non-infection conditions (i.e., in environmental niches) to avoid fitness loss. This effect can be counter-balanced and modulated by other H-NS-like proteins such as Hha(YmoA) which modify H-NS promoted silencing by heterocomplex formation under certain growth conditions. Furthermore, other DNA-bending modulators such as IHF and Fis counteract H-NS repression and support virulence gene activation by the virulence cluster/PAI-specific activator proteins.

Many of the contributing TCSs, global regulators and regulatory systems, and gene silencer/modulator proteins are conserved between *Yersinia*, *Salmonella* and many other *Enterobacteriaceae*. However, their interactions, arrangement within the complex regulatory network and composition of the control elements vary between the related bacteria. In fact, small variations of the content and organization of the genetic information of the pathogens over time, e.g., point mutations, gene rearrangements, deletions, and insertion of foreign DNA can lead to rapid and fundamental changes. These genetic modifications are the primary forces, which bring out phenotypic differences leading to (i) distinct pathogenic properties and evolution of distinct species, and (ii) adaptation to different environments and alterations of their life-style. In fact, results from our previous work clearly demonstrated that even very small variations provoke major differences in the virulence properties of related pathogens. For instance, a 20 bp insertion can transfer the *Yersinia* adhesin YadA into an invasin (Heise and Dersch, 2006). Moreover, only three amino acid substitutions can switch the thermotolerant regulatory protein SlyA of *Salmonella* into a protein thermometer similar to the orthologous *Yersinia* regulator RovA (Quade et al., 2012). Variations found between the *Salmonella* Hha and the *Yersinia* YmoA protein render YmoA but not Hha susceptible to proteases at 37°C (Jackson et al., 2004), and a 2 nt exchange can modify a 5′-UTR into an RNA-thermometer (Steinmann and Dersch, 2013). These alterations of crucial control elements can switch thermotolerant gene regulatory circuits into a thermo-responsive control system, which allows pathogens such as *Yersinia* to inhabit a wide range of environmental, insect-vector- and mammalian host-associated niches.

Another benefit of the complex arrangement of the different regulatory factors is that some regulatory circuits enable heterogenous/bistable expression of certain virulence genes. As a result, a genetically identical bacterial population can consist of subpopulations that express (ON state) or do not express (OFF state) certain virulence factors. This so-called phenotypic heterogeneity can confer clonal subsets of pathogens with different virulence properties, metabolic functions, and/or physiological features. Recent work by Davis et al. (2015) has identified three subpopulations of *Y. pseudotuberculosis* within microcolonies at a single tissue site. The most peripherally localized bacteria which contact neutrophils express anti-phagocytic virulence factors, e.g., YopE, the bacteria growing on the exterior of the microcolony induce the nitric oxide (NO)-detoxifying gene *hmp* and prevents NO diffusion and elimination of the interior bacterial population and illustrates a sophisticated form of division of labor during infection (Davis et al., 2015). Another paradigmatic example of cooperative virulence is the bistable expression of the *Salmonella* Spi-1 injectisome and flagellar genes. This results in the generation of one subpopulation that expresses Spi-1 genes. A fraction of theses cells invade the intestinal epithelium and induce inflammation, whereas other motile subpopulations consume host products released during inflammation, This enables *Salmonella* to outcompete the intestinal microbiota and establish a productive infection in a cooperative manner (Ackermann et al., 2008; Diard et al., 2013). The invasive subpopulation is also characterized by a lower growth rate, which is associated with tolerance to certain stresses including the exposure to antibiotics (Diard et al., 2013). This property is referred to as bet-hedging; one subpopulation expresses features optimized for the present environment allowing it to survive and proliferate, whereas another part of the population expresses a phenotype less well adapted to the momentary niche, but adjusted to a state the environment might turn into. This behavior allows a population to survive in an unpredictable and frequently fluctuating environment, as experienced by the pathogen during the different stages of an infection. Future research addressing virulence gene expression in different tissues throughout the infection will elucidate whether phenotypic heterogeneity is restricted to special regulatory circuits or a more general control scheme. Existence of pathogen subsets cannot only hamper the control of infection by certain antibiotic therapies, but it can result in the failure of novel anti-virulence strategies directed against crucial virulence traits, e.g., T3SSs to combat bacterial diseases. An attractive alternative in this context would be the development of anti-virulence strategies directed against major global virulence regulators, which are essential for virulence as they coordinate multiple virulence traits with crucial metabolic and physiological functions of the pathogen.

Acknowledgments

Our work mentioned in the review was funded by support from the DFG Priority Program SPP1316 (DE/5-2), SPP1617 (DE616/6). PD is supported by the German Centre for Infection Research (DZIF). ME acknowledges funding from the Helmholtz Association young investigator grant VH-NG-932, and the People Programme (Marie Curie Actions) of the European Union Seventh Framework Programme (grant 334030).

References

Abby, S. S., and Rocha, E. P. (2012). The non-flagellar type III secretion system evolved from the bacterial flagellum and diversified into host-cell adapted systems. *PLoS Genet.* 8:e1002983. doi: 10.1371/journal.pgen.1002983

Abu Kwaik, Y., and Bumann, D. (2013). Microbial quest for food in vivo: 'nutritional virulence' as an emerging paradigm. *Cell. Microbiol.* 15, 882–890. doi: 10.1111/cmi.12138

Ackermann, M., Stecher, B., Freed, N. E., Songhet, P., Hardt, W. D., and Doebeli, M. (2008). Self-destructive cooperation mediated by phenotypic noise. *Nature* 454, 987–990. doi: 10.1038/nature07067

Aldridge, C., Poonchareon, K., Saini, S., Ewen, T., Soloyva, A., Rao, C. V., et al. (2010). The interaction dynamics of a negative feedback loop regulates flagellar number in *Salmonella enterica* serovar Typhimurium. *Mol. Microbiol.* 78, 1416–1430. doi: 10.1111/j.1365-2958.2010.07415.x

Altier, C., Suyemoto, M., Ruiz, A. I., Burnham, K. D., and Maurer, R. (2000). Characterization of two novel regulatory genes affecting *Salmonella* invasion gene expression. *Mol. Microbiol.* 35, 635–646. doi: 10.1046/j.1365-2958.2000.01734.x

Anderson, D. M., Ramamurthi, K. S., Tam, C., and Schneewind, O. (2002). YopD and LcrH regulate expression of *Yersinia enterocolitica* YopQ by a posttranscriptional mechanism and bind to yopQ RNA. *J. Bacteriol.* 184, 1287–1295. doi: 10.1128/JB.184.5.1287-1295.2002

Anderson, J. K., Smith, T. G., and Hoover, T. R. (2010). Sense and sensibility: flagellum-mediated gene regulation. *Trends Microbiol.* 18, 30–37. doi: 10.1016/j.tim.2009.11.001

Atkinson, S., Chang, C. Y., Patrick, H. L., Buckley, C. M., Wang, Y., Sockett, R. E., et al. (2008). Functional interplay between the *Yersinia pseudotuberculosis* YpsRI and YtbRI quorum sensing systems modulates swimming motility by controlling expression of flhDC and fliA. *Mol. Microbiol.* 69, 137–151. doi: 10.1111/j.1365-2958.2008.06268.x

Atkinson, S., Throup, J. P., Stewart, G. S., and Williams, P. (1999). A hierarchical quorum-sensing system in *Yersinia pseudotuberculosis* is involved in the regulation of motility and clumping. *Mol. Microbiol.* 33, 1267–1277. doi: 10.1046/j.1365-2958.1999.01578.x

Autenrieth, I. B., and Firsching, R. (1996). Penetration of M cells and destruction of Peyer's patches by *Yersinia enterocolitica*: an ultrastructural and histological study. *J. Med. Microbiol.* 44, 285–294. doi: 10.1099/00222615-44-4-285

Axler-Diperte, G. L., Miller, V. L., and Darwin, A. J. (2006). YtxR, a conserved LysR-like regulator that induces expression of genes encoding a putative ADP-ribosyltransferase toxin homologue in *Yersinia enterocolitica*. *J. Bacteriol.* 188, 8033–8043. doi: 10.1128/JB.01159-06

Badger, J. L., and Miller, V. L. (1998). Expression of invasin and motility are coordinately regulated in *Yersinia enterocolitica*. *J. Bacteriol.* 180, 793–800.

Bajaj, V., Hwang, C., and Lee, C. A. (1995). hilA is a novel ompR/toxR family member that activates the expression of *Salmonella typhimurium* invasion genes. *Mol. Microbiol.* 18, 715–727. doi: 10.1111/j.1365-2958.1995.mmi_18040715.x

Bajaj, V., Lucas, R. L., Hwang, C., and Lee, C. A. (1996). Co-ordinate regulation of *Salmonella typhimurium* invasion genes by environmental and regulatory factors is mediated by control of hilA expression. *Mol. Microbiol.* 22, 703–714. doi: 10.1046/j.1365-2958.1996.d01-1718.x

Baxter, M. A., Fahlen, T. F., Wilson, R. L., and Jones, B. D. (2003). HilE interacts with HilD and negatively regulates hilA transcription and expression of the *Salmonella enterica* serovar Typhimurium invasive phenotype. *Infect. Immun.* 71, 1295–1305. doi: 10.1128/IAI.71.3.1295-1305.2003

Bearson, B. L., Bearson, S. M., Lee, I. S., and Brunelle, B. W. (2010). The *Salmonella enterica* serovar Typhimurium QseB response regulator negatively regulates bacterial motility and swine colonization in the absence of the QseC sensor kinase. *Microb. Pathog.* 48, 214–219. doi: 10.1016/j.micpath.2010.03.005

Belden, W. J., and Miller, S. I. (1994). Further characterization of the PhoP regulon: identification of new PhoP-activated virulence loci. *Infect. Immun.* 62, 5095–5101.

Bijlsma, J. J., and Groisman, E. A. (2005). The PhoP/PhoQ system controls the intramacrophage type three secretion system of *Salmonella enterica*. *Mol. Microbiol.* 57, 85–96. doi: 10.1111/j.1365-2958.2005.04668.x

Bleves, S., Marenne, M. N., Detry, G., and Cornelis, G. R. (2002). Up-regulation of the *Yersinia enterocolitica* yop regulon by deletion of the flagellum master operon flhDC. *J. Bacteriol.* 184, 3214–3223. doi: 10.1128/JB.184.12.3214-3223.2002

Bliska, J. B., Wang, X., Viboud, G. I., and Brodsky, I. E. (2013). Modulation of innate immune responses by Yersinia type III secretion system translocators and effectors. *Cell. Microbiol.* 15, 1622–1631. doi: 10.1111/cmi.12164

Boddicker, J. D., Knosp, B. M., and Jones, B. D. (2003). Transcription of the *Salmonella* invasion gene activator, hilA, requires HilD activation in the absence of negative regulators. *J. Bacteriol.* 185, 525–533. doi: 10.1128/JB.185.2.525-533.2003

Böhme, K., Steinmann, R., Kortmann, J., Seekircher, S., Heroven, A. K., Berger, E., et al. (2012). Concerted actions of a thermo-labile regulator and a unique intergenic RNA thermosensor control *Yersinia* virulence. *PLoS Pathog.* 8:e1002518. doi: 10.1371/journal.ppat.1002518

Bolin, I., Forsberg, A., Norlander, L., Skurnik, M., and Wolf-Watz, H. (1988). Identification and mapping of the temperature-inducible, plasmid-encoded proteins of Yersinia spp. *Infect. Immun.* 56, 343–348.

Bottone, E. J. (1999). *Yersinia enterocolitica*: overview and epidemiologic correlates. *Microbes Infect.* 1, 323–333. doi: 10.1016/S1286-4579(99)80028-8

Bozue, J., Mou, S., Moody, K. L., Cote, C. K., Trevino, S., Fritz, D., et al. (2011). The role of the phoPQ operon in the pathogenesis of the fully virulent CO92 strain of *Yersinia pestis* and the IP32953 strain of *Yersinia pseudotuberculosis*. *Microb. Pathog.* 50, 314–321. doi: 10.1016/j.micpath.2011.02.005

Brown, N. F., Rogers, L. D., Sanderson, K. L., Gouw, J. W., Hartland, E. L., and Foster, L. J. (2014). A horizontally acquired transcription factor coordinates *Salmonella* adaptations to host microenvironments. *MBio* 5, e1727–14. doi: 10.1128/mBio.01727-14

Brown, S. A., Palmer, K. L., and Whiteley, M. (2008). Revisiting the host as a growth medium. *Nat. Rev. Microbiol.* 6, 657–666. doi: 10.1038/nrmicro 1955

Brzostek, K., Skorek, K., and Raczkowska, A. (2012). OmpR, a central integrator of several cellular responses in *Yersinia enterocolitica*. *Adv. Exp. Med. Biol.* 954, 325–334. doi: 10.1007/978-1-4614-3561-7_40

Buchmeier, N., Bossie, S., Chen, C. Y., Fang, F. C., Guiney, D. G., and Libby, S. J. (1997). SlyA, a transcriptional regulator of *Salmonella typhimurium*, is required for resistance to oxidative stress and is expressed in the intracellular environment of macrophages. *Infect. Immun.* 65, 3725–3730.

Bücker, R., Heroven, A. K., Becker, J., Dersch, P., and Wittmann, C. (2014). The pyruvate-tricarboxylic acid cycle node: a focal point of virulence control in the enteric pathogen *Yersinia pseudotuberculosis*. *J. Biol. Chem.* 289, 30114–30132. doi: 10.1074/jbc.M114.581348

Bustamante, V. H., Martinez, L. C., Santana, F. J., Knodler, L. A., Steele-Mortimer, O., and Puente, J. L. (2008). HilD-mediated transcriptional cross-talk between SPI-1 and SPI-2. *Proc. Natl. Acad. Sci. U.S.A.* 105, 14591–14596. doi: 10.1073/pnas.0801205105

Campoy, S., Jara, M., Busquets, N., De Rozas, A. M., Badiola, I., and Barbe, J. (2002). Intracellular cyclic AMP concentration is decreased in *Salmonella typhimurium* fur mutants. *Microbiology* 148, 1039–1048. doi: 10.1099/00221287-148-4-1039

Cathelyn, J. S., Ellison, D. W., Hinchliffe, S. J., Wren, B. W., and Miller, V. L. (2007). The RovA regulons of *Yersinia enterocolitica* and *Yersinia pestis* are distinct: evidence that many RovA-regulated genes were acquired more recently than the core genome. *Mol. Microbiol.* 66, 189–205. doi: 10.1111/j.1365-2958.2007.05907.x

Chadsey, M. S., Karlinsey, J. E., and Hughes, K. T. (1998). The flagellar anti-sigma factor FlgM actively dissociates *Salmonella typhimurium* sigma28 RNA polymerase holoenzyme. *Genes Dev.* 12, 3123–3136. doi: 10.1101/gad.12.19.3123

Chavez, R. G., Alvarez, A. F., Romeo, T., and Georgellis, D. (2010). The physiological stimulus for the BarA sensor kinase. *J. Bacteriol.* 192, 2009–2012. doi: 10.1128/JB.01685-09

Chen, Y., and Anderson, D. M. (2011). Expression hierarchy in the *Yersinia* type III secretion system established through YopD recognition of RNA. *Mol. Microbiol.* 80, 966–980. doi: 10.1111/j.1365-2958.2011.07623.x

Chevance, F. F., and Hughes, K. T. (2008). Coordinating assembly of a bacterial macromolecular machine. *Nat. Rev. Microbiol.* 6, 455–465. doi: 10.1038/nrmicro1887

Choi, E., Kim, H., Lee, H., Nam, D., Choi, J., and Shin, D. (2014). The iron-sensing Fur regulator controls expression timing and levels of *Salmonella* pathogenicity island 2 genes in the course of environmental acidification. *Infect. Immun.* 82, 2203–2210. doi: 10.1128/IAI.01625-13

Chubiz, J. E., Golubeva, Y. A., Lin, D., Miller, L. D., and Slauch, J. M. (2010). FliZ regulates expression of the *Salmonella* pathogenicity island 1 invasion locus by controlling HilD protein activity in *Salmonella enterica* serovar typhimurium. *J. Bacteriol.* 192, 6261–6270. doi: 10.1128/JB.00635-10

Coombes, B. K., Wickham, M. E., Lowden, M. J., Brown, N. F., and Finlay, B. B. (2005). Negative regulation of *Salmonella* pathogenicity island 2 is required for contextual control of virulence during typhoid. *Proc. Natl. Acad. Sci. U.S.A.* 102, 17460–17465. doi: 10.1073/pnas.0505401102

Cornelis, G. R., Boland, A., Boyd, A. P., Geuijen, C., Iriarte, M., Neyt, C., et al. (1998). The virulence plasmid of *Yersinia*, an antihost genome. *Microbiol. Mol. Biol. Rev.* 62, 1315–1352.

Cornelis, G., Sluiters, C., De Rouvroit, C. L., and Michiels, T. (1989). Homology between virF, the transcriptional activator of the *Yersinia* virulence regulon, and AraC, the *Escherichia coli* arabinose operon regulator. *J. Bacteriol.* 171, 254–262.

Dandekar, T., Fieselmann, A., Fischer, E., Popp, J., Hensel, M., and Noster, J. (2014). *Salmonella*-how a metabolic generalist adopts an intracellular lifestyle during infection. *Front. Cell. Infect. Microbiol.* 4:191. doi: 10.3389/fcimb.2014.00191

Daniels, J. J., Autenrieth, I. B., Ludwig, A., and Goebel, W. (1996). The gene slyA of *Salmonella typhimurium* is required for destruction of M cells and intracellular survival but not for invasion or colonization of the murine small intestine. *Infect. Immun.* 64, 5075–5084.

Davis, K. M., Mohammadi, S., and Isberg, R. R. (2015). Community behavior and spatial regulation within a bacterial microcolony in deep tissue sites serves to protect against host attack. *Cell Host Microbe* 17, 21–31. doi: 10.1016/j.chom.2014.11.008

Diard, M., Garcia, V., Maier, L., Remus-Emsermann, M. N., Regoes, R. R., Ackermann, M., et al. (2013). Stabilization of cooperative virulence by the expression of an avirulent phenotype. *Nature* 494, 353–356. doi: 10.1038/nature11913

Dillon, S. C., and Dorman, C. J. (2010). Bacterial nucleoid-associated proteins, nucleoid structure and gene expression. *Nat. Rev. Microbiol.* 8, 185–195. doi: 10.1038/nrmicro2261

Ding, L., Wang, Y., Hu, Y., Atkinson, S., Williams, P., and Chen, S. (2009). Functional characterization of FlgM in the regulation of flagellar synthesis and motility in *Yersinia pseudotuberculosis*. *Microbiology* 155, 1890–1900. doi: 10.1099/mic.0.026294-0

Dorman, C. J. (2007). H-NS, the genome sentinel. *Nat. Rev. Microbiol.* 5, 157–161. doi: 10.1038/nrmicro1598

Dorman, C. J., Chatfield, S., Higgins, C. F., Hayward, C., and Dougan, G. (1989). Characterization of porin and ompR mutants of a virulent strain of *Salmonella typhimurium*: ompR mutants are attenuated in vivo. *Infect. Immun.* 57, 2136–2140.

Dorrell, N., Li, S. R., Everest, P. H., Dougan, G., and Wren, B. W. (1998). Construction and characterisation of a *Yersinia enterocolitica* O:8 ompR mutant. *FEMS Microbiol. Lett.* 165, 145–151. doi: 10.1111/j.1574-6968.1998.tb13139.x

Ellermeier, C. D., Ellermeier, J. R., and Slauch, J. M. (2005). HilD, HilC and RtsA constitute a feed forward loop that controls expression of the SPI1 type three secretion system regulator hilA in *Salmonella enterica* serovar Typhimurium. *Mol. Microbiol.* 57, 691–705. doi: 10.1111/j.1365-2958.2005.04737.x

Ellermeier, C. D., and Slauch, J. M. (2003). RtsA and RtsB coordinately regulate expression of the invasion and flagellar genes in *Salmonella enterica* serovar Typhimurium. *J. Bacteriol.* 185, 5096–5108. doi: 10.1128/JB.185.17.5096-5108.2003

Ellermeier, J. R., and Slauch, J. M. (2007). Adaptation to the host environment: regulation of the SPI1 type III secretion system in *Salmonella enterica* serovar Typhimurium. *Curr. Opin. Microbiol.* 10, 24–29. doi: 10.1016/j.mib.2006.12.002

Ellermeier, J. R., and Slauch, J. M. (2008). Fur regulates expression of the *Salmonella* pathogenicity island 1 type III secretion system through HilD. *J. Bacteriol.* 190, 476–486. doi: 10.1128/JB.00926-07

Ellison, D. W., and Miller, V. L. (2006). H-NS represses inv transcription in *Yersinia enterocolitica* through competition with RovA and interaction with YmoA. *J. Bacteriol.* 188, 5101–5112. doi: 10.1128/JB.00862-05

Ellison, D., Young, B., Nelson, K., and Miller, V. L. (2003). YmoA negatively regulates expression of invasin from *Yersinia enterocolitica*. *J. Bacteriol.* 185, 7153–7159. doi: 10.1128/JB.185.24.7153-7159.2003

Fabrega, A., and Vila, J. (2013). *Salmonella enterica* serovar Typhimurium skills to succeed in the host: virulence and regulation. *Clin. Microbiol. Rev.* 26, 308–341. doi: 10.1128/CMR.00066-12

Fahlen, T. F., Mathur, N., and Jones, B. D. (2000). Identification and characterization of mutants with increased expression of hilA, the invasion gene transcriptional activator of *Salmonella typhimurium*. *FEMS Immunol. Med. Microbiol.* 28, 25–35. doi: 10.1111/j.1574-695X.2000.tb01453.x

Fass, E., and Groisman, E. A. (2009). Control of *Salmonella* pathogenicity island-2 gene expression. *Curr. Opin. Microbiol.* 12, 199–204. doi: 10.1016/j.mib.2009.01.004

Feng, X., Oropeza, R., and Kenney, L. J. (2003). Dual regulation by phospho-OmpR of ssrA/B gene expression in *Salmonella* pathogenicity island 2. *Mol. Microbiol.* 48, 1131–1143. doi: 10.1046/j.1365-2958.2003.03502.x

Flamez, C., Ricard, I., Arafah, S., Simonet, M., and Marceau, M. (2007). Two-component system regulon plasticity in bacteria: a concept emerging from phenotypic analysis of *Yersinia pseudotuberculosis* response regulator mutants. *Adv. Exp. Med. Biol.* 603, 145–155. doi: 10.1007/978-0-387-72124-8_12

Franchi, L., Amer, A., Body-Malapel, M., Kanneganti, T. D., Ozoren, N., Jagirdar, R., et al. (2006). Cytosolic flagellin requires IpaF for activation of caspase-1 and interleukin 1beta in *salmonella*-infected macrophages. *Nat. Immunol.* 7, 576–582. doi: 10.1038/ni1346

Fraser, G. M., Bennett, J. C., and Hughes, C. (1999). Substrate-specific binding of hook-associated proteins by FlgN and FliT, putative chaperones for flagellum assembly. *Mol. Microbiol.* 32, 569–580. doi: 10.1046/j.1365-2958.1999.01372.x

Fredriksson-Ahomaa, M. (2012). Isolation of enteropathogenic Yersinia from non-human sources. *Adv. Exp. Med. Biol.* 954, 97–105. doi: 10.1007/978-1-4614-3561-7_12

Fredriksson-Ahomaa, M., Stolle, A., and Korkeala, H. (2006). Molecular epidemiology of *Yersinia enterocolitica* infections. *FEMS Immunol. Med. Microbiol.* 47, 315–329. doi: 10.1111/j.1574-695X.2006.00095.x

Garcia-Calderon, C. B., Casadesus, J., and Ramos-Morales, F. (2007). Rcs and PhoPQ regulatory overlap in the control of *Salmonella enterica* virulence. *J. Bacteriol.* 189, 6635–6644. doi: 10.1128/JB.00640-07

Gerlach, R. G., Jackel, D., Geymeier, N., and Hensel, M. (2007). *Salmonella* pathogenicity island 4-mediated adhesion is coregulated with invasion genes in *Salmonella enterica*. *Infect. Immun.* 75, 4697–4709. doi: 10.1128/IAI.00228-07

Goguen, J. D., Yother, J., and Straley, S. C. (1984). Genetic analysis of the low calcium response in *Yersinia pestis* mu d1(Ap lac) insertion mutants. *J. Bacteriol.* 160, 842–848.

Golubeva, Y. A., Sadik, A. Y., Ellermeier, J. R., and Slauch, J. M. (2012). Integrating global regulatory input into the *Salmonella* pathogenicity island 1 type III secretion system. *Genetics* 190, 79–90. doi: 10.1534/genetics.111.132779

Gorke, B., and Stulke, J. (2008). Carbon catabolite repression in bacteria: many ways to make the most out of nutrients. *Nat. Rev. Microbiol.* 6, 613–624. doi: 10.1038/nrmicro1932

Grabenstein, J. P., Fukuto, H. S., Palmer, L. E., and Bliska, J. B. (2006). Characterization of phagosome trafficking and identification of PhoP-regulated genes important for survival of *Yersinia pestis* in macrophages. *Infect. Immun.* 74, 3727–3741. doi: 10.1128/IAI.00255-06

Grabenstein, J. P., Marceau, M., Pujol, C., Simonet, M., and Bliska, J. B. (2004). The response regulator PhoP of *Yersinia pseudotuberculosis* is important for replication in macrophages and for virulence. *Infect. Immun.* 72, 4973–4984. doi: 10.1128/IAI.72.9.4973-4984.200472/9/4973

Grassl, G. A., Bohn, E., Muller, Y., Buhler, O. T., and Autenrieth, I. B. (2003). Interaction of *Yersinia enterocolitica* with epithelial cells: invasin beyond invasion. *Int. J. Med. Microbiol.* 293, 41–54. doi: 10.1078/1438-4221-00243

Groisman, E. A. (2001). The pleiotropic two-component regulatory system PhoP-PhoQ. *J. Bacteriol.* 183, 1835–1842. doi: 10.1128/JB.183.6.1835-1842.2001

Groisman, E. A., and Mouslim, C. (2006). Sensing by bacterial regulatory systems in host and non-host environments. *Nat. Rev. Microbiol.* 4, 705–709. doi: 10.1038/nrmicro1478

Grutzkau, A., Hanski, C., Hahn, H., and Riecken, E. O. (1990). Involvement of M cells in the bacterial invasion of Peyer's patches: a common mechanism shared by *Yersinia enterocolitica* and other enteroinvasive bacteria. *Gut* 31, 1011–1015. doi: 10.1136/gut.31.9.1011

Gu, W., Wang, X., Qiu, H., Cui, B., Zhao, S., Zheng, H., et al. (2013). Comparison of cytokine immune responses to Brucella abortus and *Yersinia enterocolitica* serotype O:9 infections in BALB/c mice. *Infect. Immun.* 81, 4392–4398. doi: 10.1128/IAI.00856-13

Gueguen, E., Durand, E., Zhang, X. Y., D'amalric, Q., Journet, L., and Cascales, E. (2013). Expression of a type VI secretion system is responsive to envelope stresses through the OmpR rranscriptional activator. *PLoS ONE* 8:e66615. doi: 10.1371/journal.pone.0066615

Hanamura, A., and Aiba, H. (1991). Molecular mechanism of negative autoregulation of *Escherichia coli* crp gene. *Nucleic Acids Res.* 19, 4413–4419. doi: 10.1093/nar/19.16.4413

Handley, S. A., Newberry, R. D., and Miller, V. L. (2005). *Yersinia enterocolitica* invasin-dependent and invasin-independent mechanisms of systemic dissemination. *Infect. Immun.* 73, 8453–8455. doi: 10.1128/IAI.73.12.8453-8455.2005

Hayashi, F., Smith, K. D., Ozinsky, A., Hawn, T. R., Yi, E. C., Goodlett, D. R., et al. (2001). The innate immune response to bacterial flagellin is mediated by Toll-like receptor 5. *Nature* 410, 1099–1103. doi: 10.1038/35074106

Heeb, S., and Haas, D. (2001). Regulatory roles of the GacS/GacA two-component system in plant-associated and other gram-negative bacteria. *Mol. Plant Microbe Interact.* 14, 1351–1363. doi: 10.1094/MPMI.2001.14.12.1351

Heise, T., and Dersch, P. (2006). Identification of a domain in *Yersinia* virulence factor YadA that is crucial for extracellular matrix-specific cell adhesion and uptake. *Proc. Natl. Acad. Sci. U.S.A.* 103, 3375–3380. doi: 10.1073/pnas.0507749103

Herbst, K., Bujara, M., Heroven, A. K., Opitz, W., Weichert, M., Zimmermann, A., et al. (2009). Intrinsic thermal sensing controls proteolysis of *Yersinia* virulence regulator RovA. *PLoS Pathog.* 5:e1000435. doi: 10.1371/journal.ppat.1000435

Heroven, A. K., Bohme, K., and Dersch, P. (2012a). The Csr/Rsm system of *Yersinia* and related pathogens: a post-transcriptional strategy for managing virulence. *RNA Biol.* 9, 379–391. doi: 10.4161/rna.19333

Heroven, A. K., Sest, M., Pisano, F., Scheb-Wetzel, M., Steinmann, R., Bohme, K., et al. (2012b). Crp induces switching of the CsrB and CsrC RNAs in *Yersinia pseudotuberculosis* and links nutritional status to virulence. *Front. Cell. Infect. Microbiol.* 2:158. doi: 10.3389/fcimb.2012.00158

Heroven, A., Bohme, K., Rohde, M., and Dersch, P. (2008). A Csr-type regulatory system, including small non-coding RNAs, regulates the global virulence regulator RovA of *Yersinia pseudotuberculosis* through RovM. *Mol. Microbiol.* 68, 1179–1195. doi: 10.1111/j.1365-2958.2008.06218.x

Heroven, A., Bohme, K., Tran-Winkler, H., and Dersch, P. (2007). Regulatory elements implicated in the environmental control of invasin expression in enteropathogenic *Yersinia*. *Adv. Exp. Med. Biol.* 603, 156–166. doi: 10.1007/978-0-387-72124-8_13

Heroven, A., Nagel, G., Tran, H. J., Parr, S., and Dersch, P. (2004). RovA is autoregulated and antagonizes H-NS-mediated silencing of invasin and rovA expression in *Yersinia pseudotuberculosis*. *Mol. Microbiol.* 53, 871–888. doi: 10.1111/j.1365-2958.2004.04162.x

Heroven, A. K., and Dersch, P. (2006). RovM, a novel LysR-type regulator of the virulence activator gene rovA, controls cell invasion, virulence and motility of *Yersinia pseudotuberculosis*. *Mol. Microbiol.* 62, 1469–1483. doi: 10.1111/j.1365-2958.2006.05458.x

Heroven, A. K., and Dersch, P. (2014). Coregulation of host-adapted metabolism and virulence by pathogenic yersiniae. *Front. Cell. Infect. Microbiol.* 4:146. doi: 10.3389/fcimb.2014.00146

Hoe, N. P., and Goguen, J. D. (1993). Temperature sensing in *Yersinia pestis*: translation of the LcrF activator protein is thermally regulated. *J. Bacteriol.* 175, 7901–7909.

Hoelzer, K., Moreno Switt, A. I., and Wiedmann, M. (2011). Animal contact as a source of human non-typhoidal salmonellosis. *Vet. Res.* 42, 34. doi: 10.1186/1297-9716-42-34

Horne, S. M., and Pruss, B. M. (2006). Global gene regulation in *Yersinia enterocolitica*: effect of FliA on the expression levels of flagellar and plasmid-encoded virulence genes. *Arch. Microbiol.* 185, 115–126. doi: 10.1007/s00203-005-0077-1

Hu, Y., Wang, Y., Ding, L., Lu, P., Atkinson, S., and Chen, S. (2009). Positive regulation of flhDC expression by OmpR in *Yersinia pseudotuberculosis*. *Microbiology* 155, 3622–3631. doi: 10.1099/mic.0.030908-0

Hughes, K. T., Gillen, K. L., Semon, M. J., and Karlinsey, J. E. (1993). Sensing structural intermediates in bacterial flagellar assembly by export of a negative regulator. *Science* 262, 1277–1280. doi: 10.1126/science.8235660

Iriarte, M., Stainier, I., Mikulskis, A. V., and Cornelis, G. R. (1995). The fliA gene encoding sigma 28 in *Yersinia enterocolitica*. *J. Bacteriol.* 177, 2299–2304.

Isberg, R. R., Swain, A., and Falkow, S. (1988). Analysis of expression and thermoregulation of the *Yersinia pseudotuberculosis* inv gene with hybrid proteins. *Infect. Immun.* 56, 2133–2138.

Ishizuka, H., Hanamura, A., Inada, T., and Aiba, H. (1994). Mechanism of the down-regulation of cAMP receptor protein by glucose in *Escherichia coli*: role of autoregulation of the crp gene. *EMBO J.* 13, 3077–3082.

Iyoda, S., Kamidoi, T., Hirose, K., Kutsukake, K., and Watanabe, H. (2001). A flagellar gene fliZ regulates the expression of invasion genes and virulence phenotype in *Salmonella enterica* serovar Typhimurium. *Microb. Pathog.* 30, 81–90. doi: 10.1006/mpat.2000.0409

Jackson, M., Silva-Herzog, E., and Plano, G. V. (2004). The ATP-dependent ClpXP and Lon proteases regulate expression of the *Yersinia pestis* type III secretion system via regulated proteolysis of YmoA, a small histone-like protein. *Mol. Microbiol.* 54, 1364–1378. doi: 10.1111/j.1365-2958.2004.04353.x

Jonas, K., Edwards, A. N., Ahmad, I., Romeo, T., Romling, U., and Melefors, O. (2010). Complex regulatory network encompassing the Csr, c-di-GMP and motility systems of *Salmonella Typhimurium*. *Environ. Microbiol.* 12, 524–540. doi: 10.1111/j.1462-2920.2009.02097.x

Jones, B. D., Ghori, N., and Falkow, S. (1994). *Salmonella typhimurium* initiates murine infection by penetrating and destroying the specialized epithelial M cells of the Peyer's patches. *J. Exp. Med.* 180, 15–23. doi: 10.1084/jem.180.1.15

Kage, H., Takaya, A., Ohya, M., and Yamamoto, T. (2008). Coordinated regulation of expression of *Salmonella* pathogenicity island 1 and flagellar type III secretion systems by ATP-dependent ClpXP protease. *J. Bacteriol.* 190, 2470–2478. doi: 10.1128/JB.01385-07

Kapatral, V., Campbell, J. W., Minnich, S. A., Thomson, N. R., Matsumura, P., and Pruss, B. M. (2004). Gene array analysis of *Yersinia enterocolitica* FlhD and FlhC: regulation of enzymes affecting synthesis and degradation of carbamoylphosphate. *Microbiology* 150, 2289–2300. doi: 10.1099/mic.0.26814-0150/7/2289

Kapatral, V., Olson, J. W., Pepe, J. C., Miller, V. L., and Minnich, S. A. (1996). Temperature-dependent regulation of *Yersinia enterocolitica* Class III flagellar genes. *Mol. Microbiol.* 19, 1061–1071. doi: 10.1046/j.1365-2958.1996.452978.x

Karlinsey, J. E., Tanaka, S., Bettenworth, V., Yamaguchi, S., Boos, W., Aizawa, S. I., et al. (2000). Completion of the hook-basal body complex of the *Salmonella typhimurium* flagellum is coupled to FlgM secretion and fliC transcription. *Mol. Microbiol.* 37, 1220–1231. doi: 10.1046/j.1365-2958.2000.02081.x

Kelly, A., Goldberg, M. D., Carroll, R. K., Danino, V., Hinton, J. C., and Dorman, C. J. (2004). A global role for Fis in the transcriptional control of metabolism and type III secretion in *Salmonella enterica* serovar Typhimurium. *Microbiology* 150, 2037–2053. doi: 10.1099/mic.0.27209-0

Komeda, Y., Suzuki, H., Ishidsu, J. I., and Iino, T. (1976). The role of cAMP in flagellation of *Salmonella typhimurium*. *Mol. Gen. Genet.* 142, 289–298.

Koornhof, H. J., Smego, R. A. Jr., and Nicol, M. (1999). Yersiniosis. II: the pathogenesis of *Yersinia* infections. *Eur. J. Clin. Microbiol. Infect. Dis.* 18, 87–112.

Kopaskie, K. S., Ligtenberg, K. G., and Schneewind, O. (2013). Translational regulation of *Yersinia enterocolitica* mRNA encoding a type III secretion substrate. *J. Biol. Chem.* 288, 35478–35488. doi: 10.1074/jbc.M113.504811

Kröger, C., Colgan, A., Srikumar, S., Handler, K., Sivasankaran, S. K., Hammarlof, D. L., et al. (2013). An infection-relevant transcriptomic compendium for *Salmonella enterica* Serovar Typhimurium. *Cell Host Microbe* 14, 683–695. doi: 10.1016/j.chom.2013.11.010

Kröger, C., Dillon, S. C., Cameron, A. D., Papenfort, K., Sivasankaran, S. K., Hokamp, K., et al. (2012). The transcriptional landscape and small RNAs of *Salmonella enterica* serovar Typhimurium. *Proc. Natl. Acad. Sci. U.S.A.* 109, E1277–E1286. doi: 10.1073/pnas.1201061109

Kutsukake, K. (1994). Excretion of the anti-sigma factor through a flagellar substructure couples flagellar gene expression with flagellar assembly in *Salmonella typhimurium*. *Mol. Gen. Genet.* 243, 605–612.

Kutsukake, K., Ikebe, T., and Yamamoto, S. (1999). Two novel regulatory genes, fliT and fliZ, in the flagellar regulon of *Salmonella*. *Genes Genet. Syst.* 74, 287–292. doi: 10.1266/ggs.74.287

Lapouge, K., Schubert, M., Allain, F. H., and Haas, D. (2007). Gac/Rsm signal transduction pathway of gamma-proteobacteria: from RNA recognition to regulation of social behaviour. *Mol. Microbiol.* 67, 241–253. doi: 10.1111/j.1365-2958.2007.06042.x

Lawhon, S. D., Frye, J. G., Suyemoto, M., Porwollik, S., Mcclelland, M., and Altier, C. (2003). Global regulation by CsrA in *Salmonella typhimurium*. *Mol. Microbiol.* 48, 1633–1645. doi: 10.1046/j.1365-2958.2003.03535.x

Lawhon, S. D., Maurer, R., Suyemoto, M., and Altier, C. (2002). Intestinal short-chain fatty acids alter *Salmonella typhimurium* invasion gene expression and virulence through BarA/SirA. *Mol. Microbiol.* 46, 1451–1464. doi: 10.1046/j.1365-2958.2002.03268.x

Lee, A. K., Detweiler, C. S., and Falkow, S. (2000). OmpR regulates the two-component system SsrA-ssrB in *Salmonella* pathogenicity island 2. *J. Bacteriol.* 182, 771–781. doi: 10.1128/JB.182.3.771-781.2000

Lee, C. A., Jones, B. D., and Falkow, S. (1992). Identification of a *Salmonella typhimurium* invasion locus by selection for hyperinvasive mutants. *Proc. Natl. Acad. Sci. U.S.A.* 89, 1847–1851. doi: 10.1073/pnas.89.5.1847

LeGrand, K., Petersen, S., Zheng, Y., Liu, K. K., Ozturk, G., Chen, J. Y., et al. (2015). CsrA impacts survival of *Yersinia enterocolitica* by affecting a myriad of physiological activities. *BMC Microbiol.* 15:31. doi: 10.1186/s12866-015-0343-6

Li, Y., Hu, Y., Francis, M. S., and Chen, S. (2015). RcsB positively regulates the *Yersinia* Ysc-Yop type III secretion system by activating expression of the master transcriptional regulator LcrF. *Environ. Microbiol.* 17, 1219–1233. doi: 10.1111/1462-2920.12556

Libby, S. J., Goebel, W., Ludwig, A., Buchmeier, N., Bowe, F., Fang, F. C., et al. (1994). A cytolysin encoded by *Salmonella* is required for survival within macrophages. *Proc. Natl. Acad. Sci. U.S.A.* 91, 489–493. doi: 10.1073/pnas.91.2.489

Lim, S., Kim, B., Choi, H. S., Lee, Y., and Ryu, S. (2006). Fis is required for proper regulation of ssaG expression in *Salmonella enterica* serovar Typhimurium. *Microb. Pathog.* 41, 33–42. doi: 10.1016/j.micpath.2006.03.005

Liu, J., Thanikkal, E. J., Obi, I. R., and Francis, M. S. (2012). Elevated CpxR~P levels repress the Ysc-Yop type III secretion system of *Yersinia pseudotuberculosis*. *Res. Microbiol.* 163, 518–530. doi: 10.1016/j.resmic.2012.07.010

Liu, X., and Matsumura, P. (1994). The FlhD/FlhC complex, a transcriptional activator of the *Escherichia coli* flagellar class II operons. *J. Bacteriol.* 176, 7345–7351.

Liu, X., and Matsumura, P. (1995). An alternative sigma factor controls transcription of flagellar class-III operons in *Escherichia coli*: gene sequence, overproduction, purification and characterization. *Gene* 164, 81–84. doi: 10.1016/0378-1119(95)00480-T

Lopez-Garrido, J., Puerta-Fernandez, E., and Casadesus, J. (2014). A eukaryotic-like 3' untranslated region in *Salmonella enterica* hilD mRNA. *Nucleic Acids Res.* 42, 5894–5906. doi: 10.1093/nar/gku222

Lucas, R. L., Lostroh, C. P., Dirusso, C. C., Spector, M. P., Wanner, B. L., and Lee, C. A. (2000). Multiple factors independently regulate hilA and invasion gene expression in *Salmonella enterica* serovar typhimurium. *J. Bacteriol.* 182, 1872–1882. doi: 10.1128/JB.182.7.1872-1882.2000

Madrid, C., Balsalobre, C., Garcia, J., and Juarez, A. (2006). The novel Hha/YmoA family of nucleoid-associated proteins: use of structural mimicry to modulate the activity of the H-NS family of proteins. *Mol. Microbiol.* 63, 7–14. doi: 10.1111/j.1365-2958.2006.05497.x

Mangan, M. W., Lucchini, S., Danino, V., Croinin, T. O., Hinton, J. C., and Dorman, C. J. (2006). The integration host factor (IHF) integrates stationary-phase and virulence gene expression in *Salmonella enterica* serovar Typhimurium. *Mol. Microbiol.* 59, 1831–1847. doi: 10.1111/j.1365-2958.2006.05062.x

Marra, A., and Isberg, R. R. (1997). Invasin-dependent and invasin-independent pathways for translocation of *Yersinia pseudotuberculosis* across the Peyer's patch intestinal epithelium. *Infect. Immun.* 65, 3412–3421.

Martinez, L. C , Yakhnin, H., Camacho, M. I., Georgellis, D., Babitzke, P., Puente, J. L., et al. (2011). Integration of a complex regulatory cascade involving the SirA/BarA and Csr global regulatory systems that controls expression of the *Salmonella* SPI-1 and SPI-2 virulence regulons through HilD. *Mol. Microbiol.* 80, 1637–1656. doi: 10.1111/j.1365-2958.2011.07674.x

Mercante, J., Edwards, A. N., Dubey, A. K., Babitzke, P., and Romeo, T. (2009). Molecular geometry of CsrA (RsmA) binding to RNA and its implications for regulated expression. *J. Mol. Biol.* 392, 511–528. doi: 10.1016/j.jmb.2009.07.034

Miao, E. A., Alpuche-Aranda, C. M., Dors, M., Clark, A. E., Bader, M. W., Miller, S. I., et al. (2006). Cytoplasmic flagellin activates caspase-1 and secretion of interleukin 1beta via IpaF. *Nat. Immunol.* 7, 569–575. doi: 10.103 8/ni1344

Miao, E. A., Freeman, J. A., and Miller, S. I. (2002). Transcription of the SsrAB regulon is repressed by alkaline pH and is independent of PhoPQ and magnesium concentration. *J. Bacteriol.* 184, 1493–1497. doi: 10.1128/JB.184.5.1493-1497.2002

Michiels, T., Vanooteghem, J. C., Lambert De Rouvroit, C., China, B., Gustin, A., Boudry, P., et al. (1991). Analysis of virC, an operon involved in the secretion of Yop proteins by *Yersinia enterocolitica*. *J. Bacteriol.* 173, 4994–5009.

Miller, H. K., Kwuan, L., Schwiesow, L., Bernick, D. L., Mettert, E., Ramirez, H. A., et al. (2014). IscR is essential for *Yersinia pseudotuberculosis* type III secretion and virulence. *PLoS Pathog.* 10:e1004194. doi: 10.1371/journal.ppat.1004194

Miller, S. I., Kukral, A. M., and Mekalanos, J. J. (1989). A two-component regulatory system (phoP phoQ) controls *Salmonella typhimurium* virulence. *Proc. Natl. Acad. Sci. U.S.A.* 86, 5054–5058. doi: 10.1073/pnas.86.13.5054

Mouslim, C., and Hughes, K. T. (2014). The effect of cell growth phase on the regulatory cross-talk between flagellar and Spi1 virulence gene expression. *PLoS Pathog.* 10:e1003987. doi: 10.1371/journal.ppat.1003987

Mulder, D. T., Mcphee, J. B., Reid-Yu, S. A., Stogios, P. J., Savchenko, A., and Coombes, B. K. (2015). Multiple histidines in the periplasmic domain of the *Salmonella enterica* sensor kinase SsrA enhance signaling in response to extracellular acidification. *Mol. Microbiol.* 95, 678–691. doi: 10.1111/mmi.12895

Nagel, G., Lahrz, A., and Dersch, P. (2001). Environmental control of invasin expression in *Yersinia pseudotuberculosis* is mediated by regulation of RovA, a transcriptional activator of the SlyA/Hor family. *Mol. Microbiol.* 41, 1249–1269. doi: 10.1046/j.1365-2958.2001.02522.x

Navarre, W. W., Porwollik, S., Wang, Y., Mcclelland, M., Rosen, H., Libby, S. J., et al. (2006). Selective silencing of foreign DNA with low GC content by the H-NS protein in *Salmonella*. *Science* 313, 236–238. doi: 10.1126/science.1128794

Nizet, V., and Johnson, R. S. (2009). Interdependence of hypoxic and innate immune responses. *Nat. Rev. Immunol.* 9, 609–617. doi: 10.1038/nri2607

Nuss, A. M., Heroven, A. K., Waldmann, B., Reinkensmeier, J., Jarek, M., Beckstette, M., et al. (2015). Transcriptomic profiling of *Yersinia pseudotuberculosis* reveals reprogramming of the Crp regulon by temperature and uncovers Crp as a master regulator of small RNAs. *PLoS Genet.* 11:e1005087. doi: 10.1371/journal.pgen.1005087

Nuss, A. M., Schuster, F., Kathrin Heroven, A., Heine, W., Pisano, F., and Dersch, P. (2014). A direct link between the global regulator PhoP and the Csr regulon in *Y. pseudotuberculosis* through the small regulatory RNA CsrC. *RNA Biol.* 11, 580–593. doi: 10.4161/rna.28676

O Croinin, T., Carroll, R. K., Kelly, A., and Dorman, C. J. (2006). Roles for DNA supercoiling and the Fis protein in modulating expression of virulence genes during intracellular growth of *Salmonella enterica* serovar Typhimurium. *Mol. Microbiol.* 62, 869–882. doi: 10.1111/j.1365-2958.2006.05416.x

Ohl, M. E., and Miller, S. I. (2001). *Salmonella*: a model for bacterial pathogenesis. *Annu. Rev. Med.* 52, 259–274. doi: 10.1146/annurev.med.52.1.259

Ohnishi, K., Kutsukake, K., Suzuki, H., and Lino, T. (1992). A novel transcriptional regulation mechanism in the flagellar regulon of *Salmonella typhimurium*: an antisigma factor inhibits the activity of the flagellum-specific sigma factor, sigma F. *Mol. Microbiol.* 6, 3149–3157. doi: 10.1111/j.1365-2958.1992.tb01771.x

Okada, N., Oi, Y., Takeda-Shitaka, M., Kanou, K., Umeyama, H., Haneda, T., et al. (2007). Identification of amino acid residues of *Salmonella* SlyA that are critical for transcriptional regulation. *Microbiology* 153, 548–560. doi: 10.1099/mic.0.29259-0

Olekhnovich, I. N., and Kadner, R. J. (2002). DNA-binding activities of the HilC and HilD virulence regulatory proteins of *Salmonella enterica* serovar Typhimurium. *J. Bacteriol.* 184, 4148–4160. doi: 10.1128/JB.184.15.4148-4160.2002

Olekhnovich, I. N., and Kadner, R. J. (2006). Crucial roles of both flanking sequences in silencing of the hilA promoter in *Salmonella enterica*. *J. Mol. Biol.* 357, 373–386. doi: 10.1016/j.jmb.2006.01.007

Olekhnovich, I. N., and Kadner, R. J. (2007). Role of nucleoid-associated proteins Hha and H-NS in expression of *Salmonella enterica* activators HilD, HilC, and RtsA required for cell invasion. *J. Bacteriol.* 189, 6882–6890. doi: 10.1128/JB.00905-07

Pegues, D. A., Hantman, M. J., Behlau, I., and Miller, S. I. (1995). PhoP/PhoQ transcriptional repression of *Salmonella typhimurium* invasion genes: evidence for a role in protein secretion. *Mol. Microbiol.* 17, 169–181. doi: 10.1111/j.1365-2958.1995.mmi_17010169.x

Pepe, J. C., Badger, J. L., and Miller, V. L. (1994). Growth phase and low pH affect the thermal regulation of the *Yersinia enterocolitica* inv gene. *Mol. Microbiol.* 11, 123–135. doi: 10.1111/j.1365-2958.1994.tb00295.x

Perez, J. C., and Groisman, E. A. (2009). Transcription factor function and promoter architecture govern the evolution of bacterial regulons. *Proc. Natl. Acad. Sci. U.S.A.* 106, 4319–4324. doi: 10.1073/pnas.0810343106

Petersen, S., and Young, G. M. (2002). Essential role for cyclic AMP and its receptor protein in *Yersinia enterocolitica* virulence. *Infect. Immun.* 70, 3665–3672. doi: 10.1128/IAI.70.7.3665-3672.2002

Pettersson, J., Nordfelth, R., Dubinina, E., Bergman, T., Gustafsson, M., Magnusson, K. E., et al. (1996). Modulation of virulence factor expression by pathogen target cell contact. *Science* 273, 1231–1233. doi: 10.1126/science.273.5279.1231

Pisano, F., Heine, W., Rosenheinrich, M., Schweer, J., Nuss, A. M., and Dersch, P. (2014). Influence of PhoP and intra-species variations on virulence of *Yersinia pseudotuberculosis* during the natural oral infection route. *PLoS ONE* 9:e103541. doi: 10.1371/journal.pone.0103541

Poncet, S., Milohanic, E., Maze, A., Abdallah, J. N., Ake, F., Larribe, M., et al. (2009). Correlations between carbon metabolism and virulence in bacteria. *Contrib. Microbiol.* 16, 88–102. doi: 10.1159/000219374

Quade, N., Mendonca, C., Herbst, K., Heroven, A. K., Ritter, C., Heinz, D. W., et al. (2012). Structural basis for intrinsic thermosensing by the master virulence regulator RovA of *Yersinia*. *J. Biol. Chem.* 287, 35796–35803. doi: 10.1074/jbc.M112.379156

Raczkowska, A., Brzostkowska, M., Kwiatek, A., Bielecki, J., and Brzostek, K. (2011a). Modulation of inv gene expression by the OmpR two-component response regulator protein of *Yersinia enterocolitica*. *Folia Microbiol. (Praha)* 56, 313–319. doi: 10.1007/s12223-011-0054-9

Raczkowska, A., Skorek, K., Bielecki, J., and Brzostek, K. (2011b). OmpR controls *Yersinia enterocolitica* motility by positive regulation of flhDC expression. *Antonie Van Leeuwenhoek* 99, 381–394. doi: 10.1007/s10482-010-9503-8

Raczkowska, A., Trzos, J., Lewandowska, O., Nieckarz, M., and Brzostek, K. (2015). Expression of the AcrAB components of the AcrAB-TolC multidrug efflux pump of *Yersinia enterocolitica* is subject to dual regulation by OmpR. *PLoS ONE* 10:e0124248. doi: 10.1371/journal.pone.0124248

Reines, M., Llobet, E., Llompart, C. M., Moranta, D., Perez-Gutierrez, C., and Bengoechea, J. A. (2012). Molecular basis of *Yersinia enterocolitica* temperature-dependent resistance to antimicrobial peptides. *J. Bacteriol.* 194, 3173–3188. doi: 10.1128/JB.00308-12

Revell, P. A., and Miller, V. L. (2000). A chromosomally encoded regulator is required for expression of the *Yersinia enterocolitica* inv gene and for virulence. *Mol. Microbiol.* 35, 677–685. doi: 10.1046/j.1365-2958.2000.01740.x

Rohde, J. R., Fox, J. M., and Minnich, S. A. (1994). Thermoregulation in *Yersinia enterocolitica* is coincident with changes in DNA supercoiling. *Mol. Microbiol.* 12, 187–199. doi: 10.1111/j.1365-2958.1994.tb01008.x

Rohde, J. R., Luan, X. S., Rohde, H., Fox, J. M., and Minnich, S. A. (1999). The *Yersinia enterocolitica* pYV virulence plasmid contains multiple intrinsic DNA bends which melt at 37 degrees C. *J. Bacteriol.* 181, 4198–4204.

Rohmer, L., Hocquet, D., and Miller, S. I. (2011). Are pathogenic bacteria just looking for food? Metabolism and microbial pathogenesis. *Trends Microbiol.* 19, 341–348. doi: 10.1016/j.tim.2011.04.003

Romeo, T., Vakulskas, C. A., and Babitzke, P. (2012). Post-transcriptional regulation on a global scale: form and function of Csr/Rsm systems. *Environ. Microbiol.* 15, 313–324. doi: 10.1111/j.1462-2920.2012.02794.x

Romeo, T., Vakulskas, C. A., and Babitzke, P. (2013). Post-transcriptional regulation on a global scale: form and function of Csr/Rsm systems. *Environ. Microbiol.* 15, 313–324. doi: 10.1111/j.1462-2920.2012.02794.x

Saini, S., Slauch, J. M., Aldridge, P. D., and Rao, C. V. (2010). Role of cross talk in regulating the dynamic expression of the flagellar *Salmonella* pathogenicity island 1 and type 1 fimbrial genes. *J. Bacteriol.* 192, 5767–5777. doi: 10.1128/JB.00624-10

Sato, Y., Takaya, A., Mouslim, C., Hughes, K. T., and Yamamoto, T. (2014). FliT selectively enhances proteolysis of FlhC subunit in FlhD4C2 complex by an ATP-dependent protease, ClpXP. *J. Biol. Chem.* 289, 33001–33011. doi: 10.1074/jbc.M114.593749

Schaake, J., Drees, A., Gruning, P., Uliczka, F., Pisano, F., Thiermann, T., et al. (2014). Essential role of invasin for colonization and persistence of *Yersinia enterocolitica* in its natural reservoir host, the pig. *Infect. Immun.* 82, 960–969. doi: 10.1128/IAI.01001-13

Schaake, J., Kronshage, M., Uliczka, F., Rohde, M., Knuuti, T., Strauch, E., et al. (2013). Human and animal isolates of *Yersinia enterocolitica* show significant serotype-specific colonization and host-specific immune defense properties. *Infect. Immun.* 81, 4013–4025. doi: 10.1128/IAI.00572-13

Schechter, L. M., and Lee, C. A. (2001). AraC/XylS family members, HilC and HilD, directly bind and derepress the *Salmonella typhimurium* hilA promoter. *Mol. Microbiol.* 40, 1289–1299.

Silphaduang, U., Mascarenhas, M., Karmali, M., and Coombes, B. K. (2007). Repression of intracellular virulence factors in *Salmonella* by the Hha and YdgT nucleoid-associated proteins. *J. Bacteriol.* 189, 3669–3673. doi: 10.1128/JB.00002-07

Singer, H. M., Erhardt, M., and Hughes, K. T. (2013). RflM functions as a transcriptional repressor in the autogenous control of the *Salmonella* flagellar master operon flhDC. *J. Bacteriol.* 195, 4274–4282. doi: 10.1128/JB.00728-13

Singer, H. M., Kuhne, C., Deditius, J. A., Hughes, K. T., and Erhardt, M. (2014). The *Salmonella* Spi1 virulence regulatory protein HilD directly activates transcription of the flagellar master operon flhDC. *J. Bacteriol.* 196, 1448–1457. doi: 10.1128/JB.01438-13

Skorek, K., Raczkowska, A., Dudek, B., Mietka, K., Guz-Regner, K., Pawlak, A., et al. (2013). Regulatory protein OmpR influences the serum resistance of *Yersinia enterocolitica* O:9 by modifying the structure of the outer membrane. *PLoS ONE* 8:e79525. doi: 10.1371/journal.pone.0079525

Stainier, I., Iriarte, M., and Cornelis, G. R. (1997). YscM1 and YscM2, two *Yersinia enterocolitica* proteins causing downregulation of yop transcription. *Mol. Microbiol.* 26, 833–843. doi: 10.1046/j.1365-2958.1997.6281995.x

Stecher, B., Barthel, M., Schlumberger, M. C., Haberli, L., Rabsch, W., Kremer, M., et al. (2008). Motility allows S. Typhimurium to benefit from the mucosal defence. *Cell. Microbiol.* 10, 1166–1180. doi: 10.1111/j.1462-5822.2008.01118.x

Stecher, B., Hapfelmeier, S., Muller, C., Kremer, M., Stallmach, T., and Hardt, W. D. (2004). Flagella and chemotaxis are required for efficient induction of *Salmonella enterica* serovar Typhimurium colitis in streptomycin-pretreated mice. *Infect. Immun.* 72, 4138–4150. doi: 10.1128/IAI.72.7.4138-4150.2004

Steinmann, R., and Dersch, P. (2013). Thermosensing to adjust bacterial virulence in a fluctuating environment. *Future Microbiol.* 8, 85–105. doi: 10.2217/fmb.12.129

Stoebel, D. M., Free, A., and Dorman, C. J. (2008). Anti-silencing: overcoming H-NS-mediated repression of transcription in Gram-negative enteric bacteria. *Microbiology* 154, 2533–2545. doi: 10.1099/mic.0.2008/020693-0

Straley, S. C., Plano, G. V., Skrzypek, E., Haddix, P. L., and Fields, K. A. (1993). Regulation by Ca2+ in the Yersinia low-Ca2+ response. *Mol. Microbiol.* 8, 1005–1010. doi: 10.1111/j.1365-2958.1993.tb01644.x

Takaya, A., Erhardt, M., Karata, K., Winterberg, K., Yamamoto, T., and Hughes, K. T. (2012). YdiV: a dual function protein that targets FlhDC for ClpXP-dependent degradation by promoting release of DNA-bound FlhDC complex. *Mol. Microbiol.* 83, 1268–1284. doi: 10.1111/j.1365-2958.2012.08007.x

Takeuchi, K., Kiefer, P., Reimmann, C., Keel, C., Dubuis, C., Rolli, J., et al. (2009). Small RNA-dependent expression of secondary metabolism is controlled by Krebs cycle function in Pseudomonas fluorescens. *J. Biol. Chem.* 284, 34976–34985. doi: 10.1074/jbc.M109.052571

Teixido, L., Carrasco, B., Alonso, J. C., Barbe, J., and Campoy, S. (2011). Fur activates the expression of *Salmonella enterica* pathogenicity island 1 by directly interacting with the hilD operator in vivo and in vitro. *PLoS ONE* 6:e19711. doi: 10.1371/journal.pone.0019711

Teplitski, M., Goodier, R. I., and Ahmer, B. M. (2003). Pathways leading from BarA/SirA to motility and virulence gene expression in *Salmonella*. *J. Bacteriol.* 185, 7257–7265. doi: 10.1128/JB.185.24.7257-7265.2003

Teplitski, M., Goodier, R. I., and Ahmer, B. M. (2006). Catabolite repression of the SirA regulatory cascade in *Salmonella enterica*. *Int. J. Med. Microbiol.* 296, 449–466. doi: 10.1016/j.ijmm.2006.06.001

Tran, H. J., Heroven, A. K., Winkler, L., Spreter, T., Beatrix, B., and Dersch, P. (2005). Analysis of RovA, a transcriptional regulator of *Yersinia pseudotuberculosis* virulence that acts through antirepression and direct transcriptional activation. *J. Biol. Chem.* 280, 42423–42432. doi: 10.1074/jbc.M504464200

Troxell, B., Sikes, M. L., Fink, R. C., Vazquez-Torres, A., Jones-Carson, J., and Hassan, H. M. (2011). Fur negatively regulates hns and is required for the expression of HilA and virulence in *Salmonella enterica* serovar Typhimurium. *J. Bacteriol.* 193, 497–505. doi: 10.1128/JB.00942-10

Tsai, C. S., and Winans, S. C. (2011). The quorum-hindered transcription factor YenR of *Yersinia enterocolitica* inhibits pheromone production and promotes motility via a small non-coding RNA. *Mol. Microbiol.* 80, 556–571. doi: 10.1111/j.1365-2958.2011.07595.x

Tsolis, R. M., Young, G. M., Solnick, J. V., and Baumler, A. J. (2008). From bench to bedside: stealth of enteroinvasive pathogens. *Nat. Rev. Microbiol.* 6, 883–892. doi: 10.1038/nrmicro2012

Uliczka, F., and Dersch, P. (2012). Unique virulence properties of *Yersinia enterocolitica* O:3. *Adv. Exp. Med. Biol.* 954, 281–287. doi: 10.1007/978-1-4614-3561-7_35

Vakulskas, C. A., Potts, A. H., Babitzke, P., Ahmer, B. M., and Romeo, T. (2015). Regulation of bacterial virulence by Csr (Rsm) systems. *Microbiol. Mol. Biol. Rev.* 79, 193–224. doi: 10.1128/MMBR.00052-14

Valentin-Weigand, P., Heesemann, J., and Dersch, P. (2014). Unique virulence properties of *Yersinia enterocolitica* O:3–an emerging zoonotic pathogen using pigs as preferred reservoir host. *Int. J. Med. Microbiol.* 304, 824–834. doi: 10.1016/j.ijmm.2014.07.008

Wada, T., Hatamoto, Y., and Kutsukake, K. (2012). Functional and expressional analyses of the anti-FlhD4C2 factor gene ydiV in *Escherichia coli*. *Microbiology* 158, 1533–1542. doi: 10.1099/mic.0.056036-0

Wada, T., Morizane, T., Abo, T., Tominaga, A., Inoue-Tanaka, K., and Kutsukake, K. (2011a). EAL domain protein YdiV acts as an anti-FlhD4C2 factor responsible for nutritional control of the flagellar regulon in *Salmonella enterica* Serovar Typhimurium. *J. Bacteriol.* 193, 1600–1611. doi: 10.1128/JB.01494-10

Wada, T., Tanabe, Y., and Kutsukake, K. (2011b). FliZ acts as a repressor of the ydiV gene, which encodes an anti-FlhD4C2 factor of the flagellar regulon in *Salmonella enterica* serovar typhimurium. *J. Bacteriol.* 193, 5191–5198. doi: 10.1128/JB.05441-11

Walthers, D., Carroll, R. K., Navarre, W. W., Libby, S. J., Fang, F. C., and Kenney, L. J. (2007). The response regulator SsrB activates expression of diverse *Salmonella* pathogenicity island 2 promoters and counters silencing by the nucleoid-associated protein H-NS. *Mol. Microbiol.* 65, 477–493. doi: 10.1111/j.1365-2958.2007.05800.x

Wang, H., Liu, B., Wang, Q., and Wang, L. (2013). Genome-wide analysis of the *Salmonella* Fis regulon and its regulatory mechanism on pathogenicity islands. *PLoS ONE* 8:e64688. doi: 10.1371/journal.pone.0064688

Wang, Q., Zhao, Y., Mcclelland, M., and Harshey, R. M. (2007). The RcsCDB signaling system and swarming motility in *Salmonella enterica* serovar typhimurium: dual regulation of flagellar and SPI-2 virulence genes. *J. Bacteriol.* 189, 8447–8457. doi: 10.1128/JB.01198-07

Wang, S., Fleming, R. T., Westbrook, E. M., Matsumura, P., and Mckay, D. B. (2006). Structure of the *Escherichia coli* FlhDC complex, a prokaryotic heteromeric regulator of transcription. *J. Mol. Biol.* 355, 798–808. doi: 10.1016/j.jmb.2005.11.020

Watson, P. R., Paulin, S. M., Bland, A. P., Libby, S. J., Jones, P. W., and Wallis, T. S. (1999). Differential regulation of enteric and systemic salmonellosis by slyA. *Infect. Immun.* 67, 4950–4954.

Wiedemann, A., Virlogeux-Payant, I., Chausse, A. M., Schikora, A., and Velge, P. (2014). Interactions of *Salmonella* with animals and plants. *Front. Microbiol.* 5:791. doi: 10.3389/fmicb.2014.00791

Wilson, R. L., Libby, S. J., Freet, A. M., Boddicker, J. D., Fahlen, T. F., and Jones, B. D. (2001). Fis, a DNA nucleoid-associated protein, is involved in *Salmonella typhimurium* SPI-1 invasion gene expression. *Mol. Microbiol.* 39, 79–88. doi: 10.1046/j.1365-2958.2001.02192.x

Winter, S. E., Thiennimitr, P., Winter, M. G., Butler, B. P., Huseby, D. L., Crawford, R. W., et al. (2010). Gut inflammation provides a respiratory electron acceptor for *Salmonella Nature* 467, 426–429. doi: 10.1038/nature09415

Wozniak, C. E., Lee, C., and Hughes, K. T. (2009). T-POP array identifies EcnR and PefI-SrgD as novel regulators of flagellar gene expression. *J. Bacteriol.* 191, 1498–1508. doi: 10.1128/JB.01177-08

Yamamoto, S., and Kutsukake, K. (2006). FliT acts as an anti-FlhD2C2 factor in the transcriptional control of the flagellar regulon in *Salmonella enterica* serovar typhimurium. *J. Bacteriol.* 188, 6703–6708. doi: 10.1128/JB.00799-06

Yanagihara, S., Iyoda, S., Ohnishi, K., Iino, T., and Kutsukake, K. (1999). Structure and transcriptional control of the flagellar master operon of *Salmonella typhimurium*. *Genes Genet. Syst.* 74, 105–111. doi: 10.1266/ggs.74.105

Young, G. M., Badger, J. L., and Miller, V. L. (2000). Motility is required to initiate host cell invasion by *Yersinia enterocolitica*. *Infect. Immun.* 68, 4323–4326. doi: 10.1128/IAI.68.7.4323-4326.2000

Young, G. M., Smith, M. J., Minnich, S. A., and Miller, V. L. (1999). The *Yersinia enterocolitica* motility master regulatory operon, flhDC, is required for flagellin production, swimming motility, and swarming motility. *J. Bacteriol.* 181, 2823–2833.

Zhang, W., Wang, Y., Song, Y., Wang, T., Xu, S., Peng, Z., et al. (2013). A type VI secretion system regulated by OmpR in *Yersinia pseudotuberculosis* functions to maintain intracellular pH homeostasis. *Environ. Microbiol.* 15, 557–569. doi: 10.1111/1462-2920.12005

Zhang, X., Kelly, S. M., Bollen, W. S., and Curtiss, R. III (1997). Characterization and immunogenicity of *Salmonella typhimurium* SL1344 and UK-1 delta crp and delta cdt deletion mutants. *Infect. Immun.* 65, 5381–5387.

Zhang, Y. J., and Rubin, E. J. (2013). Feast or famine: the host-pathogen battle over amino acids. *Cell. Microbiol.* 15, 1079–1087. doi: 10.1111/cmi.12140

17

Host FIH-Mediated Asparaginyl Hydroxylation of Translocated *Legionella pneumophila* Effectors

Christopher Price [1], Michael Merchant [2], Snake Jones [1], Ashley Best [1], Juanita Von Dwingelo [1], Matthew B. Lawrenz [1,3], Nawsad Alam [4], Ora Schueler-Furman [4] and Yousef A. Kwaik [1,3]**

[1] Department of Microbiology and Immunology, College of Medicine, University of Louisville, Louisville, KY, USA, [2] Department of Medicine-Renal, College of Medicine, University of Louisville, Louisville, KY, USA, [3] Center for Predictive Medicine, College of Medicine, University of Louisville, Louisville, KY, USA, [4] Department of Microbiology and Molecular Genetics, Institute for Medical Research Israel-Canada (IMRIC), Faculty of Medicine, Hadassah Medical School, The Hebrew University of Jerusalem, Jerusalem, Israel

Edited by:
Elizabeth L. Hartland,
University of Melbourne, Australia

Reviewed by:
Li Xu,
Cornell University, USA
Paul Race,
University of Bristol, UK

***Correspondence:**
Christopher Price
christopher.price@louisville.edu
Yousef A. Kwaik
abukwaik@louisville.edu

FIH-mediated post-translational modification through asparaginyl hydroxylation of eukaryotic proteins impacts regulation of protein-protein interaction. We have identified the FIH recognition motif in 11 *Legionella pneumophila* translocated effectors, YopM of *Yersinia*, IpaH4.5 of *Shigella* and an ankyrin protein of *Rickettsia*. Mass spectrometry analyses of the AnkB and AnkH effectors of *L. pneumophila* confirm their asparaginyl hydroxylation. Consistent with localization of the AnkB effector to the *Legionella*-containing vacuole (LCV) membrane and its modification by FIH, our data show that FIH and its two interacting proteins, Mint3 and MT1-MMP are acquired by the LCV in a Dot/Icm type IV secretion-dependent manner. Chemical inhibition or RNAi-mediated knockdown of FIH promotes LCV-lysosomes fusion, diminishes decoration of the LCV with polyubiquitinated proteins, and abolishes intra-vacuolar replication of *L. pneumophila*. These data show acquisition of the host FIH by a pathogen-containing vacuole and that asparaginyl-hydroxylation of translocated effectors is indispensable for their function.

Keywords: FIH, asparagine hydroxylation, ankyrin, Dot/Icm, AnkB, hypoxia-inducible factor (HIF), bacterial pathogenesis, *Legionella*

INTRODUCTION

Exploitation of host post-translational machineries by translocated bacterial effector proteins is a growing paradigm for effector sub-cellular localization and function in the host cell. Bacterial effectors that are injected into the host cell can be modified by host post-translational modifications including ubiquitination, sumoylation, phosphorylation, and various lipidations including prenylation, palmitoylation, S-acylation, and myristoylation (Kubori and Galan, 2003; Reinicke et al., 2005; Lewis et al., 2008; Patel et al., 2009; Ivanov et al., 2010; Price et al., 2010b; Gaus et al., 2011; Hicks et al., 2011; Hayashi et al., 2013; Beyer et al., 2014). For example, the injected SopE and SptP effectors of *Salmonella* and YopE of *Yersinia* are ubiquitinated by the host resulting in their turnover by proteasome-mediated degradation, while ubiquitination of the SopB effector of *Salmonella* and farnesylation of AnkB of *L. pneumophila* direct sub-cellular localization and impacts function of these injected effectors (Kubori and Galan, 2003; Patel et al., 2009; Ivanov et al., 2010; Price et al., 2010b; Gaus et al., 2011). Therefore, uncovering the host-mediated post-translational modifications of injected bacterial effectors is crucial in understanding function.

Although intracellular bacterial pathogens have been shown to exploit various host post-translational machineries, their exploitation of the host asparaginyl hydroxylation post-translational modification has never been described. The 2-oxoglutarate dioxygenase, designated as factor inhibiting HIF1 (FIH), is a key eukaryotic enzyme, which selectively hydroxylates an asparagine residue within the L(X)5[D/E]φNφ motif (φ represents aliphatic amino acids) in eukaryotic proteins (Hewitson et al., 2002; Lando et al., 2002a,b; Cockman et al., 2009). The addition of the strongly electronegative oxygen atom increases both polarity of a protein and can act as a hydrogen bond donor and acceptor. Therefore, hydroxylation can function as a "molecular switch" for protein-protein interactions (Loenarz and Schofield, 2011). FIH plays a key role in various cellular processes and in particular, it regulates the activity of hypoxia-inducible factor (HIF1), which is the master transcriptional regulator of hypoxia (Webb et al., 2009). During normoxia, HIF1 is hydroxylated by FIH on an asparagine residue and this modification acts as a molecular switch to prevent interaction with its co-activator p300/CBP, blocking transcription of hundreds of HIF1-regulated genes involved in oxygen homeostasis, energy production and immune responses (Hewitson et al., 2002; Lando et al., 2002a,b). In addition, FIH catalyzes asparaginyl hydroxylation of approximately 20 ankyrin repeat domain-containing (ARD) proteins such as p105 and IκBα (Cockman et al., 2009). FIH-dependent hydroxylation of the ARD protein, ASPP2, is required for binding of this protein to its target Par-3 (Janke et al., 2013). Therefore, asparaginyl hydroxylation acts as a molecular switch to promote or reduce protein-protein interactions between HIF1-p300/CBP and ASPP2-Par3 (Hewitson et al., 2002; Lando et al., 2002a,b; Janke et al., 2013). Additionally, FIH hydroxylates the deubiquitinase OTUB, which appears to regulate cellular metabolism (Scholz et al., 2016). A recent study has revealed a complex FIH interactome with many proteins that may serve as substrates for FIH enzyme activity, thus greatly expanding the number of eukaryotic proteins modified by asparaginyl hydroxylation (Rodriguez et al., 2016). However, the biological consequence of asparaginyl hydroxylation of eukaryotic proteins largely remains unclear.

When *L. pneumophila* invades amoebae or human macrophages, it evades the default endosomal-lysosomal degradation pathway and remodels its phagosome into a specialized ER-derived vacuole via intercepting ER-to-golgi vesicular traffic (Isberg et al., 2009; Al-Quadan et al., 2012; Price et al., 2014). This is achieved by the translocation of ~300 effector proteins via the Dot/Icm type IVB secretion system (de Felipe et al., 2008; Isberg et al., 2009; Zhu et al., 2011). These effectors modulate a myriad of eukaryotic processes including host signaling, vesicular trafficking, protein synthesis, apoptosis, prenylation, ubiquitination, and proteasomal degradation (Al-Quadan et al., 2012; Price et al., 2014). Surprisingly, very few of these effectors are essential for intracellular replication of *L. pneumophila*, suggesting specific requirements for different effectors in different environmental hosts.

The AnkB translocated effector is essential for proliferation of *L. pneumophila* within the two evolutionarily-distant hosts, mammalian and protozoan cells, and for intrapulmonary bacterial proliferation and manifestation of pulmonary disease in the mouse model (Al-Khodor et al., 2008; Price et al., 2009, 2010a,b, 2011; Lomma et al., 2010). Recent characterization of the crystal structure of AnkB has confirmed that it is a non-canonical F-box protein with three ankyrins repeats domain (Price et al., 2009; Lomma et al., 2010; Wong et al., 2017). The crystal structure has also confirmed that the F-box domain of AnkB interacts with the host SCF1 ubiquitin ligase, which explains show AnkB functions as a platform for the docking of polyubiquitinated proteins to the *Legionella*-containing vacuolar (LCV) membrane within macrophages and amoebae (Price et al., 2009; Lomma et al., 2010; Wong et al., 2017). The AnkB-assembled polyubiquitinated proteins are predominately Lys48-linked that are ultimately degraded by the host proteasome machinery, which generates higher levels of cellular amino acids that are the main sources of carbon and energy to power replication of *L. pneumophila* (Price et al., 2011). This enables intracellular bacteria to overcome host limitation of essential nutrients and favorable sources of carbon and energy, such as amino acids (Price et al., 2011; Abu Kwaik and Bumann, 2013).

Here we show that 11 *L. pneumophila* type IVB-translocated effectors including, AnkB and AnkH, harbor the recognition motif for FIH-dependent asparaginyl-hydroxylation. Furthermore, the FIH recognition motif is found in translocated effectors from other intracellular microbial pathogens including YopM from *Yersinia pestis*, IpaH4.5 of *Shigella flexneri* and a putative translocated ARD-protein of *Rickettsia felis*. We show that the AnkH and AnkB effectors are modified by asparaginyl hydroxylation. The LCV recruits FIH, which is indispensable for intra-vacuolar proliferation of *L. pneumophila* and plays a partial role in the ability of the LCV to evade lysosomal fusion and is needed for AnkB-dependent assembly of polyubiquitinated proteins on the LCV. This is the first example of an injected microbial effectors post-translationally modified by asparaginyl hydroxylation.

MATERIALS AND METHODS

Bacterial Strains and Cell Cultures

L. pneumophila strain AA100/130b (ATCC BAA-74), the isogenic mutants, *dotA* and *ankB* or complemented *ankB* mutants were grown as described previously (Al-Khodor et al., 2008). Maintenance of HEK293T cells was performed as previously described (Price et al., 2009). Human monocyte-derived macrophages (hMDMs) were isolated from healthy donors as described previously (Price et al., 2009). Substitutions in the *ankB* gene were performed using standard molecular biology techniques, and the resulting alleles were used to complement the AA100 *ankB* mutant. All methods were carried out and approved in accordance to the University of Louisville Institutional Review Board guidelines and blood donors gave informed consent as required by the University of Louisville Institutional Review Board (IRB # 04.0358).

qPCR of HIF1-Dependent Genes

To analyze the effect of the FIH inhibitor, N-oxalyl-D-alanine on FIH enzyme activity, the expression level of four HIF1-dependent genes were chosen as a readout. A total of 3 × 10^6 hMDMs were seeded into 6-well plates and incubated with and without 8 mM N-oxalyl-D-alanine or 1 mM dimethyloxalylglycine (Enzo) for 8 h. Total RNA was isolated from the hMDMs using a RNeasy Mini Kit (Qiagen) according to the manufacturer's instruction. Purified RNA was converted to first strand cDNA using Superscript III (Invitrogen) and then analyzed by qPCR. Primers specific to GLUT1 (5′-AACTCTTCAGCCAGGGTCC AC-3′, 5′- CACAGTGAAGATGATGAAGAC-3′), GLUT3 (5′-ACTTTGACGGACAAGGGAAATG-3′, 5′- ACCAGTGACAG CCAACAGG-3′), LDHA (5′- ACCCAGTTTCCACCATGATT-3′, 5′- CCCAAAATGCAAGGAACACT-3′) and PGK1 (5′- AT GGATGAGGTGGTGAAAGC-3′, 5′- CAGTGCTCACATGGCT GACT-3′) were analyzed relative to the house-keeping control ACTB (5′- GACAGGATGCAGAAGGAGATCACT-3′, 5′- TGA TCCACATCTGCTGGAAGGT-3′). qPCR was performed using PerfeCTa Sybr supermix (Quanta Biosciences) using a StepOne Plus qPCR machine (Applied Biosystems). Determination of fold change in gene expression was calculated using REST-XL software (Pfaffl et al., 2002).

Co-Immunoprecipitation of Skp1 and *ankB* Mutants

To assess interaction of the AnkB mutants with the known host target protein, Skp1, HEK293T cells were co-transfected with p3XFLAG CMV 10-AnkB or its mutant alleles and pHA-Skp1. Briefly, a total of 1 × 10^6 HEK293T cells were seeded into 6-well plates, and the following day transfected with plasmid DNA using polyethyleneimine for 24 h. Following transfection, 3XFLAG AnkB was immunoprecipitated using anti-FLAG M2 magnetic beads (Sigma) according to the manufacturer's instructions. Purified 3XFLAG-AnkB was then subjected to western blot analysis using anti-FLAG M2 antibody (Sigma). Blots were then stripped and reprobed with anti-HA antibody (Santa Cruz), to determine if HA-Skp1 co-immunoprecipitated with AnkB using standard procedures.

Translocation Assay

To assess translocation of the AnkB mutant alleles by *L. pneumophila* during infection of host cells, adenylate cyclase fusions were generated using standard molecular biology techniques. A total of 1 × 10^6 hMDMs were infected with *L. pneumophila* harboring plasmids expressing various adenylate cyclase fusions at an MOI of 20 for 1 h. Cells were then lysed and processed to assess cAMP concentration by ELISA using the Direct cAMP ELISA kit (Enzo) according to the manufacturer's instructions.

Localization of FIH, Mint3, MT1-MMP, and Polyubiquitin to the LCV

Dependent upon the experiment, the wild type strain, the isogenic mutants *dotA* and *ankB*, and complemented *ankB* mutants were grown on BCYE agar for 3 days at 37°C prior to infection. A total of 5 X 10^5 hMDMs were seeded into 24-well plates containing sterile glass coverslips and infected with *L. pneumophila* at an MOI of 10 for 1 h, resulting in cells infected with a single bacterium. The hMDMs were treated for 1 h with gentamicin to kill remaining extracellular bacteria and then fixed and permeabilized with −20°C methanol for 5 min. The hMDM monolayers were labeled with rabbit anti-FIH, Mint3, or MT1-MMP antibody (1/200 dilution, Santa Cruz) or mouse FK1 antibody (1/100 dilution, Enzo). *L. pneumophila* were labeled with mouse or rabbit anti-*L. pneumophila* antiserum (1/1,000 dilution) for 1 h. Corresponding Alexa-fluor conjugated secondary antibodies were used for visualization (1/4,000 dilution, Invitrogen). Monolayers were analyzed by confocal microscopy using an Olympus FV1000 scanning fluorescence confocal microscope. On average, 8–15 0.2 um serial Z sections of each image were captured and stored for further analyses, using Adobe Photoshop CS5. For "polyubiquitin cloud" area measurements, Z-stack images of ubiquitin positive LCVs were analyzed using FV10-ASW 3.1 software (Olympus). Co-localization was assessed through visual inspection of Z-stack images using FV10-ASW 3.1 software (Olympus).

Intracellular Replication of *L. pneumophila* in hMDMs

The wild type strain and the isogenic mutant *dotA* were grown on BCYE agar for 3 days at 37°C prior to infection. A total of 1 × 10^5 hMDMs were plated in 96-wells and treated with and without increasing concentrations of the FIH inhibitor, N-oxalyl-D-alanine (Enzo) for 2 h prior to infection. The hMDM monolayers were infected with *L. pneumophila* at an MOI of 10 for 1 h and then treated for 1 h with gentamicin to kill remaining extracellular bacteria At 2, 24, and 48 h post-infection the hMDMs were lysed with sterile water and *L. pneumophila* CFUs were determined by plating serial dilutions onto BCYE agar. The inhibitor was present throughout the course of the infection and did not affect viability of the hMDMs or *L. pneumophila*. Experiments were performed in triplicate.

FIH RNAi

HEK293T cells seeded into 24-well plates with glass coverslips were transfected with and without Silencer Select Negative Control #1 or FIH-specific siRNA (sense 5′-GAUAAAAGGUU ACAAACGAtt-3′, antisense 5′-UCGUUUGUAACCUUUUAU Ctg-3′) (Ambion) using Lipofectamine RNAiMax (Invitrogen) following the manufacturers' instructions for 24 h prior to infection with *L. pneumophila*. To assess knockdown of FIH in transfected cells, cell monolayers were lysed using M-PER reagent (Pierce) and then run on SDS-PAGE gels (BioRad) and analyzed by immunoblotting using a goat anti-FIH antibody (1/200 dilution) (Santa Cruz) and detected using an anti-goat HRP-antibody conjugate (Thermo Scientific). Immunoblots were then immediately stripped and reprobed with an anti-β-tubulin antibody (1/1,000 dilution) (Santa Cruz) to show equal protein loading between lanes.

Replicative Vacuole Analysis of *L. pneumophila*

A total of 2 × 10^5 RNAi treated HEK293T cells as described above or hMDMs were added to 24-well plates containing glass coverslips. The cell monolayers were infected with wild type *L. pneumophila* at an MOI of 20 for 1 h (HEK293T cells) or the wild type, *dotA* or *ankB* mutants, or complemented *ankB* mutants (hMDMs, MOI 10), and then the monolayers were treated with gentamicin for 1 h to kill remaining extracellular bacteria. Following extensive washing to remove gentamicin, the infection proceeded for 10 h. At 10 h the monolayers were permeabilized and fixed using 100% methanol held at −20°C for 5 min, and then labeled with rabbit anti-*Legionella* antiserum (1/1,000 dilution) and counter-labeled with Alexa-Fluor 488 anti-rabbit antibody (1/4,000 dilution, Invitrogen) and DAPI to stain the nuclei. Monolayers were examined by confocal microscopy. A total of 100 replicative vacuoles from 100 individual cells were analyzed for each experimental condition and performed in triplicate.

Vesicular Trafficking of *L. pneumophila* in hMDMs

The wild type strain was grown on BCYE agar for 3 days at 37°C. For killed bacteria, bacterial cells were resuspended in PBS with 3.7% formalin and incubated for 30 min at room temperature and then washed 3 times with PBS to remove residual formalin. A total of 5 × 10^5 hMDMs were seeded into 24-well plates containing sterile glass coverslips and treated with and without increasing concentrations of the FIH inhibitor, N-oxalyl-D-alanine (Enzo) for 2 h prior to infection. The hMDM monolayers were infected with live and formalin killed *L. pneumophila* at an MOI of 10 for 1 h. A total of 2 × 10^5 HEK293T cells transfected with control or FIH RNAi were seeded into 24-well plates containing sterile glass coverslips were infected with wild type bacteria at an MOI of 20 for 1 h. For both hMDMs and HEK293T cells, the monolayers were treated for 1 h with gentamicin to kill remaining extracellular bacteria and then the cells were fixed and permeabilized with −20°C methanol for 5 min. The cell monolayers were labeled with rabbit anti-*L. pneumophila* antiserum (1/1,000 dilution) and mouse anti-LAMP2, CathD, or KDEL (1/2,000, 1/100, and 1/200 dilutions respectively, Transduction Labs, Stressgen). Anti-mouse IgG Alexa-fluor 555 and anti-rabbit IgG Alexa-fluor 488 secondary antibodies (1/4,000 dilution, Invitrogen) were used to visualize vacuolar markers *L. pneumophila* respectively. A total of 100 LCVs from 100 individual cells were analyzed by confocal microscopy for each experimental condition and performed in triplicate.

Mass Spectrometry

The analysis for the presence and extent of asparagine (N) hydroxylation was conducted as described by Petkowski et al. (2013) with some modifications (Petkowski et al., 2013) as described below. A colloidal Coomassie blue stained, 1D-SDS-PAGE gel band was excised, de-stained and equilibrated into 0.1 M triethylammonium bicarbonate, pH 8.5 and digested with 80 ng mass spectrometry grade trypsin (Promega, Madison, WI) per gel plug. The digest supernatant was transferred to a new tube and the gel plugs extracted using a modification of Shevchenko et al. (1996). A 5 ul aliquot of this peptide solution was loaded onto a Dionex Acclaim PepMap 100 75 μm × 2 cm, nanoViper (C18, 3 μm, 100Å) trap, and resolved on a Dionex Acclaim PepMap RSLC 50 μm × 15 cm, nanoViper (C18, 2 μm, 100Å). The sample was eluted using a linear 2 to 60% acetonitrile gradient using an EASY nano1000 UHPLC and introduced into a LTQ-Orbitrap ELITE mass spectrometer (ThermoElectron, Waltham, MA) for accurate mass measurements using a Nanospray Flex Ion Source (ThermoElectron, Waltham, MA), a stainless steel emitter with a capillary temperature set to 225 C and a spray voltage of 1.6 kV and lock mass enabled (0% lock mass abundance) for the 371.101236 m/z polysiloxane peak as an internal calibrant (Cox et al., 2011). Data dependent tandem mass spectra were collected using HCD and ETD fragmentation using an Nth Order Double Play with ETD Decision Tree method (Swaney et al., 2008) was created in Xcalibur v2.2 and included a parent mass list for +2 or +3 charge states of tryptic peptides (0, 1, or 2 miss cleavages) based on the protein sequence and putative hydroxylation. Targeted and data dependent spectra were acquired and searched using MASCOT ver 2.1 and Sequest through Proteome Discoverer 1.4 using the 2/13/2013 version of the UniprotKB *Homo sapiens* reference proteome (canonical and isoform sequences) appended with the *Legionella* sp proteins considering up to two missed cleavages, and N-hydroxylation as variable modifications. The false discovery rate was controlled by use of (1) a decoy database generated from this database with the program decoy.pl (matrixscience.com) and use of Peptide & Protein Prophet algorithm to report FDRs below 1%.

RESULTS

Identification of the FIH Consensus Motif in Translocated Effectors of *L. pneumophila* and Other Bacterial Pathogens

To date only HIF1 and ~20 human ARD proteins are known to be hydroxylated by FIH (Hewitson et al., 2002; Lando et al., 2002a,b; Cockman et al., 2009). *L. pneumophila* translocates at least 11 ARD proteins by its Dot/Icm type IVB translocation system into the host cell upon invasion, including those that contribute to intracellular replication (AnkB, H, J) (Al-Khodor et al., 2008; Habyarimana et al., 2008; Pan et al., 2008; Price et al., 2011). Therefore, we examined the ARD proteins of *L. pneumophila* for the presence of the FIH consensus motif. The AnkB, AnkH, and AnkN effectors all harbor the FIH consensus motif in the ARD regions, suggesting that these effectors are potential targets for FIH asparaginyl hydroxylation upon their translocation into the host cell (Table S1). *In silico* analysis of the whole *L. pneumophila* genome identified 8 additional effector proteins that harbor the FIH recognition motif (Table S1).

To assess whether some of the 11 *L. pneumophila* translocated effectors and YopM from *Y. pestis* that harbor the FIH consensus sequence are hydroxylated, LC-MS MS1 analysis was performed. Purification of translocated effector proteins endogenously produced by intracellular bacteria during infection of host cells results in insufficient material for analysis. To overcome this limitation, the effector proteins were expressed and purified from transiently transfected HEK293T cells. Of the twelve translocated effectors, only AnkN, SdeC, SdcA, LepB, AnkH, AnkB, and YopM could be reliably expressed in HEK293T cells and purified in sufficient quantities for mass spectrometry analysis of post-translational modification. LC-MS MS1 analysis of AnkH protein purified from HEK293T cells assigned a site of hydroxylation on the 92N residue within the predicted FIH recognition motif in the second ankyrin repeat of AnkH (Table S1, **Figure 1A**). LC-MS MS1 analysis of AnkB revealed three hydroxylation sites, on residues 62N, 111N, and 126N (**Figure 1B**, Figure S1), while detection of hydroxylation sites within YopM was not reproducible in three attempts. Although the hydroxylated asparagine residues in AnkB were not part of the predicted FIH consensus motif, they were within the three-ankyrin repeat containing domain (Wong et al., in press) (Table S1, **Figures 1B,C** and Figure S1). LC-MS MS1 analyses of the AnkN, SdeC, SdcA, LepB effectors did not reveal asparaginyl hydroxylation modification. Therefore, our data shows that at least 2 out of the 7 effectors analyzed are modified by asparaginyl hydroxylation.

Acquisition of FIH, Mint3, and MT1-MMP by the LCV

The AnkB effector is localized to the LCV membrane through host-mediated farnesylation (Price et al., 2010b). In addition, the effectors LepB, SdeC, and SdcA which are potential candidates for FIH-mediated asparaginyl hydroxylation are also LCV-localized (Chen et al., 2004, 2007; Luo and Isberg, 2004; Bardill et al., 2005; Ingmundson et al., 2007; Tan et al., 2011). Since FIH is a cytosolic enzyme that can be sequestered to membranous structures such as the Golgi apparatus through interaction with Mint3 and MT1-MMP in macrophages (Sakamoto and Seiki, 2009, 2010) and the LCV intercepts ER-Golgi vesicular traffic (Isberg et al., 2009; Al-Quadan et al., 2012; Price et al., 2014), we determined if FIH and its two interacting partners (Mint3 and MT1-MMP) were recruited to the LCV within hMDMs. The data showed that by 2 h of infection, 75, 65, and 56% of the LCVs harboring wild type bacteria co-localized with FIH, Mint3, and MT1-MMP, respectively (**Figures 2A–D**). In contrast, only 27, 18.2, and 0% of the LCVs harboring the *dotA* translocation-deficient mutant co-localized with FIH, Mint3, and MT1-MMP, respectively (**Figures 2A–D**) and this was significantly reduced relative to co-localization observed for wild type LCVs (unpaired *t*-test, $p < 0.01$). This indicates recruitment of these proteins to the LCV is dependent on the Dot/Icm T4SS apparatus.

Role of 62N, 111N, and 126N on *ankB* Function

The AnkB F-box effector is required for recruitment of polyubiquitinated proteins to the LCV and intra-vacuolar replication of *L. pneumophila* (Al-Khodor et al., 2008; Price et al., 2009, 2011; Lomma et al., 2010). To determine the functional importance of the three hydroxylated asparagine residues identified in AnkB, the *ankB* mutant was complemented with *ankB* alleles in which each of the three asparagine residues were substituted with alanine and the effect on intra-vacuolar replication and polyubiquitinated protein recruitment to the LCV were determined. Intra-vacuolar growth in human monocyte-derived macrophages (hMDMs) was assessed by enumerating replicative vacuoles at 10 h post-infection using confocal microscopy. The data showed that expression of *ankB* alleles with a single alanine substitution at 62N, 111N, or 126N resulted in a partial defect in intra-vacuolar replication compared to the wild type allele (**Figures 3A,B**). Strikingly, expression of *ankB* alleles with multiple substitutions significantly reduced the number of LCVs harboring > 7 bacteria compared to wild type LCVs (unpaired *t*-test, $p < 0.01$), approaching that observed for the *ankB* null mutant (**Figures 3A,B**).

Since AnkB is intimately involved in recruitment of polyubiquitinated proteins to the LCV we next determined the impact of alanine substitutions at 62N, 111N, or 126N on this process. Following 2 h infection of hMDMs, ~70% of LCVs harboring wild type bacteria were decorated with polyubiquitinated proteins compared to only 29% for LCVs harboring the *ankB* mutant (**Figures 3C,D**). Complementation of the *ankB* mutant with the wild type *ankB* gene restored decoration of the LCV with polyubiquitinated proteins to wild type levels (~70%) (**Figures 3C,D**). In contrast, decoration of LCVs harboring the *ankB* mutant complemented with *ankB* 62N, 111N, or 126N substitution alleles was significantly reduced to ~50% (unpaired *t*-test, $p < 0.01$) compared to wild type LCVs. Furthermore, when multiple substitutions were made, decoration of the LCV with polyubiquitinated proteins was further reduced to ~40% (unpaired *t*-test, $p < 0.01$) compared to wild type, approaching that observed for the *ankB* mutant (**Figure 3C,D**). Importantly, substitution of these N residues did not impact translocation of the altered AnkB effectors into host cells compared to the wild type protein and did not impact interaction of this effector with its Skp1 component of the E3 ubiquitin ligase (Figures S2A,B). Taken together, this indicates that these three asparagine residues are important for AnkB function.

Role for FIH in Recruitment Polyubiquitinated Proteins to the LCV

Assembly of polyubiquitinated proteins at the LCV (Dorer et al., 2006) is mediated by the AnkB effector (Price et al., 2009, 2011; Lomma et al., 2010). Since AnkB is hydroxylated on three asparagine residues and substitution of these residues reduces polyubiquitinated protein recruitment to the LCV, we determined whether FIH activity is required for the assembly of polyubiquitinated proteins by AnkB on the LCV. FIH was chemically inhibited by N-oxalyl-D-alanine (NODA) and the effect on recruitment of polyubiquitinated proteins to the LCV was analyzed. NODA exhibited FIH inhibitory activity in hMDMs, as shown by elevated expression of HIF1-dependent genes by qPCR (Table S2). Following 2 h infection of untreated hMDMs, 75% of the wild type-containing LCVs were decorated with polyubiquitinated proteins (**Figures 4A,B**). Compared to untreated cells, blocking FIH activity with NODA, caused a dose-dependent inhibition of recruitment of polyubiquitinated proteins with only 38% of the LCVs decorated when 12 mM NODA was used (**Figures 4A,B**) (unpaired *t*-test, $p <$ 0.01 compared to untreated cells), mimicking the effect of substituting the three hydroxylated asparagine residues of AnkB (**Figures 3C,D**). In addition to the reduced number of LCVs positive for polyubiquitinated protein in hMDMs treated with NODA, microscopic examination indicated that the area of the

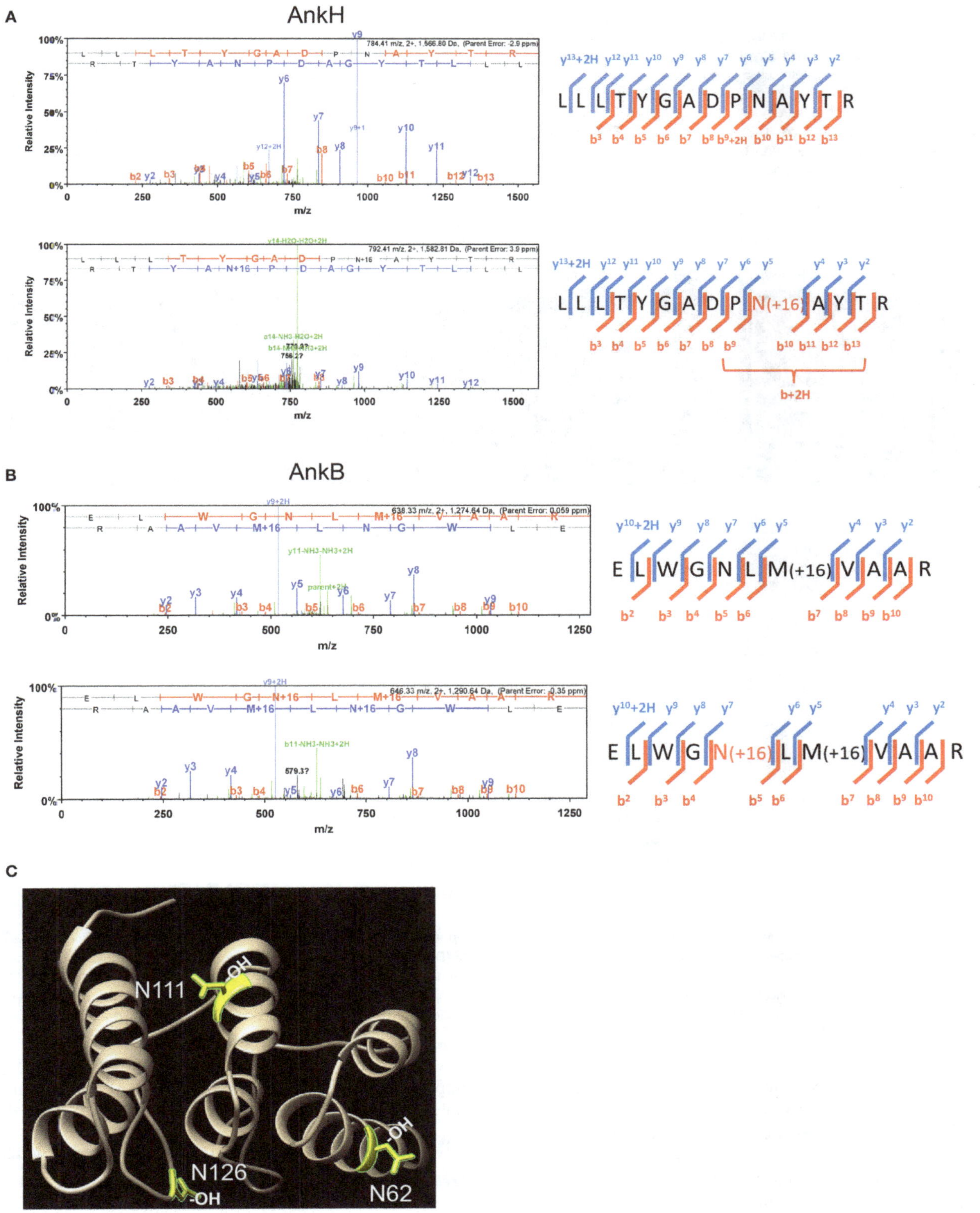

FIGURE 1 | AnkH and AnkB are modified by host asparaginyl hydroxylation. High resolution LCMS analysis 1D-LC-LTQ-Orbitrap-ELITE-MS of AnkH and AnkB protein expressed in HEK293T cells identifies hydroxylated asparagine residues. **(A)** AnkH: (Top) HCD fragmentation spectrum for +2 charged ion with a monoisotopic m/z: 1,566.81 Da; (Bottom) HCD fragmentation spectrum for +2 charged ion with a monoisotopic m/z: 1,582.81 Da; ProteomeDiscover v1.3 analysis

(Continued)

FIGURE 1 | Continued
of upper MS/MS data set identifies a tryptic peptide (Mascot Ion Score 77.8, Sequest Xcorr 3.94, and deltaCn 0.49) with the sequence LLLTYGADPNATYTR. Analysis of lower MS/MS data set identifies a tryptic peptide (Mascot Ion Score 19.7, Sequest Xcorr 2.33, and deltaCn 0.30) with the sequence LLLTYGADPN(OH)ATYTR. Both peptides were observed with sub-5 ppm mass accuracy and with near complete b-ion (red hash; includes b-, b*-, and b+2H+- ions) and y-ion (blue hash; includes y-. y*-, and y+2H+-) coverage of parent ions. **(B)** AnkB: (Top) CID fragmentation spectrum for +2 charged ion with a monoisotopic m/z: 1,274.64 Da; (Bottom) CID fragmentation spectrum for +2 charged ion with a monoisotopic m/z: 1,290.64 Da; ProteomeDiscover v1.4 analysis of upper MS/MS data set identifies a tryptic peptide (Mascot Ion Score 81.7) with the sequence ELWGNLMVAAR. Analysis of lower MS/MS data set identifies a tryptic peptide (Mascot Ion Score 41.1) with the sequence ELWGN(OH)LMVAAR. Both peptides were observed with <1 ppm mass accuracy and with near complete b-ion (red hash; includes b- ions) and y-ion (blue hash; includes y-. y*-, and y+2H+-) coverage of parent ions. **(C)** X-ray crystal structure of the AnkB ARD region illustrating the hydroxylated asparagine residues shown in yellow.

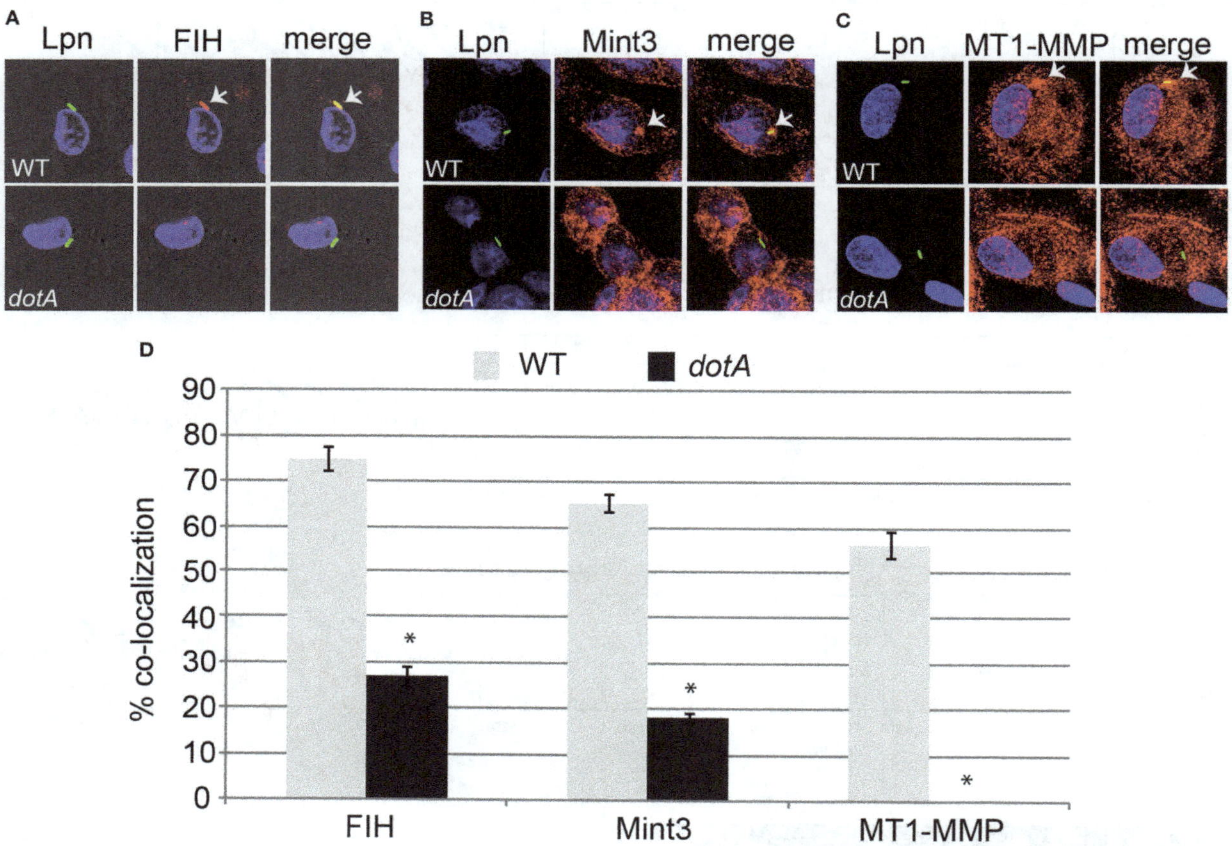

FIGURE 2 | FIH, Mint3, and MT1-MMP are recruited to the LCV in a Dot/Icm-dependent manner. hMDMs were infected with wild type *L. pneumophila* or the isogenic *dotA* mutant for 2 h to determine co-localization of the LCV with FIH **(A)**, Mint3 **(B)**, and MT1-MMP **(C)**. Bacteria were labeled with anti-Lpn antiserum (green) and FIH, Mint3, or MT1-MMP was labeled with specific antibodies (red) and then analyzed by confocal microscopy. The arrows indicate co-localization of FIH, Mint3, or MT1-MMP with the LCV. Quantification of % co-localization is shown in **(D)** and data represents (mean ± SD, $n = 100$ LCVs) of the frequency of acquisition of FIH, Mint3, or MT1-MMP by the LCV. *above the *dotA* bars indicates statistically significant difference in % co-localization compared to the corresponding wild type LCVs (unpaired *t*-test, $p < 0.01$). The data is representative of three independent experiments.

polyubiquitin "cloud" surrounding the LCV that were positive was reduced. To semi-quantitate the polyubiquitin cloud, the area of the cloud was measured using Z-stack confocal images. The average polyubiquitinated cloud surrounding LCVs in untreated hMDMs was 14.5 μm^2 while those in hMDMs treated with 12 mM NODA was significantly reduced to only 7 μm^2 (unpaired *t*-test, $p < 0.01$ compared to untreated cells) with many LCVs showing little extension of the "cloud" away from the periphery of the LCV (**Figure 4C**, Figure S3). Taken together, these data clearly show that inhibition of FIH had a negative impact on the biological function of AnkB in assembly of polyubiquitinated proteins at the LCV.

FIH is Required for Intracellular Replication of *L. pneumophila*

Since FIH is recruited to the LCV in a Dot/Icm-dependent manner, we determined the biological relevance of host FIH on intra-vacuolar replication of *L. pneumophila*. FIH was chemically inhibited by NODA and the effect on intra-vacuolar replication in hMDMs was analyzed. The data showed that NODA inhibited

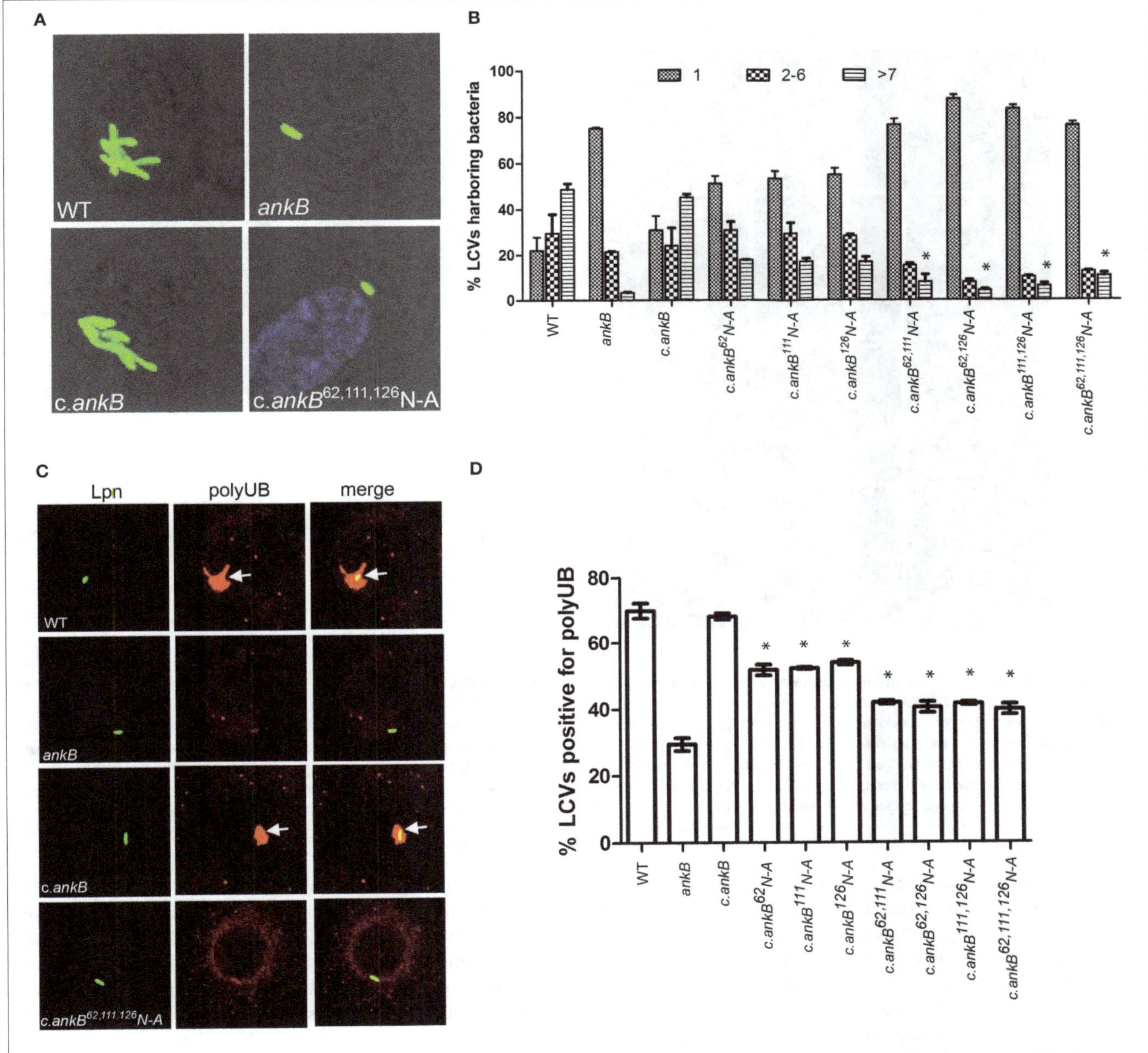

FIGURE 3 | Amino acid residues 62N, 111N, and 126N are required for AnkB function. (A) To determine intra-vacuolar replication at 10 h post-infection, hMDMs were infected with either wild type *L. pneumophila* (WT), *ankB* mutant or complemented *ankB* mutants (c.*ankB*) and then labeled with anti-Lpn antiserum (green). Images are representative of over 100 infected cells. **(B)** Quantitation of replicative vacuole analysis in hMDMs. The numbers of bacteria were determined by analyzing Z-stack confocal images of infected cells, bars represent (mean ± SD, $n = 100$ LCVs) the % total of bacteria per LCV. *above the >7 bacteria bars indicates statistically significant decrease in bacterial numbers relative to that observed for wild type (unpaired *t*-test, $p < 0.01$). **(C)** Representative confocal microscopy images of *L. pneumophila* infected hMDMs at 2 h post-infection to determine % co-localization of polyubiquitinated protein with the LCV. **(D)** Quantitation of the number of LCVs decorated by polyubiquitinated proteins in hMDMs at 2 h post-infection. Bars represent (mean ± SD, $n = 200$ LCVs) and are representative of three independent experiments *above the bars indicates statistically significant decrease in polyubiquitin recruitment relative to that observed for wild type (unpaired *t*-test, $p < 0.01$).

intra-vacuolar replication of *L. pneumophila* in a dose-dependent manner. At 2 mM, NODA had a minor negative effect but at 12 mM, it essentially blocked intracellular replication of *L. pneumophila* compared to untreated cells (**Figure 5A**). *L. pneumophila* numbers increased by 3.2 log units over 48 h of infection, but in cells treated with 12 mM NODA, only a 1.3 log unit increase was observed and this difference was statistically significant (unpaired *t*-test, $p < 0.01$) (**Figure 5A**). The addition of 12 mM NODA did not affect viability of the hMDMs or *L. pneumophila* during the course of the experiment and did not affect *L. pneumophila in vitro* growth in broth culture (data not shown). To confirm the effect of chemical inhibition of FIH on

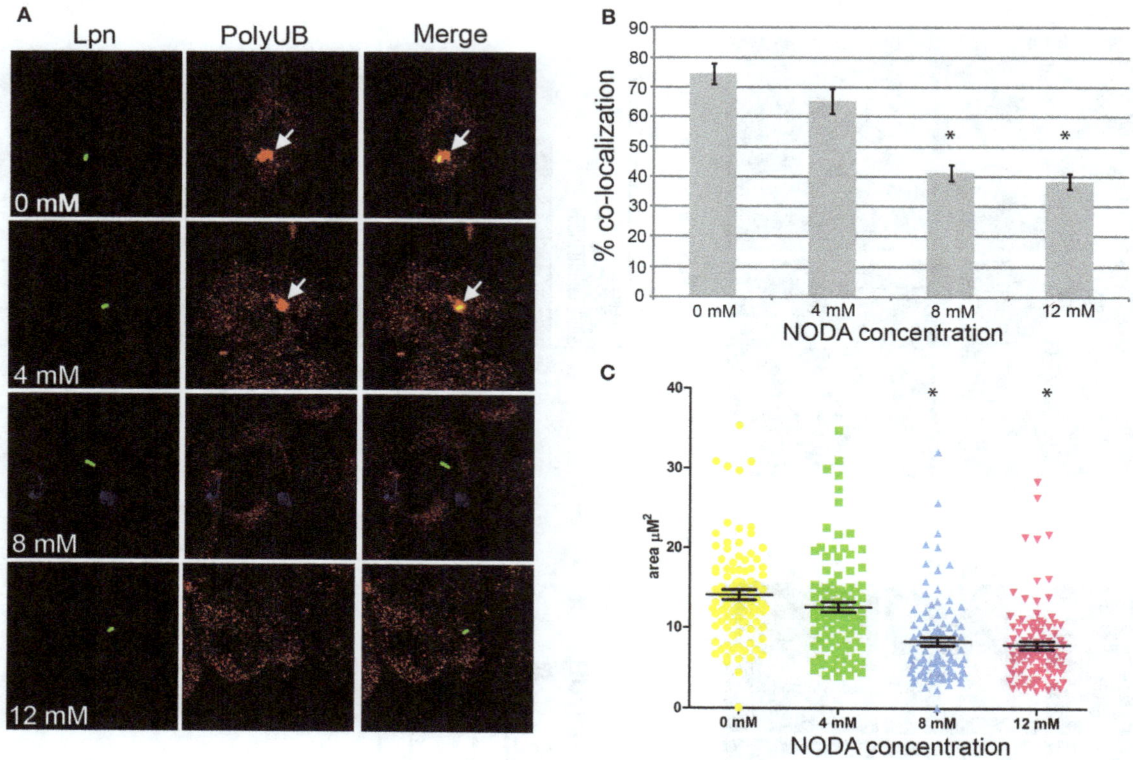

FIGURE 4 | Docking of polyubiquitinated proteins to the LCV in hMDMs requires FIH activity. (A) hMDMs pretreated with the FIH inhibitor, NODA (0–12 mM), were infected with wild type *L. pneumophila* for 2 h. Representative confocal microscopy images of infected hMDMs show co-localization of polyubiquitinated protein with the LCV. Quantification of % co-localization is shown in **(B)** and data represents (mean ± SD, $n = 100$ LCVs) of the frequency of acquisition of polyubiquitinated proteins by the LCV, *above the bars indicates statistically significant decrease in polyubiquitin recruitment relative to that observed for untreated cells (unpaired *t*-test, $p < 0.01$). **(C)** Semi-quantitative analysis of the polyubiquitin "cloud" on the LCVs in infected hMDMs treated with increasing concentrations of the FIH inhibitor, 2 h post-infection. Data represents the distribution of polyubiquitin recruitment size (mean ± SD, $n = 100$ LCVs) and is representative of three independent experiments. *above the data points indicates statistically significant decrease in polyubiquitin "cloud"size relative to that observed for untreated cells (unpaired *t*-test, $p < 0.01$).

intracellular replication, expression of FIH was knocked down in HEK293T cells using specific RNAi (**Figure 5B**). Intracellular growth was assessed by enumerating replicative vacuoles at 10 h post-infection using confocal microscopy (**Figures 5C,D**). The data showed that replication of *L. pneumophila* was significantly reduced in HEK293T cells treated with FIH specific RNAi, with only 38% of LCVs harboring replicative LCVs (three or more bacteria/LCV), compared to 69 and 77% of LCVs in untreated or control RNAi-treated cells, respectively (unpaired *t*-test, $p < 0.01$ compared to untreated cells) (**Figure 5D**). Knockdown of FIH in HEK293T cells did not affect cellular viability during the course of the experiments (data not shown). Taken together, the data clearly show that the function of FIH is indispensable for intra-vacuolar replication of *L. pneumophila*.

The Role of FIH in Biogenesis of the LCV

Since blocking FIH had a dose-dependent negative impact on intracellular replication of *L. pneumophila*, we determined whether this growth defect was associated with alterations in the biogenesis of the LCV that is rough ER-derived, and evades lysosomal fusion (Isberg et al., 2009; Al-Quadan et al., 2012; Price et al., 2014). The infected hMDMs were fixed at 2 h post-infection and labeled with antibodies specific for the late endosomal/lysosomal marker, LAMP2, and lysosomal enzyme, cathepsin D, and the ER marker, KDEL. In untreated hMDMs, only 10 and 18% of LCVs harboring live wild type bacteria co-localized with LAMP2 and cathepsin D respectively (**Figures 6A,B,D**). In contrast, 49 and 66% of LCVs harboring live wild type bacteria in 12 mM NODA-treated hMDMs co-localized with LAMP2 and cathepsin D, and this increase was significant (unpaired *t*-test, $p < 0.01$ compared to untreated cells) (**Figures 6A,B,D**). As expected, over 90% of LCVs containing formalin-killed bacteria co-localized with LAMP2 and cathepsin D (**Figures 6A,B,D**). Furthermore, only 28% of LCVs co-localized with the KDEL ER marker in NODA-treated cells compared to 80% in untreated cells (unpaired *t*-test, $p < 0.01$ compared to untreated cells) (**Figures 6C,D**). No LCVs harboring formalin killed bacteria were positive for KDEL co-localization (**Figures 6C,D**).

To confirm the effect of chemical inhibition of FIH-dependent evasion of lysosomal fusion and remodeling of the LCV by the ER, expression of FIH was silenced using specific RNAi. Similar to NODA, knockdown of FIH significantly increased the number of LCVs co-localized with LAMP2 and cathepsin D, 55 and 57%

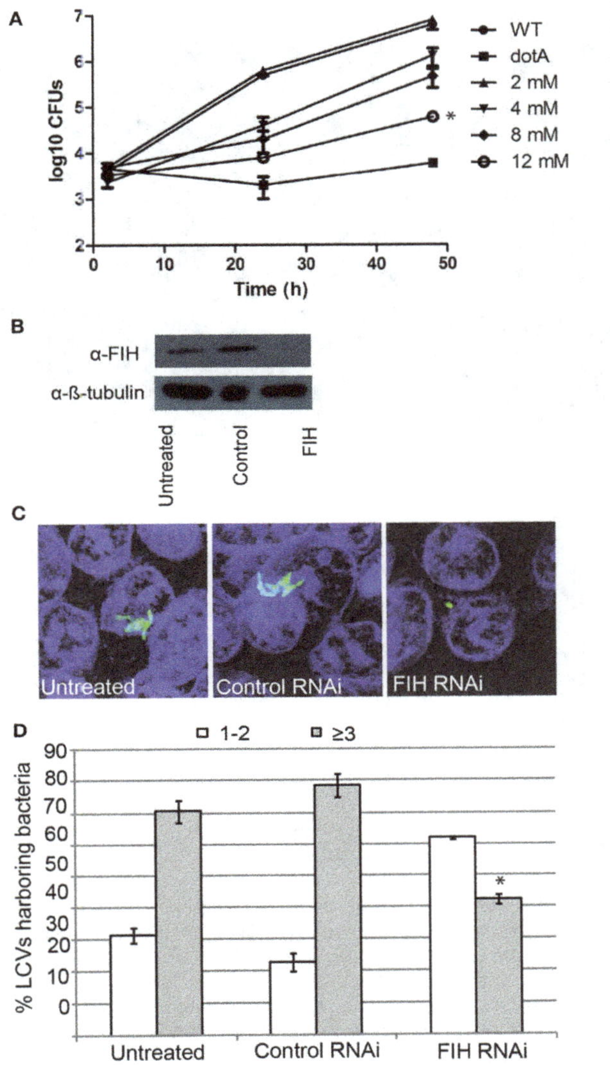

FIGURE 5 | FIH is indispensable for intracellular proliferation of *L. pneumophila*. (A) Growth kinetics of *L. pneumophila* in hMDMs treated with increasing concentrations of the FIH inhibitor, (NODA). The hMDMS were infected with the wild type strain or the isogenic *dotA* mutant and then CFUs were determined at 2, 24, and 48 h post-infection. *indicates statistically significant reduced CFUs at 48 h post-infection relative to that observed for wild type (unpaired *t*-test, $p < 0.01$). Data represents (mean ± *SD*, $n = 3$ infected hMDM monolayers). **(B)** Immunoblots showing specific knockdown of FIH protein expression in HEK293T cells transfected with control or FIH specific RNAi following 24 h. Cell lysates were run on SDS-PAGE, and FIH proteins were detected on a western blot probed with an anti-FIH antibody. The immunoblot was stripped and reprobed using an anti-β-tubulin antibody to show equivalent protein loading between lanes. **(C)** Representative confocal images of *L. pneumophila* replicative vacuoles in untreated HEK293T cells or cells treated with control or FIH specific RNAi at 10 h post-infection. Bacteria are labeled with anti-Lpn anti-serum (green), while the cell nucleus is stained with DAPI (blue). Images are representative of over 100 infected cells. **(D)** Replicative vacuole analysis quantitation in HEK293T cells treated with FIH RNAi. The numbers of bacteria were determined by analyzing Z-stack confocal images of infected cells, bars represent (mean ± SD, $n = 100$ LCVs) the % total of bacteria per LCV. *above the ≥3 bacteria bars indicates statistically significant decrease bacterial numbers relative to that observed for wild type (unpaired *t*-test, $p < 0.01$ and is representative of three independent experiments.

respectively, compared to untreated or control RNAi treated cells (unpaired *t*-test, $p < 0.01$) (**Figures 7A,B,D**). In addition, only 33% of LCVs co-localized with the KDEL ER marker compared to 80 and 78% in untreated and control RNAi HEK293T cells, respectively (unpaired *t*-test, $p < 0.01$) (**Figures 7C,D**). Confirmation of specific and complete FIH knockdown is shown in **Figure 5B**, as assessment of LCV trafficking and impact on intra-vacuolar replication of *L. pneumophila* was performed in parallel. Taken together, these data clearly show that FIH is partially required for Dot/Icm-dependent lysosomal evasion and ER-mediated remodeling of the LCV.

DISCUSSION

An emerging theme clearly shows that not only do translocated effectors of bacterial pathogens mimic eukaryotic protein functions, but they are also modified by various eukaryotic post-translational modification machineries (Ivanov and Roy, 2013; Kim et al., 2014). Our data show for the first time that injected effectors are modified by host FIH-dependent asparaginyl hydroxylation.

Our search for bacterial effectors that are potentially modified by host-asparaginyl hydroxylation was based upon the canonical FIH hydroxylation motif that was defined by using sequence data derived from the HIF1 and ARD hydroxylation sites (Hewitson et al., 2002; Lando et al., 2002a,b; Cockman et al., 2009). For AnkH, the motif was identified within the first ankyrin repeat and the predicted asparagine residue was modified by hydroxylation. Until recently, the biological consequence of asparaginyl hydroxylation within ARDs was unclear, however studies have now shown this modification regulates protein-protein interaction of ARD containing proteins, which subsequently affects their functions (Janke et al., 2013; Peng et al., 2014). Though AnkH contributes to intra-vacuolar replication of *L. pneumophila*, its functional role is unknown (Habyarimana et al., 2008, 2010). Further studies will be needed to determine the impact of asparaginyl hydroxylation on the ability of AnkH to mediate potential protein-protein interactions with its substrate, which may in turn affect its function. We also identified a single FIH hydroxylation motif within the ARD region of AnkB, but in contrast to AnkH, the predicted asparagine was not modified. However, we identified three additional asparaginyl hydroxylation sites on AnkB that occurred at distinct sites identified by the motif defined by sequence analysis. Nevertheless, they are located within the three-ankyrin repeat containing domain of AnkB (Wong et al., 2017). Interestingly, one of the hydroxylated residues (N126) occurs in a loop structure adjacent to a putative substrate interacting residue (Y127) (Wong et al., 2017) and therefore, we predict that alteration in local electronegative charge may enhance or interfere with substrate binding. FIH has not been shown to hydroxylate asparagine residues outside of the predicted motif (Wilkins et al., 2012) and therefore it will be interesting to determine if substrate specificity of FIH is wider than expected or that other host enzymes contribute to this post-translational modification. AnkB plays a central role for *L. pneumophila* by promoting the degradation of polyubiquitinated proteins

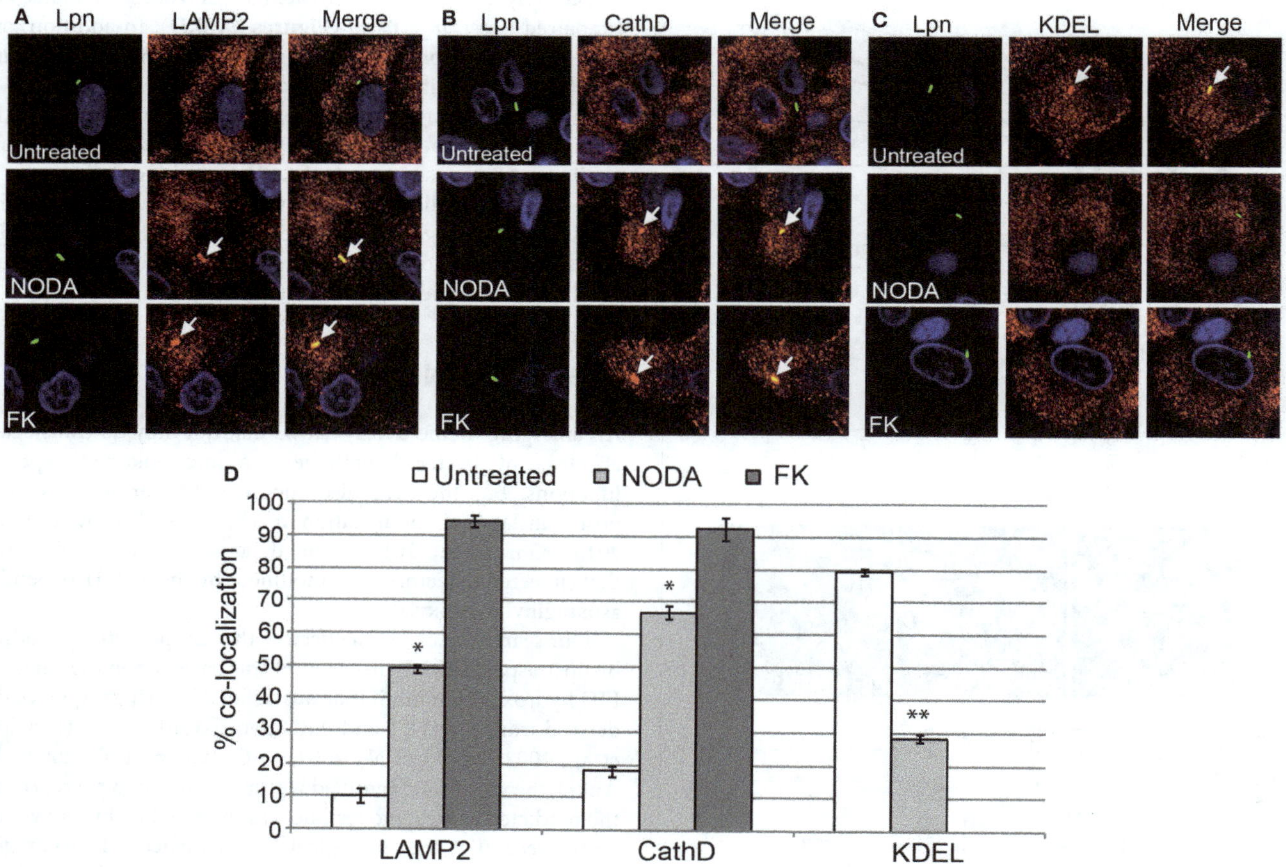

FIGURE 6 | FIH activity is required for biogenesis of the LCV. hMDMs untreated or pretreated with the FIH inhibitor, NODA, were infected with live or formalin-killed (FK) wild type *L. pneumophila* for 2 h. Representative confocal microscopy images of infected cells are shown indicating % co-localization of the LCV in hMDMS or HEK293T cells with **(A)** LAMP2, **(B)** CathD, and **(C)** KDEL. Bacteria were labeled with anti-Lpn antiserum (green) and LAMP2, CathD, and KDEL was labeled with specific antibodies (red) and analyzed by confocal microscopy. The arrows indicate co-localization of LAMP2, CathD, and KDEL with the LCV. **(D)** Quantification of the frequency of co-localization of LAMP2, CathD or KDEL with LCVs. Data represents (mean ± SD, $n = 100$ LCVs). *represents statistically significant increase in NODA-treated cells vs. untreated cells (unpaired *t*-test, $p < 0.01$). **represents statistically significant decrease in NODA-treated cells vs. untreated cells (unpaired *t*-test, $p < 0.01$). The data is representative of three independent experiments.

which allows this organism to access essential amino acids that are used for both energy and a carbon source (Price et al., 2009, 2011). Substitution of the three hydroxylated asparagine residues significantly impacts the ability of AnkB to recruit polyubiquitinated proteins to the LCV and concomitantly fails to restore intra-vacuolar replication of an *ankB* mutant strain of *L. pneumophila*. Furthermore, blocking host FIH activity results in a similar phenotype to the AnkB substitutions, and taken together suggests that asparaginyl hydroxylation of AnkB contributes to the function of this effector.

Blocking FIH activity results in a dose-dependent inhibition of intra-vacuolar replication of *L. pneumophila*. Interestingly however, only ~55% of LCVs trafficked to a lysosomal compartment, indicating that the FIH-mediated block in intra-vacuolar replication of *L. pneumophila* has both lysosomal evasion-independent and -dependent mechanisms. Both AnkB and AnkH are needed for intra-vacuolar proliferation of *L. pneumophila* but neither impacts the normal trafficking and biogenesis of the LCV (Al-Khodor et al., 2008; Habyarimana et al., 2008, 2010; Price et al., 2009, 2010a,b, 2011; Lomma et al., 2010). However, since both AnkB and AnkH are needed for intra-vacuolar replication, it is more likely that change in the asparaginyl hydroxylation status of these injected effectors contributes to the intra-vacuolar growth defect in a lysosomal evasion-independent manner. Indeed, we observed that substituting the hydroxylated residues in AnkB significantly impairs its function which correlates to the data obtained using chemical inhibition and RNAi depletion of FIH activity. The lysosomal evasion-dependent mechanism may involve both injected effectors, though no single injected effector to date has been shown to be required for the ability of the LCV to evade the lysosomes (de Felipe et al., 2008; Isberg et al., 2009; Zhu et al., 2011). A number of host proteins involved in vesicular trafficking have been identified on the LCV (Urwyler et al., 2009; Hoffmann et al., 2014; Bruckert and Abu Kwaik, 2015), and a preliminary bioinformatic analysis of the human

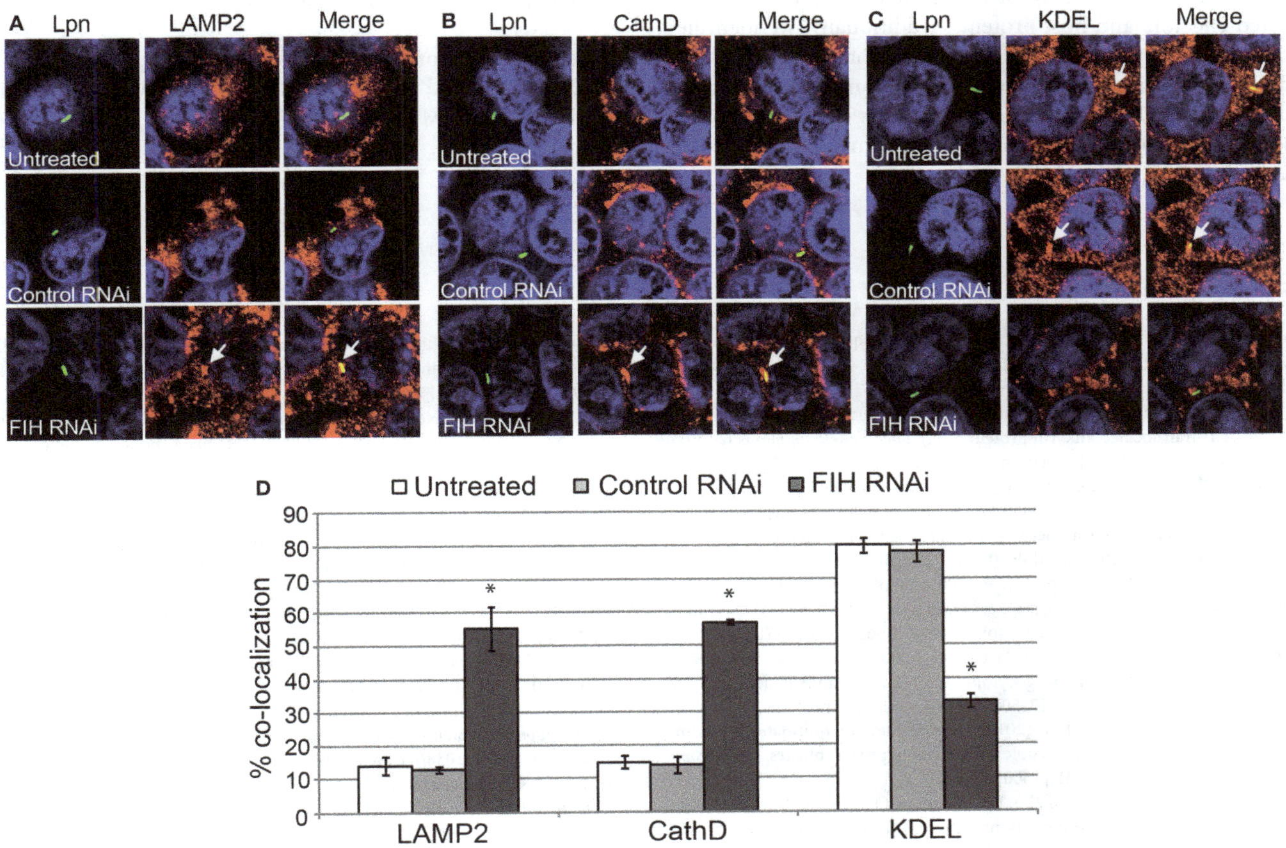

FIGURE 7 | FIH protein is required for biogenesis of the LCV. HEK293T cells were transfected with control or FIH-specific RNAi and infected with wild type *L. pneumophila* for 2 h. Representative confocal microscopy images of infected cells are shown indicating % co-localization of the LCV in hMDMS or HEK293T cells with **(A)** LAMP2, **(B)** CathD, and **(C)** KDEL. Bacteria were labeled with anti-Lpn antiserum (green) and LAMP2, CathD, and KDEL was labeled with specific antibodies (red) and analyzed by confocal microscopy. The arrows indicate co-localization of LAMP2, CathD, and KDEL with the LCV. **(D)** Quantification of the frequency of co-localization of LAMP2, CathD, or KDEL with LCVs. Data represents (mean ± SD, $n = 100$ LCVs) *represents statistical significance in FIH RNAi-treated cells vs. untreated or control RNAi-treated cells (unpaired *t*-test, $p < 0.01$). The data is representative of three independent experiments.

genome suggests that the FIH hydroxylation motif is found in a number of these including AP1G1, CopB2, ATP6AP1, CLTC, and PICALM (data not shown). It is possible that alterations in the hydroxylation status of these proteins and others may impact trafficking of the LCV, misdirecting it to the lysosome. It is clear that blocking FIH has a broad effect both on injected *L. pneumophila* effectors and host proteins, culminating in a complete block in *L. pneumophila* intra-vacuolar replication.

A central documented role of FIH is in regulation of HIF1, the key transcription factor that modulates expression of numerous genes involved in oxygen homeostasis, metabolism and immune function (Webb et al., 2009). It is possible that the LCV recruits FIH through interaction with Mint3 and MT1-MMP, which are localized to the LCV, analogous to FIH Golgi-localization observed in macrophages (Sakamoto and Seiki, 2009, 2010). In macrophages, membrane-associated FIH is inactive, at least in terms of HIF1 hydroxylation activity, but through its binding to Mint3 it enables HIF1 to promote transcription of glycolytic genes that are needed by the macrophage to generate ATP (Sakamoto and Seiki, 2009, 2010). *L. pneumophila* uses host amino acids as the primary source of carbon and energy by AnkB-dependent proteasomal degradation, but exogenous pyruvate alone can compensate for proteasomal degradation to enable intra-vacuolar replication of *L. pneumophila* (Price et al., 2011). This indicates that host pyruvate is an additional metabolite scavenged by intra-vacuolar *L. pneumophila*. Therefore, a consequence of FIH recruitment to the LCV may be increased HIF1 activity, which will ultimately increase availability of pyruvate that the bacteria can scavenge from the intracellular environment to use as an energy source and building block of macromolecules.

The biological consequence of FIH-dependent asparaginyl hydroxylation of many ARD-containing proteins remains unclear, but is likely to impact function and/or sub-cellular localization of these proteins in the host cell. For *L. pneumophila*, host FIH is essential for intra-vacuolar proliferation, and this involves lysosomal evasion-dependent and -independent mechanisms, which likely involves hydroxylation of both injected effectors and host proteins. Additionally, we identified effectors from other bacterial pathogens that harbor type III and IV secretion systems that potentially undergo asparagine hydroxylation (Table S1), suggesting this post-translational modification is a potential general paradigm in host-pathogen interaction.

AUTHOR CONTRIBUTIONS

Conceptualization, CP, OSF, and YK; Investigation, CP, MM, SJ, AB, JV, and NA; Resources, ML and OSF; Writing—Review & Editing, CP, MM, SJ, AB, JV, ML, NA, OSF, and YK; Funding Acquisition, YK.

REFERENCES

Abu Kwaik, Y., and Bumann, D. (2013). Microbial quest for food *in vivo*: 'nutritional virulence' as an emerging paradigm. *Cell. Microbiol.* 15, 882–890. doi: 10.1111/cmi.12138

Al-Khodor, S., Price, C. T., Habyarimana, F., Kalia, A., and Abu Kwaik, Y. (2008). A Dot/Icm-translocated ankyrin protein of *Legionella pneumophila* is required for intracellular proliferation within human macrophages and protozoa. *Mol. Microbiol.* 70, 908–923. doi: 10.1111/j.1365-2958.2008.06453.x

Al-Quadan, T., Price, C. T., and Abu Kwaik, Y. (2012). Exploitation of evolutionarily conserved amoeba and mammalian processes by Legionella. *Trends Microbiol.* 20, 299–306. doi: 10.1016/j.tim.2012.03.005

Bardill, J. P., Miller, J. L., and Vogel, J. P. (2005). IcmS-dependent translocation of SdeA into macrophages by the *Legionella pneumophila* type IV secretion system. *Mol. Microbiol.* 56, 90–103. doi: 10.1111/j.1365-2958.2005.04539.x

Beyer, A. R., Truchan, H. K., May, L. J., Walker, N. J., Borjesson, D. L., and Carlyon, J. A. (2014). The anaplasma phagocytophilum effector AmpA hijacks host cell SUMOylation. *Cell. Microbiol.* 17, 504–519. doi: 10.1111/cmi.12380

Bruckert, W. M., and Abu Kwaik, Y. (2015). Complete and ubiquitinated proteome of the legionella-containing vacuole within human macrophages. *J. Proteome Res.* 14, 236–248. doi: 10.1021/pr500765x

Chen, J., de Felipe, K. S., Clarke, M., Lu, H., Anderson, O. R., Segal, G., et al. (2004). Legionella effectors that promote nonlytic release from protozoa. *Science* 303, 1358–1361. doi: 10.1126/science.1094226

Chen, J., Reyes, M., Clarke, M., and Shuman, H. A. (2007). Host cell-dependent secretion and translocation of the LepA and LepB effectors of *Legionella pneumophila*. *Cell. Microbiol.* 9, 1660–1671. doi: 10.1111/j.1462-5822.2007.00899.x

Cockman, M. E., Webb, J. D., and Ratcliffe, P. J. (2009). FIH-dependent asparaginyl hydroxylation of ankyrin repeat domain-containing proteins. *Ann. N. Y. Acad. Sci.* 1177, 9–18. doi: 10.1111/j.1749-6632.2009.05042.x

Cox, J., Michalski, A., and Mann, M. (2011). Software lock mass by two-dimensional minimization of peptide mass errors. *J. Am. Soc. Mass Spectrom.* 22, 1373–1380. doi: 10.1007/s13361-011-0142-8

de Felipe, K. S., Glover, R. T., Charpentier, X., Anderson, O. R., Reyes, M., Pericone, C. D., et al. (2008). *Legionella eukaryotic*-like type IV substrates interfere with organelle trafficking. *PLoS Pathog.* 4:e1000117. doi: 10.1371/journal.ppat.1000117

Dorer, M. S., Kirton, D., Bader, J. S., and Isberg, R. R. (2006). RNA interference analysis of Legionella in Drosophila cells: exploitation of early secretory apparatus dynamics. *PLoS Pathog.* 2:e34. doi: 10.1371/journal.ppat.0020034

Gaus, K., Hentschke, M., Czymmeck, N., Novikova, L., Trulzsch, K., Valentin-Weigand, P., et al. (2011). Destabilization of YopE by the ubiquitin-proteasome pathway fine-tunes Yop delivery into host cells and facilitates systemic spread of *Yersinia enterocolitica* in host lymphoid tissue. *Infect. Immun.* 79, 1166–1175. doi: 10.1128/IAI.00694-10

Habyarimana, F., Al-Khodor, S., Kalia, A., Graham, J. E., Price, C. T., Garcia, M. T., et al. (2008). Role for the Ankyrin eukaryotic-like genes of *Legionella pneumophila* in parasitism of protozoan hosts and human macrophages. *Environ. Microbiol.* 10, 1460–1474. doi: 10.1111/j.1462-2920.2007.01560.x

Habyarimana, F., Price, C. T., Santic, M., Al-Khodor, S., and Kwaik, Y. A. (2010). Molecular characterization of the Dot/Icm-translocated AnkH and AnkJ eukaryotic-like effectors of *Legionella pneumophila*. *Infect. Immun.* 78, 1123–1134. doi: 10.1128/IAI.00913-09

Hayashi, T., Morohashi, H., and Hatakeyama, M. (2013). Bacterial EPIYA effectors–where do they come from? What are they? Where are they going? *Cell. Microbiol.* 15, 377–385. doi: 10.1111/cmi.12040

Hewitson, K. S., McNeill, L. A., Riordan, M. V., Tian, Y. M., Bullock, A. N., Welford, R. W., et al. (2002). Hypoxia-inducible factor (HIF) asparagine hydroxylase is identical to factor inhibiting HIF (FIH) and is related to the cupin structural family. *J. Biol. Chem.* 277, 26351–26355. doi: 10.1074/jbc.C200273200

Hicks, S. W., Charron, G., Hang, H. C., and Galan, J. E. (2011). Subcellular targeting of *Salmonella virulence* proteins by host-mediated S-palmitoylation. *Cell Host Microbe* 10, 9–20. doi: 10.1016/j.chom.2011.06.003

Hoffmann, C., Finsel, I., Otto, A., Pfaffinger, G., Rothmeier, E., Hecker, M., et al. (2014). Functional analysis of novel Rab GTPases identified in the proteome of purified Legionella-containing vacuoles from macrophages. *Cell. Microbiol.* 16, 1034–1052. doi: 10.1111/cmi.12256

Ingmundson, A., Delprato, A., Lambright, D. G., and Roy, C. R. (2007). *Legionella pneumophila* proteins that regulate Rab1 membrane cycling. *Nature* 450, 365–369. doi: 10.1038/nature06336

Isberg, R. R., O'Connor, T. J., and Heidtman, M. (2009). The *Legionella pneumophila* replication vacuole: making a cosy niche inside host cells. *Nat. Rev. Microbiol.* 7, 13–24. doi: 10.1038/nrmicro1967

Ivanov, S. S., Charron, G., Hang, H. C., and Roy, C. R. (2010). Lipidation by the host prenyltransferase machinery facilitates membrane localization of *Legionella pneumophila* effector proteins. *J. Biol. Chem.* 285, 34686–34698. doi: 10.1074/jbc.M110.170746

Ivanov, S. S., and Roy, C. (2013). Host lipidation: a mechanism for spatial regulation of Legionella effectors. *Curr. Top. Microbiol. Immunol.* 376, 135–154. doi: 10.1007/82_2013_344

Janke, K., Brockmeier, U., Kuhlmann, K., Eisenacher, M., Nolde, J., Meyer, H. E., et al. (2013). Factor inhibiting HIF-1 (FIH-1) modulates protein interactions of apoptosis-stimulating p53 binding protein 2 (ASPP2). *J. Cell Sci.* 126, 2629–2640. doi: 10.1242/jcs.117564

Kim, M., Otsubo, R., Morikawa, H., Nishide, A., Takagi, K., Sasakawa, C., et al. (2014). Bacterial effectors and their functions in the ubiquitin-proteasome system: insight from the modes of substrate recognition. *Cells* 3, 848–864. doi: 10.3390/cells3030848

Kubori, T., and Galan, J. E. (2003). Temporal regulation of salmonella virulence effector function by proteasome-dependent protein degradation. *Cell* 115, 333–342. doi: 10.1016/S0092-8674(03)00849-3

Lando, D., Peet, D. J., Gorman, J. J., Whelan, D. A., Whitelaw, M. L., and Bruick, R. K. (2002a). FIH-1 is an asparaginyl hydroxylase enzyme that regulates the transcriptional activity of hypoxia-inducible factor. *Genes Dev.* 16, 1466–1471. doi: 10.1101/gad.991402

Lando, D., Peet, D. J., Whelan, D. A., Gorman, J. J., and Whitelaw, M. L. (2002b). Asparagine hydroxylation of the HIF transactivation domain a hypoxic switch. *Science* 295, 858–861. doi: 10.1126/science.1068592

Lewis, J. D., Abada, W., Ma, W., Guttman, D. S., and Desveaux, D. (2008). The HopZ family of Pseudomonas syringae type III effectors require myristoylation for virulence and avirulence functions in *Arabidopsis thaliana*. *J. Bacteriol.* 190, 2880–2891. doi: 10.1128/JB.01702-07

Loenarz, C., and Schofield, C. J. (2011). Physiological and biochemical aspects of hydroxylations and demethylations catalyzed by human 2-oxoglutarate oxygenases. *Trends Biochem. Sci.* 36, 7–18. doi: 10.1016/j.tibs.2010.07.002

Lomma, M., Dervins-Ravault, D., Rolando, M., Nora, T., Newton, H. J., Sansom, F. M., et al. (2010). The *Legionella pneumophila* F-box protein Lpp2082 (AnkB) modulates ubiquitination of the host protein parvin B and promotes intracellular replication. *Cell. Microbiol.* 12, 1272–1291. doi: 10.1111/j.1462-5822.2010.01467.x

Luo, Z. Q., and Isberg, R. R. (2004). Multiple substrates of the *Legionella pneumophila* Dot/Icm system identified by interbacterial protein transfer. *Proc. Natl. Acad. Sci. U.S.A.* 101, 841–846. doi: 10.1073/pnas.0304916101

Pan, X., Luhrmann, A., Satoh, A., Laskowski-Arce, M. A., and Roy, C. R. (2008). Ankyrin repeat proteins comprise a diverse family of bacterial type IV effectors. *Science* 320, 1651–1654. doi: 10.1126/science.1158160

Patel, J. C., Hueffer, K., Lam, T. T., and Galan, J. E. (2009). Diversification of a Salmonella virulence protein function by ubiquitin-dependent differential localization. *Cell* 137, 283–294. doi: 10.1016/j.cell.2009.01.056

Peng, H., Kaplan, N., Yang, W., Getsios, S., and Lavker, R. M. (2014). FIH-1 disrupts an LRRK1/EGFR complex to positively regulate keratinocyte migration. *Am. J. Pathol.* 184, 3262–3271. doi: 10.1016/j.ajpath.2014.08.014

Petkowski, J. J., Bonsignore, L. A., Tooley, J. G., Wilkey, D. W., Merchant, M. L., Macara, I. G., et al. (2013). NRMT2 is an N-terminal monomethylase that primes for its homologue NRMT1. *Biochem. J.* 456, 453–462. doi: 10.1042/BJ20131163

Pfaffl, M. W., Horgan, G. W., and Dempfle, L. (2002). Relative expression software tool (REST) for group-wise comparison and statistical analysis of relative expression results in real-time PCR. *Nucleic Acids Res.* 30:e36. doi: 10.1093/nar/30.9.e36

Price, C. T., Al-Khodor, S., Al-Quadan, T., and Abu Kwaik, Y. (2010a). Indispensable role for the eukaryotic-like ankyrin domains of the ankyrin B effector of *Legionella pneumophila* within macrophages and amoebae. *Infect. Immun.* 78, 2079–2088. doi: 10.1128/IAI.01450-09

Price, C. T., Al-Khodor, S., Al-Quadan, T., Santic, M., Habyarimana, F., Kalia, A., et al. (2009). Molecular mimicry by an F-box effector of *Legionella pneumophila* hijacks a conserved polyubiquitination machinery within macrophages and protozoa. *PLoS Pathog.* 5:e1000704. doi: 10.1371/journal.ppat.1000704

Price, C. T., Al-Quadan, T., Santic, M., Jones, S. C., and Abu Kwaik, Y. (2010b). Exploitation of conserved eukaryotic host cell farnesylation machinery by an F-box effector of *Legionella pneumophila*. *J. Exp. Med.* 207, 1713–1726. doi: 10.1084/jem.20100771

Price, C. T., Al-Quadan, T., Santic, M., Rosenshine, I., and Abu Kwaik, Y. (2011). Host proteasomal degradation generates amino acids essential for intracellular bacterial growth. *Science* 334, 1553–1557. doi: 10.1126/science.1212868

Price, C. T., Richards, A. M., Von Dwingelo, J. E., Samara, H. A., and Abu Kwaik, Y. (2014). Amoeba host-Legionella synchronization of amino acid auxotrophy and its role in bacterial adaptation and pathogenic evolution. *Environ. Microbiol.* 16, 350–358. doi: 10.1111/1462-2920.12290

Reinicke, A. T., Hutchinson, J. L., Magee, A. I., Mastroeni, P., Trowsdale, J., and Kelly, A. P. (2005). A *Salmonella typhimurium* effector protein SifA is modified by host cell prenylation and S-acylation machinery. *J. Biol. Chem.* 280, 14620–14627. doi: 10.1074/jbc.M500076200

Rodriguez, J., Pilkington, R., Garcia Munoz, A., Nguyen, L. K., Rauch, N., Kennedy, S., et al. (2016). Substrate-trapped interactors of PHD3 and FIH cluster in distinct signaling pathways. *Cell Rep.* 14, 2745–2760. doi: 10.1016/j.celrep.2016.02.043

Sakamoto, T., and Seiki, M. (2009). Mint3 enhances the activity of hypoxia-inducible factor-1 (HIF-1) in macrophages by suppressing the activity of factor inhibiting HIF-1. *J. Biol. Chem.* 284, 30350–30359. doi: 10.1074/jbc.M109.019216

Sakamoto, T., and Seiki, M. (2010). A membrane protease regulates energy production in macrophages by activating hypoxia-inducible factor-1 via a non-proteolytic mechanism. *J. Biol. Chem.* 285, 29951–29964. doi: 10.1074/jbc.M110.132704

Scholz, C. C., Rodriguez, J., Pickel, C., Burr, S., Fabrizio, J. A., Nolan, K. A., et al. (2016). FIH regulates cellular metabolism through hydroxylation of the deubiquitinase OTUB1. *PLoS Biol.* 14:e1002347. doi: 10.1371/journal.pbio.1002347

Shevchenko, A., Jensen, O. N., Podtelejnikov, A. V., Sagliocco, F., Wilm, M., Vorm, O., et al. (1996). Linking genome and proteome by mass spectrometry: large-scale identification of yeast proteins from two dimensional gels. *Proc. Natl. Acad. Sci. U.S.A.* 93, 14440–14445. doi: 10.1073/pnas.93.25.14440

Swaney, D. L., McAlister, G. C., and Coon, J. J. (2008). Decision tree-driven tandem mass spectrometry for shotgun proteomics. *Nat. Methods* 5, 959–964. doi: 10.1038/nmeth.1260

Tan, Y., Arnold, R. J., and Luo, Z. Q. (2011). *Legionella pneumophila* regulates the small GTPase Rab1 activity by reversible phosphorylcholination. *Proc. Natl. Acad. Sci. U.S.A.* 108, 21212–21217. doi: 10.1073/pnas.1114023109

Urwyler, S., Nyfeler, Y., Ragaz, C., Lee, H., Mueller, L. N., Aebersold, R., et al. (2009). Proteome analysis of Legionella vacuoles purified by magnetic immunoseparation reveals secretory and endosomal GTPases. *Traffic* 10, 76–87. doi: 10.1111/j.1600-0854.2008.00851.x

Webb, J. D., Coleman, M. L., and Pugh, C. W. (2009). Hypoxia, hypoxia-inducible factors (HIF), HIF hydroxylases and oxygen sensing. *Cell. Mol. Life Sci.* 66, 3539–3554. doi: 10.1007/s00018-009-0147-7

Wilkins, S. E., Karttunen, S., Hampton-Smith, R. J., Murchland, I., Chapman-Smith, A., and Peet, D. J. (2012). Factor inhibiting HIF (FIH) recognizes distinct molecular features within hypoxia-inducible factor-alpha (HIF-alpha) versus ankyrin repeat substrates. *J. Biol. Chem.* 287, 8769–8781. doi: 10.1074/jbc.M111.294678

Wong, K., Perpich, J. D., Kozlov, G., Cygler, M., Abu Kwaik, Y., and Gehring, K. (2017). Structural mimicry by a bacterial F-box effector hijacks the host ubiquitin-proteasome system. *Structure* 25, 376–383. doi: 10.1016/j.str.2016.12.015

Zhu, W., Banga, S., Tan, Y., Zheng, C., Stephenson, R., Gately, J., et al. (2011). Comprehensive identification of protein substrates of the Dot/Icm Type IV transporter of *Legionella pneumophila*. *PLoS ONE* 6:e17638. doi: 10.1371/journal.pone.0017638

18

From Many Hosts, One Accidental Pathogen: The Diverse Protozoan Hosts of *Legionella*

David K. Boamah[1], Guangqi Zhou[2], Alexander W. Ensminger[2,3,4] and Tamara J. O'Connor[1]**

[1] Department of Biological Chemistry, Johns Hopkins University School of Medicine, Baltimore, MD, United States, [2] Department of Biochemistry, University of Toronto, Toronto, ON, Canada, [3] Department of Molecular Genetics, University of Toronto, Toronto, ON, Canada, [4] Public Health Ontario, Toronto, ON, Canada

Edited by:
Hayley J. Newton,
University of Melbourne, Australia

Reviewed by:
Ascel Samba-Louaka,
University of Poitiers, France
Ombeline Rossier,
Université Paris-Sud, France

***Correspondence:**
Tamara J. O'Connor
toconno7@jhmi.edu
Alexander W. Ensminger
alex.ensminger@utoronto.ca

The 1976 outbreak of Legionnaires' disease led to the discovery of the intracellular bacterial pathogen *Legionella pneumophila*. Given their impact on human health, *Legionella* species and the mechanisms responsible for their replication within host cells are often studied in alveolar macrophages, the primary human cell type associated with disease. Despite the potential severity of individual cases of disease, *Legionella* are not spread from person-to-person. Thus, from the pathogen's perspective, interactions with human cells are accidents of time and space—evolutionary dead ends with no impact on *Legionella*'s long-term survival or pathogenic trajectory. To understand *Legionella* as a pathogen is to understand its interaction with its natural hosts: the polyphyletic protozoa, a group of unicellular eukaryotes with a staggering amount of evolutionary diversity. While much remains to be understood about these enigmatic hosts, we summarize the current state of knowledge concerning *Legionella*'s natural host range, the diversity of *Legionella*-protozoa interactions, the factors influencing these interactions, the importance of avoiding the generalization of protozoan-bacterial interactions based on a limited number of model hosts and the central role of protozoa to the biology, evolution, and persistence of *Legionella* in the environment.

Keywords: *Legionella*, amoebae, protozoa, host range, environment, *Acanthamoebae*, *Hartmannella*, *Naegleria*

PREDATOR VS. PREY: *Legionella* AND ITS NATURAL PROTOZOAN HOSTS

In the environment, bacteria are targets of predation by grazing protozoa (Hahn and Höfle, 2001; Molmeret et al., 2005). In response to predation, many bacteria have developed strategies to either avoid predation or survive, and in some cases, replicate within protozoa. As bacteria are destined to encounter a large number of protozoa species in nature, their fitness will be determined by the breadth and diversity of protozoa within which they are able to grow. Though many types of bacteria are able to replicate within protozoa (Greub and Raoult, 2004), this behavior is best characterized in the bacterial pathogen *Legionella,* in particular *Legionella pneumophila*, which will be the major focus of this review.

L. pneumophila IN THE ENVIRONMENT

L. pneumophila is ubiquitous in nature (Fliermans, 1996; van Heijnsbergen et al., 2015). While various species of *Legionella* have been isolated from soil and marine environments, freshwater systems serve as the major reservoirs of *L. pneumophila* (Fliermans, 1996; van Heijnsbergen et al., 2015). *L. pneumophila* can exist in a planktonic form however, it is more often found within mixed community biofilms (Mampel et al., 2006). *L. pneumophila* intercalates into existing biofilms (Lau and Ashbolt, 2009; Stewart et al., 2012) where it acquires nutrients by forming synergistic relationships with other members of the biofilm (Tison et al., 1980; Pope et al., 1982; Bohach and Snyder, 1983; Wadowsky and Yee, 1983; Stout et al., 1986; Stewart et al., 2012; Koide et al., 2014). *L. pneumophila* is also capable of surviving in nutrient-poor conditions by necrotrophic growth on dead cell masses (Temmerman et al., 2006). Although, its interactions with other bacteria promote *L. pneumophila* survival in oligotrophic environments, intracellular growth within protozoa is likely the predominant mechanism of *L. pneumophila* proliferation in its natural habitat (Rowbotham, 1980).

THE IMPACT OF NATURAL HOSTS ON *Legionella* PERSISTENCE IN THE ENVIRONMENT AND PATHOGENESIS

Protozoa function as natural reservoirs of *L. pneumophila* and promote disease in humans. The intracellular environment of the host cell protects *L. pneumophila* from harsh environmental conditions while providing a nutrient rich replicative niche (Greub and Raoult, 2004; Abdel-Nour et al., 2013). The ability of *L. pneumophila* to survive within amoebae also protects the bacteria from killing by water disinfection procedures (Plouffe et al., 1983; King et al., 1988; Kilvington and Price, 1990; Biurrun et al., 1999; Storey et al., 2004; Bouyer et al., 2007; García et al., 2008; Cervero-Aragó et al., 2014, 2015), a reciprocal relationship that also enhances survival of the host (García et al., 2007). As a consequence, *L. pneumophila* are commonly found in man-made potable water supply and distribution systems (Ikedo and Yabuuchi, 1986; Breiman et al., 1990; Yamamoto et al., 1992; Fields et al., 2002; Lasheras et al., 2006; Brousseau et al., 2013; Thomas et al., 2014). Although, there is one reported case of probable human-to-human transmission of *Legionella* (Correia et al., 2016), the vast majority of evidence suggests a non-communicable disease. Instead, human exposure predominantly occurs through the inhalation of contaminated water aerosols (Fields, 1996), which can lead to pneumonic respiratory disease. *L. pneumophila* passaged through amoebae are more virulent in animal models of infection compared to bacteria grown in broth culture (Cirillo et al., 1994, 1999; Barker et al., 1995; Brieland et al., 1996; Garduño et al., 2002). The earliest description of *L. pneumophila*'s interaction with amoebae even proposed that an important route of human infection may be the inhalation of the pathogen in an amoebal-encapsulated state (Rowbotham, 1980). Thus, the interaction of *L. pneumophila* with protozoa is a critical determinant in both the persistence of *Legionella* in environmental and man-made reservoirs, and the incidence and severity of disease.

THE BROAD HOST RANGE OF *L. pneumophila*

Many bacterial pathogens become highly specialized for growth in one or a small subset of hosts but few are able to grow in multiple hosts. Host jumping has been observed for some pathogens but often comes at a price, the inability to grow in the previous host (Ma et al., 2006). In contrast, *L. pneumophila* exhibits an extensive host range replicating within a diverse array of protozoan hosts that span multiple phyla, from Amoebozoa (amoebae) to Percolozoa (excavates) to Ciliophora (ciliated protozoa) (Rowbotham, 1980; Fields, 1996). The ability to maintain such a broad host range is due to the assembly of a large cohort of genes that allow *L. pneumophila* to adapt to variations between hosts (O'Connor et al., 2011). Moreover, the ability to continually evolve and alter the composition of its virulence gene repertoire allows *L. pneumophila* to adapt to shifts in protozoan populations in their natural habitats (O'Connor et al., 2011). Since the discovery that *L. pneumophila* can survive and replicate within free-living amoeba (Rowbotham, 1980), the relationship between *L. pneumophila* and its protozoa hosts has garnered significant attention, largely due to the important role of protozoa in the epidemiology of this pathogen. In this review, we expand on the early works of Rowbotham and Fields (Rowbotham, 1980, 1986; Fields, 1996) to summarize the current knowledge of the host range of *L. pneumophila* in environmental reservoirs and the factors that impact the outcome of *Legionella*-protozoa interactions.

THE DIFFERENT FATES OF *L. pneumophila* WITHIN PROTOZOAN HOSTS

While *L. pneumophila* has an extensive host range, the fate of the bacterium once it enters the host cell can vary greatly. Several protozoa are able to efficiently deliver *L. pneumophila* to the lysosome for degradation, resulting in the death of the bacterium (Amaro et al., 2015). *L. pneumophila* predation by protozoa does not seem to be restricted to one particular group. While members of the Cercozoa phylum seem to be especially adept at digesting *L. pneumophila* (Amaro et al., 2015), distantly related members of the Amoebozoa phylum (*Cashia limacoides*, *Vannella platypodia*, and *Vexillifera bacillipedes*) are also efficient at killing *L. pneumophila* (Rowbotham, 1986). In contrast, many protozoa serve as hosts for *L. pneumophila* replication. In these cases, the *Legionella*-protozoa interaction is detrimental to the host: the bacteria multiply to high numbers and then kill the host as they exit the cell (Rowbotham, 1983). Alternatively, *L. pneumophila* can be toxic to the host in the absence of replication, a protist version of food-poisoning (Amaro et al., 2015). *L. pneumophila* within amoebae has been shown to inhibit both amoebae proliferation (Mengue et al., 2016) and chemotactic motility (Simon et al., 2014). The fates of the two organisms are not solely defined by this "it's you or me" relationship, as a number of intermediate outcomes have been observed. In response to extreme stress, amoebae undergo encystation, transforming into a dormant, highly resistant cyst form. While encystation restricts bacterial replication (Rowbotham, 1986; Ohno et al., 2008), *L. pneumophila* is able to survive the encystation process until more favorable conditions arise (Kilvington and Price, 1990; Greub and Raoult, 2003). Similarly, for some *Legionella*-protozoa pairs, *L. pneumophila* is resistant to grazing by the protozoan and thus survives within the host cell but fails to replicate (Smith-Somerville et al., 1991). Alternatively, *L. pneumophila* can be packaged into multi-membrane vesicles that are distinct from the replication vacuole and expelled into the extracellular environment (Rowbotham, 1983; Berk et al., 1998; Hojo et al., 2012; Amaro et al., 2015). The release of *Legionella*-containing pellets has been observed in both the ciliated protozoa *Tetrahymena* spp. (Faulkner et al., 2008; Hojo et al., 2012) and the amoebal hosts *Acanthamoeba castellanii* and *Acanthamoeba astronyxis* (Bouyer et al., 2007; Amaro et al., 2015), and does not appear to coincide with bacterial replication. Whether this process is driven by the bacterium or the host is still unclear. The pellet compartment can protect *L. pneumophila* from environmental stress (Bouyer et al., 2007; Koubar et al., 2011) which would be beneficial during its transition between host cells and thus a potential mechanism to ensure its survival. Consistent with this idea, a functional Type IVb secretion system, a major *L. pneumophila* virulence factor required for lysosome avoidance and intracellular replication, appears to be important for the release of *L. pneumophila* in pellets (Berk et al., 2008). Alternatively, the inability to digest the bacteria may simply trigger a host response that involves bacterial expulsion, as a similar phenomenon is observed with non-pathogenic *Escherichia coli, Bacillus subtilis,* and *Mycobacterium luteus* (Hojo et al., 2012; Denoncourt et al., 2014). Whether *L. pneumophila* resists predation or is expelled in pellets, the host is considered to be only partially restrictive due to the survival of *L. pneumophila* and its potential to transition to other host cells. Indeed, one might speculate that such intermediate host-bacterial interactions (resistance to protozoan predation in the absence of replication) might resemble the first evolutionary step toward becoming an intracellular pathogen.

METHODS FOR DEFINING PROTOZOAN HOSTS OF *Legionella*

Protozoan hosts of *Legionella* are defined by two main techniques: co-culture and co-isolation. When combined with microscopy, co-culture techniques allow for the direct visualization of *Legionella* within host cells, and by analyzing infected cells over time, bacterial replication within a particular host provides direct experimental evidence of *Legionella* survival and replication. When combined with plating assays to monitor bacterial numbers, co-culture methods allow bacterial growth rates, maximum growth and the impact of bacterial dose and various external conditions on the interaction to be analyzed. However, while *Legionella* may be able to replicate in a given host under specific laboratory conditions, the experimental system may not reflect conditions encountered in the environment and thus, biologically relevant interactions that commonly occur in nature. Co-isolation studies attempt to address this issue by examining the co-existence of protozoa and *Legionella* in environmental samples. In rare cases, protozoa harboring *Legionella* have been isolated from environmental samples providing direct evidence of their interaction in the environment (Thomas et al., 2006; Hsu et al., 2011; Kao et al., 2013). More commonly, *Legionella* are identified by 16S sequencing of DNA extracts from bacteria isolated by *Legionella*-selective culture methods on bacteriological medium (Salloum et al., 2002; Sheehan et al., 2005) or enrichment through co-culture of environmental samples with amoebae (Pagnier et al., 2008). Protozoa may be identified microscopically by fluorescence *in situ* hybridization (FISH) or the morphological appearance of trophozoites (Jacquier et al., 2013; Muchesa et al., 2014), or by 18S sequencing of DNA extracts following an amoebal enrichment step in which individual isolates are cultured on lawns of bacteria permissive to amoebal grazing (Greub and Raoult, 2004; Delafont et al., 2013; Muchesa et al., 2014). Thus, while most co-isolation studies do not provide direct evidence of *Legionella* growth within the protozoa identified, they can be used to predict environmentally relevant interactions, to substantiate experimental findings from co-culture techniques and are likely to implicate new protozoan species as potential hosts of *Legionella.*

EXPERIMENTALLY DEFINED PROTOZOAN HOSTS OF *L. pneumophila*

The initial discovery that *L. pneumophila* is capable of surviving and replicating in protozoa fostered a number of independent investigations to examine the host range of this bacterium (**Table 1**). Co-culture methods in combination with various microscopy techniques demonstrated growth of *L. pneumophila* in diverse protozoan hosts encompassing several species of *Acanthamoeba* (*A. castellanii*, *Acanthamoeba polyphaga,* and *Acanthamoeba palestinensis*), *Hartmannella* (*Vermamoeba vermiformis,* formerly *Hartmannella vermiformis* and *Hartmannella cantabridiensis*) and *Naegleria* (*Naegleria gruberi, Naegleria lovaniensis,* and *Naegleria jadini*) as well as *Tetrahymena pyrofomis, Echinamoeba exudans,* and *Tetramitus jugosus* (formerly *Vahlkampfia jugosus*) (Rowbotham, 1980, 1986; Tyndall and Domingue, 1982; Anand et al., 1983; Barbaree et al., 1986). While the list of hosts was dominated by three particular genera (*Acanthamoeba, Hartmannella,* and *Naegleria*), collectively it represented three different phyla Amoebozoa, Ciliophora, and Percolozoa and amongst them, four distantly related classes of protozoa, Discosea (*Acanthamoebae*), Tubulinea (*Echinamoeba* and *Hartmannella*), Heterolobosea (*Naegleria* and *Tetramitus*), and Oligohymenophorea (*Tetrahymena*) (**Figure 1**).

Subsequent studies to investigate *L. pneumophila* pathogenesis have progressively expanded the list of protozoan hosts of this

TABLE 1 | Experimentally defined protozoan hosts of *L. pneumophila*.

Protozoan species	Protozoan strain	*L. pneumophila* serogroup (Sg): strain	Fate of *L. pneumophila*	Experimental evidence	References
Acanthamoeba spp.	AMI137, AMI116, AMI073, AMI191, Humidifier strain	Sg1: Lens Sg2: Togus-1 Sg3: Bloomington-2 Sg5: Cambridge-2	Intracellular multiplication	CFU counting, Phase-contrast microscopy	Rowbotham, 1980; Dupuy et al., 2016
Acanthamoeba sp. 155		Sg1	Intracellular multiplication	CFU counting, Epifluorescence microscopy	Cervero-Aragó et al., 2014, 2015
Acanthamoeba astronyxis	Isolate C37C6	Sg1: Philadelphia-1	Live cells are packaged in expelled pellets	Electron microscopy	Marciano-Cabral and Cabral, 2003; Amaro et al., 2015
Acanthamoeba castellanii	ATCC® 30234™, CCAP 1534/2, L1501/2A, L501/2A, Neff	Sg1: JR32, Lens, Paris, Philadelphia-1, Philadelphia-2, Pontiac-1 Sg2: Togus-1 Sg3: Bloomington-2 Sg4: Los Angeles Sg6: Oxford-1	Intracellular multiplication	CFU counting, Electron microscopy	Rowbotham, 1980; Holden et al., 1984; Moffat and Tompkins, 1992; Hilbi et al., 2001; Bouyer et al., 2007; Tyson et al., 2013; Mengue et al., 2016
	Neff	Sg5: Dallas 1E	Live cells are packaged in expelled pellets	Electron microscopy	Berk et al., 1998
Acanthamoeba lenticulata	PD2	Sg1: AX71, Philadelphia-1, SC94, SC97 Sg2: AX2 Sg3: AX52, AX54, AX82	Intracellular multiplication	CFU counting	Molmeret et al., 2001
Acanthamoeba palestinensis		Sg1	Intracellular multiplication	CFU counting, Electron microscopy, Epifluorescence microscopy, Phase contrast microscopy	Anand et al., 1983; Harf et al., 1997
Acanthamoeba polyphaga	Ap-1, L1501/3A, Puschkarew	Sg1: AA100, Corby, Nottingham-8, Leeds 1A SAP, Leeds-4, Lp02, Philadelphia-2, Pontiac-1 Sg2: Oxford-2, Togus-1 Sg3: Bloomington-2 Sg4: Los Angeles-1 Sg5: Cambridge-2 Sg6 Sg7: Dallas-5, Chicago-8 Sg8: York-1, Concord-3	Intracellular multiplication	CFU counting, Electron microscopy, Phase-contrast microscopy	Rowbotham, 1980, 1986; Kilvington and Price, 1990; Gao et al., 1997; Buse and Ashbolt, 2011
	Puschkarew	Sg5: Dallas 1E	Intracellular Survival, Live cells are packaged in expelled pellets	CFU counting, Electron microscopy	Berk et al., 1998; Buse and Ashbolt, 2011

(Continued)

TABLE 1 | Continued

Protozoan species	Protozoan strain	*L. pneumophila* serogroup (Sg): strain	Fate of *L. pneumophila*	Experimental evidence	References
Acanthamoeba royreba		Sg4: Los Angeles	Intracellular multiplication	Bacteria cell count, Epifluorescence microscopy	Tyndall and Domingue, 1982
Balamuthia mandrillaris	CDC-V039	Sg1: JR32, 130b	Intracellular multiplication	CFU counting, Phase-contrast microscopy	Shadrach et al., 2005
Ciliophrya sp.		Sg1: Corby	Intracellular survival	Epifluorescence microscopy	Rasch et al., 2016
Dictyostelium discoideum	AX2, AX2-214, AX3	Sg1: Benidorm 030E, Corby, Philadelphia-1	Intracellular multiplication	CFU counting, Electron microscopy	Hägele et al., 2000; Solomon et al., 2000
Echinamoeba exudans	SH274	Sg1: RI-243	Intracellular multiplication	Electron microscopy	Fields et al., 1989
Hartmannella cantabrigiensis		Sg2: PR-1 Sg5: Leeds-10 Sg7: Chicago-8, Dallas-5 Sg8: York-1	Intracellular multiplication	Electron microscopy	Rowbotham, 1986
Naegleria spp.	AMI242, AMI117, AMI135, AMI161	Sg1: Lens	Intracellular multiplication	CFU counting	Dupuy et al., 2016
Naegleria fowleri	Lee	Sg1: Lp02	Intracellular multiplication	CFU counting, Electron microscopy	Newsome et al., 1985; Buse and Ashbolt, 2011
		Sg3: Bloomington-2 Sg6: Chicago-2 Sg5: Dallas 1E	Intracellular survival	CFU counting	Buse and Ashbolt, 2011
Naegleria gruberi	1518/1E	Sg2: Togus-1 Sg3: Bloomington-2 Sg5: Cambridge-2	Intracellular multiplication	Phase-contrast microscopy	Rowbotham, 1980
Naegleria jadini	B1518/2	Sg2: Togus-1 Sg3: Bloomington-2 Sg5: Cambridge-2	Intracellular multiplication	Phase-contrast microscopy	Rowbotham, 1980
Naegleria lovaniensis	TS	Sg1: Philadelphia-1, 130b Sg4: Los Angeles	Intracellular multiplication	Confocal microscopy, CFU counting, Bacteria cell count, Epifluorescence microscopy	Tyndall and Domingue, 1982; Declerck et al., 2005; Tyson et al., 2013, 2014
Oxytricha bifaria		Sg1: Corby	Intracellular survival	Epifluorescence microscopy	Rasch et al., 2016
Paramecium caudatum	RB-1	Sg1: Philadelphia-1	Intracellular multiplication	Fluorescence microscopy	Watanabe et al., 2016
Stylonychia mytilus		Sg1: Corby	Intracellular survival	Epifluorescence microscopy	Rasch et al., 2016

(Continued)

TABLE 1 | Continued

Protozoan species	Protozoan strain	*L. pneumophila* serogroup (Sg): strain	Fate of *L. pneumophila*	Experimental evidence	References
Tetrahymena sp.		Sg1	Intracellular multiplication	CFU counting, Epifluorescence microscopy	Barbaree et al., 1986; Berk et al., 2008
		Sg1: Lp02	Live cells are packaged in expelled pellets	Electron microscopy, Fluorescence microscopy	Berk et al., 2008
Tetrahymena pyriformis	No. 500	Sg1: Philadelphia-1, 130b Sg3: SC-6-C3	Intracellular multiplication	CFU counting, Electron microscopy	Fields et al., 1984, 1986; Cianciotto and Fields, 1992
Tetrahymena thermophila	Mating type IV	Sg1: Philadelphia-1	Intracellular multiplication	CFU counting, Light microscopy Electron microscopy	Kikuhara et al., 1994
		Sg1: Philadelphia-2	Intracellular survival	CFU counting, Light microscopy Electron microscopy	Kikuhara et al., 1994
	Inbred strain B, SB021	Sg1: JR32	Intracellular multiplication	Electron microscopy; Live cells are packaged in expelled pellets	Hojo et al., 2012
Tetrahymena tropicalis		Sg1: Lens, Philadelphia-1	Live cells are packaged in expelled pellets	Electron microscopy	Faulkner et al., 2008; Koubar et al., 2011
Tetrahymena vorax	V2S	Sg1: Philadelphia-1	Intracellular survival	Electron microscopy, Fluorescence microscopy	Smith-Somerville et al., 1991
Tetramitus jugosus[b] (*Vahlkampfia jugosa*)		Sg1: Leeds 4	Intracellular multiplication	Electron microscopy	Rowbotham, 1986
Vermamoeba vermiformis[a] (*Hartmannella vermiformis*)	ATCC® 50256™, CDC-19	Sg1: AA100, Lens, 130b Philadelphia-1, RI-243 Sg5: E-52, E-62 Sg6: E-66, E-67	Intracellular multiplication	CFU counting, Electron microscopy	Rowbotham, 1986; King et al., 1991; Wadowsky et al., 1995; Abu Kwaik, 1996; Buse and Ashbolt, 2011; Tyson et al., 2013; Dupuy et al., 2016
		Sg1: Lp02 Sg3: Bloomington-2 Sg5: Dallas 1E Sg6: Chicago-2, Sg7: Dallas-5, PR-3	Intracellular survival	CFU counting	Buse and Ashbolt, 2011
Willaertia magna	c2c Maky, T5[S]44, Z503	Sg1: Lens, Paris, Philadelphia-1, 130b	Intracellular multiplication	CFU counting, Electron microscopy	Dey et al., 2009; Tyson et al., 2014

[a]*Vahlkampfia jugosa has been renamed Tetramitus jugosus (De Jonckheere and Brown, 2005).*
[b]*Hartmannella vermiformis has been renamed Vermamoeba vermiformis (Smirnov et al., 2011).*

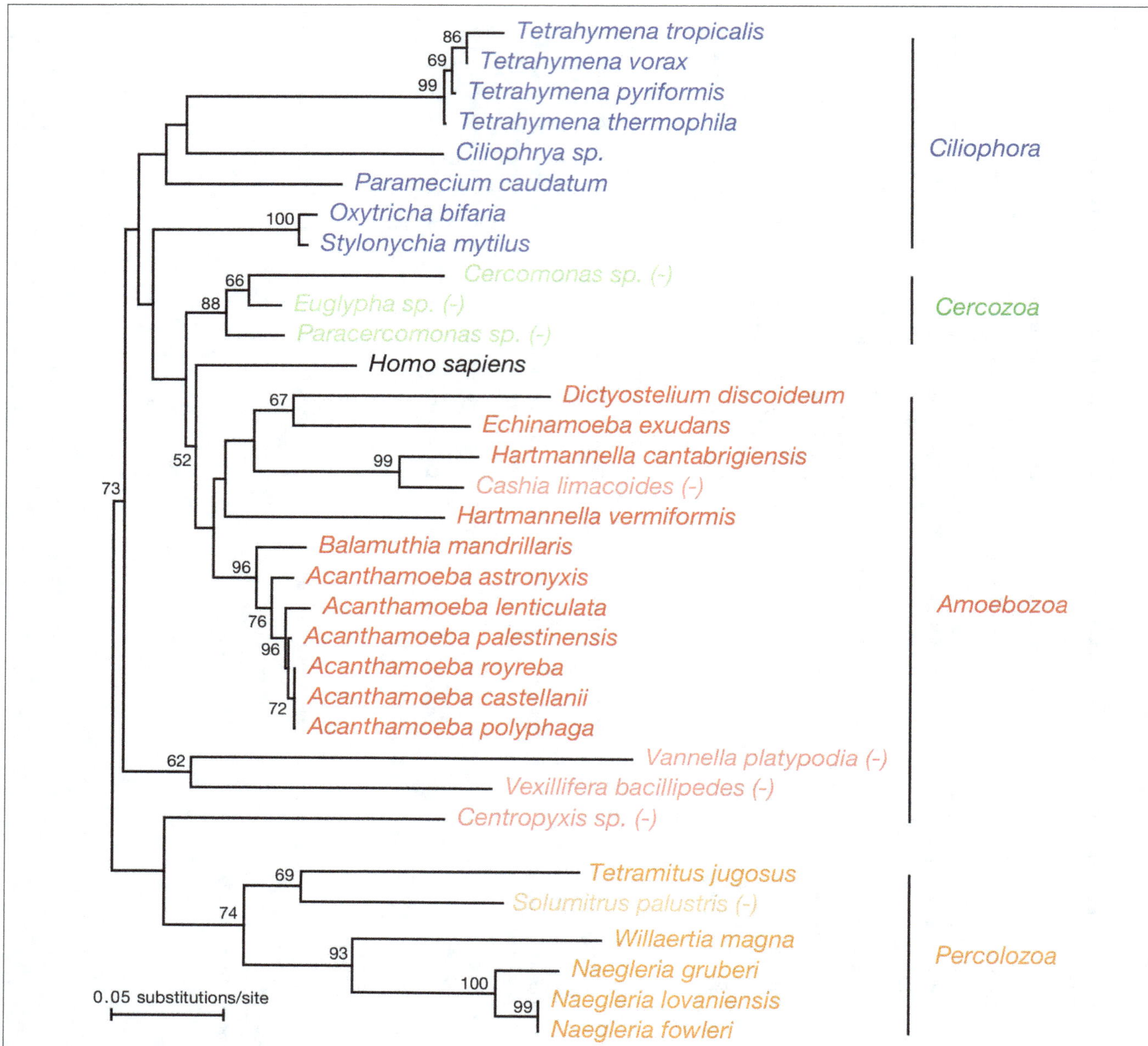

FIGURE 1 | An 18S phylogenetic tree of the experimentally defined hosts of *L. pneumophila*. Evolutionary history was inferred using the Neighbor-Joining method based on an alignment of 18S rRNA sequences. Evolutionary analyses were performed using MEGA7 (Kumar et al., 2016). Restrictive host species that do not support *L. pneumophila* replication or survival are indicated by lighter shading and the annotation "(–)". Taxonomic designations are based on the classification system outlined in Ruggiero et al. (2015).

bacterium (**Table 1** and **Figure 1**), including additional species of *Acanthamoeba* (*Acanthamoeba lenticulata* and *Acanthamoeba royreba*) and *Naegleria* (*Naegleria fowleri*) as well as more distantly related genera from their respective phyla such as *Dictyostelium discoideum* (Hägele et al., 2000; Solomon et al., 2000) and *Balamuthia mandrillaris* (Amoebozoa) (Shadrach et al., 2005) and *Willertia magna* (Percolozoa) (Dey et al., 2009; Tyson et al., 2014). Similarly, a number of additional ciliated protozoa were identified that were permissive for *L. pneumophila* survival, including *Tetrahymena* spp. (*Tetrahymena tropicalis* and *Tetrahymena vorax*), *Oxytricha bifaria*, *Stylonychia mytilus*, *Paramecium caudatum* and a member of the *Ciliophrya* genus, and in one case *L. pneumophila* replication (*Tetrahymena thermophila*), greatly expanding representation from this group (Kikuhara et al., 1994; Rasch et al., 2016; Watanabe et al., 2016). The beneficial interaction of *L. pneumophila* with these organisms appears to be specific as members from each of the representative phyla were also identified that were highly restrictive to *L. pneumophila* survival (**Figure 1**): *T. vorax* (Ciliophora), *A. astronyxis,* and *Cashia limocoides* (Amoebozoa) and *Solumitrus palustris* (Percolozoa) (Rowbotham, 1986; Smith-Somerville et al., 1991; Amaro et al., 2015). In addition, *L. pneumophila* was unable to grow in *V. platypodia* and *V. bacillipedes* (Rowbotham, 1986), which form a distantly related clade of the Amoebozoa phyla (**Figure 1**). Similarly, of the members of the Cercozoa phylum examined so far, *Cercomonas* sp., *Euglypha* sp., and *Paracercomonas* sp., all three are restrictive for *L. pneumophila* growth (Amaro et al., 2015; Rasch et al., 2016), suggesting that distinct orders and families within this class may be more restrictive than others. Thus, while the host range of *L. pneumophila* is vast, it does appear to have its limitations.

SUGGESTED ENVIRONMENTAL HOSTS OF *L. pneumophila*

Protozoa in both natural and man-made environments can alter the composition of microbial communities by eliminating bacteria through predation or augmenting populations of bacteria that are capable of replicating within these organisms (Yamamoto et al., 1992). Co-isolation techniques have been used to describe the composition of these communities within natural fresh water systems such as hot springs, thermal spas, lakes, ponds, streams, and anthropogenic reservoirs, such as cooling towers, industrial and private water networks and compost facilities. *L. pneumophila* is capable of surviving an array of physical conditions including temperatures ranging from 6 to 63°C (Fliermans et al., 1981). Thermal springs have been of particular interest as they boast characteristically high water temperatures, providing optimal conditions for *L. pneumophila* growth (Hsu et al., 2011; Ji et al., 2014; Rasch et al., 2016). Artificial aquatic reservoirs are of considerable epidemiological significance and typically support higher numbers of bacteria compared to natural water systems (Yamamoto et al., 1992), likely due to higher average water temperatures (Ikedo and Yabuuchi, 1986; Fields et al., 2002; Lasheras et al., 2006). The results of these population level analyses have validated many of the co-culture defined hosts of *L. pneumophila* while identifying several additional potential hosts (**Table 2**).

There is tremendous concordance between co-culture-confirmed *Legionella*-protozoa interactions and the results of co-isolation studies (**Tables 1, 2**). With the exception of *Balamuthia* and *Dictyostelium,* all protozoan genera shown to support intracellular growth in laboratory co-culture studies reside with *L. pneumophila* in the environment (**Table 2**). While this is not surprising for *Acanthamoeba, Hartmannella,* and *Naegleria,* as these are some of the most abundant protozoa in nature, in many cases co-isolation studies identified the same species of these genera. In particular, three of the protozoa identified, *A. palestinensis, N. lovaniensis,* and *V. vermiformis* that had been shown to support *L. pneumophila* replication in co-culture experiments (Anand et al., 1983; Rowbotham, 1986; Declerck et al., 2005; Thomas et al., 2006) were isolated from water samples harboring *L. pneumophila* (Kao et al., 2013). Similarly, amoebal enrichment assays resulted in the isolation of *Acanthamoeba jacobsi* harboring *L. pneumophila* directly from a thermal spring water sample (Hsu et al., 2011). These results identify *A. jacobsi* as a new host of *L. pneumophila* and provide direct evidence of an interaction between *L. pneumophila* and these four protozoan hosts in the environment. The lack of co-isolation of *L. pneumophila* with either *Balamuthia* or *Dictyostelium* species is likely because these protozoa are typically found in soil and the majority of samples analyzed were isolated from aquatic environments (Dunnebacke et al., 2004; Vadell and Cavender, 2007). The high degree of correlation between the co-culture and co-isolation studies supports the role of these organisms as natural hosts of *L. pneumophila* in environmental reservoirs.

Co-isolation studies predict a number of additional phyla and classes of protozoa may support *L. pneumophila* survival or growth (**Table 2**). In addition to the Amoebozoa, Ciliophora, and Percolozoa phyla, protozoa from Apusozoa (*Diphylleia rotans*), Cercozoa (*Euglypha* sp.), Euglenozoa (*Bodonidae* sp.), and Opsithokonta (*Rhinosporidium* sp.) were identified. Two additional classes of protozoa from previously identified phyla are also represented, Variosea (*Flamella balnearia*) and Oligohymenophorea with representatives encompassing four different families spanning three orders within this group. For those classes of protozoa already identified as hosts by co-culture experiments, three additional orders, Thecamoebida (*Stenamoeba limacina*), Arcellinida (*Centropyxis* sp.), and Sporadotricina (*Aspidiscidae* family) and five genera (*Comandonia operculata, C. limacoides, Paravahlkampfia ustiana, Learamoeba waccamawenis,* and *Singhamoeba horticola*) were identified. Finally, of the known hosts of *L. pneumophila* from co-culture experiments, additional species of *Acanthamoeba* (*A. jacobsi*), *Naegleria* (*Naegleria pagei* and *Naegleria australiensis*), *Tetramitus* (*Tetramius enterica*), and *Vahlkampfia* (*Valkampfia avara*) were also isolated. Combined, co-isolation and co-culture experiments represent 7 of the 8 phyla of the protozoa kingdom, 12 of the 41 classes within these phyla and 21 of the 82 defined orders, demonstrating the tremendous diversity amongst *L. pneumophila* hosts.

Protozoa more commonly found associated with *L. pneumophila* in environmental reservoirs may indicate that they are more likely to be true hosts of the bacterium. While the *Acanthamoeba* spp., *Naegleria* spp., *Vahlkampfia* spp., and *Hartmannella* spp. (including *Vermamoeba vermifomis*) are commonly found in multiple sources (**Table 2**), particular protozoa appear to co-reside with *L. pneumophila* in more than one environmental sample (**Table 2**). *A. hatchetti, A. polyphaga, H. cantabrigensis, N. fowleri, N. lovaniensis, Neoparamoeabe* sp., and *Willertia* sp. have been isolated from both natural and man-made water sources (**Table 2**), suggesting that these protozoa may function as hosts of *L. pneumophila* in both natural reservoirs and potable water. Both *E. exudans* and *Echinamoeba thermarum* have been identified in more than one potable water sample (**Table 2**), suggesting these amoebae may play more prominent roles in the epidemiology of *L. pneumophila*. A higher incidence of specific protozoa with *L. pneumophila* may indicate a stronger likelihood that these protozoa are responsible

TABLE 2 | Suggested protozoan hosts of *L. pneumophila*.

Protozoa	Environment source	Identification method used	References
Acanthamoebidae	Cooling towers	Identified morphologically via microscopy	Yamamoto et al., 1992
Acanthamoeba spp.	Compost facilities	Sequence analysis	Conza et al., 2013, 2014
	Cooling towers	Identified morphologically via microscopy	Kurtz et al., 1982
		Sequence analysis	Declerck et al., 2007
	Drinking water systems	Sequence analysis	Marciano-Cabral et al., 2010; Valster et al., 2011; Ji et al., 2014
	Hospital water networks	Identified morphologically via microscopy	Rohr et al., 1998; Steinert et al., 1998
	Industrial water networks	Identified morphologically via microscopy; Sequence analysis	Scheikl et al., 2014
	Natural water systems	Sequence analysis	Declerck et al., 2007; Hsu et al., 2011; Ji et al., 2014
Acanthamoeba castellanii	Compost facilities	Sequence analysis	Conza et al., 2013
Acanthamoeba hatchetti	Compost facilities	Sequence analysis	Conza et al., 2013, 2014
	Hospital water network	Identified morphologically via microscopy	Breiman et al., 1990
	Natural water systems	Sequence analysis	Hsu et al., 2015
Acanthamoeba jacobsi	Natural water systems	Sequence analysis	Hsu et al., 2011
Acanthamoeba lenticulata	Compost facilities	Sequence analysis	Conza et al., 2013
Acanthamoeba palestinensis	Natural water systems	Sequence analysis	Kao et al., 2013
Acanthamoeba polyphaga	Compost facilities	Sequence analysis	Conza et al., 2013, 2014
	Cooling towers	Not specified	Rowbotham, 1986
	Natural water systems	Sequence analysis	Hsu et al., 2009
Amoebidae	Cooling towers	Identified morphologically via microscopy	Yamamoto et al., 1992
Aspidiscidae	Cooling towers	Identified morphologically via microscopy	Yamamoto et al., 1992
Bodonidae	Cooling towers	Identified morphologically via microscopy	Yamamoto et al., 1992
Cashia limacoides	Cooling towers	Not specified	Rowbotham, 1986
Centropyxis sp.	Natural water systems	Identified morphologically via microscopy	Rasch et al., 2016
Ciliophrya sp.	Natural water systems	Identified morphologically via microscopy	Rasch et al., 2016
Colpodidae	Cooling towers	Identified morphologically via microscopy	Yamamoto et al., 1992
Comandonia operculata	Hospital water network	Identified morphologically via microscopy	Breiman et al., 1990
Cyclidium spp.	Cooling towers	Identified morphologically via microscopy	Barbaree et al., 1986
Diphylleia rotans	Sewage treatment systems	Sequence analysis	Valster et al., 2010
Echinamoeba spp.	Hospital water networks	Identified morphologically via microscopy	Rohr et al., 1998
Echinamoeba exudans	Drinking water systems	Sequence analysis	Valster et al., 2011
	Hospital water networks	Identified morphologically via microscopy	Fields et al., 1989
Echinamoeba thermarum	Drinking water systems	Sequence analysis	Valster et al., 2011
	Cooling towers	Sequence analysis	Valster et al., 2010
Euglypha sp.	Natural water systems	Identified morphologically via microscopy	Rasch et al., 2016
Filamoeba nolandi	Hospital water networks	Identified morphologically via microscopy	Breiman et al., 1990
Flamella balnearia	Compost facilities	Sequence analysis	Conza et al., 2013

(Continued)

TABLE 2 | Continued

Protozoa	Environment source	Identification method used	References
Hartmannellidae	Cooling towers	Identified morphologically via microscopy	Yamamoto et al., 1992
Hartmannella spp.	Cooling towers	Sequence analysis	Declerck et al., 2007
		Identified morphologically via microscopy	Kurtz et al., 1982
	Hospital water networks	Identified morphologically via microscopy	Fields et al., 1989; Breiman et al., 1990; Nahapetian et al., 1991
	Natural water systems	FISH; Identified morphologically via microscopy	Zbikowska et al., 2014
		Sequence analysis	Declerck et al., 2007
Hartmannella car tabrigiensis	Hospital water networks	Identified morphologically via microscopy	Rowbotham, 1986; Fields et al., 1989
Learamoeba waccamawenis	Compost facilities	Sequence analysis	Conza et al., 2013, 2014
Mayorella spp.	Hospital water networks	Identified morphologically via microscopy	Steinert et al., 1998
Naegleria spp.	Cooling towers	Identified morphologically via microscopy	Barbaree et al., 1986
		Sequence analysis	Declerck et al., 2007
	Compost facilities	Sequence analysis	Conza et al., 2013, 2014
	Drinking water systems	Sequence analysis	Marciano-Cabral et al., 2010; Ji et al., 2014
	Hospital water networks	Identified morphologically via microscopy	Nahapetian et al., 1991; Rohr et al., 1998
	Industrial water networks	Identified morphologically via microscopy	Scheikl et al., 2014
	Natural water systems	Sequence analysis	Declerck et al., 2007; Hsu et al., 2011; Ji et al., 2014
		FISH; Identified morphologically via microscopy	Zbikowska et al., 2014
Naegleria australiensis	Compost facilities	Sequence analysis	Conza et al., 2013
	Natural water systems	Sequence analysis	Huang and Hsu, 2010
Naegleria fowleri	Thermal saline bath	FISH; Identified morphologically via microscopy	Zbikowska et al., 2013
	Natural water systems	FISH; Identified morphologically via microscopy	Zbikowska et al., 2014
Naegleria gruberi	Compost facilities	Sequence analysis	Conza et al., 2013
	Natural water systems	Sequence analysis	Hsu et al., 2015
Naegleria lovaniensis	Natural water systems	Sequence analysis	Huang and Hsu, 2010; Kao et al., 2013
Naegleria pagei	Natural water systems	Sequence analysis	Huang and Hsu, 2010
Neoparamoeba spp.	Drinking water systems	Sequence analysis	Valster et al., 2011
	Natural water systems	Sequence analysis	Valster et al., 2010
Oxytricha bifaria	Natural water systems	Identified morphologically via microscopy	Rasch et al., 2016
Paravahlkampfia ustiana[a] (*Vahlkampfia ustiana*)	Hospital water networks	Identified morphologically via microscopy	Breiman et al., 1990
Pleuronematidae	Cooling towers	Identified morphologically via microscopy	Yamamoto et al., 1992
Rhinosporidium sp.	Tap water system	Sequence analysis	Valster et al., 2010
Saccamoeba spp.	Hospital water networks	Identified morphologically via microscopy	Rohr et al., 1998
Singhamoeba horticola	Compost facilities	Sequence analysis	Conza et al., 2013, 2014
Stenamoeba spp.	Compost facilities	Sequence analysis	Conza et al., 2013, 2014

(Continued)

TABLE 2 | Continued

Protozoa	Environment source	Identification method used	References
Stenamoeba limacina	Compost facilities	Sequence analysis	Conza et al., 2014
Stylonychia mytilus	Natural water systems	Identified morphologically via microscopy	Rasch et al., 2016
Tetrahymenidae	Cooling towers	Identified morphologically via microscopy	Yamamoto et al., 1992
Tetrahymena spp.	Cooling towers	Identified morphologically via microscopy	Barbaree et al., 1986
Tetramitus spp.	Compost facilities	Sequence analysis	Conza et al., 2013
Tetramitus enterica[b] (*Vahlkampfia enterica*)	Compost facilities	Sequence analysis	Conza et al., 2013
Vahlkampfia spp.	Compost facilities	Sequence analysis	Conza et al., 2014
	Cooling towers	Sequence analysis	Declerck et al., 2007
	Drinking water systems	Sequence analysis	Marciano-Cabral et al., 2010
	Hospital water networks	Identified morphologically via microscopy	Breiman et al., 1990; Rohr et al., 1998; Steinert et al., 1998
	Natural water systems	Sequence analysis	Declerck et al., 2007; Hsu et al., 2011
Vahlkampfia avara	Compost facilities	Sequence analysis	Conza et al., 2013, 2014
Vannella spp.	Hospital water networks	Identified morphologically via microscopy	Rohr et al., 1998
Vannella platypodia	Cooling towers	Not specified	Rowbotham, 1986
Vermamoeba vermiformis[c] (*Hartmannella vermiformis*)	Compost facilities	Sequence analysis	Conza et al., 2013, 2014
	Drinking water systems	Sequence analysis	Valster et al., 2011; Ji et al., 2014
	Hospital water networks	Identified morphologically via microscopy	Rowbotham, 1986; Fields et al., 1989; Breiman et al., 1990; Rohr et al., 1998
		Sequence analysis	Thomas et al., 2006
	Industrial water networks	Identified morphologically via microscopy	Scheikl et al., 2014
	Natural water systems	Sequence analysis	Hsu et al., 2011, 2015; Ji et al., 2014
		Sequence analysis	Kao et al., 2013
		Sequence analysis	Valster et al., 2010
	Tap water systems	Sequence analysis	Valster et al., 2010
Vexillifera bacillipedes	Cooling towers	Not specified	Rowbotham, 1986
Vorticellidae	Cooling towers	Identified morphologically via microscopy	Yamamoto et al., 1992
Willaertia spp.	Cooling towers	Sequence analysis	Declerck et al., 2007
	Natural water systems	Sequence analysis	Declerck et al., 2007
Willaertia magna	Compost facilities	Sequence analysis	Conza et al., 2013

[a] *Vahlkampfia ustiana has been renamed Paravahlkampfia ustiana.*
[b] *Vahlkampfia enterica has been renamed Tetramitus enterica.*
[c] *Hartmannella vermiformis has been renamed Vermamoeba vermiformis* (Smirnov et al., 2011).

for the persistence of *L. pneumophila* in environmental reservoirs.

Not all protozoa species isolated from the same environmental source are hosts of *L. pneumophila*. Of several species of free-living amoeba collected from a cooling tower, only *A. polyphaga* supported intracellular growth of *L. pneumophila* whereas *L. pneumophila* failed to replicate within *C. limacoides*, *V. platypodia*, and *V. bacillipedes* (Rowbotham, 1986). Similarly, of several ciliated protozoa species in biofilm samples isolated from a thermal spa, *L. pneumophila* was able to infect *Ciliophrya* sp., *O. bifaria*, and *S. mytilus*, but no intracellular bacteria were detected within *Euglypha* sp. or *Centropyxis* sp. (Rasch et al.,

2016). Thus, *L. pneumophila* is able to persist in environments comprised of both *L. pneumophila*-restrictive and permissive protozoan hosts. The relative abundancy of *L. pneumophila* in different environmental niches may reflect mixed populations of these two types of protozoa. Alternatively, in some circumstances *L. pneumophila* may deplete entire populations of permissive hosts, enriching for resistant species of protozoa that remain. Thus, the absence of certain types of protozoa may not necessarily rule them out as contributors to *L. pneumophila* growth and persistence in the environment.

The distribution of protozoa between the types of water sources examined (natural water reservoirs, cooling towers, potable water distribution system, and compost sites; **Table 2**) was relatively uniform with a few notable exceptions. Amoebozoa and Percolozoa, making up the majority of the protozoa identified, were found in all water sources. Amoebozoa were more predominant in cooling towers and potable water systems. The lower abundance of Percolozoa in cooling towers coincided with a higher abundance of Ciliophora (ciliated protozoa) whereas in potable water, an enrichment in organisms from the Tubulinea class of Amoebozoa, in particular *Echinamoeba* was observed. In contrast, fewer members of the Discosea class were reported and in particular, no members of the Centramoebida order despite their presence in all other sites. The perseverance of *L. pneumophila* within various water environments despite variation in the protozoa composition demonstrates the highly adaptive nature of this bacterium to fluctuations in host population dynamics.

METAGENOMICS

Although co-isolation studies provide valuable insights into the microbial communities that support *L. pneumophila*, these methods cannot adequately define the full diversity of these communities (Kunin et al., 2008). While enrichment steps are often necessary to identify low abundance organisms, they create experimental bottlenecks and biases by selecting against protozoa that cannot be cultured using standard protocols (Hugenholtz and Tyson, 2008; Gomez-Alvarez et al., 2012) and *Legionella* isolates with host specificities that do not overlap with amoebal species commonly used in these techniques (Evstigneeva et al., 2009). Metagenome-based analyses may circumvent the limitations inherent to culture-based approaches and provide a more comprehensive, unbiased profile of these communities (Hugenholtz and Tyson, 2008; Gomez-Alvarez et al., 2009). For example, metagenomic studies of samples from three separate watersheds showed both a high level of diversity in the population of *Legionella* (encompassing 15 different species) and a correlation between the levels of Amoebozoa present in the water and the abundance of *Legionella* isolates (Peabody et al., 2017). Monitoring the abundance of *Legionella*, *Hartmannella*, and *Naegleria* from two environmental water sources over the course of a standard water purification procedure suggested a correlation between the abundance of *Legionella* and *Naegleria*, but not *Hartmannella* (Lin et al., 2014). In general however, metagenomics studies have been somewhat difficult to interpret. Often individual sites are dominated by one or a few amoebal species and the relative abundance of *L. pneumophila* is extremely low compared to other bacteria (Liu et al., 2012; Delafont et al., 2013): these features make it difficult to correlate the presence of *L. pneumophila* with specific protozoa. As the sensitivity and depth of metagenomics analysis improves, metagenomics will most certainly be a source of tremendous insight into the full repertoire of protozoan hosts of *L. pneumophila*.

FACTORS AFFECTING THE OUTCOME OF *Legionella*-PROTOZOA INTERACTIONS

The outcome of the interaction between *L. pneumophila* and protozoa can be influenced by a number of factors; the identity of the host cell, variations in the predatory behavior or feeding preferences of the host, the strain or species of the bacterium, the relative abundance of the two organisms, the external environment, and other microorganisms.

The identity of the host cell can greatly impact the outcome of the infection. While some hosts are permissive for *L. pneumophila* replication, others are restrictive, either impeding bacterial growth or in extreme cases, survival (Amaro et al., 2015). The maximum amount and rate of *L. pneumophila* growth between hosts can vary significantly (Declerck et al., 2005). For example, *L. pneumophila* can achieve up to 10,000-fold growth in *A. castellanii* but only 10-fold growth in *N. lovaniensis* over the same time period (Declerck et al., 2005). Similarly, *L. pneumophila* strain Paris grows robustly in *A. castellanii* and *V. vermiformis* but is defective for growth in *W. magna* (Dey et al., 2009). Moreover, the differential growth of *L. pneumophila* Paris varies between different strains of *W. magna*, with robust growth in strain T5[S]44 (Tyson et al., 2014) but failure to grow in strains c2c Maky or Z502 (Dey et al., 2009). Thus, some hosts are more optimal than others for *L. pneumophila* survival and replication.

The predatory behavior and feeding preferences of the host can also influence *Legionella*-protozoa interactions. For example, the *L. pneumophila* auto-inducer LAI-1 disrupts chemotactic migration of *D. discoideum* (Simon et al., 2015) and promotes *L. pneumophila* uptake in both *D. discoideum* and *A. castellanii* (Tiaden et al., 2010). By restricting amoebal movement, *L. pneumophila* may localize feeding to the site of the bacteria—such modulation may also enrich for specific types of amoebae that support *L. pneumophila* replication. The LAI-1 biosynthesis genes are not conserved in all *Legionella* species (Burstein et al., 2016) suggesting that individual species may differentially promote their interaction with amoebae or do so via different mechanisms. Consistent with this idea, the host cell receptors that mediate *L. pneumophila* adhesion to *V. vermiformis*, *A. castellanii*, *A. polyphaga*, and *N. lovaniensis* and the underlying mechanisms governing bacterial uptake vary between these amoebal hosts (Venkataraman et al., 1997; Harb et al., 1998; Declerck et al., 2005, 2007). As a consequence, bacterial uptake can vary between protozoa. Indeed, *A. castellanii* has been shown to ingest *L. pneumophila* with much greater efficiency than *N. lovaniensis* (Declerck et al., 2005). Variations in sensing, targeting, adhesion and phagocytosis of bacteria can influence the affinity, specificity, frequency and duration with which *L. pneumophila* interacts with specific protozoa and thus, the impact of their cohabitation on the persistence of *L. pneumophila* in environmental reservoirs.

The genetic composition of the bacterium can greatly impact its fate within the host cell, as the survival and replication of different strains and species of *Legionella* can vary dramatically. Despite the growth defect of *L. pneumophila* Paris in *Willertia magna*, both the *L. pneumophila* Philadelphia-1, Lens and 130b strains are able to replicate in this amoebal host (Dey et al., 2009; Tyson et al., 2014). Similarly, comparisons between clinical and environmental isolates of *L. pneumophila* showed that while one clinical isolate was highly adept at growing in *A. lenticulata* another was severely defective and the relative amounts of replication of the environmental isolates in this host were somewhere in between (Molmeret et al., 2001). Similar differences are observed between species of *Legionella*. While *L. pneumophila*, *Legionella steelei*, *Legionella dumoffii*, and *Legionella norrlandica* are able to grow within *A. castellanii*, several other species including *Legionella longbeachae*, *Legionella jordanis*, and *Legionella anisa* are unable to do so (Neumeister et al., 1997; Edelstein et al., 2012; Rizzardi et al., 2014). Thus, the fate of both the bacterium and the host cell is greatly determined by the inherent properties of each organism.

The outcome of a *Legionella*-protozoa interaction is not only influenced by their respective identities but the relative abundance of each organism. For instance, when *L. pneumophila* is present at low levels they are digested for nutrients by *Tetrahymena* sp. but when the bacteria reach a threshold concentration, they are packaged into vesicles and secreted in pellets (Berk et al., 2008; Hojo et al., 2012). The greater the number of bacteria present, the greater the production and secretion of these bacterial pellets. Similar packaging and secretion of other types of bacteria (Denoncourt et al., 2014) suggests this may be a mechanism by which protozoa compensate for over-eating, or stock-pile food (Hojo et al., 2012).

The external environment can have a profound effect on *Legionella*-protozoa interactions. For example, temperature can greatly impact the intracellular fate of *L. pneumophila*. Although, intracellular replication of *L. pneumophila* in *A. castellanii* occurs at a range of temperatures (Rowbotham, 1981), intracellular growth is significantly reduced at lower temperatures (Ohno et al., 2008). Within more restrictive hosts, such as *A. polyphaga*, intracellular replication only occurs at higher temperatures whereas below 25°C, *L. pneumophila* is readily consumed (Nagington and Smith, 1980). In contrast, in *Tetrahymena* spp. *L. pneumophila* exhibits robust intracellular growth at 35°C (Fields et al., 1984; Barbaree et al., 1986; Kikuhara et al., 1994) but at lower temperatures, *L. pneumophila* is packaged into vesicles and secreted into the environment (Faulkner et al., 2008; Koubar et al., 2011). The factors affecting intracellular growth of *L. pneumophila* are not mutually exclusive, as different combinations of the strain of *L. pneumophila*, the host cell type and temperature can significantly alter intracellular growth of the bacterium (Buse and Ashbolt, 2011).

Much of the research examining *Legionella*-protozoa interactions has focused on specific bacterial-host pairings, which cannot address the impact of other organisms on these interactions. *L. pneumophila* naturally inhabits complex microbial communities, which could have both positive and negative impacts on *L. pneumophila* survival and population dynamics. For example, *A. castellanii* harboring the endosymbiont *Neochlamydia S13* are unable to support *L. pneumophila* replication despite efficient uptake and lack of degradation in the lysosome (Ishida et al., 2014). The impact of *Neochlamydia S13* on *L. pneumophila* replication is specific because *L. pneumophila* is able to replicate in *A. castellanii* infected with the endosymbiont *Protochlamydia R18*. Moreover, curing *A. castellanii* of *Neochlamyida S13* restores intracellular growth of *L. pneumophila*, suggesting that the presence of the endosymbiont renders *A. castellani* resistant to *L. pneumophila* pathogenesis. In contrast, *L. pneumophila* has been shown to promote the intracellular growth of *Brucella neotomae* when the two pathogens share the same vacuole (Kang and Kirby, 2017). While sharing resources does not appear to affect *L. pneumophila*, it is conceivable that *L. pneumophila* may similarly benefit from the activities of other bacteria when it finds itself in more restrictive protozoan hosts.

FUTURE DIRECTIONS

A critical challenge in understanding the molecular mechanisms of *L. pneumophila* pathogenesis, evolution and environmental persistence is the staggering diversity of the protozoan hosts that support *L. pneumophila* replication. Indeed, such diversity is thought to be responsible for shaping *L. pneumophila* into a generalist pathogen with a broad host range—a feature clearly important for pathogenesis in humans. Rather than having a single, defined "natural host," *L. pneumophila* wanders from host to host and is constantly shaped by these disparate interactions. Such a lifestyle is a challenge for researchers studying these bacteria: (1) many protozoa remain poorly characterized, difficult to culture, and/or unsequenced; (2) the shear diversity of protozoa and complexity of natural interactions makes experimental analysis of phenotypes under "physiologically relevant" conditions extremely daunting (which hosts should be used and under what chemical and physical conditions should the interaction be studied?); and (3) how can non-binary interactions with mixed bacterial and host populations be examined in a reproducible and informative fashion? Given the importance of protozoa to *L. pneumophila* biology (and pathogen evolution in general), we strongly advocate efforts for the sequencing and detailed study of these organisms. While it is enticing to retreat to the comfort of studying *Legionella*-host interactions in mammalian macrophages and perhaps one or two model protozoa, an exciting, informative, frustrating, and messy reality remains largely unexplored. Perhaps once the diversity of bacterial-protozoan behaviors is better understood, a panel of model hosts could be chosen not based on ease of culture, but instead to capture the greatest breadth of this diversity.

AUTHOR CONTRIBUTIONS

TO, DB, AE, and GZ wrote the manuscript. GZ and AE generated the phylogenetic tree.

ACKNOWLEDGMENTS

We thank Jason Park, Sara Rego, Soma Ghosh, and Mohammad Hossain for thoughtful review of the manuscript. This work was supported by the National Institutes of Health, Grant 1R21AI119580-01 (TO) and the Natural Sciences and Engineering Research Council of Canada, Grant RGPIN-2014-03641 (AE).

REFERENCES

Abdel-Nour, M., Duncan, C., Low, D. E., and Guyard, C. (2013). Biofilms: the Stronghold of *Legionella pneumophila*. *Int. J. Mol. Sci.* 14, 21660–21675. doi: 10.3390/ijms141121660

Abu Kwaik, Y. (1996). The phagosome containing *Legionella pneumophila* within the protozoan *Hartmannella vermiformis* is surrounded by the rough endoplasmic reticulum. *Appl. Environ. Microbiol.* 62, 2022–2028.

Amaro, F., Wang, W., Gilbert, J. A., Anderson, O. R., and Shuman, H. A. (2015). Diverse protist grazers select for virulence-related traits in *Legionella*. *ISME J.* 9, 1607–1618. doi: 10.1038/ismej.2014.248

Anand, C. M., Skinner, A. R., Malic, A., and Kurtz, J. B. (1983). Interaction of *L. pneumophilia* and a free living amoeba (*Acanthamoeba palestinensis*). *J. Hyg.* 91, 167–178. doi: 10.1017/S0022172400060174

Barbaree, J. M., Fields, B. S., Feeley, J. C., Gorman, G. W., and Martin, W. T. (1986). Isolation of protozoa from water associated with a legionellosis outbreak and demonstration of intracellular multiplication of *Legionella pneumophila*. *Appl. Environ. Microbiol.* 51, 422–424.

Barker, J., Scaife, H., and Brown, M. R. (1995). Intraphagocytic growth induces an antibiotic-resistant phenotype of *Legionella pneumophila*. *Antimicrob. Agents Chemother.* 39, 2684–2688. doi: 10.1128/AAC.39.12.2684

Berk, S. G., Faulkner, G., Garduño, E., Joy, M. C., Ortiz-Jimenez, M. A., and Garduño, R. A. (2008). Packaging of live *Legionella pneumophila* into pellets expelled by *Tetrahymena* spp. does not require bacterial replication and depends on a Dot/Icm-mediated survival mechanism. *Appl. Environ. Microbiol.* 74, 2187–2199. doi: 10.1128/AEM.01214-07

Berk, S. G., Ting, R. S., Turner, G. W., and Ashburn, R. J. (1998). Production of respirable vesicles containing live *Legionella pneumophila* cells by two *Acanthamoeba* spp. *Appl. Environ. Microbiol.* 64, 279–286.

Biurrun, A., Caballero, L., Pelaz, C., León, E., and Gago, A. (1999). Treatment of a *Legionella pneumophila*-colonized water distribution system using copper-silver ionization and continuous chlorination. *Infect. Control. Hosp. Epidemiol.* 20, 426–428. doi: 10.1086/501645

Bohach, G. A., and Snyder, I. S. (1983). Characterization of surfaces involved in adherence of *Legionella pneumophila* to *Fischerella* species. *Infect. Immun.* 42, 318–325.

Bouyer, S., Imbert, C., Rodier, M. H., and Héchard, Y. (2007). Long-term survival of *Legionella pneumophila* associated with *Acanthamoeba castellanii* vesicles. *Environ. Microbiol.* 9, 1341–1344 doi: 10.1111/j.1462-2920.2006.01229.x

Breiman, R. F., Fields, B. S., Sanden, G. N., Volmer, L., Meier, A., and Spika, J. S. (1990). Association of shower use with Legionnaires' disease. Possible role of amoebae. *JAMA* 263, 2924–2926. doi: 10.1001/jama.1990.03440210074036

Brieland, J., McClain, M., Heath, L., Chrisp, C., Huffnagle, G., LeGendre, M., et al. (1996). Coinoculation with *Hartmannella vermiformis* enhances replicative *Legionella pneumophila* lung infection in a murine model of Legionnaires' disease. *Infect. Immun.* 64, 2449–2456.

Brousseau, N., Lévesque, B., Guillemet, T. A., Cantin, P., Gauvin, D., Giroux, J.-P., et al. (2013). Contamination of public whirlpool spas: factors associated with the presence of *Legionella* spp., *Pseudomonas aeruginosa* and *Escherichia coli*. *Int. J. Environ. Health Res.* 23, 1–15. doi: 10.1080/09603123.2012.678001

Burstein, D., Amaro, F., Zusman, T., Lifshitz, Z., Cohen, O., Gilbert, J. A., et al. (2016). Genomic analysis of 38 *Legionella* species identifies large and diverse effector repertoires. *Nat. Genet.* 48, 167–175. doi: 10.1038/ng.3481

Buse, H. Y., and Ashbolt, N. J. (2011). Differential growth of *Legionella pneumophila* strains within a range of amoebae at various temperatures associated with in-premise plumbing. *Lett. Appl. Microbiol.* 53, 217–224 doi: 10.1111/j.1472-765X.2011.03094.x

Cervero-Aragó, S., Rodríguez-Martínez, S., Puertas-Bennasar, A., and Araujo, R. M. (2015). Effect of common drinking water disinfectants, chlorine and heat, on free *Legionella* and amoebae-associated *Legionella*. *PLoS ONE* 10:e0134726. doi: 10.1371/journal.pone.0134726

Cervero-Aragó, S., Sommer, R., and Araujo, R. M. (2014). Effect of UV irradiation (253.7 nm) on free *Legionella* and *Legionella* associated with its amoebae hosts. *Water Res.* 67, 299–309. doi: 10.1016/j.watres.2014.09.023

Cianciotto, N. P., and Fields, B. S. (1992). *Legionella pneumophila* mip gene potentiates intracellular infection of protozoa and human macrophages. *Proc. Natl. Acad. Sci. U.S.A.* 89, 5188–5191. doi: 10.1073/pnas.89.11.5188

Cirillo, J. D., Cirillo, S. L., Yan, L., Bermudez, L. E., Falkow, S., and Tompkins, L. S. (1999). Intracellular growth in *Acanthamoeba castellanii* affects monocyte entry mechanisms and enhances virulence of *Legionella pneumophila*. *Infect. Immun.* 67, 4427–4434.

Cirillo, J. D., Falkow, S., and Tompkins, L. S. (1994). Growth of *Legionella pneumophila* in *Acanthamoeba castellanii* enhances invasion. *Infect. Immun.* 62, 3254–3261.

Conza, L., Casati Pagani, S., and Gaia, V. (2014). Influence of climate and geography on the occurrence of *Legionella* and amoebae in composting facilities. *BMC Res. Notes* 7:831. doi: 10.1186/1756-0500-7-831

Conza, L., Pagani, S. C., and Gaia, V. (2013). Presence of *Legionella* and free-living amoebae in composts and bioaerosols from composting facilities. *PLoS ONE* 8:e68244. doi: 10.1371/journal.pone.0068244

Correia, A. M., Ferreira, J. S., Borges, V., Nunes, A., Gomes, B., Capucho, R., et al. (2016). Probable person-to-person transmission of Legionnaires' disease. *New. Eng. J. Med.* 374, 497–498. doi: 10.1056/NEJMc1505356

De Jonckheere, J. F., and Brown, S. (2005). The identification of vahlkampfiid amoebae by ITS sequencing. *Protist* 1, 89–96. doi: 10.1016/j.protis.2004.11.001

Declerck, P., Behets, J., van Hoef, V., and Ollevier, F. (2007). Detection of *Legionella* spp. and some of their amoeba hosts in floating biofilms from anthropogenic and natural aquatic environments. *Water Res.* 41, 3159–3167. doi: 10.1016/j.watres.2007.04.011

Declerck, P., Behets, J., Delaedt, Y., Margineanu, A., and Lammertyn, E., Ollevier, F. (2005). Impact of non-*Legionella* bacteria on the uptake and intracellular replication of *Legionella pneumophila* in *Acanthamoeba castellanii* and *Naegleria lovaniensis*. *Microb. Ecol.* 50, 536–549. doi: 10.1007/s00248-005-0258-0

Delafont, V., Brouke, A., Bouchon, D., Moulin, L., and Héchard, Y. (2013). Microbiome of free-living amoebae isolated from drinking water. *Water Res.* 47, 6958–6965. doi: 10.1016/j.watres.2013.07.047

Denoncourt, A. M., Paquet, V. E., and Charette, S. J. (2014). Potential role of bacteria packaging by protozoa in the persistence and transmission of pathogenic bacteria. *Front. Microbiol.* 5:240. doi: 10.3389/fmicb.2014.00240

Dey, R., Bodennec, J., Mameri, M. O., and Pernin, P. (2009). Free-living freshwater amoebae differ in their susceptibility to the pathogenic bacterium *Legionella pneumophila*. *FEMS Microbiol. Lett.* 290, 10–17. doi: 10.1111/j.1574-6968.2008.01387.x

Dunnebacke, T. H., Schuster, F. L., Yagi, S., and Booton, G. C. (2004). *Balamuthia mandrillaris* from soil samples. *Microbiology* 150, 2837–2842. doi: 10.1099/mic.0.27218-0

Dupuy, M., Binet, M., Bouteleux, C., Herbelin, P., Soreau, S., and Héchard, Y. (2016). Permissiveness of freshly isolated environmental strains of amoebae for growth of *Legionella pneumophila*. *FEMS Microbiol. Lett.* 363:fnw022. doi: 10.1093/femsle/fnw022

Edelstein, P. H., Edelstein, M. A., Shephard, L. J., Ward, K. W., and Ratcliff, R. M. (2012). *Legionella steelei* sp. nov., isolated from human respiratory specimens in California, USA, and South Australia. *Int. J. Syst. Evol. Microbiol.* 62, 1766–1771. doi: 10.1099/ijs.0.035709-0

Evstigneeva, A., Raoult, D., Karpachevskiy, L., and La Scola, B. (2009). Amoeba co-culture of soil specimens recovered 33 different bacteria, including four new species and *Streptococcus pneumoniae*. *Microbiology* 155, 657–664. doi: 10.1099/mic.0.022970-0

Faulkner, G., Berk, S. G., Garduño, E., and Ortiz-Jiménez, M. A., Garduño, R. A. (2008). Passage through *Tetrahymena tropicalis* triggers a rapid morphological differentiation in *Legionella pneumophila*. *J. Bacteriol.* 190, 7728–7738. doi: 10.1128/JB.00751-08

Fields, B. S. (1996). The molecular ecology of Legionellae. *Trends Microbiol.* 4, 286–290. doi: 10.1016/0966-842X(96)10041-X

Fields, B. S., Barbaree, J. M., Shotts, E. B. Jr., Feeley, J. C., Morrill, W. E., Sanden, G. N., et al. (1986). Comparison of guinea pig and protozoan models for determining virulence of *Legionella* species. *Infect. Immun.* 53, 553–559.

Fields, B. S., Benson, R. F., and Besser, R. E. (2002). *Legionella* and Legionnaires' disease: 25 years of investigation. *Clin. Microbiol. Rev.* 15, 506–526. doi: 10.1128/CMR.15.3.506-526.2002

Fields, B. S., Sanden, G. N., Barbaree, J. M., Morrill, W. E., Wadowsky, R. M., White, E. H., et al. (1989). Intracellular multiplication of *Legionella pneumophila* in amoebae isolated from hospital hot water tanks. *Curr. Microbiol.* 18, 131–137. doi: 10.1007/BF01570838

Fields, B. S., Shotts, E. B., Feeley, J. C., Gorman, G. W., and Martin, W. T. (1984). Proliferation of *Legionella pneumophila* as an intracellular parasite of the ciliated protozoan *Tetrahymena pyriformis*. *Appl. Environ. Microbiol.* 47, 467–471.

Fliermans, C. B. (1996). Ecology of *Legionella*: from data to knowledge with a little wisdom. *Microb. Ecol.* 32, 203–228. doi: 10.1007/BF00185888

Fliermans, C. B., Cherry, W. B., Orrison, L. H., Smith, S. J., Tison, D. L., and Pope, D. H. (1981). Ecological distribution of *Legionella pneumophila*. *Appl. Environ. Microbiol.* 41, 9–16.

Gao, L. Y., Harb, O. S., and Abu Kwaik, Y. (1997). Utilization of similar mechanisms by *Legionella pneumophila* to parasitize two evolutionarily distant host cells, mammalian macrophages and protozoa. *Infect. Immun.* 65, 4738–4746.

García, M. T., Baladrón, B., Gil, V., Tarancon, M. L., Vilasau, A., Ibañez, A., et al. (2008). Persistence of chlorine-sensitive *Legionella pneumophila* in hyperchlorinated installations. *J. Appl. Microbiol.* 105, 837–847. doi: 10.1111/j.1365-2672.2008.03804.x

García, M. T., Jones, S., Pelaz, C., Millar, R. D., and Abu Kwaik, Y. (2007). *Acanthamoeba polyphaga* resuscitates viable non-culturable *Legionella pneumophila* after disinfection. *Environ. Microbiol.* 9, 1267–1277. doi: 10.1111/j.1462-2920.2007.01245.x

Garduño, R. A., Garduño, E., Hiltz, M., and Hoffman, P. S. (2002). Intracellular growth of *Legionella pneumophila* gives rise to a differentiated form dissimilar to stationary-phase forms. *Infect. Immun.* 70, 6273–6283. doi: 10.1128/IAI.70.11.6273-6283.2002

Gomez-Alvarez, V., Revetta, R. P., and Santo Domingo, J. W. (2012). Metagenomic analyses of drinking water receiving different disinfection treatments. *Appl. Environ. Microbiol.* 78, 6095–6102. doi: 10.1128/AEM.01018-12

Gomez-Alvarez, V., Teal, T. K., and Schmidt, T. M. (2009). Systematic artifacts in metagenomes from complex microbial communities. *ISME J.* 3, 1314–1317. doi: 10.1038/ismej.2009.72

Greub, G., and Raoult, D. (2003). Morphology of *Legionella pneumophila* according to their location within *Hartmanella vermiformis*. *Res. Microbiol.* 154, 619–621. doi: 10.1016/j.resmic.2003.08.003

Greub, G., and Raoult, D. (2004). Microorganisms resistant to free-living amoebae. *Clin. Microbiol. Rev.* 17, 413–433. doi: 10.1128/CMR.17.2.413-433.2004

Hägele, S., Köhler, R., Merkert, H., Schleicher, M., Hacker, J., and Steinert, M. (2000). *Dictyostelium discoideum*: a new host model system for intracellular pathogens of the genus *Legionella*. *Cell. Microbiol.* 2, 165–171. doi: 10.1046/j.1462-5822.2000.00044.x

Hahn, M. W., and Höfle, M. G. (2001). Grazing of protozoa and its effect on populations of aquatic bacteria. *FEMS Microbiol. Ecol.* 35, 113–121. doi: 10.1111/j.1574-6941.2001.tb00794.x

Harb, O. S., Venkataraman, C., Haack, B. J., Gao, L. Y., and Kwaik, Y. A. (1998). Heterogeneity in the attachment and uptake mechanisms of the Legionnaires' disease bacterium, *Legionella pneumophila*, by protozoan hosts. *Appl. Environ. Microbiol.* 64, 126–132.

Harf, C., Goffinet, S., Meunier, O., Monteil, H., and Colin, D. A. (1997). Flow cytometric determination of endocytosis of viable labelled *Legionella pneumophila* by *Acanthamoeba palestinensis*. *Cytometry* 27, 269–274. doi: 10.1002/(SICI)1097-0320(19970301)27:3<269::AID-CYTO9>3.0.CO;2-9

Hilbi, H., Segal, G., and Shuman, H. A. (2001). Icm/dot-dependent upregulation of phagocytosis by *Legionella pneumophila*. *Mol. Microbiol.* 42, 603–617. doi: 10.1046/j.1365-2958.2001.02645.x

Hojo, F., Sato, D., Matsuo, J., Miyake, M., Nakamura, S., Kunichika, M., et al. (2012). Ciliates expel environmental *Legionella*-laden pellets to stockpile food. *Appl. Environ. Microbiol.* 78, 5247–5257. doi: 10.1128/AEM.00421-12

Holden, E. P., Winkler, H. H., Wood, D. O., and Leinbach, E. D. (1984). Intracellular growth of *Legionella pneumophila* within *Acanthamoeba castellanii* Neff. *Infect. Immun.* 45, 18–24.

Hsu, B. M., Huang, C. C., Chen, J. S., Chen, N. H., and Huang, J. T. (2011). Comparison of potentially pathogenic free-living amoeba hosts by *Legionella* spp. in substrate-associated biofilms and floating biofilms from spring environments. *Water Res.* 45, 5171–5183. doi: 10.1016/j.watres.2011.07.019

Hsu, B. M., Lin, C. L., and Shih, F. C. (2009). Survey of pathogenic free-living amoebae and *Legionella* spp. in mud spring recreation area. *Water Res.* 43, 2817–2828. doi: 10.1016/j.watres.2009.04.002

Hsu, T. K., Wu, S. F., Hsu, B. M., Kao, P. M., Tao, C. W., Shen, S. M., et al. (2015). Surveillance of parasitic *Legionella* in surface waters by using immunomagnetic separation and amoebae enrichment. *Pathog. Glob. Health* 109, 328–335. doi: 10.1179/2047773215Y.0000000034

Huang, S. W., and Hsu, B. M. (2010). Survey of *Naegleria* and its resisting bacteria-*Legionella* in hot spring water of Taiwan using molecular method. *Parasitol. Res.* 106, 1395–1402. doi: 10.1007/s00436-010-1815-0

Hugenholtz, P., and Tyson, G. W. (2008). Microbiology: metagenomics. *Nature* 455, 481–483. doi: 10.1038/455481a

Ikedo, M., and Yabuuchi, E. (1986). Ecological studies of *Legionella* species. I. Viable counts of *Legionella pneumophila* in cooling tower water. *Microbiol. Immunol.* 30, 413–423. doi: 10.1111/j.1348-0421.1986.tb02967.x

Ishida, K., Sekizuka, T., Hayashida, K., Matsuo, J., Takeuchi, F., Kuroda, M., et al. (2014). Amoebal endosymbiont *Neochlamydia* genome sequence illuminates the bacterial role in the defense of the host amoebae against *Legionella pneumophila*. *PLoS ONE* 9:e95166. doi: 10.1371/journal.pone.0095166

Jacquier, N., Aeby, S., Lienard, J., and Greub, G. (2013). Discovery of new intracellular pathogens by amoebal coculture and amoebal enrichment approaches. *J. Vis. Exp.* 80:51055. doi: 10.3791/51055

Ji, W. T., Hsu, B. M., Chang, T. Y., Hsu, T. K., Kao, P. M., Huang, K. H., et al. (2014). Surveillance and evaluation of the infection risk of free-living amoebae and *Legionella* in different aquatic environments. *Sci. Total Environ.* 499, 212–219. doi: 10.1016/j.scitotenv.2014.07.116

Kang, Y. S., and Kirby, J. E. (2017). Promotion and rescue of intracellular *Brucella neotomae* replication during coinfection with *Legionella pneumophila*. *Infect. Immun.* 85, e00991–e00916. doi: 10.1128/IAI.00991-16

Kao, P. M., Tung, M. C., Hsu, B. M., Hsu, S. Y., Huang, J. T., Liu, J. H., et al. (2013). Differential *Legionella* spp. survival between intracellular and extracellular forms in thermal spring environments. *Environ. Sci. Pollut. Res. Int.* 20, 3098–3106. doi: 10.1007/s11356-012-1159-7

Kikuhara, H., Ogawa, M., Miyamoto, H., Nikaido, Y., and Yoshida, S. (1994). Intracellular multiplication of *Legionella pneumophila* in *Tetrahymena thermophila*. *J. UOEH* 16, 263–275. doi: 10.7888/juoeh.16.263

Kilvington, S., and Price, J. (1990). Survival of *Legionella pneumophila* within cysts of *Acanthamoeba polyphaga* following chlorine exposure. *J. Appl. Bacteriol.* 68, 519–525. doi: 10.1111/j.1365-2672.1990.tb02904.x

King, C. H., Fields, B. S., Shotts, E. B., and White, E. H. (1991). Effects of cytochalasin D and methylamine on intracellular growth of *Legionella pneumophila* in amoebae and human monocyte-like cells. *Infect. Immun.* 59, 758–763.

King, C. H., Shotts, E. B. Jr., Wooley, R. E., and Porter, K. G. (1988). Survival of coliforms and bacterial pathogens within protozoa during chlorination. *Appl. Environ. Microbiol.* 54, 3023–3033.

Koide, M., Higa, F., Tateyama, M., Cash, H. L., Hokama, A., and Fujita, J. (2014). Role of *Brevundimonas vesicularis* in supporting the growth of *Legionella* in nutrient-poor environments. *New Microbiol.* 37, 33–39.

Koubar, M., Rodier, M. H., Garduño, R. A., and Frere, J. (2011). Passage through *Tetrahymena tropicalis* enhances the resistance to stress and the infectivity of *Legionella pneumophila*. *FEMS Microbiol. Lett.* 325, 10–15. doi: 10.1111/j.1574-6968.2011.02402.x

Kumar, S., Stecher, G., and Tamura, K. (2016). MEGA7: molecular evolutionary genetics analysis version 7.0 for bigger datasets. *Mol. Biol. Evol.* 33, 1870–1874. doi: 10.1093/molbev/msw054

Kunin, V., Copeland, A., Lapidus, A., Mavromatis, K., and Hugenholtz, P. (2008). A bioinformatician's guide to metagenomics. *Microbiol. Mol. Biol. Rev.* 72, 557–578. doi: 10.1128/MMBR.00009-08

Kurtz, J. B., Bartlett, C. L., Newton, U. A., White, R. A., and Jones, N. L. (1982). *Legionella pneumophila* in cooling water systems. Report of a survey of cooling towers in London and a pilot trial of selected biocides. *J. Hyg.* 88, 369–381. doi: 10.1017/S0022172400070248

Lasheras, A., Boulestreau, H., Rogues, A. M., Ohayon-Courtes, C., Labadie, J. C., and Gachie, J. P. (2006). Influence of amoebae and physical and chemical characteristics of water on presence and proliferation of *Legionella* species in hospital water systems. *Am. J. Infect. Control* 34, 520–525. doi: 10.1016/j.ajic.2006.03.007

Lau, H. Y., and Ashbolt, N. J. (2009). The role of biofilms and protozoa in *Legionella* pathogenesis: implications for drinking water. *J. Appl. Microbiol.* 107, 368–378. doi: 10.1111/j.1365-2672.2009.04208.x

Lin, W., Yu, Z., Zhang, H., and Thompson, I. P. (2014). Diversity and dynamics of microbial communities at each step of treatment plant for potable water generation. *Water Res.* 52, 218–230. doi: 10.1016/j.watres.2013.10.071

Liu, R., Yu, Z., Guo, H., Liu, M., Zhang, H., and Yang, M. (2012). Pyrosequencing analysis of eukaryotic and bacterial communities in faucet biofilms. *Sci. Total Environ.* 435–436, 124–131. doi: 10.1016/j.scitotenv.2012.07.022

Ma, W., Dong, F. F., Stavrinides, J., and Guttman, D. S. (2006). Type III effector diversification via both pathoadaptation and horizontal transfer in response to a coevolutionary arms race. *PLoS Genet.* 2:e209. doi: 10.1371/journal.pgen.0020209

Mampel, J., Spirig, T., Weber, S. S., Haagensen, J. A., Molin, S., and Hilbi, H. (2006). Planktonic replication is essential for biofilm formation by *Legionella pneumophila* in a complex medium under static and dynamic flow conditions. *Appl. Environ. Microbiol.* 72, 2885–2895. doi: 10.1128/AEM.72.4.2885-2895.2006

Marciano-Cabral, F., and Cabral, G. (2003). *Acanthamoeba* spp. as agents of disease in humans. *Clin. Microbiol. Rev.* 16, 273–307. doi: 10.1128/CMR.16.2.273-307.2003

Marciano-Cabral, F., Jamerson, M., and Kaneshiro, E. S. (2010). Free-living amoebae, *Legionella* and *Mycobacterium* in tap water supplied by a municipal drinking water utility in the USA. *J. Water Health* 8, 71–82. doi: 10.2166/wh.2009.129

Mengue, L., Régnacq, M., Aucher, W., Portier, E., Héchard, Y., and Samba-Louaka, A. (2016). *Legionella pneumophila* prevents proliferation of its natural host *Acanthamoeba castellanii*. *Sci. Rep.* 6:36448. doi: 10.1038/srep36448

Moffat, J. F., and Tompkins, L. S. (1992). A quantitative model of intracellular growth of *Legionella pneumophila* in *Acanthamoeba castellanii*. *Infect. Immun.* 60, 296–301.

Molmeret, M., Horn, M., Wagner, M., Santic, M., and Abu Kwaik, Y. (2005). Amoebae as training grounds for intracellular bacterial pathogens. *Appl. Environ. Microbiol.* 71, 20–28. doi: 10.1128/AEM.71.1.20-28.2005

Molmeret, M., Jarraud, S., Mori, J. P., Pernin, P., Forey, F., Reyrolle, M., et al. (2001). Different growth rates in amoeba of genotypically related environmental and clinical *Legionella pneumophila* strains isolated from a thermal spa. *Epidemiol. Infect.* 126, 231–239. doi: 10.1017/S095026880 1005258

Muchesa, P., Mwamba, O., Barnard, T. G., and Bartie, C. (2014). Detection of free-living amoebae using amoebal enrichment in a wastewater treatment plant of Gauteng Province, South Africa. *Biomed. Res. Int.* 2014:575297. doi: 10.1155/2014/575297

Nagington, J., and Smith, D. J. (1980). Pontiac fever and amoebae. *Lancet* 316, 1241. doi: 10.1016/S0140-6736(80)92494-0

Nahapetian, K., Challemel, O., Beurtin, D., Dubrou, S., Gounon, P., and Squinazi, F. (1991). The intracellular multiplication of *Legionella pneumophila* in protozoa from hospital plumbing systems. *Res. Microbiol.* 142, 677–685. doi: 10.1016/0923-2508(91)90081-K

Neumeister, B., Schöniger, S., Faigle, M., Eichner, M., and Dietz, K. (1997). Multiplication of different *Legionella* species in Mono Mac 6 cells and in *Acanthamoeba castellanii*. *Appl. Environ. Microbiol.* 63, 1219–1224.

Newsome, A. L., Baker, R. L., Miller, R. D., and Arnold, R. R. (1985). Interactions between *Naegleria fowleri* and *Legionella pneumophila*. *Infect. Immun.* 50, 449–452.

O'Connor, T. J., Adepoju, Y., Boyd, D., and Isberg, R. R. (2011). Minimization of the *Legionella pneumophila* genome reveals chromosomal regions involved in host range expansion. *Proc. Nat. Acad. Sci. U.S.A.* 108, 14733–14740. doi: 10.1073/pnas.1111678108

Ohno, A., Kato, N., Sakamoto, R., Kimura, S., and Yamaguchi, K. (2008). Temperature-dependent parasitic relationship between *Legionella pneumophila* and a free-living amoeba (*Acanthamoeba castellanii*). *Appl. Environ. Microbiol.* 74, 4585–4588. doi: 10.1128/AEM.00083-08

Pagnier, I., Raoult, D., and La Scola, B. (2008). Isolation and identification of amoeba-resisting bacteria from water in human environment by using an *Acanthamoeba polyphaga* co-culture procedure. *Environ. Microbiol.* 10, 1135–1144. doi: 10.1111/j.1462-2920.2007.01530.x

Peabody, M. A., Caravas, J. A., Morrison, S. S., Mercante, J. W., Prystajecky, N. A., Raphael, B. H., et al. (2017). Characterization of *Legionella* species from watersheds in British Columbia, Canada. *mSphere* 2, e00246–e00217. doi: 10.1128/mSphere.00246-17

Plouffe, J. F., Webster, L. R., and Hackman, B. (1983). Relationship between colonization of hospital building with *Legionella pneumophila* and hot water temperatures. *Appl. Environ. Microbiol.* 46, 769–770.

Pope, D. H., Soracco, R. J., Gill, H. K., and Fliermans, C. B. (1982). Growth of *Legionella pneumophila* in two-membered cultures with green algae and cyanobacteria. *Curr. Microbiol.* 7, 319–321. doi: 10.1007/BF01566871

Rasch, J., Krüger, S., Fontvieille, S., Ünal, C. M., Michel, R., Labrosse, A., et al. (2016). *Legionella*-protozoa-nematode interactions in aquatic biofilms and influence of Mip on *Caenorhabditis elegans* colonization. *Int. J. Med. Microbiol.* 306, 443–451. doi: 10.1016/j.ijmm.2016.05.012

Rizzardi, K., Winiecka-Krusnell, J., Ramliden, M., Alm, E., Andersson, S., and Byfors, S. (2014). *Legionella norrlandica* sp. nov., isolated from the biopurification system of a wood processing plant in northern Sweden. *Int. J. Syst. Evol. Microbiol.* 65, 598–603. doi: 10.1099/ijs.0.068940-0

Rohr, U., Weber, S., Michel, R., Selenka, F., and Wilhelm, M. (1998). Comparison of free-living amoebae in hot water systems of hospitals with isolates from moist sanitary areas by identifying genera and determining temperature tolerance. *Appl. Environ. Microbiol.* 64, 1822–1824.

Rowbotham, T. J. (1980). Preliminary report on the pathogenicity of *Legionella pneumophila* for freshwater and soil amoebae. *J. Clin. Pathol.* 33, 1179–1183. doi: 10.1136/jcp.33.12.1179

Rowbotham, T. J. (1981). Pontiac fever, amoebae and Legionellae. *Lancet* 317, 40–41. doi: 10.1016/S0140-6736(81)90141-0

Rowbotham, T. J. (1983). Isolation of *Legionella pneumophila* from clinical specimens via amoebae, and the interaction of those and other isolates with amoebae. *J. Clin. Pathol.* 36, 978–986. doi: 10.1136/jcp.36.9.978

Rowbotham, T. J. (1986). Current views on the relationships between amoebae, legionellae and man. *Isr. J. Med. Sci.* 22, 678–689.

Ruggiero, M. A., Gordon, D. P., Orrell, T. M., Bailly, N., Bourgoin, T., Brusca, R. C., et al. (2015). A higher level classification of all living organisms. *PLoS ONE* 10:e0119248. doi: 10.1371/journal.pone.0119248

Salloum, G., Meugnier, H., Reyrolle, M., Grimont, F., Grimont, P. A., Etienne, J., et al. (2002). Identification of *Legionella* species by ribotyping and other molecular methods. *Res. Microbiol.* 153, 679–686. doi: 10.1016/S0923-2508(02)01381-5

Scheikl, U., Sommer, R., Kirschner, A., Rameder, A., Schrammel, B., Zweimüller, I., et al. (2014). Free-living amoebae (FLA) co-occurring with legionellae in industrial waters. *Eur. J. Protistol.* 50, 422–429. doi: 10.1016/j.ejop.2014.04.002

Shadrach, W. S., Rydzewski, K., Laube, U., Holland, G., Ozel, M., Kiderlen, A. F., et al. (2005). *Balamuthia mandrillaris*, free-living amoeba and opportunistic agent of encephalitis, is a potential host for *Legionella pneumophila* bacteria. *Appl. Environ. Microbiol.* 71, 2244–2249. doi: 10.1128/AEM.71.5.2244-2249.2005

Sheehan, K. B., Henson, J. M., and Ferris, M. J. (2005). *Legionella* species diversity in an acidic biofilm community in Yellowstone National Park. *Appl. Environ. Microbiol.* 71, 507–511. doi: 10.1128/AEM.71.1.507-511.2005

Simon, S., Schell, U., Heuer, N., Hager, D., Albers, M. F., Matthias, J., et al. (2015). Inter-kingdom signaling by the *Legionella* quorum sensing molecule LAI-1 modulates cell migration through an IQGAP1-Cdc42-ARHGEF9-dependent pathway. *PLoS Pathog.* 11:e1005307. doi: 10.1371/journal.ppat.1005307

Simon, S., Wagner, M. A., Rothmeier, E., Müller-Taubenberger, A., and Hilbi, H. (2014). Icm/Dot-dependent inhibition of phagocyte migration by *Legionella* is antagonized by a translocated Ran GTPase activator. *Cell. Microbiol.* 16, 977–992. doi: 10.1111/cmi.12258

Smirnov, A. V., Chao, E., Nassonova, E. S., and Cavalier-Smith, T. (2011). A revised classification of naked lobose Amoebae (Amoebozoa: Lobosa). *Protist* 162, 545–570. doi: 10.1016/j.protis.2011.04.004

Smith-Somerville, H. E., Huryn, V. B., Walker, C., and Winters, A. L. (1991). Survival of *Legionella pneumophila* in the cold-water ciliate *Tetrahymena vorax*. *Appl. Environ. Microbiol.* 57, 2742–2749.

Solomon, J. M., Rupper, A., Cardelli, J. A., and Isberg, R. R. (2000). Intracellular growth of *Legionella pneumophila* in *Dictyostelium discoideum*, a system for genetic analysis of host-pathogen interactions. *Infect. Immun.* 68, 2939–2947. doi: 10.1128/IAI.68.5.2939-2947.2000

Steinert, M., Ockert, G., Lück, C., and Hacker, J. (1998). Regrowth of *Legionella pneumophila* in a heat-disinfected plumbing system. *Zentralbl. Bakteriol.* 288, 331–342. doi: 10.1016/S0934-8840(98)80005-4

Stewart, C. R., Muthye, V., and Cianciotto, N. P. (2012). *Legionella pneumophila* persists within biofilms formed by *Klebsiella pneumoniae, Flavobacterium* sp., and *Pseudomonas fluorescens* under dynamic flow conditions. *PLoS ONE* 7:e50560. doi: 10.1371/journal.pone.0050560

Storey, M. V., Winiecka-Krusnell, J., Ashbolt, N. J., and Stenström, T. A. (2004). The efficacy of heat and chlorine treatment against thermotolerant *Acanthamoebae* and *Legionellae*. *Scand. J. Infect. Dis.* 36, 656–662. doi: 10.1080/00365540410020785

Stout, J. E., Best, M. G., Yu, V. L., and Rihs, J. D. (1986). A note on symbiosis of *Legionella pneumophila* and *Tatlockia micdadei* with human respiratory flora. *J. Appl. Bacteriol.* 60, 297–299. doi: 10.1111/j.1365-2672.1986.tb01736.x

Temmerman, R., Vervaeren, H., Noseda, B., Boon, N., and Verstraete, W. (2006). Necrotrophic growth of *Legionella pneumophila*. *Appl. Environ. Microbiol.* 72, 4323–4328. doi: 10.1128/AEM.00070-06

Thomas, J. M., Thomas, T., Stuetz, R. M., and Ashbolt, N. J. (2014). Your garden hose: a potential health risk due to *Legionella* spp. growth facilitated by free-living amoebae. *Environ. Sci. Technol.* 48, 10456–10464 doi: 10.1021/es502652n

Thomas, V., Herrera-Rimann, K., Blanc, D. S., and Greub, G. (2006). Biodiversity of amoebae and amoeba-resisting bacteria in a hospital water network. *Appl. Environ. Microbiol.* 72, 2428–2438. doi: 10.1128/AEM.72.4.2428-2438.2006

Tiaden, A., Spirig, T., Sahr, T., Wälti, M. A., Boucke, K., Buchrieser, C., et al. (2010). The autoinducer synthase LqsA and putative sensor kinase LqsS regulate phagocyte interactions, extracellular filaments and a genomic island of *Legionella pneumophila*. *Environ. Microbiol.* 12, 1243–1259. doi: 10.1111/j.1462-2920.2010.02167.x

Tison, D. L., Pope, D. H., Cherry, W. B., and Fliermans, C. B. (1980). Growth of *Legionella pneumophila* in association with blue-green algae (cyanobacteria). *Appl. Environ. Microbiol.* 39, 456–459.

Tyndall, R. L., and Domingue, E. L. (1982). Cocultivation of *Legionella pneumophila* and free-living amoebae. *Appl. Environ. Microbiol.* 44, 954–959.

Tyson, J. Y., Pearce, M. M., Vargas, P., Bagchi, S., Mulhern, B. J., and Cianciotto, N. P. (2013). Multiple *Legionella pneumophila* Type II secretion substrates, including a novel protein, contribute to differential infection of the amoebae *Acanthamoeba castellanii, Hartmannella vermiformis*, and *Naegleria lovaniensis*. *Infect. Immun.* 81, 1399–1410. doi: 10.1128/IAI.00045-13

Tyson, J. Y., Vargas, P., and Cianciotto, N. P. (2014). The novel *Legionella pneumophila* type II secretion substrate NttC contributes to infection of amoebae *Hartmannella vermiformis* and *Willaertia magna*. *Microbiology* 160, 2732–2744. doi: 10.1099/mic.0.082750-0

Vadell, E. M., and Cavender, J. C. (2007). Dictyostelids living in the soils of the Atlantic Forest, Iguazú Region, Misiones, Argentina: description of new species. *Mycologia* 99, 112–124. doi: 10.1080/15572536.2007.11832606

Valster, R. M., Wullings, B. A., van den Berg, R., and van der Kooij, D. (2011). Relationships between free-living protozoa, cultivable *Legionella* spp., and water quality characteristics in three drinking water supplies in the Caribbean. *Appl. Environ. Microbiol.* 77, 7321–7328. doi: 10.1128/AEM.05575-11

Valster, R. M., Wullings, B. A., and van der Kooij, D. (2010). Detection of protozoan hosts for *Legionella pneumophila* in engineered water systems by using a biofilm batch test. *Appl. Environ. Microbiol.* 76, 7144–7153. doi: 10.1128/AEM.00926-10

van Heijnsbergen, E., Schalk, J. A., Euser, S. M., Brandsema, P. S., den Boer, J. W., and de Roda Husman, A. M. (2015). Confirmed and potential sources of *Legionella* reviewed. *Environ. Sci. Technol.* 49, 4797–4815. doi: 10.1021/acs.est.5b00142

Venkataraman, C., Haack, B. J., Bondada, S., and Abu Kwaik, Y. (1997). Identification of a Gal/GalNAc lectin in the protozoan *Hartmannella vermiformis* as a potential receptor for attachment and invasion by the Legionnaires' disease bacterium. *J. Exp. Med.* 186, 537–547. doi: 10.1084/jem.186.4.537

Wadowsky, R. M., Wang, L., Laus, S., Dowling, J. N., Kuchta, J. M., States, S. J., et al. (1995). Gentamicin-containing peptone-yeast extract medium for cocultivation of *Hartmannella vermiformis* ATCC 50256 and virulent strains of *Legionella pneumophila*. *Appl. Environ. Microbiol.* 61, 4464–4467.

Wadowsky, R. M., and Yee, R. B. (1983). Satellite growth of *Legionella pneumophila* with an environmental isolate of *Flavobacterium breve*. *Appl. Environ. Microbiol.* 46, 1447–1449.

Watanabe, K., Nakao, R., Fujishima, M., Tachibana, M., Shimizu, T., and Watarai, M. (2016). Ciliate *Paramecium* is a natural reservoir of *Legionella pneumophila*. *Sci. Rep.* 6:24322. doi: 10.1038/srep24322

Yamamoto, H., Sugiura, M., Kusunoki, S., Ezaki, T., Ikedo, M., and Yabuuchi, E. (1992). Factors stimulating propagation of Legionellae in cooling tower water. *Appl. Environ. Microbiol.* 58, 1394–1397.

Zbikowska, E., Kletkiewicz, H., Walczak, M., and Burkowska, A. (2014). Coexistence of *Legionella pneumophila* bacteria and free-living amoebae in lakes serving as a cooling system of a power plant. *Water Air Soil Pollut.* 225:2066. doi: 10.1007/s11270-014-2066-y

Zbikowska, E., Walczak, M., and Krawiec, A. (2013). Distribution of *Legionella pneumophila* bacteria and *Naegleria* and *Hartmannella* amoebae in thermal saline baths used in balneotherapy. *Parasitol. Res.* 112, 77–83. doi: 10.1007/s00436-012-3106-4

Permissions

All chapters in this book were first published in BPL, by Frontiers; hereby published with permission under the Creative Commons Attribution License or equivalent. Every chapter published in this book has been scrutinized by our experts. Their significance has been extensively debated. The topics covered herein carry significant findings which will fuel the growth of the discipline. They may even be implemented as practical applications or may be referred to as a beginning point for another development.

The contributors of this book come from diverse backgrounds, making this book a truly international effort. This book will bring forth new frontiers with its revolutionizing research information and detailed analysis of the nascent developments around the world.

We would like to thank all the contributing authors for lending their expertise to make the book truly unique. They have played a crucial role in the development of this book. Without their invaluable contributions this book wouldn't have been possible. They have made vital efforts to compile up to date information on the varied aspects of this subject to make this book a valuable addition to the collection of many professionals and students.

This book was conceptualized with the vision of imparting up-to-date information and advanced data in this field. To ensure the same, a matchless editorial board was set up. Every individual on the board went through rigorous rounds of assessment to prove their worth. After which they invested a large part of their time researching and compiling the most relevant data for our readers.

The editorial board has been involved in producing this book since its inception. They have spent rigorous hours researching and exploring the diverse topics which have resulted in the successful publishing of this book. They have passed on their knowledge of decades through this book. To expedite this challenging task, the publisher supported the team at every step. A small team of assistant editors was also appointed to further simplify the editing procedure and attain best results for the readers.

Apart from the editorial board, the designing team has also invested a significant amount of their time in understanding the subject and creating the most relevant covers. They scrutinized every image to scout for the most suitable representation of the subject and create an appropriate cover for the book.

The publishing team has been an ardent support to the editorial, designing and production team. Their endless efforts to recruit the best for this project, has resulted in the accomplishment of this book. They are a veteran in the field of academics and their pool of knowledge is as vast as their experience in printing. Their expertise and guidance has proved useful at every step. Their uncompromising quality standards have made this book an exceptional effort. Their encouragement from time to time has been an inspiration for everyone.

The publisher and the editorial board hope that this book will prove to be a valuable piece of knowledge for researchers, students, practitioners and scholars across the globe.

List of Contributors

Soma Ghosh and Tamara J. O'Connor
Department of Biological Chemistry, Johns Hopkins University School of Medicine, Baltimore, MD, United States

Mary Speir, Adam Vogrin, Azadeh Seidi, Gilu Abraham and Thomas Naderer
Department of Biochemistry and Molecular Biology, Biomedicine Discovery Institute, Monash University, Clayton, VIC, Australia
Department of Immunobiology, Howard Hughes Medical Institute, Yale University School of Medicine, New Haven, CT, USA

Stéphane Hunot
Department of Immunobiology, Howard Hughes Medical Institute, Yale University School of Medicine, New Haven, CT, USA
Centre National de la Recherche Scientifique, Institut National de la Santé et de la Recherche Médicale, Institut du Cerveau et la Moelle - Hôpital Pitié-Salpêtrière, Boulevard de l'hôpital, Sorbonne Universités, UPMC Univ Paris 06, Paris, France

Qingqing Han And Richard A. Flavell
Department of Immunobiology, Howard Hughes Medical Institute, Yale University School of Medicine, New Haven, CT, USA

Gerald W. Dorn II
Department of Medicine, Center for Pharmacogenomics, Washington University School of Medicine, St. Louis, MO, USA

Seth L. Masters and James E. Vince
Walter and Eliza Hall Institute of Medical Research, Parkville, VIC, Australia
Department of Medical Biology, University of Melbourne, Parkville, VIC, Australia

Luary C. Martínez-Chavarría
Departamento de Patología, Facultad de Medicina Veterinaria y Zootecnia, Universidad Nacional Autónoma de México, México, Mexico

Viveka Vadyvaloo
Paul G. Allen School for Global Animal Health, Washington State University, Pullman, WA, USA

Giulia Oliva, Tobias Sahr and Carmen Buchrieser
Institut Pasteur, Biologie des Bactéries Intracellulaires, Paris, France
Centre National de la Recherche Scientifique, UMR 3525, Paris, France

Tomoko Kubori and Hiroki Nagai
Department of Infectious Disease Control, Research Institute for Microbial Diseases, Osaka University, Suita, Japan
Department of Microbiology, Graduate School of Medicine, Gifu University, Gifu, Japan

Xuan T. Bui and Andree Hubber
Department of Infectious Disease Control, Research Institute for Microbial Diseases, Osaka University, Suita, Japan

Sandra Appelt
Highly Pathogenic Microorganisms, Centre for Biological Threats and Special Pathogens, Robert Koch Institute, Berlin, Germany,

Klaus Heuner
Cellular Interactions of Bacterial Pathogens, Centre for Biological Threats and Special Pathogens, Robert Koch Institute, Berlin, Germany

Jiazhang Qiu
Center of Infection and Immunity, First Hospital, Jilin University, Changchun, China
Key Laboratory of Zoonosis, Ministry of Education, College of Veterinary Medicine, Jilin University, Changchun, China

Zhao-Qing Luo
Center of Infection and Immunity, First Hospital, Jilin University, Changchun, China
Key Laboratory of Zoonosis, Ministry of Education, College of Veterinary Medicine, Jilin University, Changchun, China
Department of Biological Sciences, Purdue Institute for Inflammation, Immunology and Infectious Diseases, Purdue University, West Lafayette, IN, United States

Robeena Farzand
Leicester Microbial Sciences and Infectious Diseases Network, Department of Respiratory Sciences, University of Leicester, Leicester, United Kingdom
Department of Microbiology, Kohat University of Science and Technology, Kohat, Pakistan
Kumar Rajakumar, Michael R. Barer, Primrose P. E. Freestone and Galina V. Mukamolova
Leicester Microbial Sciences and Infectious Diseases Network, Department of Respiratory Sciences, University of Leicester, Leicester, United Kingdom

Marco R. Oggioni
Department of Genetics and Genome Biology, University of Leicester, Leicester, United Kingdom

Helen M. O'Hare
Department of Molecular and Cell Biology, University of Leicester, Leicester, United Kingdom

Amrita Lama, Samuel L. Drennan, Rudd C. Johnson, Grace L. Rubenstein and Eric D. Cambronne
Department of Molecular Microbiology and Immunology, Oregon Health and Science University, Portland, OR, United States
Laam Li and Sébastien P. Faucher
Department of Natural Resource Sciences, Faculty of Agricultural and Environmental Sciences, McGill University, Montreal, QC, Canada

Kevin Bärlocher, Amanda Welin and Hubert Hilbi
Institute of Medical Microbiology, University of Zürich, Zurich, Switzerland

Gunnar N. Schroeder
Centre for Experimental Medicine, School of Medicine, Dentistry and Biomedical Sciences, Queen's University Belfast, Belfast, United Kingdom

Julia Hoppe, Can M. Ünal, Stefanie Thiem and Louisa Grimpe
Institut für Mikrobiologie, Technische Universität Braunschweig, Braunschweig, Germany

Torsten Goldmann
Pathology of the University Hospital of Lübeck and the Leibniz Research Center, Borstel, Germany,
Airway Research Center North (ARCN), Member of the German Center for Lung Research, Borstel, Germany

Nikolaus Gaßler
Institut für Pathologie, Klinikum Braunschweig, Braunschweig, Germany

Matthias Richter
Klinikum Salzdahlumerstraße Braunschweig, Braunschweig, Germany

Olga Shevchuk
Center for Proteomics, University of Rijeka, Rijeka, Croatia

Michael Steinert
Institut für Mikrobiologie, Technische Universität Braunschweig, Braunschweig, Germany
Helmholtz Center for Infection Research, Braunschweig, Germany

Catherine Daniel, Amélie Dewitte, Sabine Poiret, Michaël Marceau, Michel Simonet, Laure Marceau, Guillaume Descombes, Denise Boutillier, Nadia Bennaceur, Sébastien Bontemps-Gallo, Nadine Lemaître and Florent Sebbane
Université de Lille, CNRS, Inserm, CHU Lille, Institut Pasteur de Lille, U1019 - UMR 8204 - Center for Infection and Immunity of Lille, Lille, France

Samual C. Allgood, Barbara P. Romero Dueñas, Rebecca R. Noll, Colleen Pike, Sean Lein and M. Ramona Neunuebel
Department of Biological Sciences, University of Delaware, Newark, DE, United States

Marc Erhardt
Young Investigator Group Infection Biology of Salmonella, Helmholtz Centre for Infection Research, Braunschweig, Germany

Petra Dersch
Department of Molecular Infection Biology, Helmholtz Centre for Infection Research, Braunschweig, Germany

Christopher Price, Snake Jones, Ashley Best and Juanita Von Dwingelo
Department of Microbiology and Immunology, College of Medicine, University of Louisville, Louisville, KY, USA

Michael Merchant
Department of Medicine-Renal, College of Medicine, University of Louisville, Louisville, KY, USA

Matthew B. Lawrenz and Yousef A. Kwaik
Department of Microbiology and Immunology, College of Medicine, University of Louisville, Louisville, KY, USA
Center for Predictive Medicine, College of Medicine, University of Louisville, Louisville, KY, USA

Nawsad Alam and Ora Schueler-Furman
Department of Microbiology and Molecular Genetics, Institute for Medical Research Israel-Canada (IMRIC), Faculty of Medicine, Hadassah Medical School, The Hebrew University of Jerusalem, Jerusalem, Israel

David K. Boamah and Tamara J. O'Connor
Department of Biological Chemistry, Johns Hopkins University School of Medicine, Baltimore, MD, United States

Guangqi Zhou
Department of Biochemistry, University of Toronto, Toronto, ON, Canada

Alexander W. Ensminger
Department of Biochemistry, University of Toronto, Toronto, ON, Canada
Department of Molecular Genetics, University of Toronto, Toronto, ON, Canada
Public Health Ontario, Toronto, ON, Canada

Index

P

R

S

T

U

V

Y

www.ingramcontent.com/pod-product-compliance
Lightning Source LLC
LaVergne TN
LVHW060532210926
848485LV00005B/1187
* 9 7 8 1 6 3 9 8 9 3 2 4 9 *